SI UNITS AND SYMBOLS

BASE UNITS

QUANTITY	UNIT	SI SYMBOL	FORMULA
length	meter	m	...
mass	kilogram	kg	...
time	second	s	...
electric current	ampere	A	...
thermodynamic temperature	kelvin	K	...
amount of substance	mole	mol	...
luminous intensity	candela	cd	...

SUPPLEMENTARY UNITS:

plane angle	radian	rad	...
solid angle	steradian	sr	...

DERIVED UNITS:

acceleration	meter per second squared	...	m/s^2
angular acceleration	radian per second squared	...	rad/s^2
angular velocity	radian per second	...	rad/s
area	square meter	...	m^2
density	kilogram per cubic meter	...	kg/m^3
electric potential difference	volt	V	W/A
electric resistance	ohm	Ω	V/A
energy	joule	J	N · m
entropy	joule per kelvin	...	J/K
force	newton	N	$kg \cdot m/s^2$
frequency	hertz	Hz	l/s
magnetomotive force	ampere	A	...
power	watt	W	J/s
pressure	pascal	Pa	N/m^2
quantity of electricity	coulomb	C	A · s
quantity of heat	joule	J	N · m
radiant intensity	watt per steradian	...	W/sr
specific heat	joule per kilogram-kelvin	...	J/kg · K
stress	pascal	Pa	N/m^2
thermal conductivity	watt per meter-kelvin	...	W/m · K
velocity	meter per second	...	m/s
viscosity, dynamic	pascal-second	...	Pa · s
viscosity, kinematic	square meter per second	...	m^2/s
volume	cubic meter	...	m^3
work	joule	J	N · m

SI PREFIXES

MULTIPLICATION FACTORS	PREFIX	SI SYMBOL
$1\ 000\ 000\ 000\ 000 = 10^{12}$	tera	T
$1\ 000\ 000\ 000 = 10^{9}$	giga	G
$1\ 000\ 000 = 10^{6}$	mega	M
$1\ 000 = 10^{3}$	kilo	k
$100 = 10^{2}$	hecto	h
$10 = 10^{1}$	deka	da
$0.1 = 10^{-1}$	deci	d
$0.01 = 10^{-2}$	centi	c
$0.001 = 10^{-3}$	milli	m
$0.000\ 001 = 10^{-6}$	micro	μ
$0.000\ 000\ 001 = 10^{-9}$	nano	n
$0.000\ 000\ 000\ 001 = 10^{-12}$	pico	p
$0.000\ 000\ 000\ 000\ 001 = 10^{-15}$	femto	f
$0.000\ 000\ 000\ 000\ 000\ 001 = 10^{-18}$	atto	a

CONVERSION FACTORS
U.S.-English Units to SI Units

TO CONVERT FROM	TO	MULTIPLY BY
(Acceleration)		
foot/second^2 (ft/s^2)	meter/second^2 (m/s^2)	3.048×10^{-1} *
inch/second^2 (in./s^2)	meter/second^2 (m/s^2)	2.54×10^{-2} *
(Area)		
foot^2 (ft^2)	meter^2 (m^2)	9.2903×10^{-2}
inch^2 (in.^{2})	meter^2 (m^2)	6.4516×10^{-4} *
yard^2 (yd^2)	meter^2 (m^2)	8.3613×10^{-1}
(Density)		
pound mass/inch^3 (lbm/in.^{3})	kilogram/meter^3 (kg/m^3)	2.7680×10^4
pound mass/foot^3 (lbm/ft^3)	kilogram/meter^3 (kg/m^3)	1.6018×10
(Energy, Work)		
British thermal unit (Btu)	joule (J)	1.0544×10^3
foot-pound force (ft · lbf)	joule (J)	1.3558
kilowatt-hour (kw · h)	joule (J)	3.60×10^6 *
(Force)		
kip (1000 lbf)	newton (N)	4.4482×10^3
pound force (lbf)	newton (N)	4.4482
ounce force	newton (N)	2.7801×10^{-1}
(Length)		
foot (ft)	meter (m)	3.048×10^{-1} *
inch (in.)	meter (m)	2.54×10^{-2} *
mile (mi), (U.S. statute)	meter (m)	1.6093×10^3
mile (mi), (international nautical)	meter (m)	1.852×10^3 *
yard (yd)	meter (m)	9.144×10^{-1} *
(Mass)		
pound mass (lbm)	kilogram (kg)	4.5359×10^{-1}
slug (lbf · s^2/ft)	kilogram (kg)	1.4594×10
ton (2000 lbm)	kilogram (kg)	9.0718×10^2
(Power)		
foot-pound/minute (ft · lbf/min)	watt (W)	2.2597×10^{-2}
horsepower (550 ft · lbf/s)	watt (W)	7.4570×10^2
(Pressure, stress)		
atmosphere (std) (14.7 lbf/in.^{2})	newton/meter^2 (N/m^2 or Pa)	1.0133×10^5
pound/foot^2 (lbf/ft^2)	newton/meter^2 (N/m^2 or Pa)	4.7880×10
pound/inch^2 (lbf/in.^{2} or psi)	newton/meter^2 (N/m^2 or Pa)	6.8948×10^3
(Velocity)		
foot/minute (ft/min)	meter/second (m/s)	5.08×10^{-3} *
foot/second (ft/s)	meter/second (m/s)	3.048×10^{-1} *
knot (nautical mi/h)	meter/second (m/s)	5.1444×10^{-1}
mile/hour (mi/h)	meter/second (m/s)	4.4704×10^{-1} *
mile/hour (mi/h)	kilometer/hour (km/h)	1.6093
mile/second (mi/s)	kilometer/second (km/s)	1.6093
(Viscosity)		
foot^2/second (ft^2/s)	meter^2/second (m^2/s)	9.2903×10^{-2}
pound-mass/foot-second (lb_m/ft · s)	pascal-second (Pa · s)	1.4882
pound-force-second/foot^2 (lb_f · s/ft^2)	pascal-second (Pa · s)	4.788×10

* Exact value

Reliability in Engineering Design

K. C. Kapur

L. R. Lamberson

Department of Industrial Engineering
and Operations Research
Wayne State University
Detroit, Michigan 48202

JOHN WILEY & SONS

New York • Chichester • Brisbane • Toronto • Singapore

Library of Congress Cataloging in Publication Data:

Kapur, Kailash Chander, 1941-
Reliability in engineering design

Includes bibliographies and indexes.
1. Engineering design. 2. Reliability

(Engineering) I. Lamberson, L. R., 1937-
joint author. II. Title.
TA174.K36 620'.004'5 76-1304
ISBN 0-471-51191-9

Printed in the United States of America

20 19 18 17 16 15

To my wife Geraldine and children Anjali and Jay

K.C.K.

To my wife Yvonne and daughter Debra

L.R.L.

Preface

In writing this book we have tried to provide a basic knowledge of reliability techniques that can be used by design or reliability engineers. The primary emphasis is on the problem of quantifying reliability in product design and testing. We were encouraged to write this book by the many students from the Detroit industries who have attended our courses in reliability. Since most of these students were experienced engineers, they contributed immensely in helping us to develop our approach to the product reliability problem. We owe these students many thanks.

A book such as this can include many topics presented in various sequences. We explain here why we have chosen this particular sequence.

Chapter 1 forms an introduction to reliability. It examines the problems encountered in quantifying reliability and presents some of the elements necessary for a total reliability program. Chapter 2 discusses and defines the many terms and measures used in reliability testing. Both quantitative and nonquantitative terms are presented. Very often these terms are misused or used in a vague manner. Therefore, we have tried to provide precise and complete definitions that will encourage the student to use reliability terms correctly. When Chapter 2 is combined with Chapter 3, which presents simple static reliability models, the student will have a basic understanding of reliability terminology.

It should be recognized that Chapters 2 and 3 could be interchanged and in some teaching situations it may be advantageous to cover Chapter 3 first. We have elected to start with the more difficult material of Chapter 2 and then give the student the easily assimilated material of Chapter 3. We hope that this grace period provided by Chapter 3 will give the student time to become familiar with the concepts and definitions of Chapter 2.

Since reliability is basically a design parameter and must be incorporated in the system at the design stage, we turn our attention early in the book to the design process as it relates to reliability. We provide the student with some approaches for assessing and verifying design reliability. Design reliability is a relatively new area as far as everyday practice is concerned, and this section emphasizes the need for improvement of reliability techniques for other design aspects.

In the design of any system, the design variables and parameters are probabilistic in nature. Thus, it is obvious that the factors that determine the stress and strength of the components are also probabilistic. This means that when the reliability aspects of design are evaluated, the probabilistic nature of the variables and parameters for a system must be considered. The probabilistic approaches to design and reliability are discussed in Chapter 4. Chapters 5 and 6 expand on the methodology necessary to analyze complex designs from a reliability standpoint. Specifically, Chapter 5 discusses combining random variables in design and presents concepts useful in statistical tolerancing. This chapter exposes the student to the further statistical work needed for design reliability. Chapter 6 presents the techniques for determining design reliability after the probabilistic modeling of design variables and parameters. In Chapter 7 simple design examples are used to illustrate the design reliability methodology. The more advanced problem of time-dependent stress-strength models is developed in Chapter 8.

Dynamic reliability models are discussed in Chapter 9 along with concepts for maintainability. It is felt that at this point the student has developed sufficient statistical maturity to easily handle the material in this chapter.

In the evolution of a new product, the testing comes after the design is complete. Thus, Chapters 10 and 11 consider the problem of reliability estimation from test data. Specifically, Chapter 10 uses the exponential distribution as the underlying model and presents reliability estimation techniques for a variety of situations. For each situation, the assumptions necessary for correct application are clearly stated, and derivations are included in the appendix for perusal by the more advanced student. Similarly, in Chapter 11 the Weibull distribution is used as the model for product life. The use of both graphical and analytical techniques is explained in this chapter.

Chapter 12 is concerned with the development of sequential tests for reliability demonstration. The methods of use as well as some problems in practical applications are considered.

Chapter 13 develops the new area of Bayesian statistical inference as it applies to product testing. It should be recognized that in many cases the design engineer has considerable á priori knowledge about the performance of a system. If test programs are to be meaningful, this knowledge must be captured and quantified. In actual practice the design and test engineers unknowingly use their prior knowledge by not accepting the large sample sizes needed to run a statistically valid test. This chapter brings together these concepts in product testing, presents related implementation problems, and illustrates how this prior knowledge might be utilized.

The solution to engineering design and reliability problems requires a compromise between many alternatives. The characteristics of the design depend on both fixed and adjustable parameters. Optimization techniques as they apply to reliability and the design process are introduced in Chapter 14. This chapter brings into focus the many trade-offs that the design engineer must consider with cost surely being one factor. The design engineer is introduced to nonlinear optimization theory as necessary for understanding; however, to completely appreciate this chapter it is advantageous for the student to have had a previous introduction to optimization.

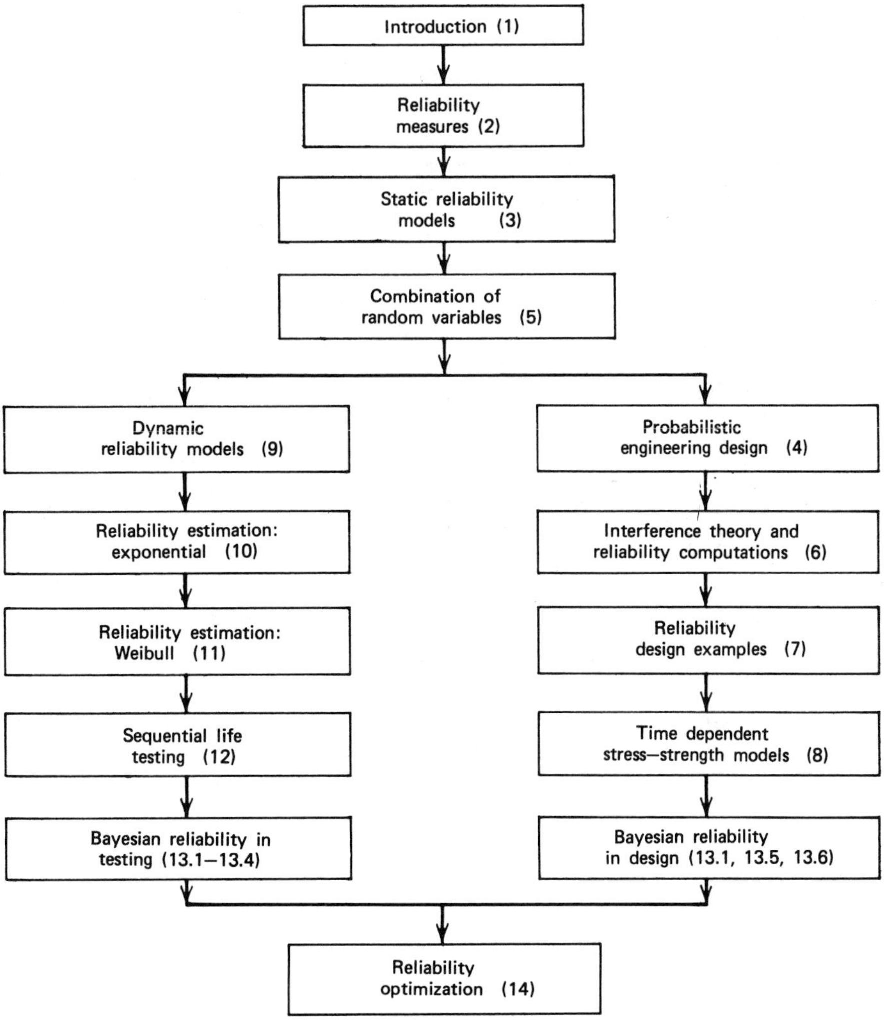

Exhibit A

We would recommend that this book be used as a two-quarter sequence in reliability for credit campus courses or as an 80 hour noncredit seminar for practicing engineers. In all cases, it must be assumed that the students have had previous training in the physical or engineering sciences and at least one, but preferably two, introductory courses in probability and statistics.

If the emphasis of the course is on design reliability, Chapters 1 through 8, Chapter 13 and Chapter 14 should be covered thoroughly. However, if the emphasis is on reliability testing, then Chapters 1 through 3, Chapter 5, and Chapters 9 through 14 can be used to teach the course. These two sequences depending upon the emphasis are shown in Exhibit A. There are undoubtedly many other possibilities for a one-quarter or one-semester course. The material covered will ultimately depend on the students' background in statistics, their educational needs, and the interest of the professor.

We have found the text material presented here suitable for students of many disciplines. We have had students in our classes with backgrounds in almost every engineering discipline as well as in mathematics and physics.

The authors must acknowledge certain organizations that have contributed to their knowledge of and experience in reliability. Our work with Chevrolet Motor Division of the General Motors Corporation, U.S. Army Tank-Automotive Command, and Ford Motor Company provided experience that was invaluable in the writing of this textbook.

We would like to thank Mr. V. Bawle, Graduate Assistant, Department of Industrial Engineering and Operations Research, for his many comments and criticisms of our original manuscript and problem sets. Also we must sincerely thank some fifty students who have read through the final text and offered many comments and criticisms. All of these contributions helped us immensely.

K. C. Kapur

L. R. Lamberson

Contents

Appendixes

Table 1 Principal SI Units

QUANTITY	UNIT	SYMBOL	FORMULA
Acceleration	Meter per second squared	...	m/s^2
Angle	Radian	rad	...
Angular acceleration	Radian per second squared	...	rad/s^2
Angular velocity	Radian per second	...	rad/s
Area	Square meter	...	m^2
Density	Kilogram per cubic meter	...	kg/m^3
Energy	Joule	J	$N \cdot m$
Force	Newton	N	$kg \cdot m/s^2$
Frequency	Hertz	Hz	s^{-1}
Length	Meter	m	...
Mass	Kilogram	kg	...
Moment of a force	Newton-meter	...	$N \cdot m$
Power	Watt	W	J/s
Pressure	Pascal	Pa	N/m^2
Stress	Pascal	Pa	N/m^2
Time	Second	s	...
Velocity	Meter per second	...	m/s
Volume, solids	Cubic meter	...	m^3
liquids	Liter	l	$10^{-3}m^3$
Work	Joule	J	$N \cdot m$

Table 2

FACTOR BY WHICH THE UNIT IS MULTIPLIED	PREFIX	
	NAME	SYMBOL
10^6	mega	M
10^3	kilo	k
10^2	hecto	h
10	deca	da
10^{-1}	deci	d
10^{-2}	centi	c
10^{-3}	milli	m
10^{-6}	micro	μ

Table 3 U.S. Customary Units and their SI Equivalents

QUANTITY	U.S. CUSTOMARY UNIT	SI EQUIVALENT
Acceleration	ft/s^2	0.3048 m/s^2
	in./s^2	0.0254 m/s^2
Area	ft^2	0.929 m^2
	in^2	645.2 mm^2
Energy	ft·lb	1.356 J
Force	kip	4.448 kN
	lb	4.448 N
	oz	0.2780 N
Length	ft	0.3048 m
	in.	25.40 mm
	mi	1.609 km
Mass	oz mass	28.35 g
	lb mass	0.4536 kg
	slug	14.59 kg
	ton	907.2 kg
Moment of a force	lb·ft	1.356 N·m
	lb·in.	0.1130 N·m
Moment of inertia:		
of an area	in^4	0.4162×10^6 mm^4
of a mass	lb·ft·s^2	1.356 kg·m^2
Power	ft·lb/s	1.356 W
	hp	745.7 W
Pressure or stress	lb/ft^2	47.88 Pa
	lb/in^2(psi)	6.895 kPa
Velocity	ft/s	0.3048 m/s
	in./s	0.0254 m/s
	mi/h (mph)	0.4470 m/s
	mi/h (mph)	1.609 km/h
Volume, solids	ft^3	0.02832 m^3
	in^3	1.609 km/h
liquids	gal	3.785 l
	qt	0.9464 l
Work	ft·lb	1.356 J

Symbol Index

Chapter 2

$\mathbf{t}$	time to failure random variable
r.v.	random variable
p.d.f.	probability density function
c.d.f.	cumulative distribution function
$E(\mathbf{x})$	expected value of random variable $\mathbf{x}$
$\mu_{\mathbf{x}}$	expected value of random variable $\mathbf{x}$
$\sigma_{\mathbf{x}}$	standard deviation of random variable $\mathbf{x}$
$F(t)$	time to failure distribution function
$\bar{F}(t)$	complement of time to failure distribution function, $\bar{F}(t)=1-F(t)$
$P(A)$	probability of event A
$R(t)$	the reliability function; $R(t)=P[\mathbf{t}\geqslant t]$
$f(t)$	probability density function for r.v. $\mathbf{t}$
$h(t)$	the hazard function
$\mathbf{z}$	the standard normal r.v.
$\phi(z)$	p.d.f. for standard normal r.v.
$\Phi(z)$	c.d.f. for standard normal r.v.
$\Gamma(n)$	gamma function evaluated at n
MTBF	mean time between failures
MTTF	mean time to failure
$\approx$	approximately equal to

Chapter 3

E_i	event that subsystem i operates successfully
R_i	reliability of subsystem i; $R_i=P(E_i)$
R_S	system reliability
Q_S	system unreliability; $Q_S=1-R_S$

Chapter 4

$\mathbf{s}$	stress random variable
$\mathcal{S}$	strength random variable
R	reliability of the component

$\bar{R}$	the unreliability, $\bar{R}=1-R$
$\mu_{\mathbf{s}}$	expected value of stress r.v. $\mathbf{s}$
$\bar{\mathbf{s}}$	estimate of expected value for r.v. $\mathbf{s}$
$\mu_{\mathbb{S}}$	expected value of strength r.v. $\mathbb{S}$
$\bar{\mathbb{S}}$	estimate of expected value for r.v. $\mathbb{S}$
$\sigma_{\mathbf{s}}$	standard deviation of stress r.v. $\mathbf{s}$
$\sigma_{\mathbb{S}}$	standard deviation of strength r.v. $\mathbb{S}$
n	factor of safety, $n=\mu_{\mathbb{S}}/\mu_{\mathbf{s}}$
$V_{\mathbb{S}}$	coefficient of variation for $\mathbb{S}$; $V_{\mathbb{S}}=\sigma_{\mathbb{S}}/\mu_{\mathbb{S}}$.
$V_{\mathbf{s}}$	coefficient of variation for $\mathbf{s}$
n_c	central factor of safety, $n_c=\bar{\mathbb{S}}/\bar{\mathbf{s}}$
$\mathbf{n}$	factor of safety r.v.; $\mathbf{n}=\mathbb{S}/\mathbf{s}$
$V(\mathbf{x})$	variance of r.v. $\mathbf{x}$

Chapter 5

$\mathrm{cov}(\mathbf{x},\mathbf{y})$	covariance of $\mathbf{x}$ and $\mathbf{y}$
ρ	correlation coefficient
$\sim$	the r.v. on the left is distributed as the distribution given on the right

Chapter 6

$f_{\mathbf{s}}(\cdot)$	p.d.f. for stress r.v. $\mathbf{s}$
$f_{\mathbb{S}}(\cdot)$	p.d.f. for strength r.v. $\mathbb{S}$
$\mathbf{y}$	the interference r.v., $\mathbf{y}=\mathbb{S}-\mathbf{s}$
$\check{\mathbf{x}}$	median value for r.v. $\mathbf{x}$
$\mathbb{S}_0$	minimum value of the strength r.v. $\mathbb{S}$
$\beta(m,n)$	beta function with arguments m and n
$\beta_x(m,n)$	incomplete beta function truncated at x with arguments m and n

Chapter 7

$\bar{\mathbf{x}}$	estimate of the expected value for r.v. $\mathbf{x}$
τ	shear stress
T	torque

Chapter 8

R_n	reliability after n cycles of loading
$\mathbf{x}_i$	stress r.v. for the ith cycle
$\mathbf{y}_i$	strength r.v. for the ith cycle
E_i	event that no failure occurs on the ith cycle
$g_i(y_i)$	p.d.f. for strength r.v. $\mathbf{y}_i$, $i=0,1,2,\ldots$
$f_i(x_i)$	p.d.f. for stress r.v. $\mathbf{x}_i$, $i=0,1,2,\ldots$
$\pi_i(t)$	probability of i cycles occurring in the time interval $[0,t]$

Chapter 9

$\mathbf{t}_i$	time to failure r.v. for ith subsystem
$R_S(t)$	system reliability at time t

$Q_S(t)$	system unreliability at time t
O.R.	operational readiness
$A(t)$	the availability function
$M(t)$	the maintainability function

Chapter 10

θ	expected value of the r.v. which is exponentially distributed; commonly termed MTBF
λ	failure rate; $\lambda = 1/\theta$
B_r	Bartlett's test statistic for goodness of fit where r is the number of failures
$\chi^2_{\alpha,n}$	$100(1-\alpha)$th percentile for chi-square with n degrees of freedom
$F_{\alpha,m,n}$	$100(1-\alpha)$th percentile for F distributed r.v. with m and n degrees of freedom
T	total accumulated test time
r	total number of failures
$\hat{\theta}$	estimator for θ
δ	minimum life

Chapter 11

β	Weibull shape parameter or slope
θ	Weibull scale parameter or characteristic life
δ	Weibull location parameter or minimum life
μ'_k	kth central moment
${}_n\mathbf{p}_j$	r.v. representing the fraction of the population failing prior to the jth ordered observation in a sample of size n

Chapter 12

H_0	the null hypothesis
H_1	the alternative hypothesis
$L_{k,n}$	the likelihood function with θ_k as the true value of the parameter; $L_{k,n} = L(x_1, x_2, \ldots, x_n \vert \theta_k)$
O.C.	operating characteristic curve

Chapter 13

$P(A\vert B)$	conditional probability of event A given that B has occurred
$f(x_1, x_2)$	joint p.d.f. of random variables $\mathbf{x}_1$ and $\mathbf{x}_2$
$k(x_1\vert x_2)$	conditional p.d.f. for $\mathbf{x}_1$ given $\mathbf{x}_2$
$\hat{\theta}$	estimator for parameter θ
T	total cumulative test time

Chapter 14

R_i	reliability of component or subsystem
n	number of subsystems
$c(\mu)$	cost function associated with parameter μ

Chapter 1 introduction

A reliability study is concerned with random occurrences of undesirable events, or failures, during the life of a physical system. In this book we use the following definition:

> The reliability of a system is the probability that, when operating under stated environmental conditions, the system will perform its intended function adequately for a specified interval of time.

With this definition, the obvious problems are: (1) the acceptance of the probabilistic notion of reliability which admits the possibility of failure, (2) the concept of adequate performance for system parameters that deteriorate slowly with time, and (3) the judgment necessary to determine the proper statement of environmental conditions.

In assessing system reliability it is first necessary to define and categorize different modes of system failure. Unfortunately, it is usually exceedingly difficult to define failure in unambiguous terms. Of course, complete and catastrophic failure is easily recognized; however, a system's performance can deteriorate gradually over time, and sometimes there is only a fine line between system success and system failure. Basically one must proceed in a logical fashion and define the various system failure modes. Once the system function and failure modes are explicitly stated, reliability can be precisely quantified by probability statements.

Before the reliability of a system is measured, certain procedural constraints must be established. For example, the amount of preventive maintenance permitted, if it is permitted at all, and the degree to which the system operator can participate in correcting failures must be specified. In other words, the manner in which the system is operated can affect the calculated reliability level.

1.1 DESIGN RELIABILITY

Reliability is an inherent attribute of a system just as is the system's capacity or power rating. The reliability level is established at the design phase, and subsequent testing and production will not raise the reliability without a basic design change. Because reliability is an abstract concept that is difficult to grasp, many organizations find themselves unable to implement a comprehensive reliability program. This is not to say that system designers or managers in the firm are not interested in a reliable product. Rather, the pressures on the designer and, very often, the organizational structure of the design department impede the development of such a program. A system design usually requires the efforts of several design groups, each of which is well versed in such things as the determination of power ratings, material selection, or tolerances. However, designers frequently are not well versed in the philosophy and principles of reliability. In addition, reliability eventually requires a systems viewpoint, which is difficult to arrive at if the system has been designed by several specialized groups.

There exist numerous engineer's handbooks to aid the designer on such things as material properties and tolerancing, and these are extensively used. However, the application of the principles of reliability at the design level are not readily summarized in a handbook fashion. This book emphasizes the methodology and philosophy necessary for anyone engaged in the reliability analysis of any system at the design stage. Various tools are developed for the assessment and verification of the design reliability.

The designer is in the best possible position to apply the philosophy of reliability; thus, he or she should be familiar with its basic principles. Designers usually can call upon several groups for professional help. For example, designers are usually well aware of the capabilities of a stress analysis group, since they have been formally trained in the concept. As a result, designers know when to consult the stress analysis group, what to expect from their efforts, and how to interpret their results. A similar relationship should exist between designers and a reliability group. Frequently, however, designers' lack of training in the principles of reliability prevents the development of this relationship.

1.2 RELIABILITY MANAGEMENT

The successful implementation of a total reliability program requires a dedicated reliability group solely responsible for the overall system reliability. The group should provide assistance in analysis, goal establishment, and progress reporting for decision making. It must also have sufficient expertise to interact during the design phase and at the systems level. Therefore, in addition to an understanding of the basic mathematics of reliability, the group must possess a good knowledge of design principles and programs, system interface problems, human factors, and cost/benefit analysis. To monitor the system's reliability performance, the group must have an efficient and well organized data collection, analysis, and reporting system, and a data base containing reliability performance data on past design configurations.

1.3 PRODUCT TESTING

Management must have faith in the benefits of a good reliability program, because it is much easier to calculate the costs of maintaining a reliability group than the benefits accruing from it. This is also true of product testing. Obviously the only way to measure system reliability is to test completed products or components, under conditions that simulate real life, until failure occurs. One simply cannot assess reliability without data, and of course, the more data available, the more confidence one will have in the estimated reliability level. Unfortunately, extensive testing is often considered undesirable because it results in expenditures of time and money. Thus, we must consider the tradeoff between the value of more confidence in estimated reliability versus the cost of more testing. It is easy to arrive at a test program's cost; the cost of not having the test program or the dollar value of different levels of confidence is difficult to calculate. For, not having a test program or reliability group does not necessarily mean unreliability; it simply means that one does not know.

1.4 RELIABILITY AND HUMAN ENGINEERING

In most cases a system is controlled and monitored by human operators. The operator must be viewed as another factor that can either degrade or improve system reliability. A deterioration in reliability can simply be a result of the man-machine interface. For example, inadequate gauge displays or improperly located controls can tax the operator's abilities and cause him to unintentionally induce failures. Factors such as operator fatigue or decreased vigilance can further complicate the situation. It should also be pointed out that human operators can improve system reliability by performing functions that were not intended in the original design but were found necessary to compensate for system weakness. So, the total system reliability picture should include the human element, and the reliability group must be able to recognize and provide input on this man-machine interface. Although this book does not consider the human factors problem, References 20, 21 provide an introduction to the extensive literature in this area.

1.5 RELIABILITY GROWTH

Reliability often improves during the initial development and testing of a system. This improvement can be attributed to a variety of factors. For example, as failures occur, the analysis of these failures provides information that is used to improve the system. Also, experience gained in constructing test components is applied to producing better components as time goes on. If the system reliability is periodically reassessed during this initial period, it will be found to be increasing in value. This phenomenon is termed "reliability growth." The tracking, prediction, and reporting of this growth phenomenon is a function of the reliability group. The reliability growth will level off as the reliability level approaches the inherent level of the design. These predictions and reports of reliability growth indicate the progress toward reliability goals. They also provide a basis for determining whether the desired reliability level will be reached on schedule. If it appears that the goals will not be achieved in time, then adjustments can be made early in the program. The reader is referred to References 5, 9, 18, 23, 24 for further thoughts on this subject of reliability growth.

1.6 DESIGN REVIEW, FAILURE MODE, AND FAULT TREE ANALYSIS

Safety hazards or costly reliability problems are never intentionally designed into a system. However, because these problems do occur, formal design review procedures applied by independent groups are desirable. Procedures such as design review, failure mode analysis (FMA) and fault tree analysis have been developed for application early in the design phase of a new product. Clearly, reliability improvements are most easily and economically accomplished before firm design commitments and production decisions are made. However, early in the design phase, reliability improvement depends heavily on the past experience of the personnel studying the product, because the work is done primarily from such things as original systems concepts, blueprints, and preliminary systems mock-ups, and no hard data is available for a quantitative assessment of reliability.

The design review, a formal and documented review of a system design, is conducted by a committee of senior company personnel experienced in such concepts as product design, reliability, and production costs. The design review normally extends over several phases of design development. In each phase previous work is updated and thus the review is based on the current information about the product. References 12, 15, 16, 17, provide excellent material on the formulation of a design review procedure.

Failure mode analysis (FMA) is a preliminary design evaluation procedure used to identify design weakness that may result in safety hazards or reliability problems. The FMA procedure might be termed a "what if" approach in that it starts at the component level and asks "what if this component fails?" The effects are then traced to the system level. Any component failures that could have a critical effect on the system are identified and either eliminated or controlled, if possible. The application and successes of FMA have been well documented [3, 4, 6, 7, 14, 22].

Fault tree analysis begins with the definition of an undesirable event and traces this event down through the system to identify basic causes. This top-down procedure can be used to identify numerous problems, including operator-induced failures. Documentation and further information on fault tree analysis can be found in References 6, 8, 13.

The value of these procedures lies in the fact that they require a total systems view. In the design process, subsystems are frequently developed by independent groups and thus failures that result from the interfacing of subsystems can go undetected.

1.7 BRIEF HISTORY OF RELIABILITY

The field of reliability analysis is an indirect outgrowth of problems with electronic systems designed during the early 1940s for use in the war effort. As the electronic systems grew in complexity, problems with electronic equipment were also compounded, prompting the Air Force, Navy, and Army to each set up committees to investigate the reliability problem. In 1952, the Department of Defense coordinated these efforts by establishing the Advisory Group on Reliability of Electronic Equipment (AGREE) [1]. In recent years the field of reliability has become an important factor in systems design and development.

1.8 SCOPE OF THE BOOK

This book emphasizes the analytical techniques useful in assessing reliability early in the life of a new product. Many terms and measures used in reliability analysis are first defined, and then various distribution and hazard models are developed. A probabilistic design methodology for assessing the reliability of a product at the design stage is presented, and relationships between this probabilistic approach and conventional design approaches are explored. Techniques for determining design reliability for both single stress and repeated stress applications are presented. A systematic approach to reliability estimation based on test data is presented, and detailed analysis is given when the underlying distributions are exponential and Weibull. In addition, the sequential method for reliability demonstration is also given. This book addresses the new area of Bayesian statistical inference as applied to design reliability and product testing. Also, reliability allocation and optimization models are developed and solved.

BIBLIOGRAPHY

1 AGREE, *Reliability of Military Electronic Equipment*, Report of the Advisory Group on Reliability of Electronic Equipment, Office of the Assistant Secretary of Defense, Washington, D. C., June 4, 1957.

2 AMCP 702-3, *Quality Assurance-Reliability Handbook*, Headquarters, U. S. Army Materiel Command, October 1968.

3 ARP-926, *Design Analysis Procedure for Failure Mode, Effects and Criticality Analysis (FMECA)*, Society of Automotive Engineers, Inc., 485 Lexington Ave., New York, New York 10017, September 1967.

4 Arnzen, H. E., "Failure Mode and Effect Analysis: A Powerful Engineering Tool for Component and System Optimization," *Annals of Reliability and Maintainability*, Fifth Reliability and Maintainability Conference, New York, New York, July 18–20, 1966.

5 Barlow, R. E. and E. M. Scheuer, "Reliability Growth During a Development Testing Program," *Technometrics*, Vol. 8, 1966.

6 Burgess, J. A., "Spotting Trouble Before it Happens," *Machine Design*, September 17, 1970.

7 Crown, P. L., "Design Effective Failure Mode and Effect Analysis," *Annals of Assurance Science*, 1969.

8 Corsetti, P. A. and R. A. Bruce, "Commercial Application of Fault Tree Analysis," *Proceedings of the Reliability and Maintainability Conference*, Vol. 9, 1970.

9 Duane, J. T., "Learning Curve Approach to Reliability Monitoring," *IEEE Transactions on Aerospace*, Vol. 2, April 1964.

10 Drenick, R. F., "Mathematical Aspects of the Reliability Problem," *Journal of the Society of Industrial and Applied Mathematics*, Vol. 8, No. 1, March 1960.

11 Drenick, R. F., "The Failure Law of Complex Equipment," *Journal of the Society for Industrial and Applied Mathematics*, Vol. 8, No. 4, December 1960.

12 Frederiksen, K. A., "Reliability Design Review Techniques," *Proceedings of the Aerospace Systems Reliability Symposium*, Salt Lake City, Utah, April 16–18, 1962.

13 Fussell, J. B., G. J. Powers and R. G. Bennetts; "Fault Trees—A State of the Art Discussion," *IEEE Transactions on Reliability*, Vol. R-23, No. 1, April 1974.

14 Greene, K. and T. J. Cunningham, "Failure Mode, Effects, and Criticality Analysis," *Annals of Assurance Science*, 1968.

15 Jacobs, R. M., "Implementing Formal Design Review," *Proceedings of the Annual Symposium on Reliability*, Washington, D. C., January 10–11, 1967.

16 Jacobs, R. M. and R. Cazanjian, "Design Review—The Why and the How," *Seventh National Symposium on Reliability and Quality Control*, January 9, 1961.

17 Jacobs, R. M., "Competitive Product Design via Design Review," *Proceedings of the Annual Symposium on Reliability*, San Francisco, California, January 25–27, 1966.

18 Littlewood, B. and J. L. Verrall, "A Bayesian Reliability Growth Model for Computer Software," *Journal of the Royal Statistical Society*, Series C, Vol. 22, No. 3, 1973.

19 Lloyd, D. K. and M. Lipow, *Reliability: Management, Methods, and Mathematics*, Prentice-Hall, Englewood Cliffs, N. J., 1962.

20 McCormick, E. J., *Human Factors Engineering*, McGraw-Hill, New York, 1970.

21 Meister, D., "A Critical Review of Human Performance Reliability Predictive Methods," *IEEE Transactions on Reliability*, Vol. R-22, No. 3, August 1973.

22 MIL-STD-1629(SHIPS), *Procedures for Performing a Failure Mode and Effects Analysis for Shipboard Equipment*, Department of the Navy, Naval Ship Engineering Center, Hyattsville, Md. 20782, November 1, 1974.

23 Read, R. R., "A Remark on the Barlow-Scheuer Reliability Growth Estimation Scheme," *Technometrics*, Vol. 13, No. 1, February 1971.

24 Weiss, H. K., "Estimation of Reliability Growth in a Complex System with a Poisson-Type Failure," *Operations Research*, Vol. 4, No. 5, October 1956.

Chapter 2 reliability measures

Any reliability analysis of a system must be based on precisely defined concepts. Since it is readily accepted that a population of supposedly identical systems operating under similar conditions fail at different points in time, then it follows that a failure phenomenon can only be described in probabilistic terms. Thus the fundamental definitions of reliability must depend heavily on concepts from probability theory.

This chapter is concerned with the development of the fundamental definitions and concepts for reliability. These concepts provide the basis for quantifying the reliability of a system, and only precise quantification allows comparison between systems or provides a logical basis for reliability improvement in a system. Terms such as reliability function, expected life, hazard function, and failure rate are defined and various examples reinforce the meaning of these concepts. The basic definitions are presented in Sections 2.1 to 2.3.

Section 2.4 examines various distribution models useful in reliability. In this section several distribution models are described and then the resulting hazard functions are derived. Section 2.5 takes the opposite approach, first presenting various hazard functions and then deriving the distribution models. This familiarity with distribution models and hazard models is important if the various models are to be applied correctly.

Since frequently a decision on the selection of a distribution or hazard model is based on test data, Section 2.6 is concerned with calculating and displaying the data for use as a guide in selecting one of the previously covered models.

The best way to select a distribution model would be to describe the physical phenomena producing the failure and then use these phenomena as a logical basis for derivation of the distribution model. Unfortunately this is a difficult process; however, Section 2.7 considers some possibilities as well as provides further familiarity with distribution models.

Recall from the previous chapter that the reliability of a system is defined as the probability that the system will adequately perform its intended function under stated environmental conditions for a specified interval of time, number of cycles of operations, or number of kilometers. Thus, reliability is a quantitative measure. Terms such as use reliability, operational reliability, and inherent reliability all mean the same thing where the function is explicitly stated. The term "probability of survival" is also used to describe reliability. Here the term survival means the continued ability of the system to perform the desired function.

The purpose of, and the need for, a particular system actually determines the kind of reliability measure that is most meaningful and most useful. In general a system may be required to perform various functions, each of which may have a different reliability. In addition, at different times (or number of cycles or any other measure of the use of the system) the system may have a different probability of successfully performing the required function under stated conditions.

In reliability, the term failure means that the system is not capable of performing the required function. The term capable is used here in the sense that either the system is capable of performing the required function or not. Generally, however, the term capable is very fuzzy and we could define various degrees of capability.

2.1 THE RELIABILITY FUNCTION

The probability of failure as a function of time can be defined by

$$P(\mathbf{t} \leqslant t) = F(t), \qquad t \geqslant 0 \tag{2.1}$$

where $\mathbf{t}$ is a random variable denoting the failure time. Then $F(t)$ is the probability that the system will fail by time t. In other words, $F(t)$ is the failure distribution function (also called the unreliability function). If we define reliability as the probability of success, or the probability that the system will perform its intended function at a certain time t, we can write

$$R(t) = 1 - F(t) = P(\mathbf{t} > t) \tag{2.2}$$

where $R(t)$ is the reliability function.

If the time to failure random variable $\mathbf{t}$ has a density function $f(t)$, then

$$R(t) = 1 - F(t) = 1 - \int_0^t f(\tau)\,d\tau = \int_t^\infty f(\tau)\,d\tau \tag{2.3}$$

For example, if the time to failure is described by an exponential density function, then

$$f(t) = \frac{1}{\theta} e^{-t/\theta}, \qquad t \geqslant 0, \quad \theta > 0 \tag{2.4}$$

and this will lead to a reliability function

$$R(t) = \int_t^\infty \frac{1}{\theta} e^{-\tau/\theta} d\tau$$

$$= e^{-(\tau/\theta)}, \qquad t \geqslant 0 \tag{2.5}$$

Or consider the Weibull distribution where the failure distribution function is given by

$$F(t) = 1 - \exp\left\{ -\left(\frac{t}{\theta}\right)^\beta \right\}, \qquad t \geqslant 0, \quad \theta > 0, \quad \beta > 0 \tag{2.6}$$

then the reliability function is

$$R(t) = e^{-(t/\theta)^\beta}, \qquad t \geqslant 0 \tag{2.7}$$

Thus given a particular failure density function or distribution function, the reliability function can be found directly. Section 2.4 will provide further insight for specific distributions.

2.2 THE EXPECTED LIFE

The expected life, or the expected time during which a component will perform successfully, is defined as

$$E(\mathbf{t}) = \int_0^\infty \tau f(\tau) d\tau \tag{2.8}$$

Another convenient method for determining the expected life is given by

$$E(\mathbf{t}) = \int_0^\infty R(t) dt \tag{2.9}$$

In order to derive Equation 2.9, perform this integration by parts using

$$\int u\, dv = uv - \int v\, du$$

Let $u = R(t)$ and $dv = dt$. Then $du = -f(t)dt$ and $v = t$. The integral becomes

$$\int_0^\infty R(t)\, dt = \left[t \cdot R(t) \right]\Big|_0^\infty + \int_0^\infty t \cdot f(t)\, dt = E(\mathbf{t}) \tag{2.10}$$

since, when $t = 0$, $t \cdot R(t) = 0$, and we assume $\lim_{t \to \infty} [t \cdot R(t)] = 0$ for a finite mean.

When the system being tested is renewed through maintenance and repairs, $E(\mathbf{t})$ is also known as the mean time between failures or mean time to failure.

Much of renewal theory is based on the assumption that the behaviors of a repaired system and a new system are identical from a failure standpoint. In general, however, perfect renewal is not possible, and in such cases, terms such as the mean time to the first failure or the mean time to the second failure become relevant.

The mean time to failure (MTTF), or the mean time between failures (MTBF), should be used when the failure distribution function is specified, because the reliability level implied by the MTBF depends on the underlying failure distribution. A careful examination of Equations 2.3 and 2.10 will show that two failure distributions can have the same MTBF and yet produce different reliability levels. We will illustrate this by a case where the MTBFs are equal, but the failure distributions are normal and exponential. The normal failure distribution is symmetrical about its mean, thus

$$R\,(\text{MTBF}) = P\,(\mathbf{z} \geqslant 0) = 0.5$$

where $\mathbf{z}$ is a standard normal deviate. When we compute for the exponential failure distribution using Equation 2.5, recognizing that θ is equal to the MTBF, the reliability at the MTBF is

$$R\,(\text{MTBF}) = \exp\left(-\frac{\text{MTBF}}{\text{MTBF}}\right) = \exp(-1) = 0.368$$

Clearly, the reliability in the case of the exponential distribution is about 75% of that for the normal failure distribution with the same MTBF.

For one-shot situations where failure does not depend on time, the concept of the probability of success may be utilized. This concept is especially useful when considering components or systems that are used once and cannot be repaired, such as ammunition or solid fuel rocket engines. This approach to reliability, independent of time, will be discussed in some design examples.

2.3 THE FAILURE RATE AND HAZARD FUNCTION

The probability of failure of a system in a given time interval $[t_1, t_2]$ can be expressed in terms of either the unreliability function as

$$\int_{t_1}^{t_2} f(t)\,dt = \int_{-\infty}^{t_2} f(t)\,dt - \int_{-\infty}^{t_1} f(t)\,dt = F(t_2) - F(t_1)$$

or in terms of the reliability function as

$$\int_{t_1}^{t_2} f(t)\,dt = \int_{t_1}^{\infty} f(t)\,dt - \int_{t_2}^{\infty} f(t)\,dt = R(t_1) - R(t_2)$$

The rate at which failures occur in a certain time interval $[t_1, t_2]$ is called the failure rate during that interval. It is defined as the probability that a failure per unit time occurs in the interval, given that a failure has not occured prior to t_1, the beginning of the interval. Thus the failure rate is

$$\frac{R(t_1) - R(t_2)}{(t_2 - t_1)R(t_1)} \tag{2.11}$$

Note that the failure rate is a function of time.

If we redefine the interval as $[t, t+\Delta t]$, the expression in Equation 2.11 becomes

$$\frac{R(t) - R(t+\Delta t)}{\Delta t \cdot R(t)} \tag{2.12}$$

The "rate" in the above definitions is expressed as failures per unit time; in reality the "time" units might be kilometers, revolutions, etc., as previously discussed.

The hazard function is defined as the limit of the failure rate as the interval approaches zero. Thus the hazard function is the instantaneous failure rate. The hazard function $h(t)$ is defined by

$$\begin{aligned} h(t) &= \lim_{\Delta t \to 0} \frac{R(t) - R(t+\Delta t)}{\Delta t \cdot R(t)} = \frac{1}{R(t)}\left[-\frac{d}{dt}R(t) \right] \\ &= \frac{f(t)}{R(t)} \end{aligned} \tag{2.13}$$

The quantity $h(t)dt$ represents the probability that a device of age t will fail in the small interval of time t to $t+dt$. The importance of the hazard function is that it indicates the change in the failure rate over the life of a population of devices. For example, two designs may provide the same reliability at a specific point in time; however, the failure rates up to this point in time can differ.The death rate, in "actuarial theory," is analogous to the failure rate as the force of mortality is analogous to the hazard function.

The hazard function can also be derived by considering a population containing N identical items with a failure distribution function $F(t)$. Let $N_s(t)$ be a random variable denoting the number of items functioning successfully at time t. Then $N_s(t)$ has a binomial distribution given by

$$P\left[N_s(t) = n\right] = \frac{N!}{n!(N-n)!}\left[R(t)\right]^n\left[1 - R(t)\right]^{N-n}, \qquad n = 0, 1, \ldots, N$$

The expected value of $N_s(t)$ is $E[N_s(t)] = NR(t) = \bar{N}(t)$, or,

$$R(t) = \frac{E[N_s(t)]}{N} = \frac{\bar{N}(t)}{N} \tag{2.14}$$

Thus the reliability at time t is the average fraction of successfully functioning units at time t. Observe that

$$F(t) = 1 - R(t) = 1 - \frac{\bar{N}(t)}{N} = \frac{N - \bar{N}(t)}{N} \tag{2.15}$$

and

$$f(t) = \frac{dF(t)}{dt} = -\frac{1}{N}\frac{d\bar{N}(t)}{dt}$$

or

$$f(t) = \lim_{\Delta t \to 0} \frac{\bar{N}(t) - \bar{N}(t+\Delta t)}{N \cdot \Delta t} \tag{2.16}$$

It is clear that the failure density function is normalized in terms of the size of the original population N. However, it is often more meaningful to normalize with respect to the average number of units successfully functioning at time t, since this indicates the failure rate for those surviving. Replacing N with $\bar{N}(t)$ in Equation 2.16 leads to the hazard function

$$h(t) = \lim_{\Delta t \to 0} \frac{\bar{N}(t) - \bar{N}(t+\Delta t)}{\bar{N}(t) \cdot \Delta t}$$

$$= \frac{N}{\bar{N}(t)} f(t) \tag{2.17}$$

or

$$h(t) = \frac{f(t)}{R(t)} \tag{2.18}$$

The hazard function and the reliability function can be related yet another way. From Equation 2.17 we have

$$h(t) = -\frac{d\bar{N}(t)}{dt} \cdot \frac{1}{\bar{N}(t)} = -\frac{d}{dt}\left[\ln \bar{N}(t)\right]$$

or

$$\ln \bar{N}(t) = -\int_0^t h(\tau)\,d\tau + c \tag{2.19}$$

where c is the integration constant. Hence, we have

$$\bar{N}(t) = e^{c} \cdot \exp\left[-\int_0^t h(\tau)\,d\tau\right] \tag{2.20}$$

Now, $\bar{N}(0) = N = e^{c}$, which, after substituting into Equation 2.20, leads to

$$\bar{N}(t) = N \cdot \exp\left[-\int_0^t h(\tau)\,d\tau\right]$$

or the reliability and hazard functions are related by

$$R(t) = \frac{\bar{N}(t)}{N} = \exp\left[-\int_0^t h(\tau)\,d\tau\right] \tag{2.21}$$

From Equations 2.18 and 2.21, we can write

$$f(t) = h(t)\exp\left[-\int_0^t h(\tau)\,d\tau\right] \tag{2.22}$$

Thus $f(t)$, $R(t)$ and $h(t)$ are all related and any one implies the other two.

The development of the hazard function can also be approached from a conditional probability standpoint. Consider a device that has survived an interval of time $[0,t]$. What is the probability that it will fail over the next interval of time (t,t_1)? That is, we are interested in the conditional probability

$$P\left[t < \mathbf{t} \leqslant t_1 \mid \mathbf{t} > t\right] = \frac{P\left[(t < \mathbf{t} \leqslant t_1) \cap (\mathbf{t} > t)\right]}{P(\mathbf{t} > t)}$$

Realizing that

$$(t < \mathbf{t} \leqslant t_1) \cap (\mathbf{t} > t) = t < \mathbf{t} \leqslant t_1$$

yields

$$P\left[t < \mathbf{t} \leqslant t_1 \mid \mathbf{t} > t\right] = \frac{P(t < \mathbf{t} \leqslant t_1)}{P(\mathbf{t} > t)} = \frac{F(t_1) - F(t)}{1 - F(t)} = \frac{F(t_1) - F(t)}{R(t)} \tag{2.23}$$

If we replace t_1 with $t + \Delta t$ in Equation 2.23, divide both sides by Δt, and take the limit as $\Delta t \to 0$, we will obtain

$$\lim_{\Delta t \to 0} \frac{P[t<\mathbf{t}\leqslant t+\Delta t|\mathbf{t}>t]}{\Delta t} = \frac{1}{R(t)} \lim_{\Delta t \to 0} \frac{F(t+\Delta t)-F(t)}{\Delta t}$$

$$= \frac{1}{R(t)} \cdot \frac{dF(t)}{dt} = \frac{f(t)}{R(t)} \tag{2.24}$$

So, we can conclude that

$$h(t) = \lim_{\Delta t \to 0} \frac{P[t<\mathbf{t}\leqslant t+\Delta t|\mathbf{t}>t]}{\Delta t} \tag{2.25}$$

Thus $h(t)$ is the rate of change of the conditional probability of failure given that the system has survived up to time t. Also worth noting is that $f(t)$ is the time rate of change of the ordinary probability of failure.

2.4 THE RELIABILITY AND HAZARD FUNCTION FOR WELL-KNOWN DISTRIBUTION FUNCTIONS

A reliability function and its related hazard function are unique. Thus each reliability function has only one hazard function and vice versa. In the following section we examine some of the common failure density functions and their related hazard functions. We will also look at several hazard models.

2.4.1 EXPONENTIAL DISTRIBUTION

The exponential distribution is widely used in reliability. It should not be used indiscriminately, since there exist numerous situations where it clearly does not apply. For this distribution, we merely reproduce Equations 2.4 and 2.5

$$f(t) = \frac{1}{\theta} e^{-t/\theta}, \qquad t \geqslant 0$$

where θ is a parameter such that $\theta > 0$, and

$$R(t) = e^{-t/\theta}, \qquad t \geqslant 0$$

The form of the exponential density function is illustrated in Figure 2.1.

The hazard function for the exponential density function is constant. It is given by

$$h(t) = \frac{f(t)}{R(t)} = \frac{1}{\theta} \tag{2.26}$$

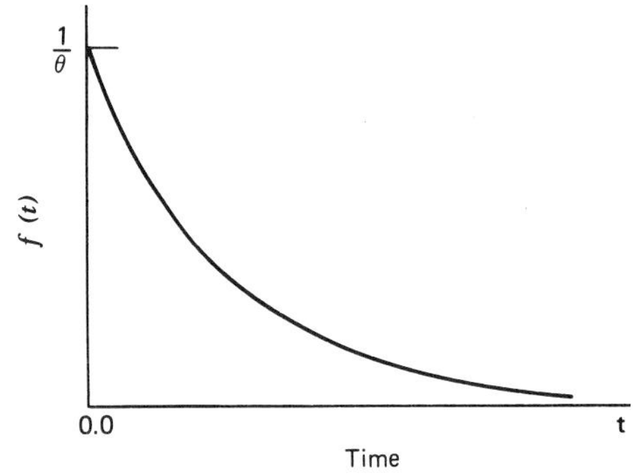

Figure 2.1 The exponential failure density function.

Chapter 10 is devoted entirely to the application of this density function in life testing.

2.4.2 NORMAL DISTRIBUTION

The normal distribution takes the well-known bell shape. This distribution is symmetrical about its mean value.

In this case, the cumulative distribution is

$$F(t) = P[\mathbf{t} \leqslant t] = \int_{-\infty}^{t} \frac{1}{\sigma\sqrt{2\pi}} \exp\left[-\frac{1}{2}\left(\frac{\tau-\mu}{\sigma}\right)^2\right] d\tau \qquad (2.27)$$

and

$$R(t) = 1 - F(t)$$

The integral cannot be evaluated in a closed form; however, tables for the standard normal denisty function are readily available and can be used to find probabilities for any normal distribution.

The probability density function (p.d.f.) for a standard normal distribution is given by

$$\phi(z) = \frac{1}{\sqrt{2\pi}} \exp(-z^2/2), \qquad -\infty < z < \infty \qquad (2.28)$$

Then the standardized cumulative distribution function (c.d.f.) is

$$\Phi(z) = \int_{-\infty}^{z} \frac{1}{\sqrt{2\pi}} \exp(-\tau^2/2)\, d\tau \qquad (2.29)$$

Then, for a normally distributed random variable **t**, with mean μ and standard deviation σ

$$P(\mathbf{t} \leqslant t) = P\left(\mathbf{z} \leqslant \frac{t-\mu}{\sigma}\right) = \Phi\left(\frac{t-\mu}{\sigma}\right) \qquad (2.30)$$

yields the relationship necessary if standard normal tables (such as those given in Appendix II) are to be used.

The hazard function for a normal distribution is a monotonically increasing function of t. This can be easily shown by proving that $h'(t) \geqslant 0$ for all t. Since

$$h(t) = \frac{f(t)}{1 - F(t)}$$

then

$$h'(t) = \frac{(1-F)f' + f^2}{(1-F)^2} \tag{2.31}$$

The denominator in Equation 2.31 is nonnegative for all t. Hence, it is sufficient to show that the numerator

$$(1-F)f' + f^2 \geqslant 0$$

The reader is encouraged to try this proof by employing the basic definition of a normal density function f.

The reliability function and the hazard function for some normal random variables are shown in Figures 2.2 and 2.3, respectively.

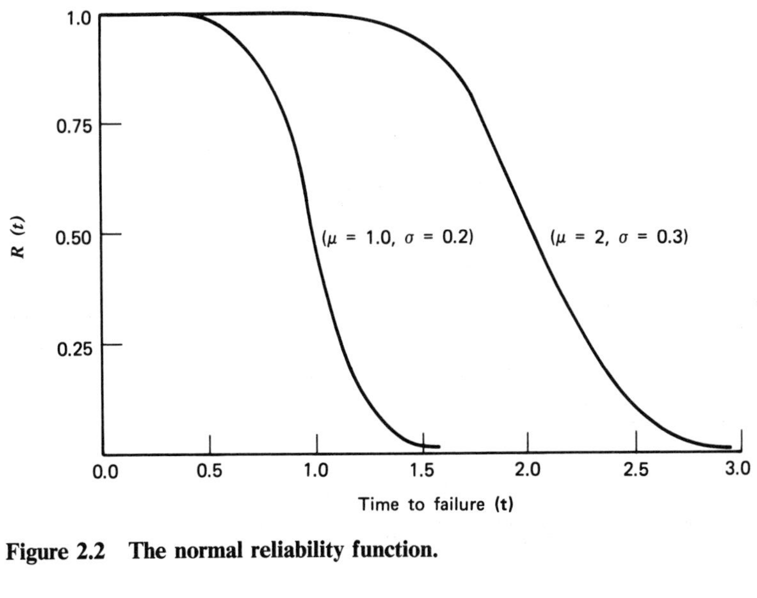

Figure 2.2 The normal reliability function.

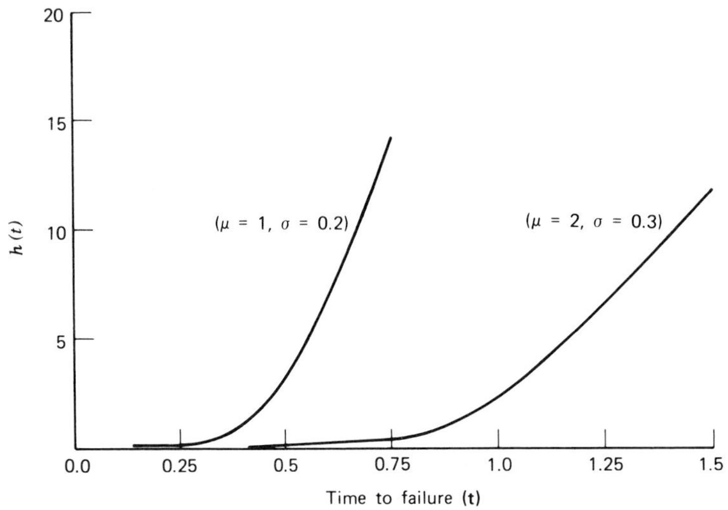

Figure 2.3 The normal hazard function.

EXAMPLE 2.1

A component has a normal distribution of failure times with $\mu=20{,}000$ cycles and $\sigma=2{,}000$ cycles. Find the reliability of the component and the hazard function at 19,000 cycles.

solution

The reliability function is related to the standard normal deviate $\mathbf{z}$ by

$$R(t)=P\left(\mathbf{z}>\frac{t-\mu}{\sigma}\right)$$

where the distribution function for $\mathbf{z}$ is given by Equation 2.29.

For this particular application,

$$R(19{,}000)=P\left[\mathbf{z}>\frac{19{,}000-20{,}000}{2{,}000}\right]$$

$$=P[\mathbf{z}>-0.5]=\Phi(0.5)$$

where the value of $\Phi(z)$ as defined in Equation 2.29 is found in Appendix II. From the tables we see that

$$R(19{,}000)=0.69146$$

The value of the hazard function is found from the relationship

$$h(t)=\frac{f(t)}{R(t)}=\frac{\phi\left(\mathbf{z}=\frac{t-\mu}{\sigma}\right)/\sigma}{R(t)}$$

where $\phi(z)$ is given by Equation 2.28 and found in Appendix I. Here

$$h(19{,}000)=\frac{\phi(-0.50)}{\sigma R(t)}=\frac{0.3521}{(2{,}000)(0.69146)}=0.000254 \quad \text{failures/cycle}$$

2.4.3 LOG NORMAL DISTRIBUTION

The log normal density function is

$$f(t)=\frac{1}{\sigma t\sqrt{2\pi}}\exp\left[-\frac{1}{2}\left(\frac{\ln t-\mu}{\sigma}\right)^2\right], \qquad t\geqslant 0 \tag{2.32}$$

where μ and σ are parameters such that $-\infty<\mu<\infty$ and $\sigma>0$. Various forms of the log normal density function are illustrated in Figure 2.4.

If a random variable $\mathbf{x}$ is defined as $\mathbf{x}=\ln \mathbf{t}$, then $\mathbf{x}$ is normally distributed with a mean of μ and standard deviation of σ. That is,

$$E(\mathbf{x})=E(\ln \mathbf{t})=\mu$$

and

$$V(\mathbf{x})=V(\ln \mathbf{t})=\sigma^2$$

Since $t=e^x$, the mean of the log normal distribution can be found by using the normal distribution. Consider that

$$E(\mathbf{t})=E(e^{\mathbf{x}})=\int_{-\infty}^{\infty}\frac{1}{\sigma\sqrt{2\pi}}\exp\left[x-\frac{1}{2}\left(\frac{x-\mu}{\sigma}\right)^2\right]dx$$

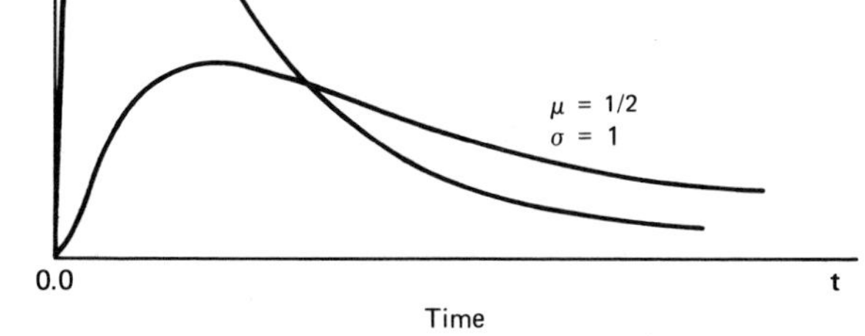

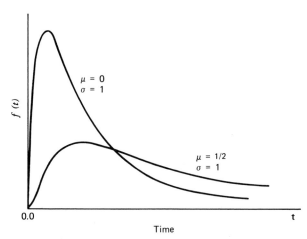

Figure 2.4 The failure density function for the log normal distribution.

and by rearrangement of the exponent this integral becomes

$$E(\mathbf{t}) = \exp\left[\mu + \frac{\sigma^2}{2}\right] \int_{-\infty}^{\infty} \frac{1}{\sigma\sqrt{2\pi}} \exp\left[-\frac{1}{2\sigma^2}\left\{x - (\mu + \sigma^2)\right\}^2\right] dx$$

Thus the mean of the log normal distribution is

$$E(\mathbf{t}) = \exp\left[\mu + \frac{\sigma^2}{2}\right]$$

Proceeding in a similar manner

$$E(\mathbf{t}^2) = E(e^{2\mathbf{x}}) = \exp\left[2(\mu + \sigma^2)\right]$$

and thus the variance for the log normal is

$$V(\mathbf{t}) = \left[e^{2\mu + \sigma^2}\right]\left[e^{\sigma^2} - 1\right]$$

The cumulative distribution function for the log normal is

$$F(t) = \int_0^t \frac{1}{\tau\sigma\sqrt{2\pi}} \exp\left[-\frac{1}{2}\left(\frac{\ln\tau - \mu}{\sigma}\right)^2\right] d\tau \tag{2.33}$$

and this can be related to the standard normal deviate **z** by

$$F(t) = P[\mathbf{t} \leqslant t] = P\left[\mathbf{z} \leqslant \frac{\ln t - \mu}{\sigma}\right]$$

The reliability function is

$$R(t) = P[\mathbf{t} > t] = P\left[\mathbf{z} > \frac{\ln t - \mu}{\sigma}\right] \tag{2.34}$$

Thus the hazard function would be

$$h(t) = \frac{f(t)}{R(t)} = \frac{\phi\left(\dfrac{\ln t - \mu}{\sigma}\right)}{t\sigma R(t)} \tag{2.35}$$

where ϕ is the standard normal probability density function and μ and σ are, respectively, the mean and the standard deviation of the natural logarithm of the random variable **t**, the time to failure. The reliability function and hazard function are shown in Figures 2.5 and 2.6 for various log normal distributions.

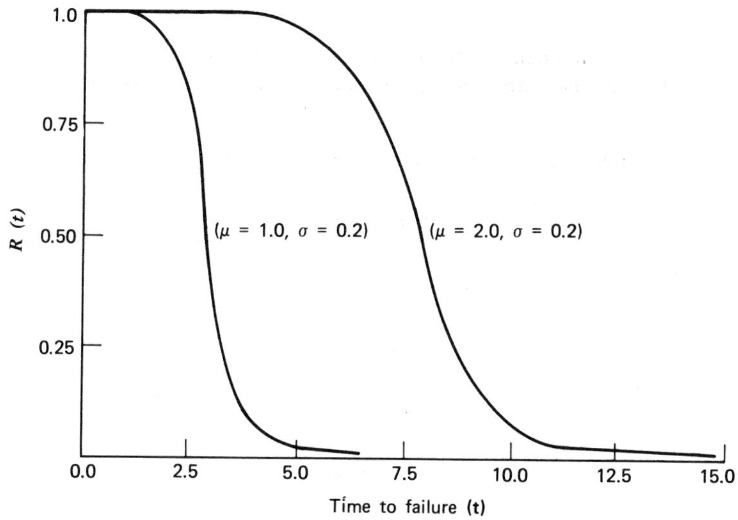

Figure 2.5 The log normal reliability function.

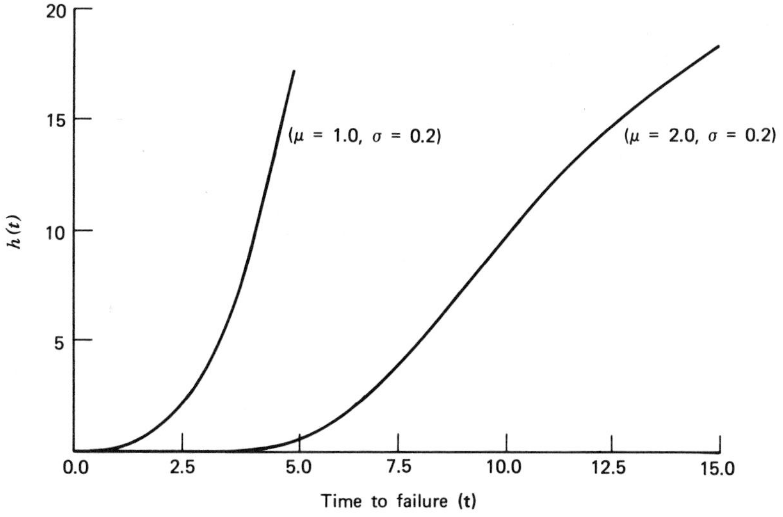

Figure 2.6 The log normal hazard function.

EXAMPLE 2.2

The failure time of a certain component is log normally distributed with $\mu = 5$ and $\sigma = 1$. Find the reliability of the component and the hazard rate for a life of 150 time units.

solution

Substituting the numerical values of μ, σ, and t into Equation 2.34, we compute

$$R(150) = P\left[\mathbf{z} > \frac{\ln 150 - 5}{1}\right] = P[\mathbf{z} > 0.01] = 0.496$$

and using Equation 2.35 for the hazard function

$$h(150) = \frac{\phi\left(\frac{\ln 150 - 5}{1}\right)}{(150)(1)(0.496)} = \frac{\phi(0.01)}{(150)(0.496)} = \frac{0.399}{(150)(0.496)}$$

$$= 0.0053 \text{ failures/unit time}$$

Thus values for the log normal distribution are easily computed by using the standard normal tables.

2.4.4 WEIBULL DISTRIBUTION

The Weibull failure density function is

$$f(t) = \frac{\beta(t-\delta)^{\beta-1}}{(\theta-\delta)^{\beta}} \exp\left[-\left(\frac{t-\delta}{\theta-\delta}\right)^{\beta}\right], \qquad t \geqslant \delta \geqslant 0 \tag{2.36}$$

where β, known as the shape parameter, and $(\theta - \delta)$, known as the scale parameter, are always positive. It can be easily shown that, for $t \geqslant \delta$,

$$R(t) = 1 - F(t) = \exp\left[-\left(\frac{t-\delta}{\theta-\delta}\right)^{\beta}\right] \tag{2.37}$$

and hence

$$h(t) = \frac{f(t)}{R(t)} = \frac{\beta(t-\delta)^{\beta-1}}{(\theta-\delta)^{\beta}} \tag{2.38}$$

Various forms of the failure density function are illustrated in Figure 2.7, where δ is taken as zero.

Figures 2.8 and 2.9 depict the varying shapes of $R(t)$ and $h(t)$, respectively, with changing β. The hazard function is decreasing for $\beta < 1$, increasing for $\beta > 1$, and constant when β is exactly one.

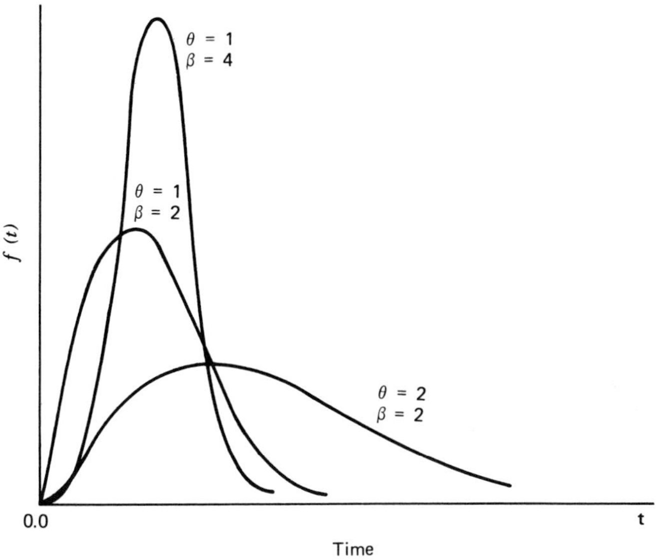

Figure 2.7 The failure density function for the Weibull distribution.

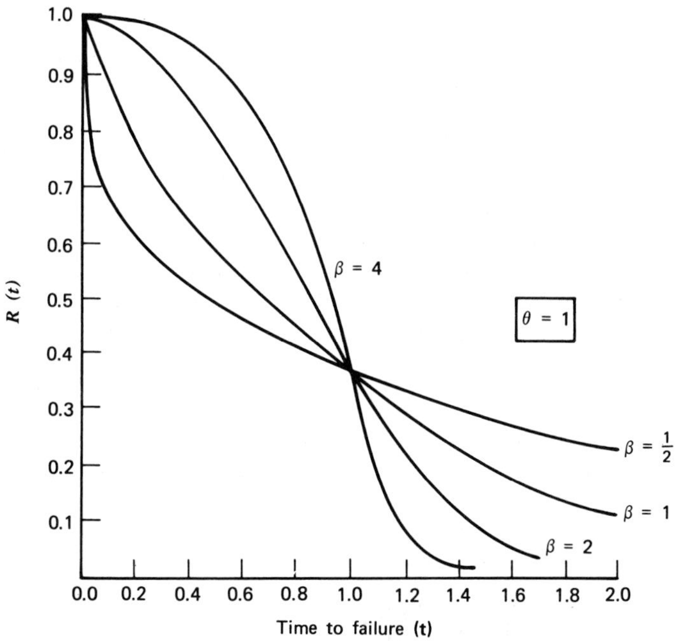

Figure 2.8 The Weibull reliability function.

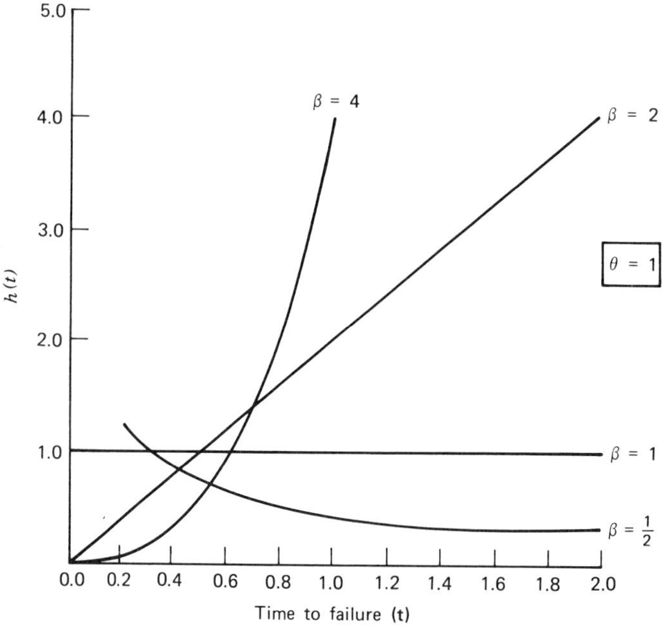

Figure 2.9 The Weibull hazard function.

In particular, consider the Weibull distribution with $\beta = 1$ and $\delta = 0$. The expression for the reliability function in Equation 2.37 reduces to

$$R(t) = \exp\left[-\frac{t}{\theta}\right] \tag{2.39}$$

and the hazard function given in Equation 2.38 reduces to $1/\theta$, a constant. Thus the exponential is a special case of the Weibull distribution.

When $\beta < 1$ the Weibull distribution takes a hyperexponential shape. For $\beta = 3.5$, the distribution is approximately symmetrical starting at δ, while for $\beta > 3.5$ the distribution moves away from δ along the t axis and becomes negatively skewed.

EXAMPLE 2.3

The failure time of a certain component has a Weibull distribution with $\beta = 4$, $\theta = 2{,}000$, and $\delta = 1{,}000$. Find the reliability of the component and the hazard rate for an operating time of 1,500 hours.

solution

A direct substitution into Equation 2.37 yields

$$R(1{,}500) = \exp\left[-\left(\frac{1500-1000}{2000-1000}\right)^4\right] = \exp[-0.0625]$$
$$= 0.939$$

Using Equation 2.38, the desired hazard function is given by

$$h(1{,}500)=\frac{4(1500-1000)^{4-1}}{(2000-1000)^{4}}=\frac{4(500)^{3}}{(1000)^{4}}=0.0005 \text{ failures/hour}$$

2.4.5 GAMMA DISTRIBUTION

The failure density function for a gamma distribution is

$$f(t)=\frac{\lambda^{\eta}}{\Gamma(\eta)}t^{\eta-1}e^{-\lambda t}, \qquad t\geqslant 0, \quad \eta>0, \quad \lambda>0 \tag{2.40}$$

where η is the shape parameter and λ is the scale parameter. Hence,

$$F(t)=\int_0^t \frac{\lambda^{\eta}}{\Gamma(\eta)}\tau^{\eta-1}e^{-\lambda\tau}\,d\tau$$

If η is an integer, it can be shown by successive integration by parts that

$$F(t)=\sum_{k=\eta}^{\infty}\frac{(\lambda t)^{k}\exp[-\lambda t]}{k!} \tag{2.41}$$

Then

$$R(t)=1-F(t)=\sum_{k=0}^{\eta-1}\frac{(\lambda t)^{k}\exp[-\lambda t]}{k!} \tag{2.42}$$

and

$$h(t)=\frac{f(t)}{R(t)}=\frac{\dfrac{\lambda^{\eta}}{\Gamma(\eta)}t^{\eta-1}e^{-\lambda t}}{\displaystyle\sum_{k=0}^{\eta-1}\frac{(\lambda t)^{k}\exp[-\lambda t]}{k!}} \tag{2.43}$$

Typical examples of $R(t)$ and $h(t)$ are shown in Figures 2.10 and 2.11, respectively. The gamma failure density function has shapes that are very similar to the Weibull distribution.

The gamma distribution can also be used to model the time to the nth failure of a system if the underlying failure distribution is exponential. That is, if $\mathbf{x}_i$ is exponentially distributed with parameter $\theta=1/\lambda$, then $\mathbf{t}=\mathbf{x}_1+\mathbf{x}_2+\cdots+\mathbf{x}_n$ is gamma distributed with parameters λ and $\boldsymbol{n}$.

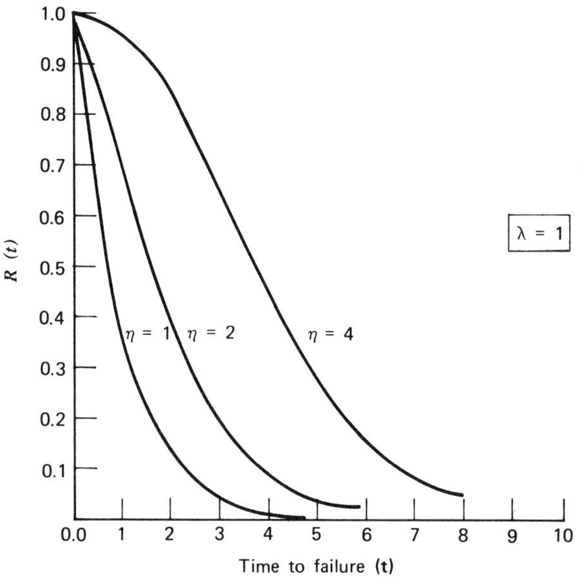

Figure 2.10 The gamma reliability function.

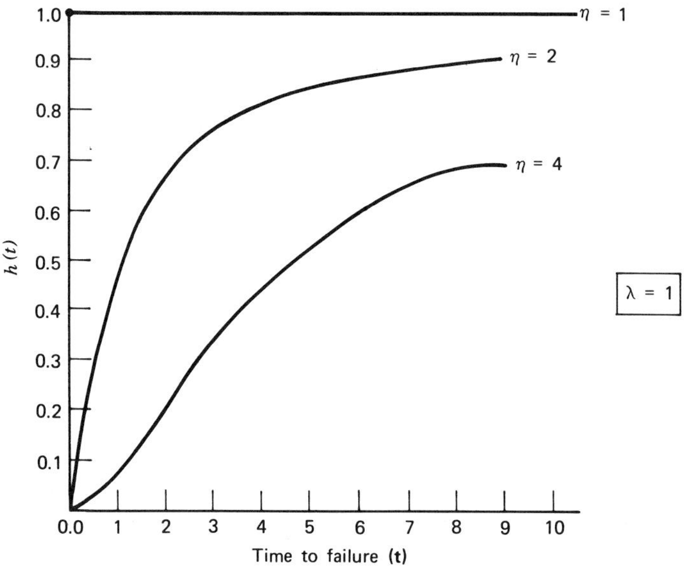

Figure 2.11 The gamma hazard function.

EXAMPLE 2.4

The time to failure of a component has a gamma distribution with $\eta=3$ and $\lambda=0.05$. Determine the reliability of the component and the hazard rate at 24 time units.

solution

Using Equation 2.42, we compute

$$R(24)=\sum_{k=0}^{2}\frac{(0.05\times 24)^k \exp[-0.05\times 24]}{k!}=0.88$$

Using Equation 2.40, we compute

$$f(24)=\frac{(0.05)^3(24)^2\exp[-0.05\times 24]}{\Gamma(3)}=0.011$$

Then

$$h(24)=\frac{f(24)}{R(24)}=\frac{0.011}{0.88}=0.0125 \text{ failures/unit time}$$

This completes the presentation of the density functions and their related reliability and hazard functions. We now look at some hazard functions.

2.5 HAZARD MODELS AND PRODUCT LIFE

The hazard function $h(t)$ will change over the lifetime of a population of products somewhat as shown in Figure 2.12. The first interval of time, t_0 to t_1, represents early failures due to material or manufacturing defects. Quality control and initial product testing usually eliminate many substandard devices, and thus avoid this higher initial failure rate. Actuarial statisticians call this phase of the curve "infant mortality."

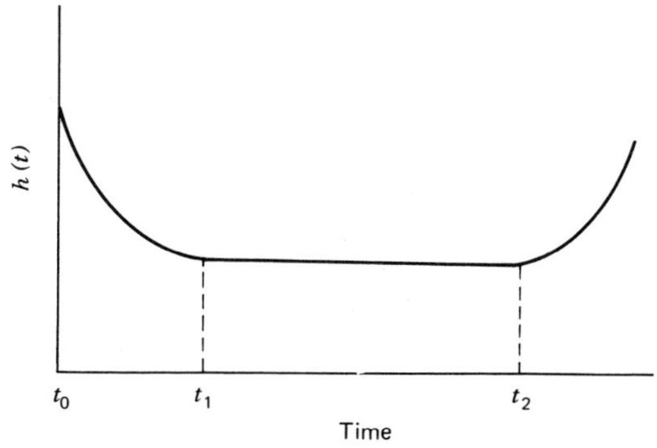

Figure 2.12 Hazard function over product life.

The second phase of the curve, from t_1 to t_2, represents chance failures caused by sudden stresses, extreme conditions, etc. In actuarial terms this could be equated to the accidents encountered by a population of individuals on a day-to-day basis. Unfortunately, in mechanical devices such as automobiles, these failures can be eliminated only by designing for the extreme conditions. But by doing so the vehicle is overdesigned for the vast majority of situations.

The portion of the curve beyond t_2 represents wearout failures. Here the hazard rate increases as equipment deteriorates. If time t_2 could be predicted with certainty, then equipment could be replaced before this wearout phase begins.

In some instances, it may be more convenient to select a hazard function model than a distribution model. So next we consider various kinds of hazard models.

2.5.1 CONSTANT HAZARD FUNCTION

We first consider a model in which the hazard function is constant. As discussed before, from Figure 2.12, the constant hazard rate represents the portion of the hazard function over product life between times t_1 and t_2. During this time chance or random failures occur, and failure cannot be attributed to deterioration over time in the strength of the components. Let the hazard function be given by

$$h(t)=\lambda \text{ failures/unit time} \tag{2.44}$$

where λ is a constant. From Equation 2.22 we have

$$f(t)=h(t)\exp\left[-\int_0^t h(\tau)\,d\tau\right]$$

which in this case gives

$$f(t)=\lambda e^{-\lambda t} \tag{2.45}$$

The reliability function is given by

$$R(t)=\frac{f(t)}{h(t)}=e^{-\lambda t} \tag{2.46}$$

Thus a constant hazard rate leads to the exponential density function for the time to failure random variable.

2.5.2 LINEARLY INCREASING HAZARD FUNCTION

The increasing hazard function represents the wearout portion of the curve in Figure 2.12 after time t_2. In general, this is a nonlinear function. Here we will assume it to be linear for the sake of simplicity. Let

$$h(t)=ct \tag{2.47}$$

where c is a positive constant. Then, by Equation 2.22, we have

$$f(t)=ct\exp\left[-\int_0^t c\tau\,d\tau\right]=ct\exp\left[-\frac{ct^2}{2}\right],\qquad t\geqslant 0 \tag{2.48}$$

The density function in Equation 2.48 corresponds to the Rayleigh density function. The reliability function is

$$R(t)=\frac{f(t)}{h(t)}=\exp\left[-\frac{ct^2}{2}\right]$$

2.5.3 PIECEWISE LINEAR BATHTUB HAZARD FUNCTION

The decreasing hazard rate corresponds to the time t_0 to t_1 in the hazard function plot and suggests initial failure rate. Generally, after a system is produced, or assembled, and placed in operation, the initial failure rate is higher than that encountered later. The initial failures may be due to various manufacturing and assembling defects that escape detection by the quality control system. As the defective parts are replaced with new ones, the reliability improves; this phenomenon is sometimes facetiously referred to as servicing-in reliability. The following model is an example of a piecewise linear bathtub hazard model as given by Reference 17. For this model

$$h(t)=\begin{cases} c_0-c_1t+\lambda, & 0\leqslant t\leqslant c_0/c_1\\ \lambda, & c_0/c_1<t\leqslant t_0\\ c(t-t_0)+\lambda, & t_0<t\end{cases} \tag{2.49}$$

where $\lambda>0$. This hazard function linearly decreases to λ at time c_0/c_1, remains constant until time t_0, and then linearly increases. The resulting density function must be defined over three regions, and using Equation 2.22 it is

$$f(t)=\begin{cases} (c_0+\lambda-c_1t)\exp\{-[(c_0+\lambda)t-c_1(t^2/2)]\}, & 0\leqslant t\leqslant c_0/c_1\\ \lambda\exp\{-(\lambda t+c_0^2/2c_1)\}, & c_0/c_1<t\leqslant t_0\\ [c(t-t_0)+\lambda]\exp\{-[(c/2)(t-t_0)^2+(c_0^2/2c_1)+\lambda t]\}, & t_0<t\end{cases} \tag{2.50}$$

Then the reliability function is defined over the three regions as

$$R(t)=\begin{cases} \exp\{-[(c_0+\lambda)t-c_1(t^2/2)]\}, & 0\leqslant t\leqslant c_0/c_1\\ \exp\{-(\lambda t+c_0^2/2c_1)\}, & c_0/c_1<t\leqslant t_0\\ \exp\{-[(c/2)(t-t_0)^2+\lambda t+(c_0^2/2c_1)]\}, & t_0<t\end{cases} \tag{2.51}$$

This hazard model is quite versatile and can be fitted to empirically calculated hazard functions as obtainable from test data.

2.5.4 POWER FUNCTION MODEL

A hazard function might be characterized by a power function. If we parameterize the power function as

$$h(t)=\frac{\beta t^{\beta-1}}{\theta^{\beta}} \tag{2.52}$$

then

$$f(t)=\frac{\beta t^{\beta-1}}{\theta^{\beta}}\exp\left[-\left(\frac{t}{\theta}\right)^{\beta}\right] \tag{2.53}$$

and

$$R(t)=\exp\left[-\left(\frac{t}{\theta}\right)^{\beta}\right] \tag{2.54}$$

This parameterization gives us the Weibull distribution. By choosing different values of β, we can obtain decreasing, constant, or increasing hazard functions.

2.5.5 EXPONENTIAL MODEL

If the hazard function increases or decreases sharply, exhibiting exponential behavior, we can use the model

$$h(t)=ce^{\alpha t} \tag{2.55}$$

Then

$$f(t)=ce^{\alpha t}e^{-(c/\alpha)(e^{\alpha t}-1)} \tag{2.56}$$

and

$$R(t)=e^{-(c/\alpha)(e^{\alpha t}-1)} \tag{2.57}$$

The density function in this case is said to be a form of the extreme value distribution. This distribution is discussed further in Section 2.7.4. The nature of the above hazard function depends on the values of the constants c and α.

2.6 ESTIMATING THE HAZARD FUNCTION, FAILURE DENSITY FUNCTION, AND RELIABILITY FUNCTION FROM EMPIRICAL DATA

We now discuss the procedures used to estimate the various reliability measures from empirical data. Both a small and a large sample are used in illustrating the procedures.

The data in Table 2.1 represent kilocycles to failure for eight springs. These data will be used to illustrate the procedure for a small sample.

Table 2.1 Failure data for eight springs

FAILURE NUMBER	KILOCYCLES TO FAILURE
1	190
2	245
3	265
4	300
5	320
6	325
7	370
8	400

For small samples, the cumulative distribution at the ith ordered failure time t_i is estimated by

$$\hat{F}(t_i) = (i - 0.3)/(n + 0.4) \tag{2.58}$$

where n is the sample size. The rationale for this estimator is developed in Section 11.2.

The reliability function is estimated by

$$\begin{aligned} \hat{R}(t_i) &= 1 - \hat{F}(t_i) \\ &= (n - i + 0.7)/(n + 0.4) \end{aligned} \tag{2.59}$$

From the interpretation of Equations 2.11 and 2.12, the failure rate is calculated by

$$\hat{h}(t_i) = \frac{\hat{R}(t_i) - \hat{R}(t_{i+1})}{(t_{i+1} - t_i)\hat{R}(t_i)} \tag{2.60}$$

Substituting the estimator for $\hat{R}(t_i)$ into this equation yields

$$\hat{h}(t_i) = 1/\left[(t_{i+1} - t_i)(n - i + 0.7)\right] \tag{2.61}$$

The failure rate $\hat{h}(t_i)$, as calculated from the data, is an estimate of the hazard function.

The ordinate heights of the failure density function at time t_i can be developed from the interpretation of Equation 2.16. The estimator is given by

$$\hat{f}(t_i) = \frac{\hat{R}(t_i) - \hat{R}(t_{i+1})}{(t_{i+1} - t_i)} \tag{2.62}$$

which simplifies to

$$\hat{f}(t_i) = 1/\left[(n+0.4)(t_{i+1} - t_i)\right] \tag{2.63}$$

The computations for the spring test data are shown in Table 2.2. Because the short interval of time between failures five and six produced a large increase in $\hat{h}(t)$, this interval was combined with the previous interval. With empirical data this kind of smoothing must frequently be done.

Table 2.2 Computation of reliability measures for the spring test data

FAILURE NUMBER	t	$\hat{F}(t)$	$\hat{R}(t)$	$t_{i+1} - t_i$	$\hat{f}(t)$	$\hat{h}(t)$
1	190	0.083	0.917	55	0.0022	0.0024
2	245	0.202	0.798	20	0.0060	0.0075
3	265	0.321	0.679	35	0.0034	0.0050
4	300	0.440	0.560	20	0.0059	0.0171*
5	320	0.560	0.440	5	0.0248	
6	325	0.679	0.321	45	0.0025	0.0082
7	370	0.798	0.202	30	0.0040	0.0198
8	400	0.917	0.083	–	–	–

*This value of the hazard rate was obtained by combining intervals four and five together and thus considering it as a single interval of 20+5=25 kilocycles.

The reliability and hazard rate estimates are plotted in Figures 2.13 and 2.14. In analyzing small samples, extreme caution must be used because one spurious observation has considerable influence. However, from Figure 2.14, the hazard function appears to be increasing, which suggests that a distribution such as the Weibull or Rayleigh be used for the theoretical failure density function.

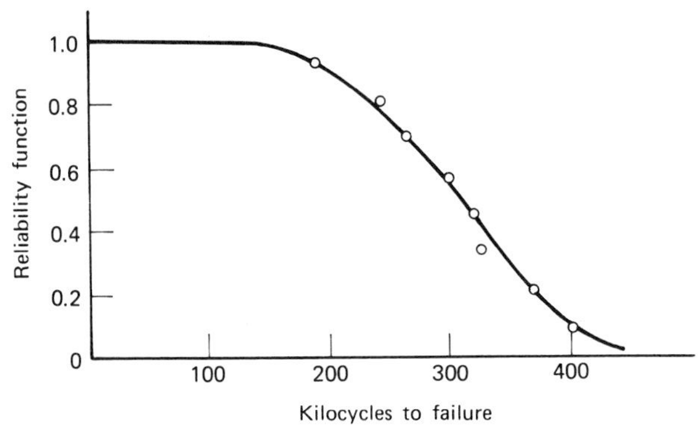

Figure 2.13 Reliability function—spring test data.

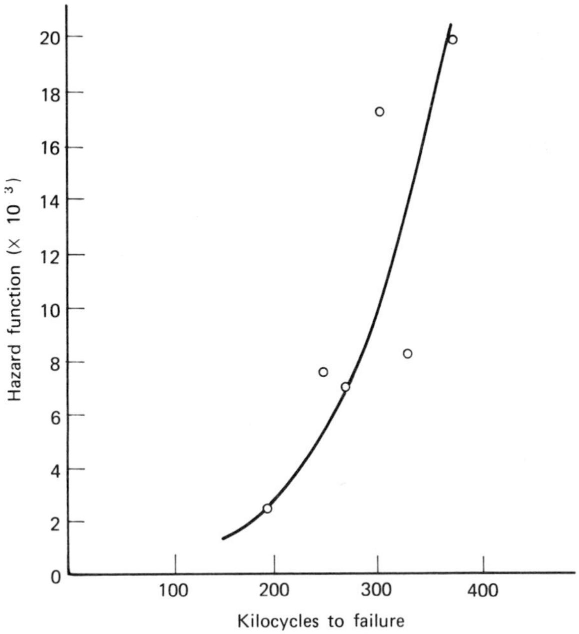

Figure 2.14 Hazard function—spring test data.

A second example will illustrate the calculations for larger samples. The data in Table 2.3 represents drive shaft failures on 46 test cars. In this particular test failure occured when the drive shaft deteriorated to such a point that excess noise was produced.

Table 2.3 Drive Shaft Failure Data

INTERVAL (km)	NUMBER OF FAILURES
$0 \leqslant m \leqslant 20{,}000$	19
$20{,}000 < m \leqslant 40{,}000$	11
$40{,}.000 < m \leqslant 60{,}000$	7
$60{,}000 < m \leqslant 80{,}000$	5
$80{,}000 < m \leqslant 100{,}000$	4
$100{,}000 < m$	0

In this example the data is grouped into class intervals of 20,000 kilometers each. The reliability function is estimated by Equation 2.14 which is

$$R(t) = \frac{\bar{N}(t)}{N}$$

where $\bar{N}(t)$ is the number of units surviving at time t and N is the total number originally put on test. The first value of t, t_o, is taken to be zero.

Estimating the reliability function for $t=20{,}000$ km yields

$$\hat{R}(20{,}000)=27/46=0.587$$

The hazard rate is calculated by a direct application of Equation 2.17 which is

$$h(t)=\frac{\bar{N}(t)-\bar{N}(t+\Delta t)}{\bar{N}(t)\Delta t}$$

where $\bar{N}(t)$ is the number of surviving components at time t and Δt is the class interval. For the first group in Table 2.4 we calculate

$$\hat{h}(t)=\frac{46-27}{46(20{,}000)}=0.207\times 10^{-4}$$

The failure density function is estimated from Equation 2.16. Thus for the second interval

$$\hat{f}(40{,}000)=\frac{27-16}{46(20{,}000)}=0.120\times 10^{-4}$$

Table 2.4 Computation of the reliability measures for the drive shaft failure data

t	$N-\bar{N}(t)$	$\hat{F}(t)$	$\hat{R}(t)$	$\hat{f}(t)\times 10^4$	$\hat{h}(t)\times 10^4$
20,000	19	0.413	0.587	0.207	0.207
40,000	30	0.652	0.348	0.120	0.204
60,000	37	0.804	0.196	0.076	0.219
80,000	42	0.913	0.087	0.055	0.278
100,000	46	1.000	0.0	0.044	0.500

The reliability measures for the drive shaft failure data are plotted in Figures 2.15, 2.16, and 2.17. From the plot of the hazard rate and the failure density function it appears that the exponential distribution would be a good choice for a theoretical model for kilometers to failure. In the plot of the hazard function the large increase in the last interval is an inevitable result of the discreteness of the calculation method.

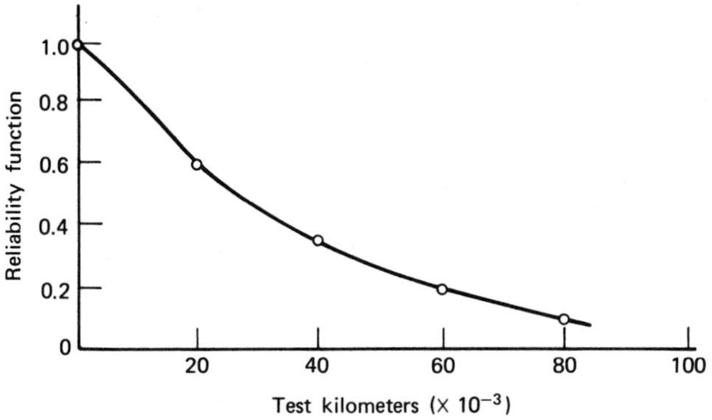

Figure 2.15 Reliability function—drive shaft test data.

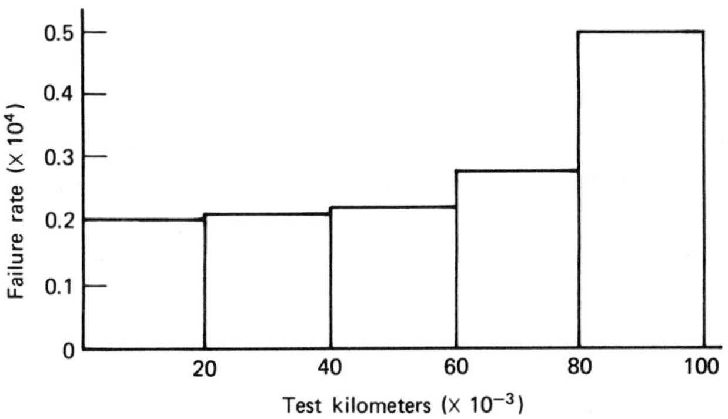

Figure 2.16 Failure rate—drive shaft test data.

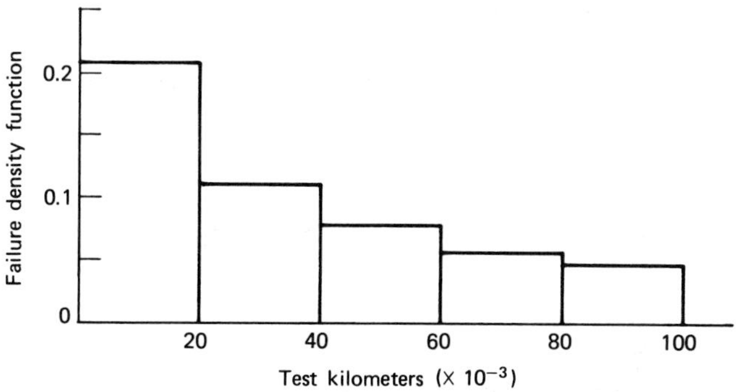

Figure 2.17 Failure density function—drive shaft test data.

2.7 SOME COMMENTS ON DISTRIBUTION SELECTION

In practice, one is usually forced to select a distribution model without having enough data to actually verify its appropriateness. The selection must be based on either previous experience or knowledge of the particular physical situation causing the failures. So, if it were possible, it would indeed be very desirable to be able to identify associations between certain physical situations and a particular model. Unfortunately the problem is complex. The following section looks into the reasons for this complexity and the manner in which some of the various distribution models are developed.

2.7.1 EXPONENTIAL, WEIBULL, AND RELATED MODELS

Consider a situation where the environment causes extreme disturbances to a system, for example, road shocks received by a vehicle due to bumps and holes, voltage surges in electrical appliances, or gusts of wind on buildings or airframes. Let us assume that

1. The disturbances occur completely at random and independently. That is, the occurrence of a disturbance does not provide any information as to when the next disturbance will occur.
2. The probability of an occurrence during any interval of time Δt is proportional to the length of the interval, the constant of proportionality being λ. That is, P[one disturbance during Δt]$=\lambda\Delta t$.
3. A function $p(t)$ denotes the probability that the system will fail if a disturbance does occur at time t.

If $R(t)$ is defined to be the probability that the system is surviving at the time t, then, for a small interval of time Δt, we have

$$R(t+\Delta t)=P\left\{\begin{array}{l}[(\text{system surviving at time } t \cap \text{no disturbances occur during } \Delta t) \cup (\text{system}\\ \text{surviving at time } t \cap \text{one disturbance occurs} \cap \text{no failure occurs}) \cup \ldots]\end{array}\right\}$$

$$=R(t)(1-\lambda\Delta t)+R(t)\lambda\Delta t\left[1-p(t)\right]+R(t)(\lambda\Delta t)^2\left[1-p(t)\right]^2+\ldots$$

Rearranging and dividing by Δt yields

$$\frac{R(t+\Delta t)-R(t)}{\Delta t}=-\lambda R(t)p(t)+\left(\text{terms of the order } (\Delta t)^n, n\geqslant 1\right)$$

Taking the limit as $\Delta t\to o$, we get

$$\frac{dR(t)}{dt}=-\lambda R(t)p(t)$$

Rearranging and solving for $R(t)$ gives

$$\int \frac{dR(t)}{R(t)} = -\int \lambda p(t)\,dt$$

or

$$\ln R(t) = -\int \lambda p(t)\,dt$$

or

$$R(t) = e^{-\int \lambda p(t)\,dt}$$

Define

$$P(t) = \int_0^t p(\tau)\,d\tau$$

Then

$$R(t) = e^{-\lambda P(t)} \tag{2.64}$$

and

$$f(t) = \lambda p(t) e^{-\lambda P(t)}, \qquad t \geqslant 0 \tag{2.65}$$

We can now observe how a disturbance must affect a system over time for different distributions to apply. A constant value of unity for $p(t)$ would mean that any disturbance would cause system failure. Here $f(t)$ would be exponential.

If $\lambda p(t)$ is a power function of t given by $(\beta/\theta)(t/\theta)^{\beta-1}$, then $f(t)$ is Weibull. In this case, as time passes, the system will deteriorate if $\beta > 1$; that is, a disturbance will have a greater chance of causing a failure with increasing age. The gamma and Rayleigh distributions are, of course, the other contenders for this situation. Specifically, a linear function with $\beta = 2$ would give rise to the Rayleigh distribution.

Let us suppose that Figure 2.18 is a typical pattern of stress in an automotive front suspension component and $\sigma_{\max}$ is the upper stress limit which, if exceeded, will result in a failure. For the time being we will assume that this upper limit is fixed. This implies that the component does not undergo any changes over time, such as work hardening due to previous stressing.

The stress is changing directions in a random manner and, for any two nonadjacent intervals of time such as Δt_1 and Δt_3, the stress level can reasonably be assumed to be independent. In adjacent short intervals this independence is probably absent in that a low stress in one interval would probably mean a low

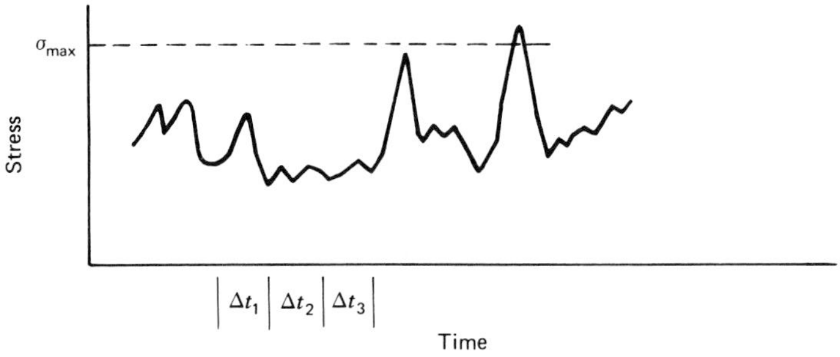

Figure 2.18 Stress changes on a component over time.

stress in the next. However, we can reasonably assume independence for nonadjacent intervals particularly since we are concerned with the failure peaks when the stress exceeds the upper limit.

Normally the above physical situation would be represented by the exponential time to failure distribution. However, if the basic load situation were to remain unchanged but gradual aging or wear were to cause σ_{max} to change over time, then a distribution such as the Weibull, gamma, or Rayleigh might prove to be more appropriate.

2.7.2 GEOMETRIC AND EXPONENTIAL DISTRIBUTIONS

Consider a particular instrument system on an airplane. Every time the plane lands, this system undergoes stresses, and an extremely rough landing can cause the system to fail. A rough landing may result from pilot error, poorly maintained runways, or weather conditions. That is, it is independent of the state of the instrument system. Let

λ_i = the probability of a rough landing occurring on the ith landing
$\overline{E}_i$ = the event that the ith landing is rough
F_n = the event that the first failure occurs on the nth landing

Then

$$F_n = E_1 \cap E_2 \cap \dots \cap E_{n-1} \cap \overline{E}_n$$

which leads to

$$P(F_n) = (1-\lambda_1)(1-\lambda_2)\dots(1-\lambda_{n-1})\lambda_n$$

If we assume that λ_i does not change from landing to landing, then

$$P(F_n) = (1-\lambda)^{n-1}\lambda$$

If we denote the random variable representing the landing on which damage occurred to the instrument system by **m**, then

$$P(m)=(1-\lambda)^{m-1}\lambda, \qquad m=1,2,\ldots$$

This is called the geometric distribution.

Now, in a reliability sense, we want to know the probability that the plane makes n landings without a failure, which is simply

$$P(m>n)=(1-\lambda)^{n} \tag{2.66}$$

This is the probability of a failure-free operation for a "time period" of n landings.

For large n and small λ we can say that

$$R(n)=P(m>n)\approx e^{-n\lambda} \tag{2.67}$$

So if n is viewed as a continuous variable, then this reliability function is precisely the one obtained from the exponential distribution. This indicates the link between a geometric situation and the exponential distribution.

2.7.3 SYSTEM FAILURE MODELS

Most mechanical or electromechanical products sold for consumer use can essentially be regarded as series systems from a reliability standpoint. This means that all components must function for the system to function. Consumers will frequently not be aware that a specific component has degraded, but they will be aware of a complete failure of operation and will take the product in for servicing.

Suppose we are interested in the system reliability $R_S(t)$ and that the ith component reliability is given by $R_i(t)$. Then for a system of m components in series where all components must function, we have

$$R_S(t)=\prod_{i=1}^{m} R_i(t) \tag{2.68}$$

where independence is assumed. Let us consider a rather general hazard function for the ith component as suggested by Reference 17 and given by $h_i(t)=\lambda_i + c_i t^k$, where λ_i, c_i, and k are constants. Then

$$R_i(t)=\exp\left[-\left(\lambda_i t+\frac{t^{k+1}}{k+1}c_i\right)\right]$$

and

$$R_S(t)=\exp\left[-\left(t\sum_{i=1}^{m}\lambda_i+\frac{t^{k+1}}{k+1}\sum_{i=1}^{m}c_i\right)\right]$$

Now let

$$\lambda^* = \sum_{i=1}^{m} \lambda_i$$

$$c^* = \sum_{i=1}^{m} c_i$$

and

$$T = \lambda^* t$$

Then

$$R_S(t) = \exp\left[-\left[T + \left(\frac{c^*}{\lambda^*}\right)\frac{1}{k+1}\left(\frac{T^{k+1}}{(\lambda^*)^k}\right)\right]\right]$$

As the number of components (m) becomes large, we assume that

$$\lambda^* \to \infty$$

and

$$\frac{c^*}{(k+1)\lambda^*}$$

is bounded. Then

$$\lim_{m \to \infty} R_S(t) = e^{-T} = e^{-\lambda^* t} \tag{2.69}$$

That is, the time to failure distribution for the system approaches the exponential. In practice, this means that the time between failures for a large complex system will tend to follow the exponential distribution.

2.7.4 THE EXTREME VALUE DISTRIBUTION

Consider a random sample of size n from an infinite population having a cumulative distribution function $F(x)$ where $\mathbf{x}$ is a continuous random variable $(-\infty < \mathbf{x} < \infty)$. Let the sample be denoted as $\mathbf{x}_1, \mathbf{x}_2, \ldots, \mathbf{x}_n$. Define the random variable

$$\mathbf{y}_n = \min(\mathbf{x}_1, \mathbf{x}_2, \ldots, \mathbf{x}_n)$$

The random variable $\mathbf{y}_n$ is termed the smallest extreme value.

Since material or equipment failure is related to the weakest point or the weakest component, the extreme value distribution for the smallest value is the one usually encountered in reliability work. The smallest extreme value distribution will be considered here; however, the concepts are easily applied to obtain the largest extreme value distribution.

The cumulative distribution for $\mathbf{y}_n$ is given by

$$P(\mathbf{y}_n > y) = P[(\mathbf{x}_1 > y) \cap (\mathbf{x}_2 > y) \cap \cdots \cap (\mathbf{x}_n > y)]$$

Since independence is assured by virtue of a random sample

$$P(\mathbf{y}_n > y) = \prod_{i=1}^{n} P(\mathbf{x}_i > y)$$

or

$$P(\mathbf{y}_n > y) = [1 - F(y)]^n$$

and then the cumulative distribution for $\mathbf{y}_n$ is

$$G_n(y) = 1 - [1 - F(y)]^n, \qquad -\infty < y < \infty \tag{2.70}$$

The p.d.f. is easily obtained by differentiating Equation 2.70 and is

$$g_n(y) = nf(y)[1 - F(y)]^{n-1}, \qquad -\infty < y < \infty \tag{2.71}$$

This is also called the p.d.f. for the first-order statistic in a sample of size n.

EXAMPLE 2.5

Let us find $G_n(y)$ and $g_n(y)$ when

$$f(x) = \lambda e^{-\lambda x}, \qquad x \geqslant 0$$

which is the exponential distribution.

For the exponential distribution the c.d.f. is

$$F(x) = 1 - e^{-\lambda x}, \qquad x \geqslant 0$$

Then substituting into Equation 2.70

$$G_n(y) = 1 - e^{-n\lambda y}, \qquad y \geqslant 0 \tag{2.72}$$

This is the cumulative distribution for the smallest extreme value, and from this the p.d.f. is

$$g_n(y) = n\lambda e^{-n\lambda y}, \qquad y \geqslant 0 \tag{2.73}$$

Here $g_n(y)$ should be recognized as the exponential distribution with parameter $n\lambda$.

If $f(x)$ is integrable, then $G_n(y)$ is easily obtained; however, $f(x)$ is not always integrable and this has led to the study of $G_n(y)$ as n becomes large.

Define the random variable $\mathbf{u}_n$ as

$$\mathbf{u}_n = nF(\mathbf{y}_n) \tag{2.74}$$

Here $F(x)$ is the c.d.f. for $\mathbf{x}$ and since $0 \leqslant F(x) \leqslant 1$, then $0 \leqslant \mathbf{u}_n \leqslant n$.

Now consider the c.d.f. for $\mathbf{u}_n$. This is obtained by

$$H_n(u) = P(\mathbf{u}_n \leqslant u) = P(nF(\mathbf{y}_n) \leqslant u) = P\left[\mathbf{y}_n \leqslant F^{-1}\left(\frac{u}{n}\right)\right]$$

$$= G_n\left[F^{-1}\left(\frac{u}{n}\right)\right]$$

or substituting into $G_n(y)$ from Equation 2.70

$$H_n(u) = 1 - \left(1 - \frac{u}{n}\right)^n, \qquad 0 \leqslant u \leqslant n \tag{2.75}$$

As $n \to \infty$ this becomes

$$H(u) = 1 - e^{-u}, \qquad u \geqslant 0 \tag{2.76}$$

and

$$h(u) = H'(u) = e^{-u}, \qquad u \geqslant 0 \tag{2.77}$$

It follows from the above result that, since the sequence of distribution functions $H_n(u)$ converges to $1 - e^{-u}$, the sequence of random variables $\mathbf{u}_n$ converges in distribution to a random variable, say $\mathbf{u}$. From Equation 2.74, it is clear that, since the sequence of random variables $\mathbf{u}_n$ converges in distribution to a random variable $\mathbf{u}$, the sequence of random variables $\mathbf{y}_n$ converges in distribution to a random variable, say $\mathbf{y}$, where

$$\mathbf{y} = F^{-1}\left(\frac{\mathbf{u}}{n}\right)$$

Thus the limiting distribution of the smallest extreme value is given by the distribution of $\mathbf{y}$. The following example explains the derivation for the exponential case.

EXAMPLE 2.6

Let $f(x) = \lambda e^{-\lambda x}$, $x \geqslant 0$, $\lambda > 0$. As before, let

$$\mathbf{u}_n = nF(\mathbf{y}_n)$$

Hence

$$\mathbf{u}_n = n(1 - e^{-\lambda \mathbf{y}_n})$$

or

$$\mathbf{y}_n = \frac{1}{\lambda}\ln\left[\frac{1}{1-(\mathbf{u}_n/n)}\right]$$

$$= \frac{1}{\lambda}\left[\frac{\mathbf{u}_n}{n} + \frac{1}{2}\left(\frac{\mathbf{u}_n}{n}\right)^2 + \frac{1}{3}\left(\frac{\mathbf{u}_n}{n}\right)^3 + \cdots\right]$$

As $n \to \infty$, $\mathbf{u}_n$ converges in distribution to $\mathbf{u}$, and, ignoring second and higher order terms, we conclude that $\mathbf{y}_n$ converges in distribution to $\mathbf{u}/\lambda n$. The asymptotic density function of $\mathbf{u}$ is of the form e^{-u} as given by Equation 2.77 , and hence we have by the transformation of random variables that

$$g_n(y) = n\lambda e^{-n\lambda y}, \qquad y \geqslant 0$$

and

$$G_n(y) = 1 - e^{-n\lambda y}$$

This result is identical to the result given in Example 2.5, where the exact distribution of $\mathbf{y}_n$ was obtained. The limiting form of the above distribution is termed a Type III asymptotic distribution of the smallest extreme value.

If the underlying density function is of such a form that it approaches zero exponentially as $\mathbf{x} \to \infty$, the limiting distribution of the smallest extreme value is referred to by Gumbel[10] as the Type I asymptotic distribution of the smallest extreme. For example, this is true if the underlying distribution of $\mathbf{x}$ is standard normal. In this situation, it has been shown by Cramer [4] that the limiting distribution of the smallest extreme value is of the form

$$G(y) = 1 - \exp\left[-\exp\left(\frac{y-\alpha}{\beta}\right)\right], \qquad -\infty < y < \infty \tag{2.78}$$

where α and $\beta > 0$ are constants.

If in Equation 2.78 we let $\mathbf{z} = (\mathbf{y} - \alpha)/\beta$, then

$$H(z) = 1 - \exp(-e^z), \qquad -\infty < z < \infty$$

This is called the reduced distribution of extreme values and has standardized parameters of $\alpha = 0$ and $\beta = 1$. The variables $\mathbf{y}$ and $\mathbf{z}$ are, of course, directly related. For, if z_R is a value of $\mathbf{z}$ such that $H(z_R) = 1 - R$, then $y_R = \alpha + \beta z_R$ is a value of $\mathbf{y}$ such that $G(y_R) = 1 - R$.

The asymptotic distribution of largest extreme value is the mirror image of the smallest extreme value distribution. The c.d.f. for the largest extreme value is

$$J(y) = \exp(-e^{-y}), \qquad -\infty < y < \infty \tag{2.79}$$

The moments of the two reduced extreme value distributions are directly related by $E(\mathbf{x}_L^k)=(-1)^k E(\mathbf{x}_S^k)$ where L and S refer to the largest and the smallest extreme value respectively. Thus, the distributions have opposite signs on their means and the same variance.

2.7.4.1. An Application of the Extreme Value Distribution One of the applications of the extreme value distribution is in the study of failures caused by corrosion. Consider automotive exhaust pipes that have various pits when new. The exhaust gases and other corrosives increase the depth of these pits and, ultimately, a failure occurs when the exhaust gases can escape through one pit that has penetrated the thickness of the pipe and has become a hole. If we assume that the time of penetration is proportional to the difference between the pipe thickness and the initial depth of the pit, and that the initial pit depths have an exponential probability distribution, then we will show that the time to failure is an extreme value distribution.

Let D be the thickness of the pipe and d_i the initial depth of the ith pit, $i=1,2,\ldots,N$. Then the $\mathbf{d}_i$'s constitute a random sample from a truncated exponential distribution with range $0 \leqslant \mathbf{d} \leqslant D$; that is,

$$f(d)=\frac{\lambda e^{-\lambda d}}{1-e^{-\lambda D}}, \qquad 0 \leqslant d \leqslant D$$

Hence

$$1-F(d)=P(\mathbf{d}_i \geqslant d)=\frac{e^{-\lambda d}-e^{-\lambda D}}{1-e^{-\lambda D}}, \qquad 0 \leqslant d \leqslant D$$

Let $\mathbf{t}_i$ be the time to failure of the ith pit. Based on our assumption, $\mathbf{t}_i=k(D-\mathbf{d}_i)$, where k is a corrosion rate constant associated with the material. Then

$$G(t)=P(\mathbf{t}_i \leqslant t)=P\left(\mathbf{d}_i \geqslant D-\frac{t}{k}\right)=\frac{e^{\lambda t/k}-1}{e^{\lambda D}-1}, \qquad 0 \leqslant t \leqslant kD$$

If $\mathbf{t}$ is the time to failure of the pipe, then

$$\mathbf{t}=\min(\mathbf{t}_i, i=1,\ldots,N)$$

and because failure occurs as soon as one pit becomes a pinhole, the distribution function for $\mathbf{t}$ is given by Equation 2.70 as

$$P(\mathbf{t} \leqslant t)=H(t)=1-(1-G(t))^N$$

If we assume that the number of pits is very large, then as $N \to \infty$, we have

$$H(t) \approx 1-e^{-NG(t)}$$

Substituting for $G(t)$,

$$H(t) \approx 1 - \exp\left[-N\frac{(e^{\lambda t/k} - 1)}{e^{\lambda D} - 1}\right]$$

Defining $\alpha = N/(e^{\lambda D} - 1)$ and $\gamma = \lambda/k$,

$$H(t) \approx 1 - e^{-\alpha(e^{\gamma t} - 1)}, \qquad t \geqslant 0$$

Differentiating,

$$h(t) = H'(t) \approx \alpha\gamma e^{\gamma t} e^{-\alpha(e^{\gamma t} - 1)}$$

This is the extreme value distribution.

EXAMPLE 2.7

The thickness D of a pipe is 1/16 inch. The constant of proportionality k is 10^6 hours per inch for the pipe. The number of initial pits is 10^4 and the mean pit depth in the beginning is 1/128 in. We can consider $\lambda = 128$ in.$^{-1}$. This is based on the observation that $P(d_i \geqslant d) \cong e^{-\lambda d}$ because we can ignore $e^{-\lambda D}$ which is very small. We are interested in a reliability of 0.90 and want to find what life will give us this reliability.

solution

$$R = 0.90 = 1 - H(t) = \exp\left[-\frac{10^4(e^{128t/10^6} - 1)}{e^8 - 1}\right]$$

or

$$0.90 = \exp\left[-3.3557(e^{0.000128t} - 1)\right]$$

Solving for t, we find

$$t = 242 \text{ hours}$$

2.7.5 SOME COMMENTS ON COMPONENT FAILURE

The failure mechanism and the time to failure pattern for various metals have been studied for several years. By far the most common mode of failure is fatigue. Basically, a fatigue failure is a progressive failure resulting when a component is subjected to a cyclical load of sufficient intensity. In a fatigue failure, one or more cracks start near the surface and progress until the stress on the remaining resisting area is sufficient to cause an abrupt fracture.

Two theories purport to explain how the fracture and the resulting fatigue failure occur. One theory claims that there exist minute flaws in the molecular structure of any metal. The fracture starts at the weakest point analogous to the weakest link in a chain. Since the number of molecules, or links, is very large, this leads to the extreme value distribution as a model for time or cycles to failure.

A second theory asserts that the interlinking structures in a metal support each other roughly like a bundle of threads acting in parallel. This produces an averaging effect because the load is distributed over all the strands. In this case the time to failure distribution is reasoned to be normal by virtue of the central limit theorem. Experimental results are not conclusive and seem to suggest the normal, the extreme value, or the log normal distribution as adequate failure density models.

The fatigue life for a material is described by an S-N diagram obtained from polished standard laboratory specimens subjected to fully reversed bending at a particular stress level until failure occurs. A theoretical S-N diagram is shown in Figure 2.19*a*, with the various portions explained in Figure 2.19*b*. In practice, various factors distort this theoretical relationship. First of all, even the idealized test specimens fail in a pattern about this theoretical line such as

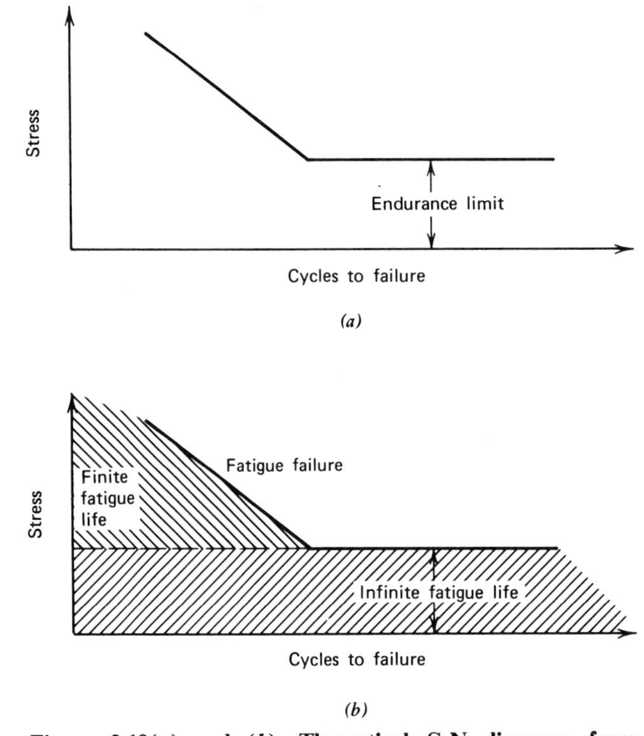

Figure 2.19(*a*) and (*b*) Theoretical S-N diagrams from fatigue tests (plotted on log scales).

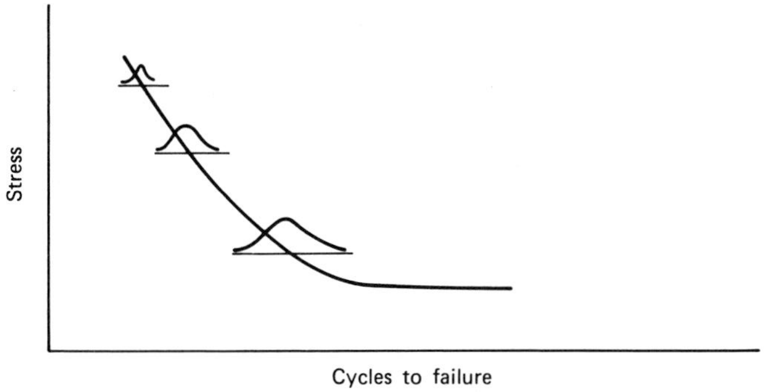

Figure 2.20 S-N diagram indicating variability at a particular stress level (plotted on a log scale).

shown in Figure 2.20. As illustrated in this figure, it is known that the distribution of the cycles to failure has a larger variance at lower stresses than at higher stresses. Unfortunately, the amount of information available to a designer on the actual variability has been somewhat limited in the past. It is also true that although the statistical methodology necessary in properly dealing with this problem has been available for quite some time, it remains largely unapplied to this situation.

We must mention one further complication. Usually the stress on a component is varying and is influenced by other factors such as corrosion, creep, and wear. Undoubtedly, there exists, in practice, a distribution of stress such as illustrated in Figure 2.21. By transformation about the S-N line this results in a

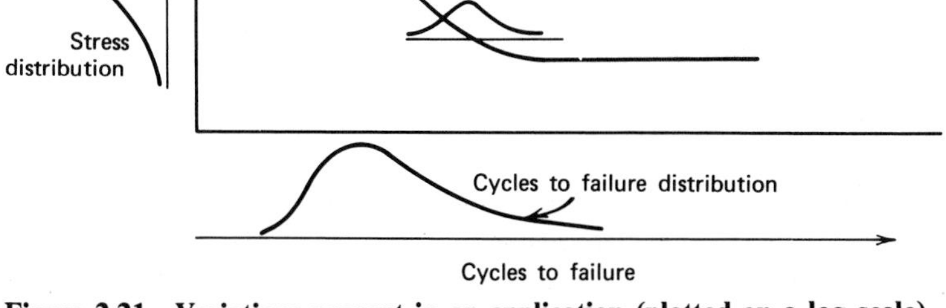

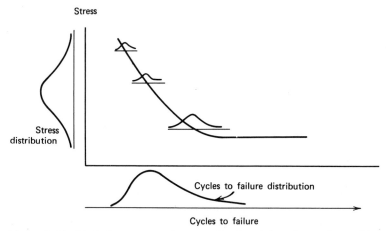

Figure 2.21 Variations present in an application (plotted on a log scale).

failure distribution. Although the plot describes the complications of an actual situation, no specific failure distribution is obvious on theoretical grounds. However, it does suggest that the exponential can be ruled out at the component level.

2.8 SUMMARY

In this chapter we have presented most of the basic formal definitions of reliability, and we apply the concepts presented here in different ways throughout the textbook. In this chapter $R(t)$, the reliability function, was expressed as a function of time, t. In preliminary design analysis it is very difficult to know how $R(t)$ varies with time. In other words, selecting a distribution or hazard model is difficult without considerable past experience gained from testing. Thus it is frequently advantageous to consider a static or constant reliability level. When this is done, it is understood that some reference time base is implied. This time base may be a warranteed life or some other logical choice. Once the time base is selected, then system and subsystem reliabilities are estimated from experience. These reliability values are used to determine the adequacy of the design from a reliability standpoint. The next chapter is devoted to this particular problem and approach.

The difficult problem associated with the material in this chapter from a practical standpoint is the selection of a distribution model. Unless one has considerable test data it is difficult to determine whether the proper model is, for instance, Weibull, log normal, or gamma. Distribution models such as these will generally "fit" well in the "middle" of the range of the random variable; however, they will differ in the "tails" of the distribution. Unfortunately, in reliability work, since the focus is usually on high reliability, the "tails" of the distribution are the most important. Even though there exist many statistical goodness-of-fit tests, in practice one usually has limited test data, and distribution selection is probably best made based on experience with similar systems. We hope that this chapter will help in the process of making a logical selection.

EXERCISES

1 If **t** is a random variable representing the hours to failure for a device where

$$f(t) = te^{-t^2/2}, \qquad t \geqslant 0$$

find (a) the reliability function, and (b) the hazard function. (c) If 50 devices are placed in operation and 27 are still in operation 1 hour later, find approximately the expected number of failures in the time interval from 1 to 1·1 hours using the hazard function.

2 The life of a population of devices is normally distributed with a mean of 800 hours and standard deviation of 25 hours. (a) Graph, to scale, the normal distribution for this situation. (b) Plot the hazard function on the same graph up to at least 800 hours. (c) At 775 hours indicate the value of the p.d.f. and the area that produces the point on the hazard function.

Now assume that 100 devices are placed in operation at time zero. (d) Use the normal distribution to determine the expected number of devices failing prior to 775 hours. (e) Repeat part (d) for 800 hours. (f) Then, using the information gained in parts (d) and (e), how many devices will fail in the interval from 775 to 800 hours? (g) Find graphically the area under the hazard function for the interval 775 to 800 hours. Then show how to use this area to determine approximately how many devices will fail over the interval. Why do the two results differ?

3 Two designs for a critical component are being studied for adoption. From extensive testing on prototypes it was found that the time to failure is Weibull distributed with a minium life of zero. Design A costs $1,200 to build and has Weibull parameters of $\beta=2$ and $\theta=100\sqrt{10}$ hours. Design B cost $1,500 to build and has Weibull parameters of $\beta=3$ and $\theta=100$ hours. (a) The component has a 10 hour guaranteed life. Which design should the manufacturer produce and why? (b) For a 15 hour guaranteed life what should the choice be?

4 Fifteen units of a certain automotive component are placed on a life test. The life is measured in kilocycles. The failures occurs at 90, 150, 240, 340, 410, 450, 510, 550, 600, 670, 710, 770, 790, 830, 880. (a) Plot the failure density, unrealiability, and hazard functions based on the above data. (b) What kind of model would you suggest for the hazard function?

5 Consider a normal variable **t** with $h(t)$ as its hazard function. Find the asymptote to the hazard function for arbitrarily large values of t.

6 Consider the piecewise linear bathtub hazard function defined over three regions of interest given below.

$$h(t)=b_1-c_1t, \qquad 0\leqslant t\leqslant t_1$$

$$h(t)=b_1-c_1t_1-c_2(t-t_1), \qquad t_1<t\leqslant t_2$$

$$h(t)=b_1-c_1t_1-c_2(t_2-t_1)+c_3(t-t_2), \qquad t_2<t<+\infty$$

The constants in the above expressions are determined so that they satisfy the normal requirements for $h(t)$ to be a hazard function. Find the reliability function based on the above hazard function.

7 Consider the following functions:

(a)e^{-at} (c)ct^5 (e)e^{2t}/t^3

(b)e^{at} (d)dt^{-3}

where a, c, and d are constants and $t \geqslant 0$.

Which of the above functions can serve as hazard models? Also, develop mathematical expressions for the density function and the reliability function.

8 Prove that for the normal density function, the hazard rate is monotonically increasing.

9 Compute the expected number of failures after 10,000 miles if 50 automobile components are placed on test with a hazard function as below:

(a)$h(t) = 10^{-5}$ (b)$h(t) = 10^{-6}$ (c)$h(t) = 10^{-4}\exp[10^{-2}t]$

10 Consider the gamma density function with respect to the parameter λ for $\eta = 2$ and plot curves similar to Figures 2.10 and 2.11.

11 Prove that $\int_0^\infty h(t)\, dt \to \infty$.

12 The failure distribution of a component has the form of the gamma distribution with parameters $\eta = 2$ and $\lambda = 0.001$. Determine the reliability of the component and the hazard function after an operation time of 100 units. What is the mean life of the component?

13* The following data represent the number of miles to the first and the succeeding major motor failures for a fleet of 191 city buses. Plot the hazard functions and the failure density functions for miles-to-first failure, miles-to-second failure, etc. Suggest possible theoretical failure distributions.

Initial bus motor failures

DISTANCE INTERVAL (THOUSANDS OF MILES)	OBSERVED NUMBER OF FAILURES
Less than 20	6
20–40	11
40–60	16
60–80	25
80–100	34
100–120	46
120–140	33
140–160	16
160–180	2
180 or more	2
Total	191

*Data taken with permission from "An Analysis of Some Failure Data," by D. J. Davis, *Journal of the American Statistical Association,* Vol. 47, No. 258, June 1952.

Second bus motor failures

DISTANCE INTERVAL (THOUSANDS OF MILES)	OBSERVED NUMBER OF FAILURES
0–20	19
20–40	13
40–60	13
60–80	15
80–100	15
100–120	18
120–160	7
160 or more	4
Total	104

Third bus motor failures

DISTANCE INTERVAL THOUSANDS OF MILES	OBSERVED NUMBER OF FAILURES
0–20	27
20–40	16
40–60	18
60–80	13
80–100	11
100 or more	16
Total	101

Fourth bus motor failures

DISTANCE INTERVAL (THOUSANDS OF MILES)	OBSERVED NUMBER OF FAILURES
0–20	34
20–40	20
40–60	15
60–80	15
80 or more	12
Total	96

Fifth bus motor failures

DISTANCE INTERVAL (THOUSANDS OF MILES)	OBSERVED NUMBER OF FAILURES
0–20	29
20–40	27
40–60	14
60–80	8
80 or more	7
Total	85

14 Assuming that $f(x)$ is a unimodal probability density function with modal value $\tilde{x}$ prove that $h'(\tilde{x}) = h^2(\tilde{x})$.

15 Given that the population distribution is uniform with p.d.f.

$$f(x) = \frac{1}{\theta}, \qquad 0 \leqslant x \leqslant \theta$$

Find (a) the c.d.f. for the smallest extreme value in a random sample of size n and (b) the p.d.f. from part (a). Use the exact theory and not the approximation for large n.

16 For Problem 15 find the expected value and variance for the smallest extreme value.

17 Rework Problem 15 assuming n large and using the asymptotic distribution of extreme values.

18 Given the probability density function

$$f(x) = \beta x^{\beta - 1}, \qquad 0 \leqslant x \leqslant 1, \quad \beta > 0$$

find the asymptotic distribution for the smallest extreme value.

19 Assume that $\mathbf{x}$ is a continuous random variable with c.d.f. $F(x)$ and $\mathbf{x}_1, \mathbf{x}_2, \ldots, \mathbf{x}_n$ is a random sample of size n from $F(x)$. Define $\mathbf{z}_n = \max(\mathbf{x}_1, \mathbf{x}_2, \ldots, \mathbf{x}_n)$ and find the c.d.f. and p.d.f. of $\mathbf{z}_n$. For $n \to \infty$ find the asymptotic distribution for $\mathbf{z}_n$. *Hint*: Let $\mathbf{u}_n = n[1 - F(\mathbf{z}_n)]$.

20 Consider the Weibull distribution as given by

$$F(x) = 1 - \exp\left[-\left(\frac{x}{\theta}\right)^{\beta}\right], \qquad x \geqslant 0;\ \beta, \theta > 0$$

and find the distribution for the smallest extreme value using both the exact and asymptotic theory.

21 A warranty reporting system reports field failures. For the rear brake drums on a particular pickup truck the following data was obtained:

KILOMETER INTERVAL	NUMBER OF FAILURES
$M < 2{,}000$	707
$2{,}000 \leqslant M < 4{,}000$	532
$4{,}000 \leqslant M < 6{,}000$	368
$6{,}000 \leqslant M < 8{,}000$	233
$8{,}000 \leqslant M < 10{,}000$	231
$10{,}000 \leqslant M < 12{,}000$	136
$12{,}000 \leqslant M < 14{,}000$	141
$14{,}000 \leqslant M < 16{,}000$	78
$16{,}000 \leqslant M < 18{,}000$	101
$18{,}000 \leqslant M < 20{,}000$	46
$20{,}000 \leqslant M < 22{,}000$	51
$22{,}000 \leqslant M < 24{,}000$	56

For the above data plot the hazard rate, failure function, and reliability function. Assume that the population size is 2,680 and the above data represents all of the failures.

22 Windshield wiper arms have a normally distributed fatigue failure life with a mean of 500,000 cycles and standard deviation of 25,000 cycles.

(a) Find the reliability of the wiper arms at 450,000 cycles.

(b) Plot the hazard function.

(c) If you desire only a 10% chance of a field failure, how often should you replace the wiper blades on your automobile?

(Note that this problem will require you to make several estimates, such as frequency of your wipers and number of rain days in your area.)

BIBLIOGRAPHY

1 Aziz, P. M., "Application of the Statistical Theory of Extreme Values to the Analysis of Maximum Pit Depth Data for Aluminum." *Corrosion*, Vol. 12, October 1956.

2 Bazovsky, I., *Reliability Theory and Practice*, Prentice-Hall, Englewood Cliffs, N. J., 1961.

3 Bompas-Smith, J. H., "The Determination of Distributions that Describe the Failure of Mechanical Components," *Annals of Assurance Science*, 1969.

4 Cramer, H., *Mathematical Methods of Statistics,* Princeton University Press, Princeton, N. J., 1946.

5 Epstein, B., "Statistical Aspects of Fracture Problems,"*Journals of Applied Physics,* Vol. 19, February 1948.

6 Epstein, B., "Elements of the Theory of Extreme Values," *Technometrics*, Vol. 2, No. 1, February 1960.

7 Freudenthal, A. M. and E. J. Gumbel, "Failure and Survival in Fatigue," *Journal of Applied Physics*, Vol. 25, No. 11, November 1954.

8 Freudenthal, A. M. and E. J. Gumbel, "Minimum Life in Fatigue,"*American Statistical Association Journal*, Vol. 49, 1954.

9 Gertsbakh, I. B. and Kh.B. Kordonskiy, *Models of Failure*, Translation from the Russian, Springer-Verlag, New York, 1969.

10 Gumbel, E. J.; *Statistics of Extremes*, Columbia University Press, New York, 1958.

11 Kececioglu, D., R. E. Smith, and E. A. Felstead, "Distributions of Cycles-to-Failure in Simple Fatigue and the Associated Reliability," *Proceedings of the 8th Reliability and Maintainability Conference*, Denver, Colorado, July 7–9, 1969.

12 Lloyd, D. K. and M. Lipow, *Reliability: Management, Methods and Mathematics*, Prentice-Hall, Englewood Cliffs, N. J., 1962.

13 Military Standard 781B, *Reliability Tests: Exponential Distribution*, Department of Defense, November 1967.

14 Miner, M. A., "Cumulative Damage in Fatigue," *American Society of Mechanical Engineers*, Vol. 12, No. 3, 1945.

15 *Reliability Handbook*, AMCP 702-3, U.S. Army Materiel Command, Washington, D. C., October 1968.

16 Sasieni, M., A. Yaspen, and L. Friedman, *Operations Research: Methods and Problems*, John Wiley & Sons, New York, 1959.

17 Shooman, M. L., *Probabilistic Reliability: An Engineering Approach*, McGraw-Hill, New York, 1968.

Chapter 3 static reliability models

In this chapter, we use static models to demonstrate reliability analysis. In modeling a system from a reliability standpoint using static models, the component or subsystem reliabilities are considered to be constants; thus some base time period is implied. Sections 3.1 and 3.2 present the basic series and parallel system reliability models and demonstrate the reliability calculations. Sections 3.3 and 3.4 then consider system reliability calculations for more complex combinations of the basic models. Section 3.5 points out some design factors that are related to reliability.

System reliability analysis using static models is a form of preliminary analysis. It is used to evaluate possible design configurations and to determine the necessary reliability levels for subsystems and components. As the design progresses to its final stages a more detailed analysis can be made and, finally, prototypes are built to verify the design reliability by testing.

The functional block diagram for a system represents the effect of subsystem failure on system performance. It is a black box sort of analysis which requires that complex systems be broken down into subsystems or components. In the static model each black box is assumed to be in one of two states: success or failure.

The structure of a block diagram for a particular system will depend on the definition of reliability for that system. As an example, consider a simple dual circuit breaker shown in Figure 3.1. Now if the primary concern is that the circuit closes when required, then the block diagram would be a series arrangement as shown in Figure 3.2a. However, if instead the main concern is that the circuit opens when required, the block diagram would take the parallel arrangement shown in Figure 3.2b. Note that the block diagrams are constructed to determine operational success and not to show electrical circuitry.

The following discussion illustrates the application of probability theory in the calculation of system reliability. We are most interested in the probability of

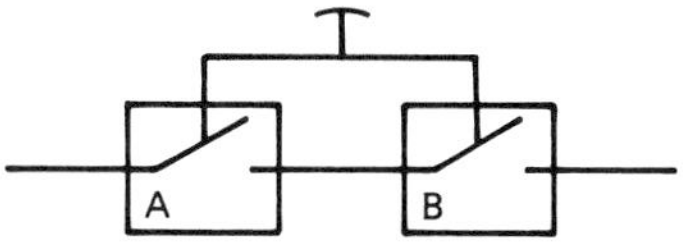

Figure 3.1 Dual breaker example.

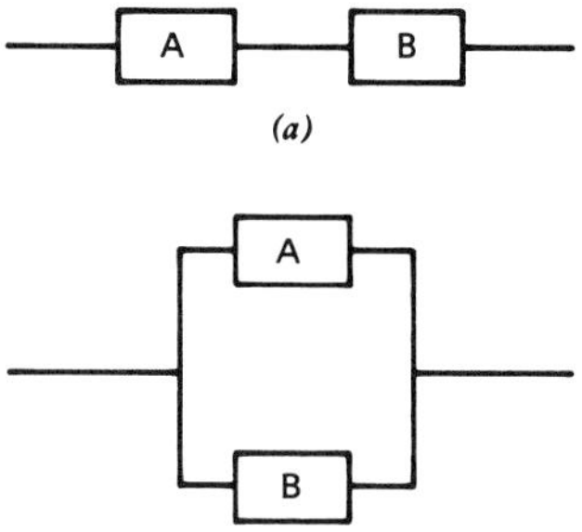

(a) *(b)*

Figure 3.2 Block diagram for dual breaker.

successful operation over a stated interval of time. For purposes of this discussion we assume that the functional diagram has been developed and the subsystems under consideration fail independently. A subsystem will be taken to mean a particular low-level grouping of components or just one component, depending on the way the system is subdivided. The following notation will be used:

E_i = event that subsystem i operates successfully
$R_i = P(E_i)$ = subsystem reliability
R_S = system reliability.

3.1 SERIES SYSTEMS

The series configuration is probably the most commonly encountered model and, fortunately, is also the simplest to analyze. In a series system all subsystems must operate successfully if the system is to function. The block diagram model of a series system is given in Figure 3.3. For the series model

$$R_S = P[E_1 \cap E_2 \cap \cdots \cap E_n]$$

and because of our assumption of independence this becomes

$$= P(E_1)P(E_2)\cdots P(E_n)$$

or

$$R_S = \prod_{i=1}^{n} R_i \tag{3.1}$$

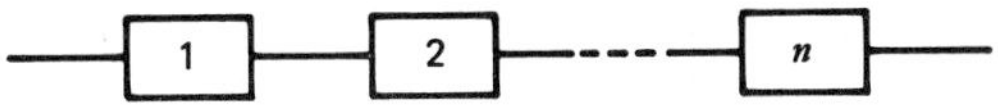

Figure 3.3 Series system block diagram.

where the right hand side indicates the product of the subsystem reliabilities.

Equation 3.1 constitutes what is commonly called the product rule in reliability. Very often a product's design configuration will dictate that this rule be used in computing reliability. This is unfortunate because the system reliability decreases rapidly as the number of series components increases, and the reliability will always be less than or equal to the least reliable component. Thus for a series system

$$R_S \leqslant \min_i \{R_i\} \tag{3.2}$$

If one wants to meet a given system reliability, a rapid approximation for the necessary level of subsystem reliability is obtained as follows. Let

$$q = \text{the probability that a subsystem will fail}$$

Then assuming that q is identical for all subsystems

$$R_S = (1-q)^n$$

and by simple application of the binomial theorem this becomes

$$1 + n(-q)^1 + \frac{n(n-1)}{2}(-q)^2 + \cdots + (-q)^n$$

Thus if we ignore higher order terms by assuming that q is small we have

$$R_S \approx 1 - nq$$

When applying this approximation it is helpful to know that if $nq = 0.1$, the results will be accurate to two places.

EXAMPLE 3.1

If we want $R_S = 0.99999$ in a system with 20 components, then

$$0.99999 = 1 - 20(q)$$

and

$$q = 0.0000005$$

or

$$R = 0.9999995$$

This is an approximate value of the subsystem reliability required to meet a specified system reliability.

An approximation for the series system reliability where the q_i's are different is given by

$$R_S \approx 1 - \sum_{i=1}^{n} q_i \tag{3.3}$$

3.2 PARALLEL SYSTEMS

A parallel system is a system that is not considered to have failed unless all components have failed. The reliability block diagram for a parallel system is given in Figure 3.4.

The system reliability is calculated as follows. If we define

$$Q_S = \text{unreliability of the system}$$

then

$$Q_S = P\left[\bar{E}_1 \cap \bar{E}_2 \cap \cdots \cap \bar{E}_n\right]$$

where $\bar{E}$ is the complementary event. Assuming independence

$$Q_S = P\left(\bar{E}_1\right) P\left(\bar{E}_2\right) \cdots P\left(\bar{E}_n\right)$$

or

$$Q_S = \prod_{i=1}^{n} (1 - R_i) \tag{3.4}$$

Then the system reliability is given by the complementary probability and

$$R_S = 1 - \prod_{i=1}^{n} (1 - R_i) \tag{3.5}$$

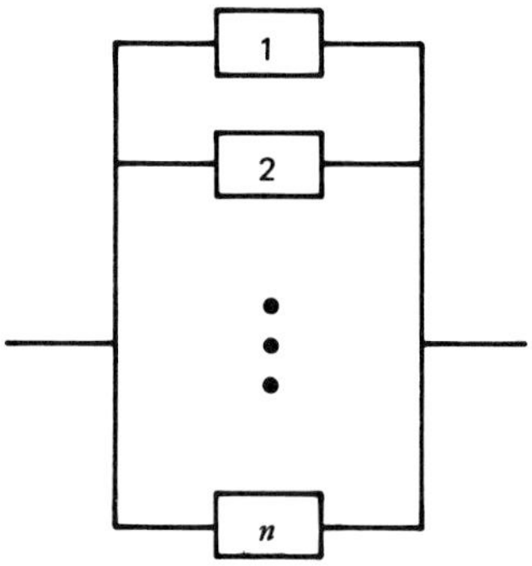

Figure 3.4 Parallel system block diagram.

In analyzing a parallel system in this fashion it is implied that all subsystems are activated when the system is activated and that failures do not influence the reliability of the surviving subsystems.

The type of parallel arrangement previously considered is termed a pure parallel situation and, as was mentioned, is not representative of many parallel arrangements. In reality the standby redundant and shared parallel arrangements are more often used in parallel systems, particularly in mechanical systems.

In the standby redundant system, the standby component is not activated unless the on-line component fails. This situation is illustrated in Figure 3.5. The switch (S) can represent an automatic sensing device or can simply mean that the operator replaces component A with component B. A good example of standby redundancy is the spare tire in an automobile. The standby redundant system should be analyzed as a dynamic model and is covered in a later chapter.

In the shared parallel system, the failure rate of surviving components increases as failures occur. An automobile wheel assembly is an example of the shared parallel arrangement; if a lug nut comes loose the remaining nuts must support an increased load, and hence the failure rate is increased with each successive failure. Thus, the shared parallel is not truly a static model.

Another form of redundancy is an r out of n system. This system has n parallel components; however, at least r components must survive if the system is to continue operating. An example of this form of redundancy are cables for a bridge where a certain minimum number of cables are necessary to support the structure.

The reliability for an r out of n system is given by

$$R_S = \sum_{x=r}^{n} \binom{n}{x} R^x (1-R)^{n-x} \tag{3.6}$$

where R is the subsystem reliability and is assumed to be equal for all subsystems, and

$$\binom{n}{x} = \frac{n!}{x!(n-x)!}$$

For unequal subsystem reliabilities direct enumeration can be used (this is covered in a later section).

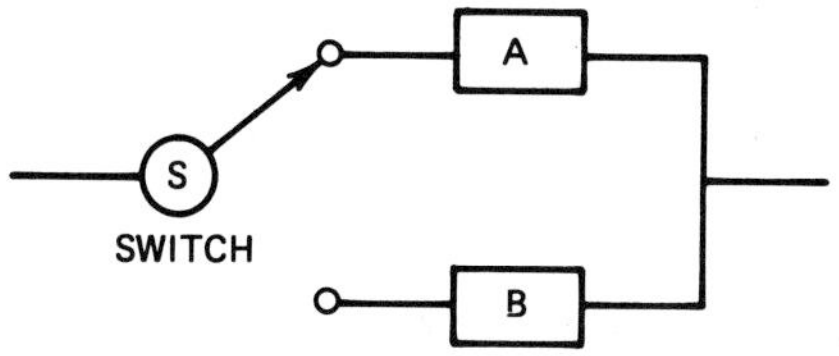

Figure 3.5 Standby redundant system.

3.3 PARALLEL AND SERIES COMBINATIONS

Simple combinations of parallel and series subsystems are easily analyzed by successively collapsing subsystems into equivalent parallel or series components. As an example consider the series arrangement of parallel subsystems shown in Figure 3.6. To calculate the reliability of this system, we first collapse the parallel subsystems into equivalent series components.

Let us say that we have determined that the reliabilities are $R_A=0.9$, $R_B=0.8$, $R_C=0.7$, and $R_D=0.6$. Then the equivalent series component reliabilities are

$$R_{AB}=1-(0.1)(0.2)=0.98$$

and

$$R_{CD}=1-(0.3)(0.4)=0.88$$

So the system reliability is

$$R_S=(0.98)(0.88)=0.8624$$

A second arrangement is shown in Figure 3.7, where the banks of series subsystems are in parallel. The procedure here is to first collapse the series subsystems into equivalent parallel components. For purposes of this example let us assume that the subsystem reliabilities are the same as those used for the previous calculations. Therefore, the equivalent parallel components are

$$R_{AC}=(0.9)(0.7)=0.63$$

and

$$R_{BD}=(0.8)(0.6)=0.48$$

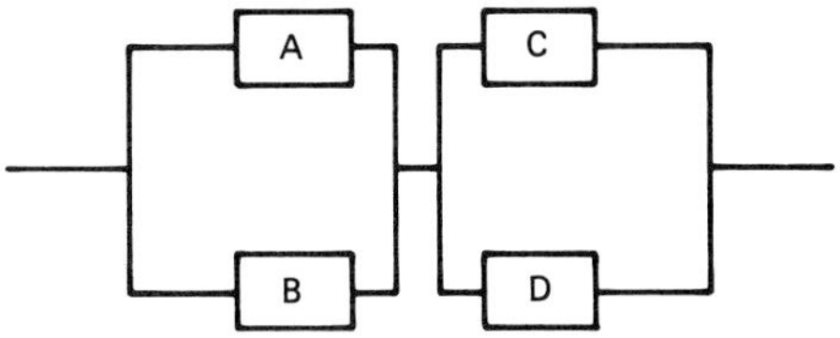

Figure 3.6 Series-parallel system.

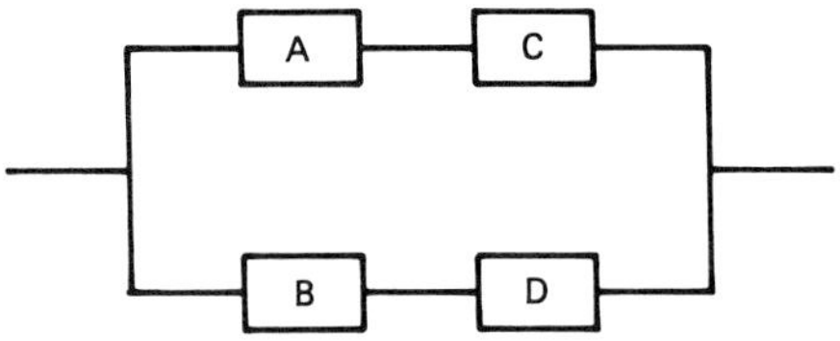

Figure 3.7 Parallel-series system.

This gives a system reliability of

$$R_S = 1 - (1 - R_{AC})(1 - R_{BD})$$
$$= 1 - (0.37)(0.52) = 1 - 0.1924$$

or

$$R_S = 0.8076$$

Note the difference in system reliability resulting from the different arrangement of subsystems.

The procedure for series and parallel combinations in a system is a rather straightforward application of the calculation methods for basic series and parallel systems. Thus, the basic formulas can be successively applied to analyze systems with combinations of series and parallel arrangements.

3.4 COMPLEX SYSTEM ANALYSIS

Certain design configurations or complex failure modes may produce systems in which pure parallel or series configurations are not appropriate. As an example consider the system shown in Figure 3.8. In this system the failure of subsystem E drops out both the E, D and E, C paths. Thus we do not have a pure parallel arrangement. Several ways to handle such situations have been proposed. The following method is not the simplest to use in some specific applications. However, it is always applicable, and it also allows one to consider the effects of different failure modes on system performance.

The procedure is simply to examine all possible mutually exclusive modes of the system. For example, let us now consider a three-component system; we label the components A, B, and C. Define the event A as the event that component A succeeds and so forth for B and C. Let $P(A) = 0.95$, $P(B) = 0.90$, and $P(C) = 0.85$. Then the probability for each mode is calculated under the assumption of independence. Thus $P(A \cap \bar{B} \cap C)$ becomes $P(A)P(\bar{B})P(C)$. These calculations are shown in Table 3.1.

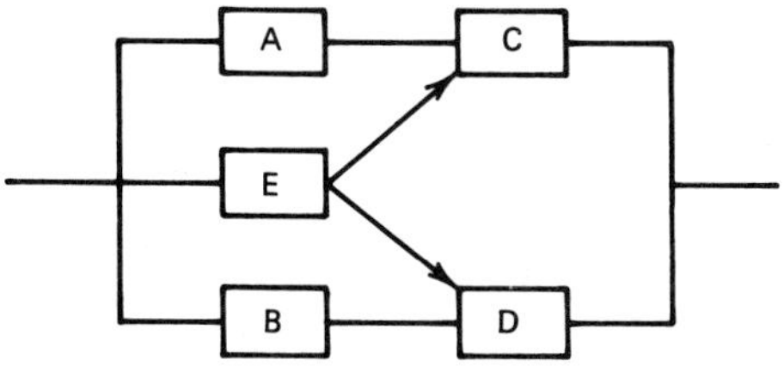

Figure 3.8 Complex system configuration.

Table 3.1 System reliability calculations

Number of Failed Components	System Modes	Probability
0	$A\cap B\cap C$	0.7268
1	$\bar{A}\cap B\cap C$	0.0382
	$A\cap \bar{B}\cap C$	0.0808
	$A\cap B\cap \bar{C}$	0.1282
2	$\bar{A}\cap \bar{B}\cap C$	0.0042
	$\bar{A}\cap B\cap \bar{C}$	0.0068
	$A\cap \bar{B}\cap \bar{C}$	0.0142
3	$\bar{A}\cap \bar{B}\cap \bar{C}$	0.0008
		$\Sigma = 1.0000$

Because each mode is a mutually exclusive event the probabilities can be summed. For instance, if only modes $A\cap \bar{B}\cap C$, $\bar{A}\cap \bar{B}\cap C$, and $\bar{A}\cap \bar{B}\cap \bar{C}$ resulted in catastrophic failure then the probability of a catastrophic failure is $0.0808+0.0042+0.0008=0.0858$. This method allows one to review and categorize the effect of each failure mode on the system.

The obvious drawback to this method is that the number of failure modes increases rapidly with the number of components. The number of modes can be easily calculated. For, if n is the total number of subsystems and x is the number of failures under consideration then $\binom{n}{x}$ is the number of failure modes containing x failures. For example, if $n=5$ and $x=2$, then there are $\binom{5}{2}=10$ failure modes containing two failed components, and the total number of different modes is given by

$$\sum_{x=0}^{n}\binom{n}{x}=\binom{5}{0}+\binom{5}{1}+\binom{5}{2}+\binom{5}{3}+\binom{5}{4}+\binom{5}{5}=32$$

The extensive calculations required by this method can be simplified by the use of computer programs that evaluate all combinations.

3.5 RELIABILITY CONSIDERATIONS IN DESIGN

The reliability level of a system is determined during the design process and subsequent production, assembly, and delivery of the system will certainly not improve upon this inherent reliability level. The design process also dictates the system configuration and the configuration chosen influences the reliability level as well as the cost of achieving this level. Thus, a preliminary reliability analysis as well as the many other design factors should be considered during the design phase.

Since the designer is the system architect he or she should be familiar with the basic reliability analysis concepts that can be used to evaluate the design. Only after the design is completed can an independent reliability group analyze and test the product. So it is important that the designer evaluate the reliability levels and costs of various designs before the final choice is made. This section will emphasize some reliability trade-offs that might be helpful to a designer in developing alternatives.

3.5.1 SERIES SYSTEMS

Because of the nature of a series system, the component reliability level and the number of components are determinants of system reliability. This relationship is shown in Figure 3.9. As can be seen, the reliability of a series system can be improved by (1) decreasing the number of series components, or (2) increasing the component reliabilities. The graph shows the effects of increasing component reliability; however, the marginal gain in system reliability becomes smaller as component reliabilities increase. The deterioration of system reliability that results from increasing the number of components is also evident.

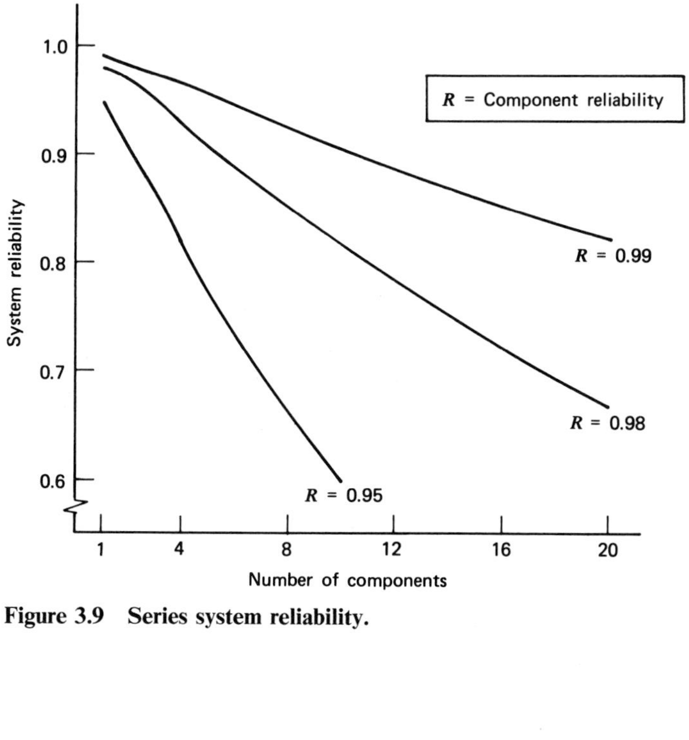

Figure 3.9 Series system reliability.

3.5.2 PARALLEL SYSTEMS

The paralleling of components is usually mentioned as a means to improve system reliability. However, the gains are not always realizable. First of all, designing a parallel system for a mechanical device is usually extremely difficult. Some forms of parallel arrangement such as providing spare parts (a standby parallel arrangement) or using a load-sharing design (a shared parallel arrangement) are probably more representative of the true situation.

The second problem with parallel arrangements is illustrated in Figure 3.10. As can be noted, for a given component reliability the marginal gain due to the addition of parallel elements slows down rapidly. As can be seen in the graph, the gain in system reliability by adding parallel components beyond the fourth is exceedingly small. Thus the marginal gain in reliability by paralleling may not be as profitable as considering an improved component.

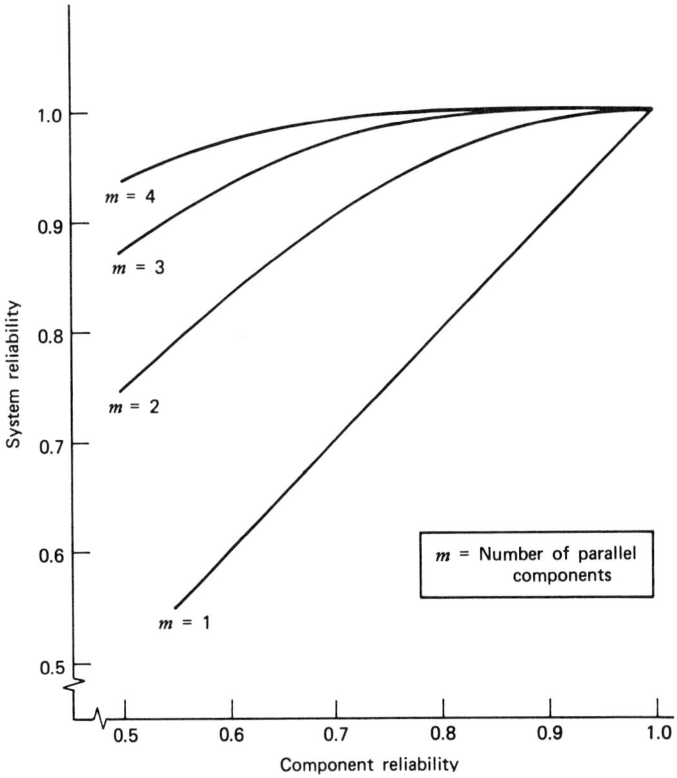

Figure 3.10 Reliability improvement by *m* parallel components.

3.5.3 SERIES-PARALLEL DESIGN CONSIDERATIONS

Now let us assume that we have an n component system. We can either provide redundant components, which gives a system block diagram as shown in Figure 3.11, or provide a total redundant system as shown in Figure 3.12. The component level redundancy we will term low-level redundancy and the system level redundancy we will term high-level redundancy. The basic question from a design standpoint is how do these two levels of redundancy compare? In the following analysis we will assume that all components have the same reliability, R.

For the low-level redundant system (Figure 3.11), the equivalent reliability for a parallel bank is

$$R_{\mathrm{EQ}} = 1 - (1 - R)^m \tag{3.7}$$

Then the R_{EQ} components are in series, which gives a system reliability of

$$R_{\mathrm{S,low}} = \left[1 - (1 - R)^m\right]^n \tag{3.8}$$

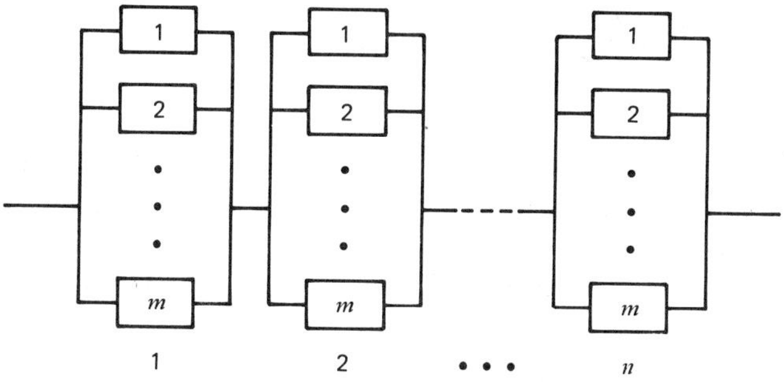

Figure 3.11 Low-level redundancy of components.

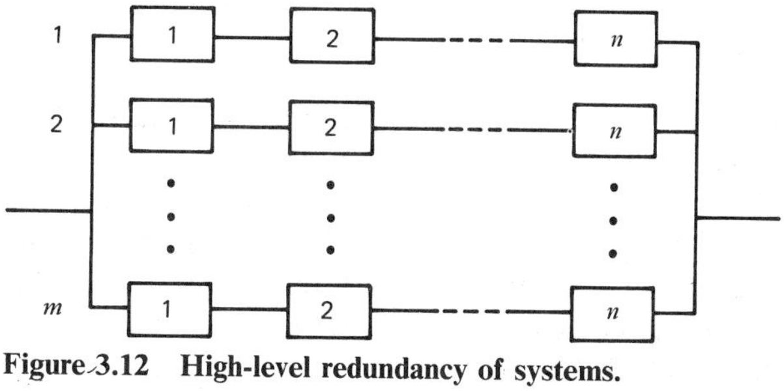

Figure 3.12 High-level redundancy of systems.

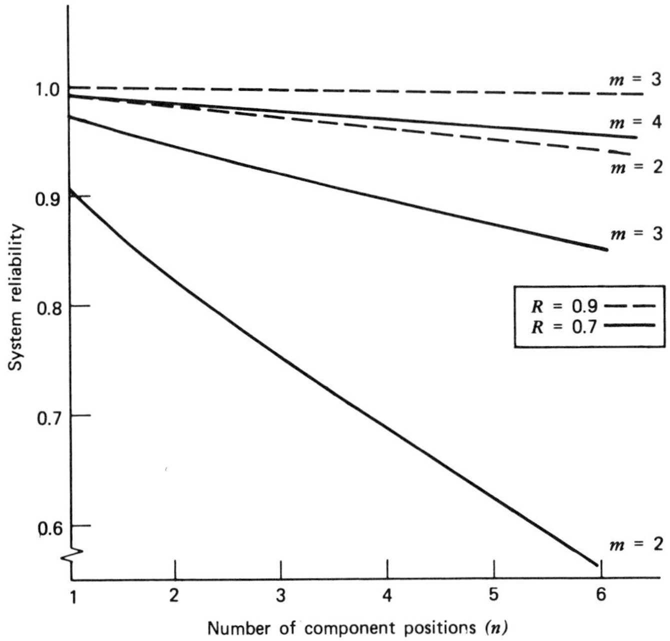

Figure 3.13 Low-level component redundancy.

This equation is plotted in Figure 3.13 for different component reliability levels. This figure shows the effects of component reliability, number of components, and number of parallel banks on the system reliability.

Now the high-level redundancy as shown in Figure 3.12 gives an equivalent path reliability of

$$R_{EQ} = R^n$$

for each parallel path. Thus, the system reliability is

$$R_{S,\text{high}} = 1 - (1 - R^n)^m \tag{3.9}$$

This is plotted in Figure 3.14. Again the effects of system configuration and component reliability can be observed.

By comparing the graphs in Figures 3.13 and 3.14 it is evident that the low-level redundancy gives a higher system reliability in all cases. However, the difference is not as pronounced if components have high reliabilities. Basically Figures 3.13 and 3.14 indicate that providing spare components will result in better overall reliability than providing a spare system. Of course, this can be applied at different levels to subsystems, depending on the possible system

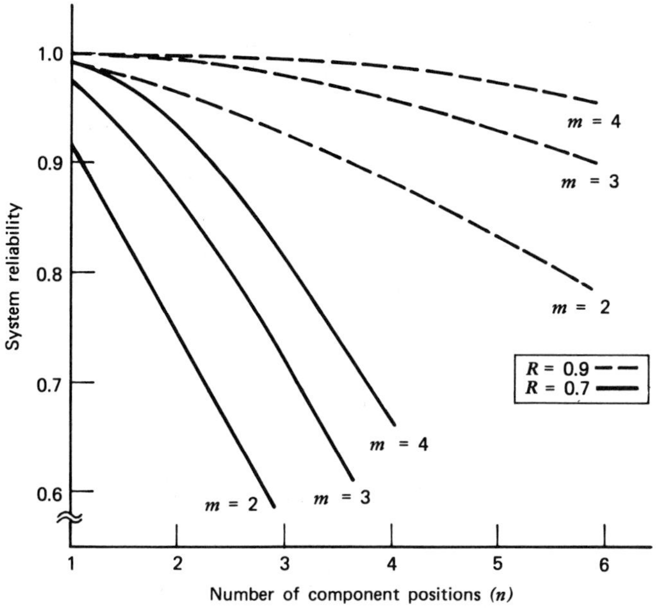

Figure 3.14 High-level component redundancy.

breakdown, for, in some instances, design or system peculiarities make it impossible to apply all of these rules. Also the total system operation must be considered. For instance, if your automotive brake system fails at 80 km/h in heavy traffic it would not do you any good to have a complete set of components in your glove compartment. So the rules must be used as guides and applied with discretion.

3.6 SUMMARY

Reliability must be considered early in the design process when changes are most easily and economically made. The concepts presented in this chapter can be used to analyze reliability levels so that necessary design improvements or reliability trade-offs can be made early in the design stage. The simple series and parallel analysis as presented here is very preliminary; however, it is important because it helps suggest a course of action that will result in a satisfactory reliability level.

To apply this analysis one must first break the system into elements small enough to yield reasonably accurate reliability estimates. For large products accomplishing this breakdown takes considerable effort. Once the breakdown is accomplished reliability calculations can be made to determine design weaknesses. The solution to a design weakness may be paralleling, providing an improved subsystem, or increasing reliability levels by limiting stresses. The best way to improve reliability ultimately depends on the cleverness of the design engineer.

EXERCISES

1 Calculate the reliability of each of the system configurations in Figure 3.15, where each component has the indicated reliability.

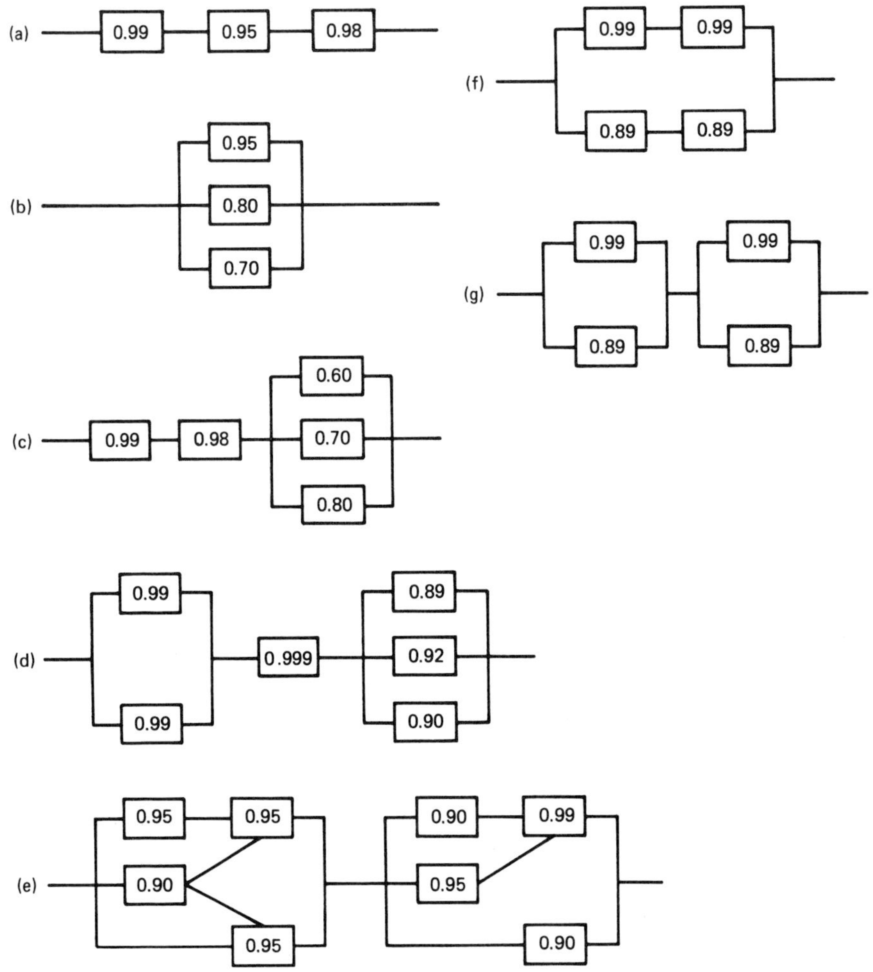

Figure 3.15 System configurations.

2 A system consists of 100 parts connected functionally in series. Each part has a 1,000 hour reliability of 0.9999. Calculate the reliability of the system.

3 A system is comprised of four major subsystems in parallel. Each subsystem has a reliability of 0.900. At least two of the four subsystems must operate if the system is to perform properly. What is the reliability of this system?

4 A system is comprised of 10 subsystems connected functionally in series. If a system reliability of 0.999 is desired, what is the minimum subsystem reliability that is needed?

5 Assume that four wheel bolts are adequate from a design standpoint. However, the wheel attachment under consideration has five bolts. If the chances of losing a wheel bolt are 0.00001, what is the reliability of this bolt system?

6 Figure 3.16 illustrates a carburetor with a return spring. In this design if the carburetor spring breaks there will be no return force on the throttle.

(a) Discuss whether this is a reliable system.

(b) Discuss alternative ways to improve the reliability of this system.

(c) Management believes that the reliability of this throttle closing system must be improved and that the obvious way is by a parallel arrangement. Management also states that only external bracketry on the carburetor can be changed, since they have no control over the carburetor design as such. Sketch a new design of the bracketry that will accommodate a second spring. Explain from a design standpoint the problems encountered in this process and why the

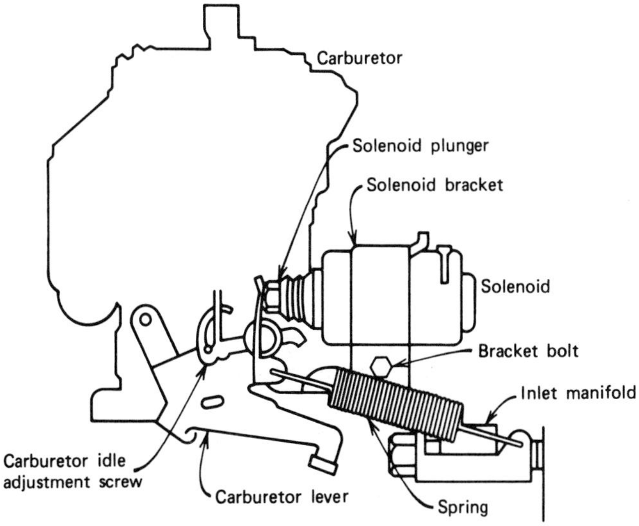

Figure 3.16 Carburetor return spring example.

theoretical effect of redundancy may not be obtained. Can you suggest alternate methods?

7 The system diagram given in Figure 3.17 describes the circuitry for a neutral start switch on a manual automobile transmission. According to the service manual, in order to start this vehicle the clutch pedal must be fully depressed and the ignition switch must be in the start position.

(a) Define reliability as it relates to this system.

(b) Draw an appropriate reliability block diagram.

(c) Assuming that each functional block in your diagram has a 0.0001 chance of failing, calculate the reliability.

(d) What is the probability that the starter cannot be energized?

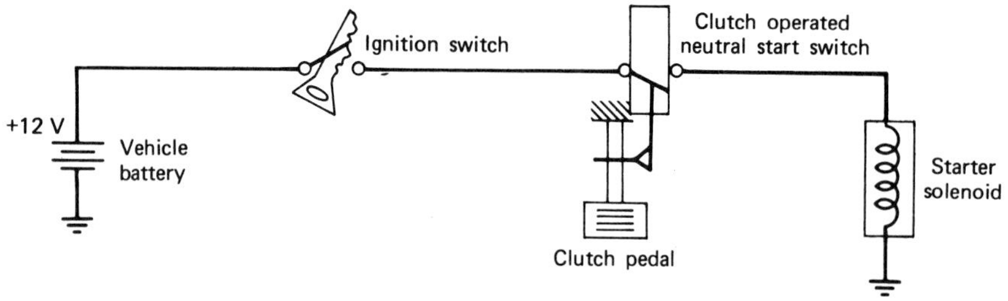

Figure 3.17 Neutral start switch circuit.

8 A detection system for the CO level in a test cell is under consideration. Specifically, there is a sensor available that will close a circuit and thereby signal the personnel if it detects a particular level of CO concentration. However, this sensor can fail in the following ways:

FAIL STATE	PROBABILITY
Signal high CO level when none is present	0.10
Not detect high CO level when it is present	0.15

Obviously, the sensor is not too reliable, and it is decided to use three of them in a dc circuit.

(a) Arrange the sensors such that the probability of detecting a high emission level if it is present, is maximized.

(b) Calculate the probability of a false signal for each arrangement considered in (a).

9 A manufacturer wishes to know the reliability of a skid protection system to be used on military tractor trailers. The system consists of:

(a) Two battery or generator powered sensors per wheel.

(b) One logic unit per sensor to predict wheel skid.

(c) A command unit, which operates an electric or an engine vacuum solenoid.

(d) The solenoids in (c) operate an actuator that controls the pressure to the brake.

The system block diagram (not reliability diagram) is shown in Figure 3.18.

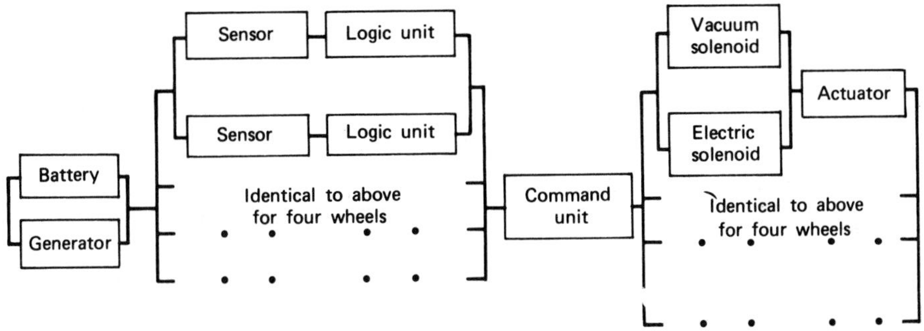

Figure 3.18 Block diagram for a skid control system.

For the following component reliabilities calculate the system reliability.

COMPONENT	RELIABILITY
Battery	0.90
Generator	0.99
Sensor	0.98
Logic unit	0.97
Command unit	0.99
Vacuum solenoid	0.97
Electric solenoid	0.98
Actuator	0.99

10 In an electrical distribution system, electronically operated circuit breakers can be activated to interrupt the current. If the current exceeds 105% of the rated line current it is required that the circuit breakers open, thereby disconnecting the supply. The probability that a circuit breaker functions correctly is 0.98, and each breaker has its own line voltage sensor. If the reliability associated with interrupting the circuit is to be at least 0.999, how many circuit breakers in series are necessary to achieve the desired reliability?

11 A dc battery has a time to failure that is normally distributed with a mean of 30 hours and a standard deviation of 4 hours. (a) What is the 25 hour reliability? (b) When should a battery be replaced to ensure that there is no more than a 10% chance of failure prior to replacement? (c) Two batteries are connected in parallel to power a light. Assuming that the light does not fail, what is the 35 hour reliability for the power source? (d) A particular battery has been in continuous use for 30 hours. What is the probability that this battery will last another 4 hours?

BIBLIOGRAPHY

1 Amstadter, B. L., *Reliability Mathematics: Fundamentals; Practices; Procedures,* McGraw-Hill, New York, 1971.

2 Balaban, H. S., "Some Effects of Redundancy on System Reliability," *Proceedings of the 6th National Symposium on Reliability and Quality Control,* Washington, D. C., January 11–13, 1960.

3 Bazovsky, I., *Reliability Theory and Practices,* Prentice-Hall, Englewood Cliffs, N. J., 1961.

4 Esary, J. D., F. Proschan, and D. W. Walkup, "Association of Random Variables, with Applications," *Annals of Mathematical Statistics,* Vol. 38, 1967.

5 Lieberman, G. J., "The Status and Impact of Reliability Methodology," *Naval Research Logistics Quarterly,* Vol. 16, No. 1, March 1969.

6 Shooman, M. L,. *Probabilistic Reliability: An Engineering Approach,* McGraw-Hill, New York, 1968.

7 Von Alven, W. H. (Ed.), *Reliability Engineering,* Prentice-Hall, Englewood Cliffs, N. J., 1964.

Chapter 4 probabilistic engineering design

In very complex systems found everywhere today, extremely grave consequences can result from the failure of a single component. Therefore, a primary goal of reliability and design engineers is to choose the best structural and mechanical designs, considering factors such as cost, reliability, weight, and volume. To attain this goal, a methodology is necessary for estimating component reliability at the design stage.

The basis of the concept of reliability is that a given component has a certain stress resisting capacity; if the stress induced by the operating conditions exceeds this capacity failure results. The conventional design approach, which is based on somewhat arbitrary multipliers such as safety factors and safety margins, gives little indication of the failure probability of the component. Some designers believe that a component failure can be entirely eliminated by using a safety factor above a certain preconceived magnitude. In reality, the failure probability may vary from a low to an intolerably high value for the same safety factor. Using a safety factor is justified only when its value is based on considerable experience with parts similar to the one under consideration. Furthermore, the design variables and parameters are often random variables, a fact completely ignored by the conventional design approach.

Clearly, the conventional design approach is not adequate from a reliability standpoint. Hence, another design methodology that does consider the probabilistic nature of the design is needed so that component reliability can be calculated at the design stage. Such a design methodology is called *probabilistic design*. It identifies explicitly all the design variables and parameters, which, in turn, determine both the stress and the strength distributions (Figure 4.1). Once these two distributions are determined, the component reliability can be easily calculated. That is, this approach expresses the component reliability as a function of the stress and strength distributions (and their parameters). The probabilistic design methodology is explained in Section 4.1. Some background references for the stress and the strength distributions are given in Section 4.2.

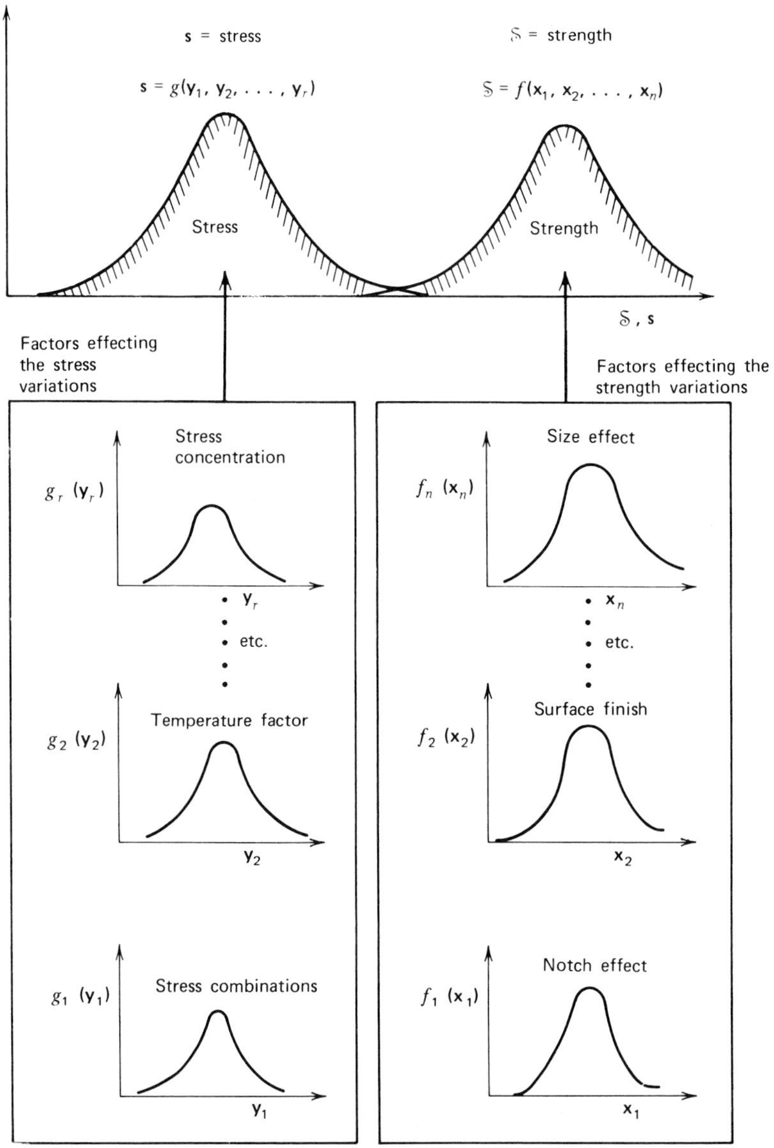

Figure 4.1 An illustration of various factors contributing to the stress and strength distribution.

In the remainder of the chapter, the relationship between the traditional design approach, which uses safety factors, and the probabilistic approach, which employs the concept of reliability, is developed. Section 4.3 shows the relationship between safety factors, coefficients of variation of stress and strength random variables and reliability. If we do not know the density functions for the stress and strength random variables, but we do know the first two moments of these random variables, then bounds on reliability as a function of safety factor and the coefficient of variation of stress and strength random variables are given in Section 4.3. These bounds help the design engineer study design reliability as a function of stress and strength variability. Reliability depends only on the interference of the stress and the strength distributions and hence only this local information is needed to compute the reliability. The design engineer may be able to estimate the tail probabilities for the stress and strength variables within certain limits based on previous experience and intimate knowledge of the design and operating environment. The lower and upper limits on these probabilities quantify the uncertainty of the estimates made by the engineer. Based on these estimates, bounds are developed on reliability in Section 4.4.

4.1 PROBABILISTIC DESIGN METHODOLOGY

Recently, engineers and designers have become increasingly concerned about problems of design adequacy in various disciplines. This has led to the widespread interest in the use of the probabilistic design approach. This approach, which originated in aerospace engineering, has now spread to the consumer products industry. Figure 4.2 shows a block diagram for this methodology. For the component under consideration, the first step is to perform environmental computations, which affect the stress and strength computations. For the strength computations, consideration must be given to the properties of the material used and the probability distributions of factors affecting the strength, such as the surface finish and surface treatment. For the stress computations, the load statistics history and the probability distributions of factors affecting the stress, such as stress concentration and temperature, must be considered. Based on these computations, the stress and strength distributions and their statistics can be obtained. These distributions are then used to compute the reliability of the component (with respect to a certain mode of failure), which is defined to be the probability that the strength of the component is greater than the stress acting on the component. For an effective application of this methodology, the design engineer must have adequate information on the probabilistic strength, the strength degradation data for the material to be used, and the design data on the statistical distribution of loads.

The steps in the process are:

1. Establishment of a preliminary design.
2. Estimation of external forces.

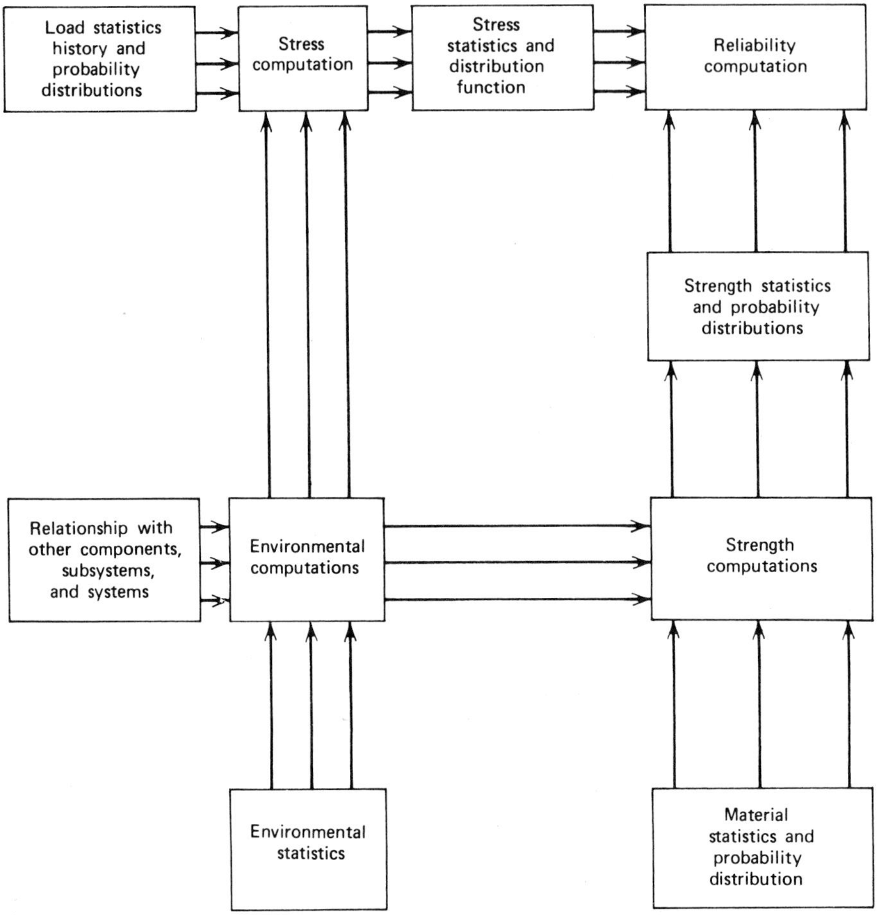

Figure 4.2 Block diagram showing a probabilistic approach.

3. Analysis of the preliminary system, including force intensity in components expressed as probability density functions.

4. Material selection based on mechanical and physical properties and economic feasibility.

5. Description of strength and failure characteristic of the material, including its probability density function.

6. Quantitative estimates of strength and failure characteristics of components. These are functions of

a. Engineering.

b. Geometric considerations.

c. Anticipated operational loads.

7. Description of collective strength and failure characteristics.

If we consider the total design reliability program, the steps will be as summarized below:

1. Define the design problem.
2. Identify the design variables and parameters involved.
3. Conduct a "failure modes, effects, and criticality analysis."
4. Verify the significant design parameter selection.
5. Formulate the relationship between the critical parameters and the failure governing criteria involved.
6. Determine the failure governing stress function.
7. Determine the failure governing stress distribution.
8. Determine the failure governing strength function.
9. Determine the failure governing strength distribution.
10. Calculate the reliability associated with these failure governing distributions for each critical failure mode.
11. Iterate the design to obtain the design reliability goal.
12. Optimize the design in terms of performance, cost, weight, etc.
13. Repeat optimization for each critical component.
14. Calculate system reliability.
15. Iterate to optimize system reliability.

4.2 STRENGTH AND STRESS DISTRIBUTIONS

A search of the existing literature in this area shows that there is a considerable body of data discussing strength and stress distributions [1–5, 16, 17, 23]. There are several ASTM technical special publications on the statistical aspects of fatigue data [2, 3, 4]. Also available [11, 18, 22, 23] are data that support and guide the use of adjustment or degradation factors to account for the differences between laboratory specimens and manufactured parts. The probability distributions for strength can be approached in two ways [6]. If it is assumed that the strength of the components is determined by the weakest point, then the distribution will be determined by the lowest value of samples taken from a distribution that describes the strength of all points in the material. This assumption leads to an extreme value distribution [6, 23]. An alternative assumption is that the weaker points receive support from the stronger points surrounding them; that is, an averaging process occurs. Accordingly, the distribution of the strength is related to the mean value of samples from the distribution of strength of all points, and this leads to a normal distribution. Some of the experimental data tends to support the latter assumption. For example, the probability distributions of the ultimate tensile, yield, and endurance strengths of steels are found to be normal [2, 6, 17, 22, 27]. Also, for a fixed cycle life the fatigue strength distribution is approximately normally distributed [22, 27]. It should be noted that the normal approximation has some limitations. Negative values for strength are inadmissible, hence the strength cannot be truly normally distributed since the normal distribution has the limits from minus infinity to plus infinity. However, when the coefficient of variation is less than 0.3, the probability of negative values is negligible for the strength data. Also, because the normal distribution is a symmetrical distribution, it cannot be used to represent any distribution known to be skewed. It is found that strength properties of structural alloy materials often tend to follow a log normal distribution [2, 12]. Lipson et al. [23] have gathered fatigue data for various materials, heat treatments, surface conditions, and so on and have converted them into strength at a given life. In their investigation of ferrous materials, they use the Weibull distribution to fit the strength distribution. For each material, the Weibull parameters have been tabulated for various treatments and surface conditions. However, in another study on nonferrous materials, they report trying several distributions such as Weibull, largest extreme value, smallest extreme value, and normal. They have investigated the effect of factors such as heat treatment, surface finish, and temperature on the fatigue strength distribution. In most cases, they found that these factors changed the distribution function. For instance, for a given material with a certain surface finish the distribution function may be Weibull but a change in the surface finish might cause the distribution to change to the largest extreme value distribution. The probability distribution of time to failure at a given stress level is sometimes found to be left-skewed. It probably follows the log normal or the gamma distribution [6, 12, 14].

On the other hand, stress distributions cannot be generalized as conveniently as strength distributions. Some loads follow nearly normal distribution; for example, rocket motor thrust [14] or the gas pressure in the cylinder heads of reciprocating engines [23]. Other load components may be skewed and/or possess relatively wide scatter. Bompas-Smith [6] discusses the expected stress and strength distributions for the various conditions and gives very useful information and reliability data for different usage and duty conditions.

4.3 SAFETY FACTORS AND RELIABILITY

Let $\mathbb{S}$ denote the strength random variable and $\mathbf{s}$ the stress random variable. The random variable $\mathbf{y} = (\mathbb{S} - \mathbf{s})$ is then related to the reliability of the component by

$$R = P(\mathbf{y} \geqslant 0) \tag{4.1}$$

When strength and stress random variables have normal density functions, $\mathbf{y}$ is normally distributed and the reliability R is given by

$$R = \frac{1}{\sqrt{2\pi}} \int_{-\frac{(\mu_{\mathbb{S}} - \mu_{\mathbf{s}})}{\sqrt{\sigma_{\mathbb{S}}^2 + \sigma_{\mathbf{s}}^2}}}^{\infty} e^{-z^2/2} dz \tag{4.2}$$

Table 4.1 Safety factors and reliability[a]

CASE NO.	MEAN STRENGTH $\mu_{\mathbb{S}}$	MEAN STRESS μ_s	STRENGTH STANDARD DEVIATION $\sigma_{\mathbb{S}}$	STRESS STANDARD DEVIATION σ_s	FACTOR OF SAFETY $\mu_{\mathbb{S}}/\mu_s$	RELIABILITY R
1	50,000	20,000	2,000	2,500	2.5	1.0
2	50,000	20,000	8,000	3,000	2.5	0.9997
3	50,000	20,000	10,000	3,000	2.5	0.9979
4	50,000	20,000	8,000	7,500	2.5	0.9965
5	50,000	20,000	12,000	6,000	2.5	0.987
6	25,000	10,000	2,000	2,500	2.5	$0.9_{(6)}4$
7	25,000	10,000	1,000	1,500	2.5	$0.9_{(16)}6$
8	50,000	10,000	20,000	5,000	5.0	0.9738
9	50,000	40,000	2,000	2,500	1.25	0.99909
10	50,000	10,000	5,000	5,000	5.0	1.0

[a]The stress and strength are assumed to be normally distributed.

where **z** is the standard normal random variable, $\mu_{\mathbb{S}}$ is the mean value of the strength, μ_s is the mean value of the stress, and $\sigma_{\mathbb{S}}$ and σ_s are the standard deviations of strength and stress, respectively. This is discussed further in Chapter 6. The reliability clearly depends on the lower limit of the integral in Equation 4.2. A higher value of reliability can be obtained by lowering the lower limit.

Table 4.1 gives the reader an idea about the variability in reliability related to different magnitudes of variability in strength and stress random variables. The factor of safety is given by $\mu_{\mathbb{S}}/\mu_s$.

Let the negative of the lower limit of the integral in Equation 4.2 be denoted by

$$z_0 = \frac{\mu_{\mathbb{S}} - \mu_s}{\sqrt{\sigma_{\mathbb{S}}^2 + \sigma_s^2}} \tag{4.3}$$

The relationship between z_0 and reliability R is shown in Figure 4.3.

Let us now study the effect of error in inputs on the computed reliability. Let $\mu_y = 40$ and $\sigma_y = 10$, where $\mathbf{y} = \mathbb{S} - \mathbf{s}$. Then the unreliability of the component is found to be 3×10^{-5}. Suppose the mean and the standard deviation of **y** are in error by 10%. Then, as the second and the third rows in Table 4.2 show, the unreliability changes rather significantly. It is evident that we need a high degree of accuracy for μ_y and σ_y to arrive at a good estimate of the reliability. This extreme sensitivity of the reliability to input errors is often frustrating to all but the most determined designer.

If the mean factor of safety is defined as

$$n = \frac{\mu_{\mathbb{S}}}{\mu_s}$$

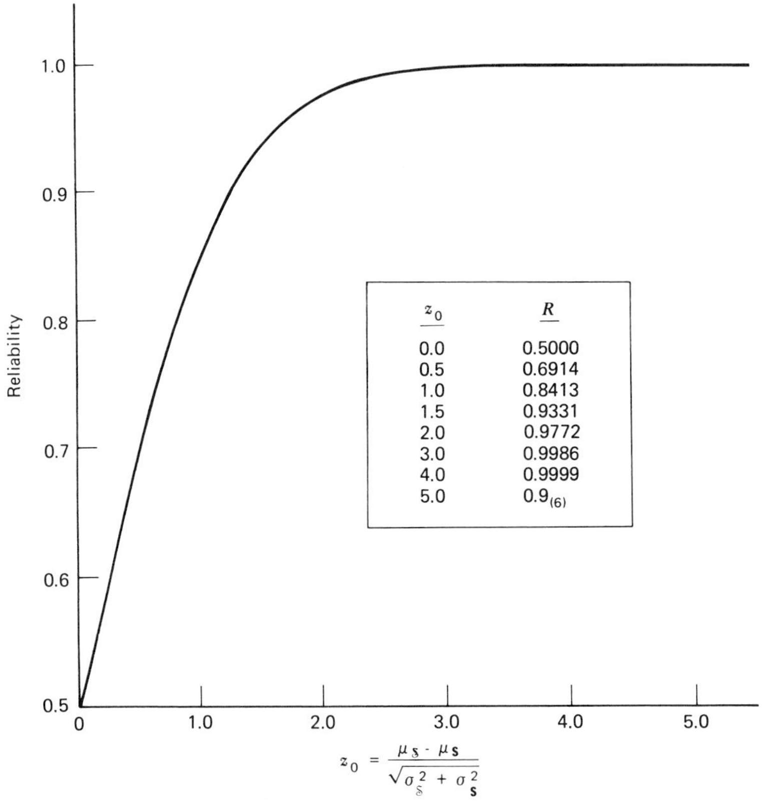

Figure 4.3 Plot of reliability vs z_0.

and if the coefficients of variation of the strength and stress, denoted by $V_{\mathcal{S}}$ and V_s respectively, are defined as

$$V_{\mathcal{S}} = \frac{\sigma_{\mathcal{S}}}{\mu_{\mathcal{S}}}$$

and

$$V_s = \frac{\sigma_s}{\mu_s}$$

Table 4.2 Sensitivity of unreliability to μ_y and σ_y

μ_y	σ_y	z_0	UNRELIABILITY
40.0	10.0	4.00	3×10^{-5}
36.0	11.0	3.27	53×10^{-5}
44.0	9.0	4.89	5×10^{-7}

then from Equation 4.3, we get

$$z_0 = \frac{\dfrac{\mu_{\mathfrak{S}}}{\mu_s} - 1}{\sqrt{\left(\dfrac{\sigma_{\mathfrak{S}}}{\mu_s}\right)^2 + \left(\dfrac{\sigma_s}{\mu_s}\right)^2}}$$

$$= \frac{n-1}{\sqrt{V_{\mathfrak{S}}^2 n^2 + V_s^2}} \tag{4.4}$$

Equation 4.4 is represented by a set of graphs shown in Figure 4.4. These graphs show the relationship among the following four design variables:

1. The factor of safety n.
2. The coefficient of strength variation $V_{\mathfrak{S}}$.

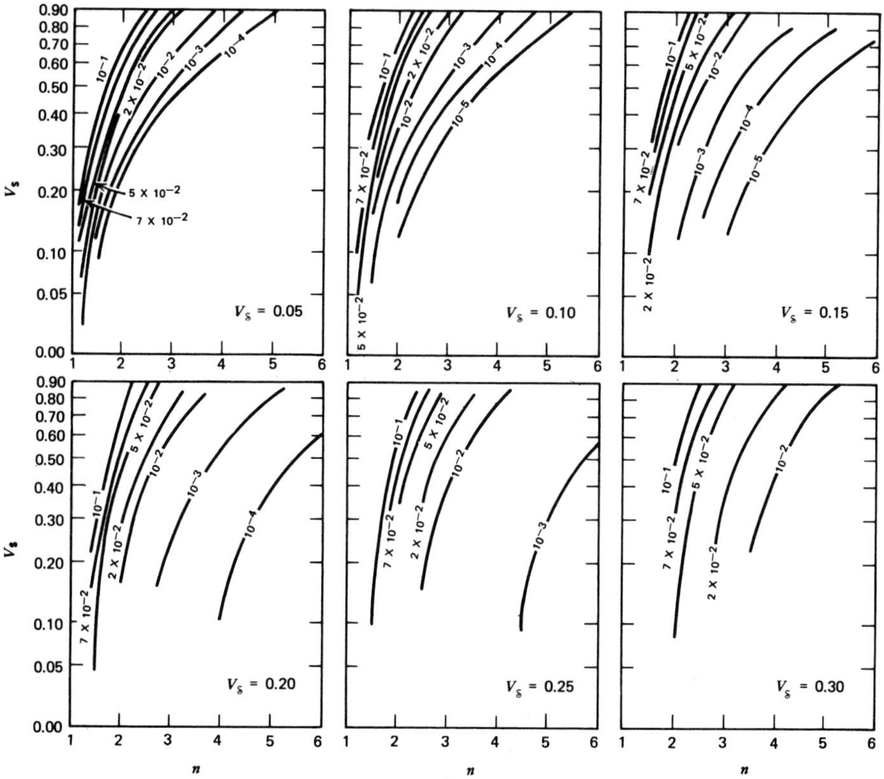

Figure 4.4 Relationships between n, $V_{\mathfrak{S}}$, V_s and R.

3. The coefficeint of stress variation V_s.

4. The reliability R.

Let us now treat the factor of safety n as a random variable. We define **n** as the ratio of the failure governing strength $\mathbb{S}$ to the failure governing stress **s**; that is,

$$\mathbf{n} = \frac{\mathbb{S}}{\mathbf{s}} \tag{4.5}$$

The following derivations are based on results given in References 24 and 9. Let $\bar{\mathbb{S}}$ and $\bar{\mathbf{s}}$ denote the mean strength and stress respectively. Let $\bar{\mathbf{n}}$ denote the expected value of the factor of safety **n**. Then we have the Bienayme—Chebyshev inequality

$$P(|\mathbf{n}-a| \leqslant \varepsilon) \geqslant 1 - \frac{E\{(\mathbf{n}-a)\}^2}{\varepsilon^2} \tag{4.6}$$

where a is any positive constant and $\varepsilon > 0$.

In order to prove Equation 4.6, we observe that

$$E\{(\mathbf{n}-a)^2\} = \int_{\mathbf{n}} (n-a)^2 f(n)\,dn \geqslant \int_{*} (n-a)^2 f(n)\,dn \tag{4.7}$$

where * indicates that the integration is over only those values of **n** for which $|\mathbf{n}-a| > \varepsilon$. But

$$\int_{*} (n-a)^2 f(n)\,dn > \varepsilon^2 \int_{*} f(n)\,dn = \varepsilon^2 P(|\mathbf{n}-a| > \varepsilon) \tag{4.8}$$

From Equations 4.7 and 4.8 we have

$$P(|\mathbf{n}-a| > \varepsilon) < \frac{1}{\varepsilon^2} E\{(\mathbf{n}-a)^2\}$$

which immediately leads to Equation 4.6.

Set $a = k\bar{\mathbf{n}}$, then we have

$$E(\mathbf{n}-a)^2 = E\{(\mathbf{n}-k\bar{\mathbf{n}})^2\} = E\{\mathbf{n}^2 - 2k\bar{\mathbf{n}}\cdot\mathbf{n} + k^2\bar{\mathbf{n}}^2\}$$

$$= E(\mathbf{n}^2) - 2k\bar{\mathbf{n}}^2 + k^2\bar{\mathbf{n}}^2 = \sigma_{\mathbf{n}}^2 + \bar{\mathbf{n}}^2 - 2k\bar{\mathbf{n}}^2 + k^2\bar{\mathbf{n}}^2$$

$$= \bar{\mathbf{n}}^2\left[\frac{\sigma_{\mathbf{n}}^2}{\bar{\mathbf{n}}^2} + (1-k)^2\right] = \bar{\mathbf{n}}^2\left[V_{\mathbf{n}}^2 + (1-k)^2\right]$$

Equation 4.6 can now be rewritten as

$$P(a-\varepsilon \leqslant \mathbf{n} \leqslant a+\varepsilon) \geqslant 1-\frac{E\left\{(\mathbf{n}-a)^2\right\}}{\varepsilon^2} \tag{4.9}$$

Let $a-\varepsilon=1$, then Equation 4.9 can be written as

$$P(1 \leqslant \mathbf{n} \leqslant 2k\bar{\mathbf{n}}-1) \geqslant 1-\frac{\bar{\mathbf{n}}^2\left[V_{\mathbf{n}}^2+(1-k)^2\right]}{(k\bar{\mathbf{n}}-1)^2} \tag{4.10}$$

By definition of reliability, we have

$$R=P(\mathbf{n} \geqslant 1) \tag{4.11}$$

From Equations 4.10 and 4.11 we have

$$R \geqslant 1-\frac{\bar{\mathbf{n}}^2\left[V_{\mathbf{n}}^2+(1-k)^2\right]}{(k\bar{\mathbf{n}}-1)^2} \tag{4.12}$$

The right hand side represents the lower bound of reliability. The highest lower bound can be obtained if we minimize

$$w=\frac{\bar{\mathbf{n}}^2\left[V_{\mathbf{n}}^2+(1-k)^2\right]}{(k\bar{\mathbf{n}}-1)^2} \tag{4.13}$$

with respect to k. Differentiating w with respect to k and equating to zero, we have

$$\frac{\partial w}{\partial k}=\frac{-2\bar{\mathbf{n}}^3\left[V_{\mathbf{n}}^2+(1-k)^2\right]}{(k\bar{\mathbf{n}}-1)^3}+\frac{-2\bar{\mathbf{n}}^2(1-k)}{(k\bar{\mathbf{n}}-1)^2}=0 \tag{4.14}$$

Solving Equation 4.14, the critical value k^* is given by

$$k^*=\frac{\bar{\mathbf{n}}(V_{\mathbf{n}}^2+1)-1}{(\bar{\mathbf{n}}-1)} \tag{4.15}$$

The second partial derivative $\partial^2 w/\partial k^2$ evaluated at this k^* is always positive, which guarantees that k^* minimizes w. Substituting the value of k^* in Equation 4.12 we obtain

$$R \geqslant 1-\frac{\bar{\mathbf{n}}^2 V_{\mathbf{n}}^2}{\left[\bar{\mathbf{n}}^2 V_{\mathbf{n}}^2+(\bar{\mathbf{n}}-1)^2\right]} \tag{4.16}$$

or

$$\bar{\mathbf{n}} \geqslant \frac{1}{1 - V_{\mathbf{n}}\sqrt{\dfrac{R}{1-R}}} \tag{4.17}$$

Equations 4.16 and 4.17 represent the desired relationships between the mean factor of safety, the coefficient of variation of the factor of safety, and the reliability. Equation 4.17 gives us the lower bound on $\bar{\mathbf{n}}$ which will ensure that the probability of finding $\mathbf{n}$ such that $(1 \leqslant \mathbf{n} \leqslant [2k^*\bar{\mathbf{n}} - 1])$ is R.

Another approach to developing bounds is by using Taylor's series. We recall from Equation 4.5 that $\mathbf{n} = \mathbb{S}/\mathbf{s}$. The expected value and the standard deviation of $\mathbf{n}$ can be obtained by considering only the first two terms of the Taylor's series as discussed in Chapter 5. We have

$$\bar{\mathbf{n}} \approx \frac{\bar{\mathbb{S}}}{\bar{\mathbf{s}}} + \frac{\bar{\mathbb{S}}}{\bar{\mathbf{s}}^3}\sigma_{\mathbf{s}}^2$$

$$= n_c\left[1 + V_{\mathbf{s}}^2\right] \tag{4.18}$$

where $n_c \triangleq \bar{\mathbb{S}}/\bar{\mathbf{s}}$, the central factor of safety. Further, we have

$$E(\mathbf{n}^2) = E(\mathbb{S}^2/\mathbf{s}^2)$$

$$\approx E(\mathbb{S}^2)\left[\frac{1}{\bar{\mathbf{s}}^2} + 3\frac{\sigma_{\mathbf{s}}^2}{\bar{\mathbf{s}}^4}\right]$$

or

$$\sigma_{\mathbf{n}}^2 + \bar{\mathbf{n}}^2 \approx \left(\sigma_{\mathbb{S}}^2 + \bar{\mathbb{S}}^2\right)\left[\frac{1}{\bar{\mathbf{s}}^2} + 3\frac{\sigma_{\mathbf{s}}^2}{\bar{\mathbf{s}}^4}\right] \tag{4.19}$$

Dividing by $\bar{\mathbf{n}}^2$ and substituting $\bar{\mathbf{n}} = n_c(1 + V_{\mathbf{s}}^2)$, we have

$$V_{\mathbf{n}}^2 + 1 \approx \left(V_{\mathbb{S}}^2 + 1\right)\left[\frac{1 + 3V_{\mathbf{s}}^2}{\left(1 + V_{\mathbf{s}}^2\right)^2}\right]$$

or

$$V_{\mathbf{n}} \approx \frac{\left[(1 + V_{\mathbb{S}}^2)(1 + 3V_{\mathbf{s}}^2) - (1 + V_{\mathbf{s}}^2)^2\right]^{1/2}}{(1 + V_{\mathbf{s}}^2)} \tag{4.20}$$

Substituting Equation 4.20 into Equation 4.12 yields

$$R \geqslant 1 - \frac{n_c^2\left[(1+V_{\mathfrak{S}}^2)(1+3V_{\mathrm{s}}^2)+k^*(k^*-2)(1+V_{\mathrm{s}}^2)^2\right]}{\left[k^* n_c(1+V_{\mathrm{s}}^2)-1\right]^2} \tag{4.21}$$

where

$$k^* = \frac{n_c\left[(1+V_{\mathfrak{S}}^2)(1+3V_{\mathrm{s}}^2)/(1+V_{\mathrm{s}}^2)\right]-1}{\left[n_c(1+V_{\mathrm{s}}^2)-1\right]} \tag{4.22}$$

The equality in Equation 4.21 gives a lower bound on reliability in terms of the central factor of safety n_c and the coefficients of variation of strength and stress. Equations 4.16, 4.17, and 4.20 can be combined to yield

$$n_c \geqslant \frac{1}{1-\left\{\left[\dfrac{\left[(1+V_{\mathfrak{S}}^2)(1+3V_{\mathrm{s}}^2)-(1+V_{\mathrm{s}}^2)^2\right]R}{(1-R)}\right]^{1/2}-V_{\mathrm{s}}^2\right\}} \tag{4.23}$$

If we neglect the third and higher order terms of $V_{\mathfrak{S}}$ and V_{s}, Equation 4.23 simplifies to

$$n_c \geqslant \frac{1}{\left[1-\left\{\dfrac{(V_{\mathfrak{S}}^2+V_{\mathrm{s}}^2)R}{(1-R)}\right\}^{1/2}+V_{\mathrm{s}}^2\right]} \tag{4.24}$$

Equation 4.24 provides the smallest value of the deterministic factor of safety with at least a probability R of finding the actual factor of safety $\mathbf{n}$ in the range

$$1 \leqslant \mathbf{n} \leqslant \left\{2k^* n_c(1+V_{\mathrm{s}}^2)-1\right\}$$

Another way to compute Equation 4.24 directly is to apply Taylor's series approximation directly to $\mathbf{n}=\mathfrak{S}/\mathbf{s}$ and we have

$$\sigma_{\mathbf{n}}^2 \approx \frac{\sigma_{\mathfrak{S}}^2\bar{\mathbf{s}}^2+\sigma_{\mathbf{s}}^2\bar{\mathfrak{S}}^2}{\bar{\mathbf{s}}^4} \tag{4.25}$$

Thus, we have

$$V_{\mathbf{n}} = \frac{\sigma_{\mathbf{n}}}{\bar{\mathbf{n}}} \approx \frac{\sqrt{\sigma_{\mathfrak{S}}^2\bar{\mathbf{s}}^2+\sigma_{\mathbf{s}}^2\bar{\mathfrak{S}}^2}}{\bar{\mathbf{s}}^2} \times \frac{\bar{\mathbf{s}}^3}{\bar{\mathfrak{S}}(\bar{\mathbf{s}}^2+\sigma_{\mathbf{s}}^2)}$$

or

$$V_{\mathbf{n}} \approx \frac{\sqrt{V_{\mathbb{S}}^2 + V_{\mathbf{s}}^2}}{\left[1 + V_{\mathbf{s}}^2\right]} \tag{4.26}$$

k^*, which will give the highest lower bound in Equation 4.12, is given by Equation 4.15 and, thus, substituting Equations 4.18 and 4.26 into Equation 4.15, we have

$$k^* = \frac{n_c\left(V_{\mathbb{S}}^2 + V_{\mathbf{s}}^2\right)}{\left(1 + V_{\mathbf{s}}^2\right)\left\{n_c\left(1 + V_{\mathbf{s}}^2\right) - 1\right\}} + 1 \tag{4.27}$$

Substituting into Equation 4.12 the values of $\bar{\mathbf{n}}$, $V_{\mathbf{n}}$, and k^*, we have

$$R \geqslant 1 - \frac{n_c^2\left(V_{\mathbb{S}}^2 + V_{\mathbf{s}}^2\right)}{n_c^2\left(V_{\mathbb{S}}^2 + V_{\mathbf{s}}^2\right) + \left\{n_c\left(1 + V_{\mathbf{s}}^2\right) - 1\right\}^2} \tag{4.28}$$

The above inequality gives a lower bound on R given n_c, $V_{\mathbb{S}}$, and $V_{\mathbf{s}}$.

If we use Equation 4.17, we have

$$n_c \geqslant \frac{1}{\left(1 + V_{\mathbf{s}}^2\right) - \sqrt{V_{\mathbb{S}}^2 + V_{\mathbf{s}}^2}\sqrt{\frac{R}{1-R}}} \tag{4.29}$$

Thus, Equation 4.29 is the same as Equation 4.24, derived before.

It should be noted that the bounds given by Equations 4.16 and 4.17 are exact while the bounds given by Equations 4.21 and 4.24 are approximate. This is due to the fact that we substituted approximate values of $\bar{\mathbf{n}}$ and $V_{\mathbf{n}}$ given by Taylor's series approximation in Equations 4.16 and 4.17.

Thus in this section we developed the bounds on the relationships between reliability, factor of safety, and variability in stress and strength random variables. Based on Equations 4.17, 4.23, 4.24, 4.28 and 4.29, we can draw various design nomographs (see Exercise 4). These can be used by the design engineer to estimate the reliability of the design and to study its relationship to factor of safety and variability of the stress and strength random variables.

EXAMPLE 4.1

The strength of a component has a Weibull distribution with the following parameters:

$\mathbb{S}_0$ = the lower bound for strength = 54 MPa

β = shape parameter = 2

θ = scale parameter = 72 MPa

The stress acting on the component follows the normal density function with mean $\bar{s} = 54$ MPa and standard deviation $\sigma_s = 1.8$ MPa. Thus, the coefficient of variation V_s for the stress is 0.033. In order to compute the coefficient of variation for the strength, we find $\bar{\mathbb{S}}$ and $\sigma_{\mathbb{S}}$. We have (see Section 6.7)

$$\bar{\mathbb{S}} = \mathbb{S}_0 + (\theta - \mathbb{S}_0)\Gamma\left(1 + \frac{1}{\beta}\right)$$

$$= 54 + (72 - 54)\Gamma\left(1 + \frac{1}{2}\right)$$

$$= 54 + 18 \times 0.88623 = 69.95 \text{ MPa}$$

$$\sigma_{\mathbb{S}}^2 = (\theta - \mathbb{S}_0)^2\left[\Gamma\left(\frac{2}{\beta} + 1\right) - \left\{\Gamma\left(\frac{1}{\beta} + 1\right)\right\}^2\right]$$

$$= (18)^2\left[1 - (0.88623)^2\right] = 69.53 \text{ MPa}$$

Thus,

$$V_{\mathbb{S}} = \frac{\sqrt{69.53}}{69.95} = 0.119$$

We have

$$n_c = \frac{\bar{\mathbb{S}}}{\bar{s}} = \frac{69.95}{54.0} = 1.295$$

Using Equation 4.28, we find that R should be greater than or equal to 0.8647. The exact reliability can be computed by the method discussed in Chapter 6, Section 6.7. In order to compute the reliability by the tables in the Appendix III, we calculate

$$A = \frac{\mathbb{S}_0 - \bar{s}}{\sigma_s} = \frac{54 - 54}{1.8} = 0.0$$

$$C = \frac{\theta - \mathbb{S}_0}{\sigma_s} = \frac{72 - 54}{1.8} = 10.0$$

which leads to a reliability of 0.9951 from the tables in Appendix III.

4.4 RELIABILITY BOUNDS IN PROBABILISTIC DESIGN

Design for reliability considers the probabilistic nature of the variables. The methodology identifies explicitly the design variables and parameters for the stress and the strength distributions. Once these distributions are determined, the reliability (for the simple stress-strength model of failure) is computed by analytic, numeric, or simulation methods (see Chapter 6). In the real world, it is difficult to know the true distributions over the complete range of the stress and the strength r.v.'s (random variables); reliability depends only on the interference of the stress and strength r.v.'s. Hence, only the local information in the interference range is needed to compute the reliability.

In many cases, it may not be possible to know the actual distributions of the stress and the strength r.v.'s in the interference interval. However, design engineers may be able to estimate the tail probabilities for the stress and strength variables within certain limits based on previous experience and intimate knowledge of the design and operating environment. The upper and lower limits on these probabilities quantify the uncertainty of the estimates made by the engineer. We assume that such upper and lower limits on the tail probabilities for various intervals are already available. Now, we develop a methodology to compute bounds on design reliability, given some bounds on the interval probabilities for the tails of the stress and the strength random variables [19].

The unreliability (failure probability) $\bar{R}$ is given by (see Equations 6.6 and 6.7)

$$\bar{R} = P\{\mathbf{s} > \mathbb{S}\} = \int_{-\infty}^{\infty} f(u) G(u)\, du = \int_{-\infty}^{\infty} g(u) \bar{F}(u)\, du \tag{4.30}$$

where $f(\cdot)$ is the p.d.f. for the r.v. $\mathbf{s}$ and $g(\cdot)$ is the p.d.f. for the r.v. $\mathbb{S}$.

Let $s_{\max}$ be the upper limit for $\mathbf{s}$ and $\mathbb{S}_{\min}$ be the lower limit for $\mathbb{S}$. Thus the interference interval is $[\mathbb{S}_{\min}, s_{\max}]$. $\mathbb{S}_{\min}$ and $s_{\max}$ are either known from the p.d.f.'s of $\mathbb{S}$ and $\mathbf{s}$, respectively, or their values are estimated according to the accuracy desired. The probability of interference r.v. having values outside this interference interval is assumed extremely small, or negligible. Based on this interference interval, Equation 4.30 can be rewritten as

$$\bar{R} = \int_{\mathbb{S}_{\min}}^{s_{\max}} f(u) G(u)\, du = \int_{\mathbb{S}_{\min}}^{s_{\max}} g(u) \bar{F}(u)\, du \tag{4.31}$$

Now break the interval $[\mathbb{S}_{\min}, s_{\max}]$ into n subintervals. Let $a_0 = \mathbb{S}_{\min}$ and $a_n = s_{\max}$ denote the end points of $[\mathbb{S}_{\min}, s_{\min}]$ and $a_1, a_2, \ldots, a_{n-1}$ the end points of the subintervals. Define

$$p_i \equiv P\{a_{i-1} < \mathbf{s} \leqslant a_i\} \qquad \text{for } i = 1, \ldots, n \tag{4.32}$$

$$q_i \equiv P\{a_{i-1} < \mathbb{S} \leqslant a_i\} \qquad \text{for } i = 1, \ldots, n \tag{4.33}$$

The unreliability $\bar{R}$ is given by Equation 4.34 and is obtained by the discrete approximation of the integrals in Equation 4.31

$$\bar{R}=\sum_{i=1}^{n}\left[p_i\left(\sum_{k=1}^{i}q_k\right)\right]=\sum_{i=1}^{n}\left[q_i\left(\sum_{k=i}^{n}p_k\right)\right] \tag{4.34}$$

The p_i's refer to the right-hand tail of the stress p.d.f. and q_i's to the left-hand tail of the strength p.d.f. Let the p_i's and q_i's be estimated in terms of lower and upper bounds given by

$$L_{p,i} \leqslant p_i \leqslant U_{p,i} \qquad \text{for } i=1,\ldots,n \tag{4.35}$$

$$L_{q,i} \leqslant q_i \leqslant U_{q,i} \qquad \text{for } i=1,\ldots,n \tag{4.36}$$

Obviously, we have

$$P\{\mathbb{S} \leqslant s_{\max}\}=\sum_{i=1}^{n} q_i,\ P\{s \geqslant \mathbb{S}_{\min}\}=\sum_{i=1}^{n} p_i \tag{4.37}$$

the bounds on which are also assumed supplied by the design engineer. They are

$$a_p \leqslant \sum_{i=1}^{n} p_i \leqslant b_p,\ a_q \leqslant \sum_{i=1}^{n} q_i \leqslant b_q \tag{4.38}$$

Based on the bounds given by Equations 4.35, 4.36, and 4.38, we wish to find bounds for $\bar{R}$ in Equation 4.34. The upper bound on $\bar{R}$ is given by maximizing $\bar{R}$ in Equation 4.34 subject to the constraints given by Equations 4.35, 4.36, and 4.38, while the lower bound on $\bar{R}$ is given by minimizing $\bar{R}$ in Equation 4.34 subject to the same constraints. In either case we face a quadratic programming problem.

EXAMPLE 4.2

We are given the bounds on p_i's and q_i's as shown by the following constraints for the case when $n=4$. In this case the unreliability $\bar{R}$ is

$$\bar{R}=p_1q_1+p_2(q_1+q_2)+p_3(q_1+q_2+q_3)+p_4(q_1+q_2+q_3+q_4)$$

and the constraints are

$0.05 \leqslant p_1 \leqslant 0.12,$ $\qquad$ $0.04 \leqslant q_1 \leqslant 0.10;$

$0.04 \leqslant p_2 \leqslant 0.08,$ $\qquad$ $0.03 \leqslant q_2 \leqslant 0.07;$

$0.02 \leqslant p_3 \leqslant 0.06,$ $\qquad$ $0.02 \leqslant q_3 \leqslant 0.05;$

$0.01 \leqslant p_4 \leqslant 0.05,$ $\qquad$ $0.01 \leqslant q_4 \leqslant 0.03;$

$0.14 \leqslant \sum_{i=1}^{4} p_i \leqslant 0.20,$ $\qquad$ $0.13 \leqslant \sum_{i=1}^{4} q_i \leqslant 0.22.$

$\bar{R}$, which is a quadratic function of p_i's and q_i's, is maximized and minimized subject to the above 10 inequality constraints. The optimal values are

	VALUES FOR $\bar{R}_{\min}$		VALUES FOR $\bar{R}_{\max}$	
i	p_i^*	q_i^*	p_i^*	q_i^*
1	0.07	0.04	0.05	0.10
2	0.04	0.03	0.04	0.07
3	0.02	0.03	0.06	0.04
4	0.01	0.03	0.05	0.01
$\bar{R}$	0.0089		0.0354	

The conclusion is that $0.0089 \leqslant \bar{R} \leqslant 0.0354$; that is, $\bar{R}$ is known to within a factor of four.

EXAMPLE 4.3

The stress acting on a component is normally distributed with a mean (μ_s) of 30 and a standard deviation (σ_s) of 3, and the coefficient of variation of 0.1. The strength of the component has a three-parameter Weibull distribution with parameters:

$$\text{Minimum strength} = \mathbb{S}_0 = 30$$
$$\text{Scale parameter } \theta = 60$$
$$\text{Shape parameter } \beta = 2, 3, \text{or } 4$$

The unreliability of the component is [Equation 6.36]

$$\bar{R} = 1 - \Phi(a) - \frac{b}{\sqrt{2\pi}} \int_0^{\infty} \exp\left[-y^{\beta} - \frac{1}{2}(by + a)^2 \right] dy$$

where

$$a \equiv (\mathbb{S}_0 - \mu_s)/\sigma_s,\ b \equiv (\theta - \mathbb{S}_0)/\sigma_s,\ y \equiv (s - \mathbb{S}_0)/(\theta - \mathbb{S}_0)$$

It is clear that $\mathbb{S}_{\min} = 30$; $s_{\max}$ was taken as $\mu_s + 6.67\sigma_s = 50$. Thus the interference interval is [30, 50], and was divided into 10 equal parts ($n = 10$). The probabilities in the interference range can be computed by using the normal and Weibull distributions. The tail probability for the stress above 50 MPa is of the order of 10^{-9} and was lumped in the last interval.

An uncertainty (of $\pm\ \alpha$ percent of the actual probability) was present in these probabilities; that is, $L_{p,i} = p_i - (\alpha/100)p_i$ and $U_{p,i} = p_i + (\alpha/100)p_i$. Similar bounds were found for the strength probabilities. The errors on the estimates of the cumulative probabilities for the total interference interval in Equation 4.37 were assumed to be $\alpha/2$ percent of the actual probabilities obtained. This assumption is valid because the analyst can have a better estimate of the cumulative probabilities for the total interference interval than for the small interval probabilities. The analysis was performed for $\alpha = 2\%$ (2%) 10%. The optimization problem for each value of α and β was solved.

Table 4.3 shows the upper and lower limits on unreliability.

Table 4.3

SHAPE PARAMETER β	UNCERTAINTY $\alpha(\%)$	$\bar{R}_{min}$	$\bar{R}_{max}$
	2	.00704	.00760
	4	.00677	.00789
2	6	.00650	.00818
	8	.00624	.00848
	10	.00598	.00878
	2	.00123	.00133
	4	.00118	.00138
3	6	.00113	.00143
	8	.00108	.00148
	10	.00104	.00154
	2	.00024	.00026
	4	.00023	.00027
4	6	.00022	.00029
	8	.00021	.00030
	10	.00020	.00031

4.5 Summary

Probabilistic design is a realistic way to develop a quantitative and statistical estimate of the performance of a design before it leaves the drawing board. The main drawback of this approach is that it requires a good knowledge of probability and statistics, and not every design engineer has this knowledge. The use of the methods discussed in this chapter can be an aid in overcoming this problem. Once carried out, the analysis will result in a set of design nomographs as discussed in Section 4.3, and these can be used by the design engineers to study the sensitivity of design performance with respect to the measures of variability in the stress and strength variables. The approach discussed in Section 4.4 can be carried out jointly by design engineers and reliability engineers working together. The probabilistic approach forces the designer to quantify uncertainty in the design variables and thus understand the inherent reliability of the design. Because this approach statistically quantifies design performance, it will help the designer to assess warranty costs, establish maintenance programs, and schedule inventories. It must be realized that this approach is still in its infancy and, as it grows, an increasing amount of statistical data will be available to designers in the future.

EXERCISES

1 Consider a simple design problem that is familiar to you. Develop a methodology for the design of this problem similar to the methodology given in Section 4.1.

2 A mechanical element in combined bending and torsion has to be designed. The strength and the stress for this element are random variables and are functions of other random variables. List some of these other random variables that affect the stress and the strength and show how you would determine their distributions.

3 Discuss the use of the extreme value distributions in developing stress and strength distributions.

4 Develop a design nomograph showing the relationship of the following four design variables in one nomograph:

a. The factor of safety $\bar{\mathbf{n}}$
b. The coefficient of strength variation $V_{\mathfrak{S}}$
c. The coefficient of stress variation $V_{\mathbf{s}}$
d. The reliability R.

5 Suppose it is known that the p.d.f. for the factor of safety **n** is unimodal. Develop a new relationship similar to Equation 4.16 that uses the unimodality of **n**. What impact does this have on the relationship between $\bar{\mathbf{n}}$, $V_{\mathbf{n}}$, and R? By what factor are the new bounds improved?

6 Using the relationship developed in Exercise 5, develop a relationship similar to Equation 4.24, which relates n_c, $V_{\mathfrak{S}}$, $V_{\mathbf{s}}$, and R. What impact does the assumption of unimodality have on the bounds developed for this case?

7 With the assumptions given in Section 4.3 is it possible to improve the bounds given by Equations 4.16 and 4.24? If so, develop the new improved bounds. What other information is available in practice to improve the bounds given by Equations 4.16 and 4.24?

8 The strength of a component is log normally distributed with a mean value of 400 MPa and a standard deviation of 50 MPa. The stress acting on the component is normally distributed with a mean value of 250 MPa and a standard deviation of 50 MPa. Compute the bounds on reliability for α equal to 5% and 10%, where α has the same meaning as given in Example 4.3.

9 The stress and the strength distributions for a component are Weibull with the following parameters:

Strength: $\mathfrak{S}_0 = 300$ MPa, $\theta_{\mathfrak{S}} = 400$ MPa, $\beta_{\mathfrak{S}} = 3.0$
Stress: $s_0 = 150$ MPa, $\theta_{\mathbf{s}} = 300$ MPa, $\beta_{\mathbf{s}} = 4.0$

Compute the bounds on reliability for α equal to 5% and 10%, where α has been defined as in Example 4.3.

BIBLIOGRAPHY

1 *Aerospace Structural Metals Handbook,* Syracuse University Press, Syracuse, N.Y., 1963.

2 *American Society for Metals, Metals Handbook, Properties and Selection,* Vol. 1, 8th Ed., 1969.

3 *American Society for Testing Materials,* "Symposium on Statistical Aspects of Fatigue," ASTM Special Technical Publication No. 121, June 19, 1951.

4 *American Society for Testing Materials,* "Symposium on Fatigue with Emphasis on Statistical Approach—II," ASTM Special Technical Publication No. 137, June 24, 1952.

5 *American Society for Testing Materials,* "A Guide for Fatigue Testing and the Statistical Analysis of Fatigue Data," ASTM Special Technical Publication No. 91-A, 2nd Ed., 1963.

6 Bompas-Smith, J.H., *Mechanical Survival: The Use of Reliability Data,* by R. H. W. Brook, Ed., McGraw-Hill, New York, 1973.

7 Bouton, I., "Fundamental Aspects of Structural Reliability," *Aerospace Engineering,* Vol. 1, No. 2116, June 1962.

8 Bratt, M. J., Reethoff, G., and Weber, G. W. "A Model for Time Varying and Interfering Stress/Strength Probability Density Distributions with Consideration for Failure Incidence and Property Degradation," *Proceedings Third Annual Aerospace Reliability and Maintainability Conference,* 1964, pp. 566–575.

9 Dao-Thien, M. and Massoud, M., "On the Relation Between the Factor of Safety and Reliability," *Journal of Engineering for Industry,* Transactions of ASME, 1973, paper No. 73-WA/DE-1.

10 Disney, R. L. and Sheth, N. J., "The Determination of the Probability of Failure by Stress/Strength Interference Theory," *Proceedings of Annual Symposium on Reliability,* 1968, pp. 417–422.

11 Forrest, P. G., *Fatigue of Metals,* Pergamon Press, New York, 1962.

12 Freudenthal, A. M. and Gumbel, E. J., "Distribution Functions for the Prediction of Fatigue Life and Fatigue Strength," *International Conference on Fatigue of Metals,* British Institute of Mechanical Engineers, London, 1966.

13 Haugen, E B., "Statistical Methods for Structural Reliability Analysis," *Proceedings of the Tenth National Symposium on Reliability and Quality Control,* 1964, pp. 97–121.

14 Haugen, E. B., "Implementing a Structural Reliability Program.," *Proceedings Eleventh National Symposium on Reliability and Quality Control,* 1965, pp. 158–168.

15 Haugen, E. B, *Probabilistic Approach to Design,* John Wiley & Sons, New York, 1968.

16 Haugen, E. B. and Mutti, D. H., *Statistical Metals Manual,* The University of Arizona, Tuscon, Arizona, 1972.

17 Haugen, E. B., *Probabilistic Mechanical Design,* The University of Arizona, Tuscon, Arizona, 1974.

18 Juvenal, R. C. and Lipson, C., *Handbook of Stress and Strength,* Macmillan, New York, 1963.

19 Kapur, K. C., "Reliability Bounds in Probabilistic Design," *IEEE* Transactions on Reliability, Vol. R-24, No. 3, August 1975.

20 Kececioglu, D. and Cormier, D., "Designing a Specified Reliability Directly into a Component," *Proceedings Third Annual Aerospace Reliability and Maintainability Conference,* 1964, pp. 546–565.

21 Kececioglu, D. and Haugen E. B., "A Unified Look at Design Safety Factors, Safety Margins and Measures of Reliability," *Annals of Assurance Sciences—Seventh Reliability and Maintainability Conference,* 1968, pp. 520–530.

22 Lipson, C., Kerawalla, J., Mitchel, L., and Krafus, A.; Engineering for Reliability, Unpublished Lecture Notes, University of Michigan, Ann Arbor, Michigan, 1962.

23 Lipson, C., Sheth, N. J. and Disney, R. L., "Reliability Prediction—Mechanical Stress/Strength Interference," Rome Air Development Center, Technical Report No. RADC-TR-66-710, March 1967.

24 Mischke, C., "A Method of Relating Factor of Safety and Reliability," *Journal of Engineering for Industry,* Transactions of ASME, August 1970, pp. 537–542.

25 Papoulis, A., *Probability, Random Variables and Stochastic Process,* McGraw-Hill, New York, 1965.

26 Shaw, L., Shooman, M. and Schatz, R., "Time Dependent Stress/Strength Models for Non-Electrical and Electrical Systems," *Annals of Reliability and Maintainability Symposium,* 1972, pp. 186–197.

27 Vincent, Liffi R. and Kececioglu, D. B., "An Approach to Reliability Determination of a Rotating Component Subjected to Complex Fatigue," *Annals of Reliability and Maintainability,* 1970, pp. 534–548.

Chapter 5 combinations of random variables in design

The reliability of an engineering design is usually a function of several design variables and parameters. As discussed in Chapter 4, most of these variables and parameters are random variables. The design performance is expressed mathematically as a function of these random variables. A complete characterization of the performance can be obtained only if sufficient data have been collected on the behavior of the system. In the case of large, complex systems this data collection may not always be possible. Often, the length of the time of operation is not adequate to give us sufficiently large statistical samples needed to estimate the system performance distribution function. However, if the total system is first broken down into subsystems and components, the statistical behavior of these individual components can be studied and then integrated into a complete description of the behavior of the original system.

In this chapter, we discuss how to combine several random variables. For instance, velocity is expressed as a combination of distance and time in the form of a quotient, or force as a combination of mass and acceleration in a product form. A voltage distribution may be derived from the resistance and current distributions. In a mechanical system, the stress distribution may be derived from the force and area distributions. We do not confine our discussion to the sum, difference, product, and division of two or more random variables; we also discuss several other functional combinations.

In the design of beams we are usually interested in computing the bending stresses. The maximum stresses occur at the top of the upper flange and bottom of the lower flange of the beam. The maximum fiber stress is given by

$$s = \frac{Mc}{I} \tag{5.1}$$

where s = maximum fiber stress, kPa

M = external bending moment, N·m

c = distance from the neutral axis to the extreme fibers, m

I = moment of inertia of the beam cross-section about the neutral axis, m^4

If we were to design a beam that was tubular in cross-section with an outside radius of r meters and a wall thickness of t meters, then, for $r \gg t$,

$$I = \pi r^3 t \tag{5.2}$$

Thus, from Equation 5.1 we have

$$s = \frac{Mc}{\pi r^3 t} \tag{5.3}$$

In real world design problems, M, c, r, and t are random variables. Hence, to compute the distribution, or any other statistical property of the stress s, we need to know how to combine the random variables **M**, **c**, **r** and **t**.

Thus the basic problem we address ourselves to in this chapter is this: given a function f of random variables $\mathbf{x}_1,\mathbf{x}_2,\ldots,\mathbf{x}_n$ as

$$\mathbf{y} = f(\mathbf{x}_1,\mathbf{x}_2,\ldots,\mathbf{x}_n) \tag{5.4}$$

how do we find certain properties of the random variable **y** as functions of the properties of random variables $\mathbf{x}_1,\mathbf{x}_2,\ldots,\mathbf{x}_n$? The random variables $\mathbf{x}_1,\ldots,\mathbf{x}_n$ may or may not be independent. The density functions of the random variables $\mathbf{x}_1,\ldots,\mathbf{x}_n$ may be known, and we may want to find the density function of the random variable **y**. Section 5.1 shows how to find the density function for **y** given the density function for only one random variable **x**. The results for several random variables can be found in References 2, 6, and 7. In general, it is quite complicated to find the density function for **y** and sometimes this may not be possible. In many design situations, only the first few moments of the random variables $\mathbf{x}_1,\mathbf{x}_2,\ldots,\mathbf{x}_n$ are known, and it is necessary to find the corresponding moments of the random variable **y**. Some of the well-known results for expectation and variance are given in Section 5.2. Section 5.3 gives the useful results for sum and difference of normal random variables. Computation of the moments of the random variable **y**, using Taylor's series approximation, is discussed in Section 5.4; error analysis in the approximation methods is also discussed. Concepts in statistical tolerancing are very useful in probabilistic design methodology and these are discussed in Section 5.5.

5.1 TRANSFORMATION OF RANDOM VARIABLES

Let us first consider the case of one random variable **x**. We are given

$$\mathbf{y} = f(\mathbf{x}) \tag{5.5}$$

Suppose we know the density function $g(x)$ of the random variable **x**. Our objective is to find the density function $h(y)$ for the random variable **y**. It is well known [2, 6] that

$$h(y)=\left|\frac{dk}{dy}\right|g(k(y)) \tag{5.6}$$

where $k(y)$ is defined to be the inverse function of f. From Equation 5.5, we can now write

$$k(y)=f^{-1}(y)=x \tag{5.7}$$

The term $|dk/dy|$ of course represents the absolute value of the derivative of the inverse function $k(y)$ with respect to $\mathbf{y}$. If the inverse function is double valued, we will have two values for x from Equation 5.7. We will denote them by x_1 and x_2. Then Equation 5.6 becomes

$$h(y)=\left|\frac{dx_1}{dy}\right|g(x_1)+\left|\frac{dx_2}{dy}\right|g(x_2) \tag{5.8}$$

In general, if the inverse function $k(y)$ has n roots, $x_1,\ldots,x_n$, Equation 5.8 will have n terms, one term for each root.

If two or more random variables are present, a similar change of variables transformation can be made except that the analysis will be more complicated. Later we consider the properties of the random variable $\mathbf{y}$ when $\mathbf{x}_1,\ldots,\mathbf{x}_n$ follow certain well-known distributions.

EXAMPLE 5.1

To compute the tensile stresses induced in a circular bar of diameter D, we need to know the cross-section $A=\frac{\pi}{4}D^2$. Suppose the diameter has a normal distribution with a mean μ and a standard deviation σ. We wish to find the density function for the random variable A.

We know that

$$A=f(D)=(\pi/4)D^2$$

Hence the inverse function is given by

$$D=f^{-1}(A)=\pm\sqrt{4A/\pi}$$

Differentiating yields

$$\left|\frac{dD}{dA}\right|=\sqrt{1/A\pi}$$

The inverse function is found to be double valued. Therefore, a substitution in Equation 5.8 yields

$$h(A)=\sqrt{1/A\pi}\left[g(\sqrt{4A/\pi})+g(-\sqrt{4A/\pi})\right] \tag{5.9}$$

where g is given to be a normal density function with mean μ and standard deviation σ. Hence,

$$h(A)=\sqrt{1/A\pi}\;\frac{1}{\sigma\sqrt{2\pi}}\left\{\exp\left[-\frac{\left(\sqrt{4A/\pi}-\mu\right)^2}{2\sigma^2}\right]+\exp\left[-\frac{\left(-\sqrt{4A/\pi}-\mu\right)^2}{2\sigma^2}\right]\right\}\tag{5.10}$$

5.2 EXPECTATION AND VARIANCE OF A FUNCTION OF RANDOM VARIABLES

Next we present some of the well-known results for the expectation and the variance of a function of random variables.

5.2.1 PROPERTIES OF EXPECTATION

Let a denote a constant. Further, let $\mathbf{x}$ and $\mathbf{y}$ denote two random variables with expectations $E(\mathbf{x})$ and $E(\mathbf{y})$, respectively. Then

$$E(a\mathbf{x})=aE(\mathbf{x})\tag{5.11}$$

$$E(a+\mathbf{x})=a+E(\mathbf{x})\tag{5.12}$$

$$E(\mathbf{x}+\mathbf{y})=E(\mathbf{x})+E(\mathbf{y})\tag{5.13}$$

$$E(\mathbf{x}-\mathbf{y})=E(\mathbf{x})-E(\mathbf{y})\tag{5.14}$$

$$E(\mathbf{x}^2)=\left[E(\mathbf{x})\right]^2+V(\mathbf{x})\tag{5.15}$$

If $\mathbf{x}$ and $\mathbf{y}$ are independent, then

$$E(\mathbf{xy})=E(\mathbf{x})\cdot E(\mathbf{y})\tag{5.16}$$

5.2.2 PROPERTIES OF VARIANCE

Again let a be a constant, and let $\mathbf{x}$ and $\mathbf{y}$ be two random variables with expected values $\mu_{\mathbf{x}}$ and $\mu_{\mathbf{y}}$ and variances $\sigma_{\mathbf{x}}^2$ and $\sigma_{\mathbf{y}}^2$, respectively. Then

$$V(a\mathbf{x})=a^2V(\mathbf{x})\tag{5.17}$$

$$V(a+\mathbf{x})=V(\mathbf{x})\tag{5.18}$$

$$V(\mathbf{x}^2)=E(\mathbf{x}^4)-\left(\mu_{\mathbf{x}}^2+\sigma_{\mathbf{x}}^2\right)^2\tag{5.19}$$

If $\mathbf{x}$ and $\mathbf{y}$ are independent, we further have

$$V(\mathbf{x}+\mathbf{y})=V(\mathbf{x})+V(\mathbf{y}) \tag{5.20}$$

$$V(\mathbf{x}-\mathbf{y})=V(\mathbf{x})+V(\mathbf{y}) \tag{5.21}$$

$$V(\mathbf{xy})=\sigma_{\mathbf{x}}^2\sigma_{\mathbf{y}}^2+\sigma_{\mathbf{y}}^2\mu_{\mathbf{x}}^2+\sigma_{\mathbf{x}}^2\mu_{\mathbf{y}}^2 \tag{5.22}$$

The covariance of $\mathbf{x}$ and $\mathbf{y}$, denoted by cov($\mathbf{x},\mathbf{y}$), is defined to be

$$\text{cov}(\mathbf{x},\mathbf{y})=E\left[(\mathbf{x}-\mu_{\mathbf{x}})(\mathbf{y}-\mu_{\mathbf{y}})\right]=E(\mathbf{xy})-E(\mathbf{x})E(\mathbf{y}) \tag{5.23}$$

We can normalize the covariance with respect to $\sigma_{\mathbf{x}}$ and $\sigma_{\mathbf{y}}$. The normalized variable is called the correlation coefficient ρ and is defined to be

$$\rho=\frac{\text{cov}(\mathbf{x},\mathbf{y})}{\sigma_{\mathbf{x}}\sigma_{\mathbf{y}}} \tag{5.24}$$

It can be shown that $-1\leqslant\rho\leqslant+1$. When $\rho=0$ and $\sigma_{\mathbf{x}}$ and $\sigma_{\mathbf{y}}$ are finite, Equation 5.24 leads to cov($\mathbf{x},\mathbf{y}$) = 0, and we have

$$E(\mathbf{xy})=E(\mathbf{x})E(\mathbf{y}) \tag{5.25}$$

5.3 SUM AND DIFFERENCE OF NORMAL RANDOM VARIABLES

Let $\mathbf{x}$ and $\mathbf{y}$ be two normal random variables with parameters $(\mu_{\mathbf{x}},\sigma_{\mathbf{x}})$ and $(\mu_{\mathbf{y}},\sigma_{\mathbf{y}})$, respectively, and with a correlation coefficient ρ. Then we have the following results:

i. If $\mathbf{z}=\mathbf{x}+\mathbf{y}$, then $\mathbf{z}$ is normally distributed with (5.26)

$$\mu_{\mathbf{z}}=\mu_{\mathbf{x}}+\mu_{\mathbf{y}} \tag{5.27}$$

and

$$\sigma_{\mathbf{z}}=\sqrt{\sigma_{\mathbf{x}}^2+\sigma_{\mathbf{y}}^2+2\rho\sigma_{\mathbf{x}}\sigma_{\mathbf{y}}} \tag{5.28}$$

ii. If $\mathbf{z}=\mathbf{x}-\mathbf{y}$, then $\mathbf{z}$ is normally distributed with

$$\mu_{\mathbf{z}}=\mu_{\mathbf{x}}-\mu_{\mathbf{y}} \tag{5.29}$$

and

$$\sigma_{\mathbf{z}}=\sqrt{\sigma_{\mathbf{x}}^2+\sigma_{\mathbf{y}}^2-2\rho\sigma_{\mathbf{x}}\sigma_{\mathbf{y}}} \tag{5.30}$$

EXAMPLE 5.2

An assembly consists of five machine parts put together such that the tolerances are additive. We want the final tolerance on the assembly, Δ, to be ± 0.25mm. Let us assume that the machining tolerances on all the components are the same and that they are independently normally distributed. Our objective is to find the component tolerance in order to achieve the required assembly tolerance.

Let us denote the component tolerance as $\pm t$. Since these component tolerances are normally distributed, we know that the assembly tolerance will also be normally distributed. Now let us assume that all tolerances are based on 3σ limits (i.e., the standard deviation for each component is $t/3$). Then the desired standard deviation for the assembly dimension is $\Delta/3$. Hence, extending Equation 5.28 to five variables, we have

$$\frac{\Delta}{3} = \sqrt{(t/3)^2 + (t/3)^2 + (t/3)^2 + (t/3)^2 + (t/3)^2}$$

$$= \frac{t}{3}\sqrt{5}$$

or

$$t = \frac{\Delta}{\sqrt{5}} = \frac{0.25}{\sqrt{5}} = 0.1118\text{mm}$$

Hence the desired machining tolerance for each component is ± 0.1118mm.

EXAMPLE 5.3

In the design of a certain electrical circuit, we want the current $I = V/R$ to be within certain specified limits. We will assume that the voltage supplied to the circuit is fairly constant; that is, the variation in the voltage is negligible. Hence the design emphasis is on controlling the resistance R to meet certain specified limits for the current. The resistor R is made up of two separate resistors R_1 and R_2 connected in series with their mean resistance values 100 Ω and 200 Ω, respectively. Based on a study of the manufacturing process, the resistance of a resistor is found to be normally distributed with a standard deviation equal to 5% of the mean value of the resistance.

The mean value of the total resistance R is

$$\mu_R = \mu_{R_1} + \mu_{R_2} = 100 + 200 = 300\ \Omega$$

and the standard deviation is

$$\sigma_R = \sqrt{\sigma_{R_1}^2 + \sigma_{R_2}^2} = \sqrt{(5)^2 + (10)^2} = 11.18\ \Omega$$

Note that the standard deviation of the total resistance R is only 3.73% of its mean value. This percentage is considerably less than that for the individual resistors. Thus, by putting two resistors in series we achieved smaller tolerances, or we have gained a "tolerance advantage."

Let us now suppose that the two resistors are to be manufactured on two separate machines. Suppose that the resistors are selectively assembled and, therefore, are correlated. Let us assume that the correlation coefficient is -0.7. Then, using Equation 5.28, we have

$$\sigma_R = \sqrt{\sigma_{R_1}^2 + \sigma_{R_2}^2 + 2\rho\sigma_{R_1}\sigma_{R_2}}$$

$$= \sqrt{(5)^2 + (10)^2 - 2(0.7)(5)(10)}$$

$$= 7.42\ \Omega$$

In this case, because of the correlation the standard deviation of the total resistance is only 2.47% of its mean value, which is significantly smaller than the original manufacturing capability of 5.0%.

5.4 COMPUTATION OF MOMENTS OF A FUNCTION OF RANDOM VARIABLES

Generally, it is very difficult to find the density function for a function of random variables. In such cases, for engineering purposes, knowledge of the moments of the transformed random variable is quite useful. The probability bounds can be obtained using Chebyshev's bound or Gauss' bound.

If **y** is a random variable with mean μ and variance σ^2, then for any positive number k the following inequality, known as Chebyshev's inequality, holds:

$$P(|\mathbf{y} - \mu| \geqslant k\sigma) \leqslant (1/k^2) \tag{5.31}$$

If the density function of **y** is continuous and has only one maximum (mode), say at $y_{\max}$, and if the expected value of **y** is $y_{\max}$, then the following inequality, known as Gauss' inequality, holds:

$$P(|\mathbf{y} - \mu| \geqslant k\sigma) \leqslant (4/9k^2) \tag{5.32}$$

5.4.1 APPROXIMATIONS FOR $E[F(\mathbf{X})]$ AND $V[F(\mathbf{X})]$

Let us first consider the case where **x** is a one-dimensional variable. The expansion of $\mathbf{y} = f(\mathbf{x})$ about the point $\mathbf{x} = \mu$ by Taylor's series up to the first three terms is

$$\mathbf{y} = f(\mathbf{x}) = f(\mu) + (\mathbf{x} - \mu)f'(\mu) + \frac{(\mathbf{x} - \mu)^2}{2!} f''(\mu) + \mathcal{R} \tag{5.33}$$

where $\mathcal{R}$ is the remainder. Taking the expectation of Equation 5.33, we have

$$E(\mathbf{y}) = E[f(\mu)] + E\{\mathbf{x}f'(\mu) - \mu f'(\mu)\} + E\left\{\frac{1}{2} f''(\mu)(\mathbf{x} - \mu)^2\right\} + E(\mathcal{R})$$

$$= f(\mu) + \{\mu f'(\mu) - \mu f'(\mu)\} + \frac{1}{2} f''(\mu)V(\mathbf{x}) + E(\mathcal{R})$$

$$\approx f(\mu) + \frac{1}{2} f''(\mu)V(\mathbf{x}) \tag{5.34}$$

Equation 5.34 is an approximation for the expected value of $\mathbf{y}$ because we have ignored the remainder terms in the Taylor's series expansion. If the variance of $\mathbf{x}$ is small, then we may further ignore the second term in Equation 5.34 to obtain

$$E(\mathbf{y}) = E[f(\mathbf{x})] \approx f(\mu) \tag{5.35}$$

To obtain the approximate value of $V(\mathbf{y})$, consider once again the Taylor's series expansion up to the first two terms:

$$\mathbf{y} = f(\mu) + (\mathbf{x} - \mu) f'(\mu) + \mathcal{R} \tag{5.36}$$

Taking the variance of Equation 5.36, we have

$$\begin{aligned} V(\mathbf{y}) &\approx V[f(\mu)] + V[(\mathbf{x} - \mu) f'(\mu)] \\ &\approx [f'(\mu)]^2 V(\mathbf{x}) \end{aligned} \tag{5.37}$$

Equation 5.37 is an approximation for the variance of y because we ignored the remainder $\mathcal{R}$, a procedure not recommended for highly nonlinear functions.

EXAMPLE 5.4

The radius of a bar has a mean value $\mu_R = 2.0\,\text{mm}$ and a standard deviation $\sigma_R = 0.10\,\text{mm}$. Find the mean and the standard deviation of the cross-sectional area.

We have

$$A = f(R) = \pi R^2$$

Hence,

$$f'(R) = 2\pi R$$

$$f''(R) = 2\pi$$

Therefore, using Equation 5.34 yields

$$\begin{aligned} E[A] &\approx f(\mu_R) + \frac{1}{2}(2\pi)\sigma_R^2 \\ &= \pi(2)^2 + \frac{1}{2}(2\pi)(0.1)^2 \\ &= 4.01\pi \text{ mm}^2 \end{aligned}$$

and from Equation 5.37,

$$\begin{aligned} V[A] &\approx (2\pi\mu_R)^2 \sigma_R^2 \\ &= (2\pi 2)^2 (0.1)^2 \end{aligned}$$

or

$$\sigma_A = \sqrt{V(A)} = 0.4\pi \text{ mm}^2$$

We will now consider the approximation for a function f of n random variables, that is

$$\mathbf{y}=f(\mathbf{x}_1,\mathbf{x}_2,\ldots,\mathbf{x}_n)=f(\mathbf{x}) \tag{5.38}$$

Let $\mu=(\mu_1,\ldots,\mu_n)$ and $\sigma=(\sigma_1,\ldots,\sigma_n)$ denote the vectors of the expected values and the standard deviations of $\mathbf{x}_1,\ldots,\mathbf{x}_n$, respectively. Then, by the Taylor's series expansion, we have

$$\begin{aligned}\mathbf{y}&=f(\mathbf{x}_1,\mathbf{x}_2,\ldots,\mathbf{x}_n)\\&=f(\mu_1,\ldots,\mu_n)+\sum_{i=1}^{n}\left.\frac{\partial f(\mathbf{x})}{\partial \mathbf{x}_i}\right|_{\mathbf{x}=\mu}(\mathbf{x}_i-\mu_i)\\&\quad+\frac{1}{2!}\sum_{j=1}^{n}\sum_{i=1}^{n}\left.\frac{\partial^2 f(\mathbf{x})}{\partial \mathbf{x}_i\partial \mathbf{x}_j}\right|_{\mathbf{x}=\mu}(\mathbf{x}_i-\mu_i)(\mathbf{x}_j-\mu_j)+\mathfrak{R}\end{aligned} \tag{5.39}$$

Taking the expectation of Equation 5.39, we have

$$\begin{aligned}E(\mathbf{y})&=f(\mu_1,\ldots,\mu_n)+\sum_{i=1}^{n}\left.\frac{\partial f(\mathbf{x})}{\partial \mathbf{x}_i}\right|_{\mathbf{x}=\mu}E(\mathbf{x}_i-\mu_i)\\&\quad+\frac{1}{2!}\sum_{j=1}^{n}\sum_{i=1}^{n}\left.\frac{\partial^2 f(\mathbf{x})}{\partial \mathbf{x}_i\partial \mathbf{x}_j}\right|_{\mathbf{x}=\mu}E\left[(\mathbf{x}_i-\mu_i)(\mathbf{x}_j-\mu_j)\right]+E(\mathfrak{R})\end{aligned}$$

If $\mathbf{x}_1,\mathbf{x}_2,\ldots,\mathbf{x}_n$ are independent random variables, then, after deleting the zero terms, we get

$$E(\mathbf{y})=f(\mu_1,\ldots,\mu_n)+\frac{1}{2}\sum_{i=1}^{n}\left.\frac{\partial^2 f(\mathbf{x})}{\partial \mathbf{x}_i^2}\right|_{\mathbf{x}=\mu}V(\mathbf{x}_i)+E(\mathfrak{R})$$

Approximating by dropping the remainder term,

$$E(\mathbf{y})\approx f(\mu_1,\ldots,\mu_n)+\frac{1}{2}\sum_{i=1}^{n}\left.\frac{\partial^2 f(\mathbf{x})}{\partial \mathbf{x}_i^2}\right|_{\mathbf{x}=\mu}V(\mathbf{x}_i) \tag{5.40}$$

If we further ignore the second term in Equation 5.40, then,

$$E[f(\mathbf{x})]\approx f(\mu_1,\ldots,\mu_n) \tag{5.41}$$

Now, considering only the first two terms in Equation 5.39 and taking the variance, we have

$$V(\mathbf{y}) \approx V\left(f(\mathbf{x})|_{\mathbf{x}=\mu}\right) + V\left[\sum_{i=1}^{n} \frac{\partial f(\mathbf{x})}{\partial \mathbf{x}_i}\bigg|_{\mathbf{x}=\mu} \cdot (\mathbf{x}_i - \mu_i)\right]$$

$$= \sum_{i=1}^{n} \left\{ \frac{\partial f(\mathbf{x})}{\partial \mathbf{x}_i}\bigg|_{\mathbf{x}=\mu} \right\}^2 V(\mathbf{x}_i) \tag{5.42}$$

EXAMPLE 5.5

The load P acting on a bar in tension has a mean value $\mu_P = 10{,}000$ N and a standard deviation $\sigma_P = 1{,}000$ N. The mean value of the cross-sectional area A is $\mu_A = 5.0\ \text{cm}^2$, and the standard deviation of A is $\sigma_A = 0.4$ cm. Find the mean and standard deviation of the tensile stress $\mathbf{s}$ on the bar.

The stress is given by

$$\mathbf{s} = f(P,A) = \frac{P}{A}$$

Hence

$$\mu_{\mathbf{s}} \approx f(\mu_P, \mu_A) = \frac{\mu_P}{\mu_A} = \frac{10{,}000}{5} = 2{,}000\ \text{N/cm}^2 = 20{,}000\ \text{kPa}$$

Differentiating, we get

$$\frac{\partial f}{\partial P} = \frac{1}{A} \quad \text{and} \quad \frac{\partial f}{\partial A} = -\frac{P}{A^2}$$

Hence,

$$\sigma_{\mathbf{s}}^2 \approx \left\{ \frac{\partial f}{\partial P}\bigg|_{\mu_P,\mu_A} \right\}^2 \sigma_P^2 + \left\{ \frac{\partial f}{\partial A}\bigg|_{\mu_P,\mu_A} \right\}^2 \sigma_A^2$$

$$= \left(\frac{1}{5}\right)^2 (1000)^2 + \left\{ -\frac{10{,}000}{(5)^2} \right\}^2 (0.4)^2$$

$$= 65{,}600$$

or

$$\sigma_{\mathbf{s}} = \sqrt{65{,}600} = 256.1\ \text{N/cm}^2 = 2{,}561\ \text{kPa}$$

EXAMPLE 5.6

An electrical circuit has two resistances in parallel as shown in Figure 5.1. The value of each resistance is a random variable. We know that

$$\mu_{R_1} = 100\ \Omega, \quad \sigma_{R_1} = 10\ \Omega$$
$$\mu_{R_2} = 200\ \Omega, \quad \sigma_{R_2} = 15\ \Omega$$

Determine μ_{R_T} and σ_{R_T}.

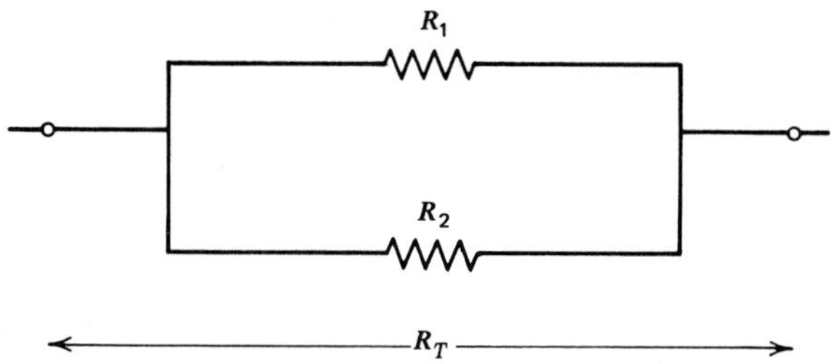

Figure 5.1 Two resistors in parallel.

We know that

$$R_T = f(R_1, R_2) = \frac{R_1 R_2}{R_1 + R_2}$$

Therefore

$$E[R_T] \approx f(\mu_{R_1}, \mu_{R_2}) = \frac{\mu_{R_1}\mu_{R_2}}{\mu_{R_1} + \mu_{R_2}} = \frac{100 \times 200}{100 + 200} = 66.7\ \Omega$$

Differentiating we have

$$\frac{\partial f}{\partial R_1} = \frac{R_2^2}{(R_1 + R_2)^2} \quad \text{and} \quad \frac{\partial f}{\partial R_2} = \frac{R_1^2}{(R_1 + R_2)^2}$$

Hence,

$$\left.\frac{\partial f}{\partial R_1}\right|_{\mu_{R_1}, \mu_{R_2}} = \frac{(200)^2}{(300)^2} = 0.444$$

$$\left.\frac{\partial f}{\partial R_2}\right|_{\mu_{R_1}, \mu_{R_2}} = \frac{(100)^2}{(300)^2} = 0.111$$

The variance is given by

$$\sigma_{R_T}^2 = (0.444)^2(10)^2 + (0.111)^2(15)^2$$
$$= 22.4858$$

or

$$\sigma_{R_T} = 4.74\ \Omega$$

If we know the covariances of the variables in Equation 5.39, we can generalize the result given in Equation 5.42. It is easy to prove that

$$V(\alpha_0 + \alpha_1 \mathbf{x}_1 + \alpha_2 \mathbf{x}_2 + \cdots + \alpha_n \mathbf{x}_n)$$
$$= \alpha_1^2 V(\mathbf{x}_1) + \cdots + \alpha_n^2 V(\mathbf{x}_n) + 2 \sum_{i=1}^{n-1} \sum_{j=i+1}^{n} \alpha_i \alpha_j \operatorname{cov}(\mathbf{x}_i, \mathbf{x}_j) \qquad (5.43)$$

Now consider the first two terms of Equation 5.39. Taking the variance, we have

$$V(\mathbf{y}) = \sum_{i=1}^{n} \left\{ \left. \frac{\partial f(\mathbf{x})}{\partial \mathbf{x}_i} \right|_{\mathbf{x}=\mu} \right\}^2 V(\mathbf{x}_i)$$

$$+2\sum_{i=1}^{n-1} \sum_{j=i+1}^{n} \left(\left. \frac{\partial f}{\partial \mathbf{x}_i} \right|_{\mathbf{x}=\mu} \right) \left(\left. \frac{\partial f}{\partial \mathbf{x}_j} \right|_{\mathbf{x}=\mu} \right) \operatorname{cov}(\mathbf{x}_i,\mathbf{x}_j) \tag{5.44}$$

Let us now examine some special cases of the function $f(\cdot)$. Let

$$\mathbf{y} = f(\mathbf{x}) = \log_{10}\mathbf{x}$$

Then

$$\sigma_{\mathbf{y}} \approx 0.4343 \frac{\sigma_{\mathbf{x}}}{\mu_{\mathbf{x}}} = 0.4343\, V_{\mathbf{x}} \tag{5.45}$$

where $V_{\mathbf{x}}$ is the coefficient of variation of $\mathbf{x}$. Thus the standard deviation of $\log \mathbf{x}$ is proportional to the coefficient of variation of $\mathbf{x}$.

Next consider the function $\mathbf{y} = \mathbf{x}^a$, for which we have

$$\sigma_{\mathbf{y}} \approx |a\mu_{\mathbf{x}}^{a}| \frac{\sigma_{\mathbf{x}}}{\mu_{\mathbf{x}}} \tag{5.46}$$

Since $\mu_{\mathbf{y}} \approx \mu_{\mathbf{x}}^{a}$, we have

$$V_{\mathbf{y}} \approx |a|\, V_{\mathbf{x}} \tag{5.47}$$

We can also easily verify that

$$V_{1/\mathbf{x}} \approx V_{\mathbf{x}} \tag{5.48}$$

and

$$V_{\sqrt{\mathbf{x}}} \approx \frac{1}{2} V_{\mathbf{x}} \tag{5.49}$$

Now consider the following general case

$$\mathbf{y} = a_0 \mathbf{x}_1^{a_1} \mathbf{x}_2^{a_2} \cdots \mathbf{x}_n^{a_n} \tag{5.50}$$

Using Equation 5.44, we can write

$$V_{\mathbf{y}}^2 \approx \sum_{i=1}^{n} a_i^2 V_{\mathbf{x}_i}^2 + 2 \sum_{i=1}^{n-1} \sum_{j=i+1}^{n} a_i a_j \rho_{ij} V_{\mathbf{x}_i} V_{\mathbf{x}_j} \tag{5.51}$$

where ρ_{ij} denotes the coefficient of correlation for $(\mathbf{x}_i,\mathbf{x}_j)$.

Two special cases of Equation 5.50 and Equation 5.51 are

(i) $$\mathbf{y} = a\mathbf{x}_1\mathbf{x}_2 \tag{5.52}$$

$$V_{\mathbf{y}}^2 \approx V_{\mathbf{x}_1}^2 + V_{\mathbf{x}_2}^2 + 2\rho_{12}V_{\mathbf{x}_1}V_{\mathbf{x}_2} \tag{5.53}$$

(ii) $$\mathbf{y} = a\mathbf{x}_1\mathbf{x}_2/\mathbf{x}_3 \tag{5.54}$$

$$V_{\mathbf{y}}^2 \approx V_{\mathbf{x}_1}^2 + V_{\mathbf{x}_2}^2 + V_{\mathbf{x}_3}^2 + 2\rho_{12}V_{\mathbf{x}_1}V_{\mathbf{x}_2} - 2\rho_{13}V_{\mathbf{x}_1}V_{\mathbf{x}_3} - 2\rho_{23}V_{\mathbf{x}_2}V_{\mathbf{x}_3} \tag{5.55}$$

5.4.2 ERROR ANALYSIS IN APPROXIMATION METHODS

Consider the Taylor's series expansion up to the third-order term for a function of two variables. We have

$$\begin{aligned}
\mathbf{y} &= f(\mathbf{x}_1, \mathbf{x}_2) \\
&= f(\mu_1, \mu_2) + \left.\frac{\partial f}{\partial \mathbf{x}_1}\right|_{\mu_1,\mu_2}(\mathbf{x}_1 - \mu_1) + \left.\frac{\partial f}{\partial \mathbf{x}_2}\right|_{\mu_1,\mu_2}(\mathbf{x}_2 - \mu_2) \\
&+ \frac{1}{2!}\left[\left.\frac{\partial^2 f}{\partial \mathbf{x}_1^2}\right|_{\mu_1,\mu_2}(\mathbf{x}_1 - \mu_1)^2 + 2\left.\frac{\partial^2 f}{\partial \mathbf{x}_1 \partial \mathbf{x}_2}\right|_{\mu_1,\mu_2}(\mathbf{x}_1 - \mu_1)(\mathbf{x}_2 - \mu_2) + \left.\frac{\partial^2 f}{\partial \mathbf{x}_2^2}\right|_{\mu_1,\mu_2}(\mathbf{x}_2 - \mu_2)^2\right] \\
&+ \frac{1}{3!}\left[\left.\frac{\partial^3 f}{\partial \mathbf{x}_1^3}\right|_{\mu_1,\mu_2}(\mathbf{x}_1 - \mu_1)^3 + 3\left.\frac{\partial^3 f}{\partial \mathbf{x}_1^2 \partial \mathbf{x}_2}\right|_{\mu_1,\mu_2}(\mathbf{x}_1 - \mu_1)^2(\mathbf{x}_2 - \mu_2)\right. \\
&\left. + 3\left.\frac{\partial^3 f}{\partial \mathbf{x}_1 \partial \mathbf{x}_2^2}\right|_{\mu_1,\mu_2}(\mathbf{x}_1 - \mu_1)(\mathbf{x}_2 - \mu_2)^2 + \left.\frac{\partial^3 f}{\partial \mathbf{x}_2^3}\right|_{\mu_1,\mu_2}(\mathbf{x}_2 - \mu_2)^3\right] + \mathcal{R}
\end{aligned} \tag{5.56}$$

Let us apply the above expansion for the following special function, which commonly occurs in design reliability. The safety factor **n** which is a random variable, is given by

$$\mathbf{n} = \frac{\mathbf{S}}{\mathbf{s}} \tag{5.57}$$

where $\mathbf{S}$ and **s** represent the strength and stress random variables, respectively.

Then we have

$$\mathbf{n}=\frac{\bar{\mathbb{S}}}{\bar{\mathbf{s}}}+\frac{1}{\bar{\mathbf{s}}}(\mathbb{S}-\bar{\mathbb{S}})+\left(-\frac{\bar{\mathbb{S}}}{\bar{\mathbf{s}}^2}\right)(\mathbf{s}-\bar{\mathbf{s}})$$

$$+\frac{1}{2!}\left[0-\frac{2}{\bar{\mathbf{s}}^2}(\mathbb{S}-\bar{\mathbb{S}})(\mathbf{s}-\bar{\mathbf{s}})+\frac{2\bar{\mathbb{S}}}{\bar{\mathbf{s}}^3}(\mathbf{s}-\bar{\mathbf{s}})^2\right]$$

$$+\frac{1}{3!}\left[0+0+\frac{6}{\bar{\mathbf{s}}^3}(\mathbb{S}-\bar{\mathbb{S}})(\mathbf{s}-\bar{\mathbf{s}})^2-\frac{6\bar{\mathbb{S}}}{\bar{\mathbf{s}}^4}(\mathbf{s}-\bar{\mathbf{s}})^3\right]+\mathcal{R} \tag{5.58}$$

$$=\frac{\bar{\mathbb{S}}}{\bar{\mathbf{s}}}+\frac{1}{\bar{\mathbf{s}}}(\mathbb{S}-\bar{\mathbb{S}})-\frac{\bar{\mathbb{S}}}{\bar{\mathbf{s}}^2}(\mathbf{s}-\bar{\mathbf{s}})-\frac{1}{\bar{\mathbf{s}}^2}(\mathbb{S}-\bar{\mathbb{S}})(\mathbf{s}-\bar{\mathbf{s}})+\frac{\bar{\mathbb{S}}}{\bar{\mathbf{s}}^3}(\mathbf{s}-\bar{\mathbf{s}})^2$$

$$+\frac{(\mathbb{S}-\bar{\mathbb{S}})(\mathbf{s}-\bar{\mathbf{s}})^2}{\bar{\mathbf{s}}^3}-\frac{\bar{\mathbb{S}}}{\bar{\mathbf{s}}^4}(\mathbf{s}-\bar{\mathbf{s}})^3+\mathcal{R} \tag{5.59}$$

If we ignore the remainder term $\mathcal{R}$ in Equation 5.59 and take the expectation, we have

$$\bar{\mathbf{n}}=E(\mathbf{n})\approx\frac{\bar{\mathbb{S}}}{\bar{\mathbf{s}}}+\frac{\bar{\mathbb{S}}}{\bar{\mathbf{s}}^3}\sigma_{\mathbf{s}}^2-\frac{\bar{\mathbb{S}}}{\bar{\mathbf{s}}^4}\mu_3^{\mathbf{s}} \tag{5.60}$$

where $\mu_3^{\mathbf{s}}$ is the third moment of s about its mean.

Now taking the variance of both sides of Equation 5.59 after ignoring the remainder term, we have

$$V(\mathbf{n})=E\left\{\left[\frac{\bar{\mathbb{S}}}{\bar{\mathbf{s}}}+\frac{1}{\bar{\mathbf{s}}}(\mathbb{S}-\bar{\mathbb{S}})-\frac{\bar{\mathbb{S}}}{\bar{\mathbf{s}}^2}(\mathbf{s}-\bar{\mathbf{s}})-\frac{1}{\bar{\mathbf{s}}^2}(\mathbb{S}-\bar{\mathbb{S}})(\mathbf{s}-\bar{\mathbf{s}})\right.\right.$$

$$\left.\left.+\frac{\bar{\mathbb{S}}}{\bar{\mathbf{s}}^3}(\mathbf{s}-\bar{\mathbf{s}})^2+\frac{(\mathbb{S}-\bar{\mathbb{S}})(\mathbf{s}-\bar{\mathbf{s}})^2}{\bar{\mathbf{s}}^3}-\frac{\bar{\mathbb{S}}}{\bar{\mathbf{s}}^4}(\mathbf{s}-\bar{\mathbf{s}})^3\right]^2\right\}$$

$$-\left[\frac{\bar{\mathbb{S}}}{\bar{\mathbf{s}}}+\frac{\bar{\mathbb{S}}}{\bar{\mathbf{s}}^3}\sigma_{\mathbf{s}}^2-\frac{\bar{\mathbb{S}}}{\bar{\mathbf{s}}^4}\mu_3^{\mathbf{s}}\right]^2 \tag{5.61}$$

If we assume that both $\mathbb{S}$ and s are normally distributed, then it is known that

$$\mu_3=0,\qquad \mu_4=3\sigma^4,\qquad \mu_5=0,\qquad \text{and}\qquad \mu_6=15\sigma^6$$

Using these values in Equations 5.60 and 5.61 and simplifying, we have

$$E(\mathbf{n}) \approx \frac{\bar{\mathfrak{S}}}{\bar{\mathbf{s}}} + \frac{\bar{\mathfrak{S}}}{\bar{\mathbf{s}}^3}\sigma_{\mathbf{s}}^2 \tag{5.62}$$

and

$$V(\mathbf{n}) \approx \frac{\sigma_{\mathbf{s}}^2}{\bar{\mathbf{s}}^2} + \frac{\sigma_{\mathbf{s}}^2}{\bar{\mathbf{s}}^4}\left(\bar{\mathfrak{S}}^2 + 3\sigma_{\mathfrak{S}}^2\right) + \frac{\sigma_{\mathfrak{S}}^4}{\bar{\mathbf{s}}^6}\left(3\sigma_{\mathfrak{S}}^2 + 8\bar{\mathfrak{S}}^2\right) + 15\frac{\bar{\mathfrak{S}}\sigma_{\mathbf{s}}^6}{\bar{\mathbf{s}}^8} \tag{5.63}$$

Table 5.1 shows the differences in the values of the mean and the variance of the safety factor **n** as approximated, first by using Equations 5.40 and 5.42, and then by using Equations 5.62 and 5.63, respectively. Equations 5.62 and 5.63 provide better approximations because they consider higher order terms.

Table 5.1 Error analysis of Taylor's series approximation for $\mathbf{n} = \dfrac{\mathfrak{S}}{\mathbf{s}}$

				FIRST-ORDER approx.		SECOND-ORDER approx.		THIRD-ORDER approx.	
$\mu_{\mathfrak{S}}$	$\sigma_{\mathfrak{S}}$	$\mu_{\mathbf{s}}$	$\sigma_{\mathbf{s}}$	$\bar{\mathbf{n}}$	$V(\mathbf{n})$	$\bar{\mathbf{n}}$	$V(\mathbf{n})$	$\bar{\mathbf{n}}$	$V(\mathbf{n})$
100.0	9.0	10.0	3.0	10.0000	9.8100	10.9000	11.5029	10.9000	17.6219
120.0	11.0	20.0	4.0	6.0000	1.7425	6.2400	1.8698	6.2400	2.2756
140.0	12.0	30.0	5.0	4.6666	0.7649	4.7962	0.8029	4.7962	0.9200
150.0	13.0	40.0	7.0	3.7500	0.5362	3.8648	0.5659	3.8648	0.6578
180.0	15.0	50.0	8.0	3.6000	0.4218	3.6921	0.4410	3.6921	0.5000
200.0	16.0	60.0	10.0	3.3333	0.3798	3.4259	0.3988	3.4259	0.4580

Next we offer another example to show the size of error to be expected by the Taylor's series approximation method.

Let $\mathbf{y} = e^{a\mathbf{x}}$, where $\mathbf{x}$ is a normal random variable with mean μ and standard deviation σ. We can easily derive the following exact expressions:

$$E(e^{a\mathbf{x}}) = \exp\left(a\mu + \frac{a^2\sigma^2}{2}\right) \tag{5.64}$$

$$V(e^{a\mathbf{x}}) = \exp(2a^2\sigma^2 + 2a\mu) - \left\{\exp\left(\frac{a^2\sigma^2}{2} + a\mu\right)\right\}^2 \tag{5.65}$$

Let us now compute the first two moments of $\mathbf{y}$ using the Taylor's series. We see that

$$\mathbf{y} = e^{a\mathbf{x}} = e^{a\mu} + ae^{a\mu}(\mathbf{x} - \mu) + \frac{1}{2!}a^2e^{a\mu}(\mathbf{x} - \mu)^2$$
$$+ \frac{1}{3!}(a^3e^{a\mu})(\mathbf{x} - \mu)^3 + \mathfrak{R} \tag{5.66}$$

Hence,

$$\bar{y} \approx e^{a\mu} + 0 + \frac{1}{2} a^2 e^{a\mu} \sigma^2 + 0 \tag{5.67}$$

Let us consider the first three terms of Equation 5.66 and compute the variance. We have

$$V(\mathbf{y}) = E\left[\left\{ e^{a\mu} + a e^{a\mu} (\mathbf{x} - \mu) + \frac{1}{2} a^2 e^{a\mu} (\mathbf{x} - \mu)^2 \right\}^2\right]$$

$$- \left[E\left\{ e^{a\mu} + a e^{a\mu} (\mathbf{x} - \mu) + \frac{1}{2} a^2 e^{a\mu} (\mathbf{x} - \mu)^2 \right\} \right]^2$$

$$= a^2 e^{2a\mu} \sigma^2 \left(1 + \frac{1}{2} a^2 \sigma^2 \right) \tag{5.68}$$

If we consider all the terms in Equation 5.66, we have

$$V(\mathbf{y}) = a^2 e^{2a\mu} \sigma^2 \left(1 + \frac{3}{2} a^2 \sigma^2 + \frac{5}{12} a^4 \sigma^4 \right) \tag{5.69}$$

Table 5.2 has been developed based on the above relationships. It shows the differences between the exact values of the mean and the variance of the random variable **y** and their approximate values obtained by using up to the second-order, and the third-order terms, in the Taylor's series expansion.

5.5 STATISTICAL TOLERANCING

The specifications for a dimension or some other design variable are usually given as a nominal value plus or minus the tolerance. For example, Figure 5.2 shows the length of a component as 2.500 ± 0.003 m. Here 2.500 m is the nominal length and 0.003 m is the tolerance. The dimension is a random variable with a certain distribution; it possesses the property that no more than a certain fraction of the components will fall outside the lower and the upper

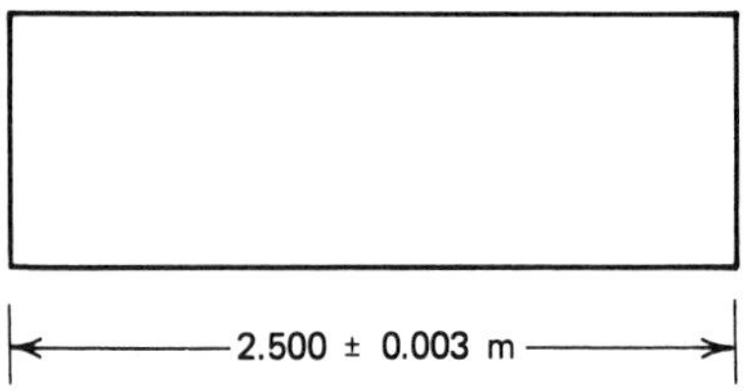

Figure 5.2 Example of a tolerance specification.

Table 5.2 Approximation errors in μ_y and σ_y^2 for $y = e^{ax}$ using Taylor's series expansion

			EXACT VALUES		APPROXIMATE VALUES USING FIRST TWO TERMS		APPROXIMATE VALUES USING ONLY FIRST THREE TERMS	
a	μ	σ	$E(e^{ax})$	$V(e^{ax})$	$E(e^{ax})$	$V(e^{ax})$	$E(e^{ax})$	$V(e^{ax})$
1.00	0.00	1.00	1.64872	4.67077	1.50000	1.50000	1.50000	2.91667
1.00	1.00	1.00	4.48169	34.51262	4.07742	11.08358	4.07742	21.55141
1.00	1.00	0.50	3.08022	2.69476	3.05807	2.07817	3.05807	2.58809
1.00	2.00	1.00	12.18249	255.01555	11.08358	81.89722	11.08358	159.24458
2.00	0.00	1.00	7.38906	2926.35986	3.00000	12.00000	3.00000	54.66666
2.00	0.00	0.50	1.64872	4.67077	1.50000	1.50000	1.50000	2.91667
2.00	1.00	1.00	54.59814	159773.812	22.16716	655.17773	22.16716	2984.69824
2.00	1.00	0.50	12.18249	255.01555	11.08358	81.89722	11.08358	159.24458
3.00	0.00	1.00	90.01714	65651856.0	5.50000	49.50000	5.50000	434.24976
3.00	0.00	0.50	3.08022	80.52939	2.12500	4.78125	2.12500	14.58984
3.00	2.00	0.50	1242.64819	13106545.0	857.28589	778171.375	857.28589	2374566.000
3.00	2.00	0.25	534.45630	215676.562	516.89282	117297.875	516.89282	180864.000

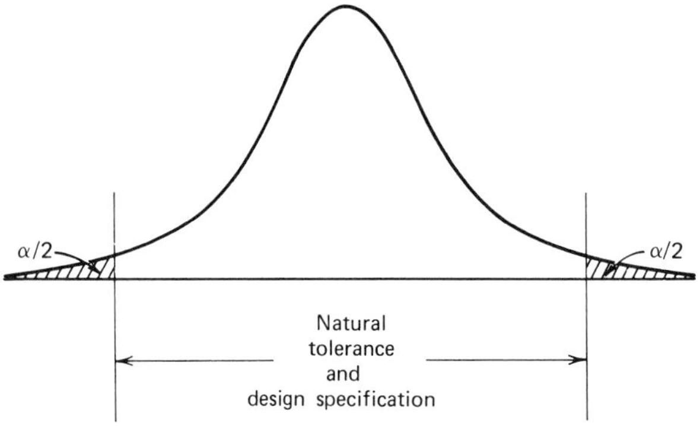

Figure 5.3 Design specification for a given α.

specification limits. Thus, if the dimension is normally distributed with a mean of 2.500 m and a standard deviation of 0.001 m and a design specification of 2.500 ± 0.003 m, then no more than 27 components out of 10,000 will fall outside these specification limits. Figure 5.3 shows the density function for the dimensional random variable and natural or design specification. This means that no more than $\alpha/2$ percent of the components will be above the upper limit and no more than $\alpha/2$ percent of the components will be below the lower limit. Conversely, if we are given the distribution of the dimension and an α as the allowable percent defective, we can compute the design specifications as shown in Figure 5.3.

Let η_i denote the nominal dimension of the ith component and t_i its tolerance. Then the dimension of the component may be written as $\eta_i \pm t_i$. In a statistical tolerance analysis, we usually approximate t_i by $3\sigma_i$.

Let x_i denote the actual dimension of the ith component and x_A that of the assembly. Then the mean value of x_i will be given by η_i and its standard deviation by $t_i/3$. Hence, for the case of linear tolerancing, we have, for an assembly of k components,

$$\eta_A = \eta_1 \pm \eta_2 \pm \eta_3 \pm \cdots \pm \eta_k$$

$$\sigma_A = \sqrt{\sigma_1^2 + \sigma_2^2 + \cdots + \sigma_k^2}$$

$$= \sqrt{(t_1/3)^2 + (t_2/3)^2 + \cdots + (t_k/3)^2}$$

or

$$t_A = 3\sigma_A = \sqrt{t_1^2 + t_2^2 + \cdots + t_k^2}$$

EXAMPLE 5.7
Consider the assembly of a shaft and a bearing. Let x_s denote the actual dimension of the shaft and x_b that of the bearing. Let x_c denote the clearance between the shaft and the bearing. Then

$$x_c = x_b - x_s$$

The tolerance for the clearance is ± 0.05 mm. Determine the tolerances for the shaft and bearing, given that they are equal.

Using the relationship

$$t_c = \sqrt{t_b^2 + t_s^2}$$

we have

$$0.05 = \sqrt{t^2 + t^2} = \sqrt{2}\, t$$

or

$$t = t_b = t_c = \frac{0.05}{\sqrt{2}} = 0.035 \text{ mm}$$

5.6 SUMMARY

In probabilistic design for reliability all the design variables are treated as random variables. This methodology results in a reflection of a spectrum of possible values of the variable. Thus, the design performance, which is a function of these random variables, is also a random variable. This chapter discussed several approaches for combining random variables. In general it is difficult to find the density function for the random variable **y** as a function of the density functions of the random variables $\mathbf{x}_1, \ldots, \mathbf{x}_n$. In most of the design computations, we have information available only about the first few moments of the random variables $\mathbf{x}_1, \ldots, \mathbf{x}_n$. This information is used to compute the approximate values of the moments of the random variable **y**. Chapter 7 uses the results given in this chapter extensively for various simple design problems. The section on error analysis gives us some idea about the error in computing the first two moments of the random variable **y** by the Taylor's series approximation method. The design analysis in Chapter 7 is based on computing the first two moments of the random variable **y** by using Equations 5.41 and 5.42.

EXERCISES

1 The parts of a contact assembly for a relay are shown in Figure 5.4. The dimension x represents the amount of intentional overtravel (called "wipe") of the upper contact that would occur if the upper contact was clamped to the part at left. Find the nominal dimension x and its tolerance.

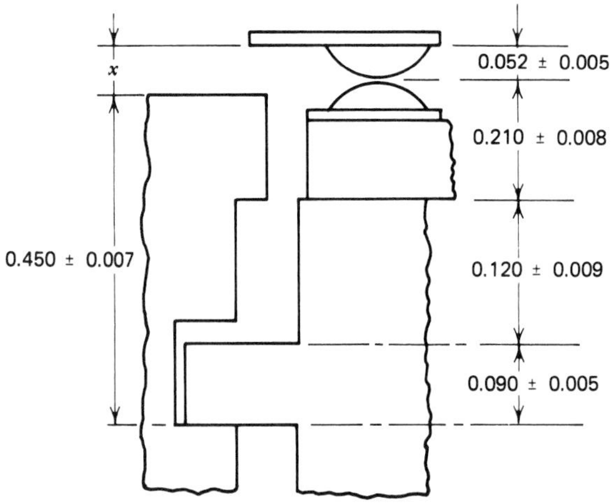

Figure 5.4 A contact relay.

2 A shaft and ring assembly is dimensioned as shown in Figure 5.5. The dimension x represents the clearance. Find the nominal dimension x and its tolerance.

3 A partially finished connecting rod is shown in Figure 5.6. Each radius has a tolerance of ± 0.002. The tolerance for the distance L between the centers of the holes is ± 0.004. Find the tolerance for the dimension h.

4 Determine the tolerance for the calculated engine brake fuel consumption (BFC) Z when tolerances on both the fuel rate X and the brake horsepower (BHP) Y at a specific injector rack setting are known. By definition, we have BFC = pounds of fuel per hour/BHP, or $Z = X/Y$. The tolerance on the BFC specified for this particular diesel engine is $\pm 2\sigma$ limits. The needed values are:

$$\mu_X = 2020 \text{ lb/hr}$$

$$\sigma_X = 25 \text{ lb/hr}$$

$$\mu_Y = 2500 \text{ BHP}$$

$$\sigma_Y = 24 \text{ BHP}$$

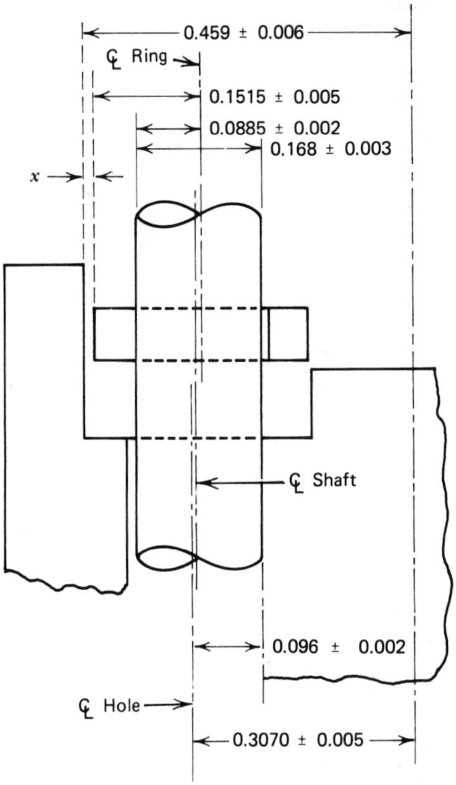

Figure 5.5 A shaft assembly.

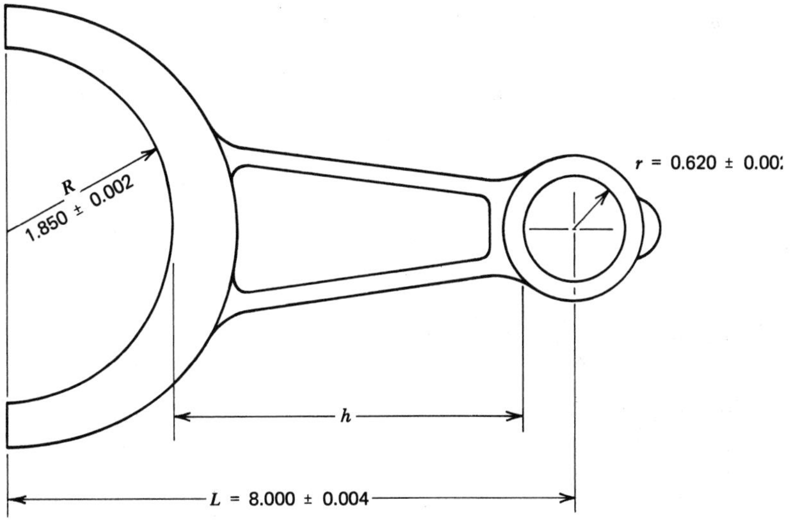

Figure 5.6 A connecting rod.

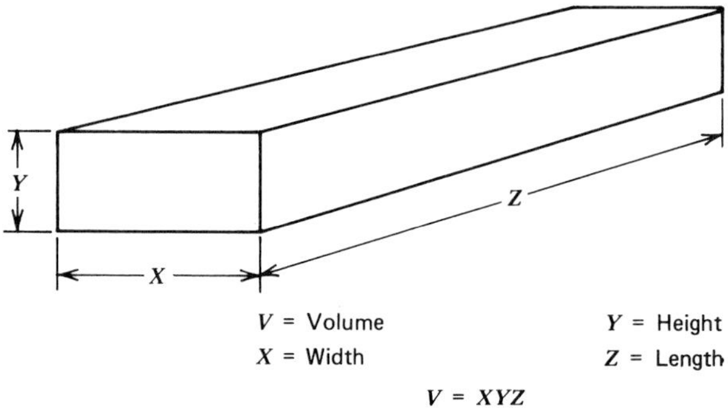

Figure 5.7 Rectangular solid bar.

5 Determine the tolerance for the volume of the rectangular solid bar shown in Figure 5.7. The three sides are given by

$$X = 2 \pm 0.002 \text{ m}$$

$$Y = 1 \pm 0.001 \text{ m}$$

$$Z = 4 \pm 0.008 \text{ m}$$

6 Determine the tolerance for the volume of a cylinder having the following tolerances for the diameter D and its length L. The diameter of the cylinder is produced independently of the length.

$$D = 2.5 \pm 0.002 \text{ m}$$

$$L = 4.0 \pm 0.005 \text{ m}$$

7 The head of a screw is shown in Figure 5.8. The various dimensions are formed in such a manner that there is no association between them; that is, they

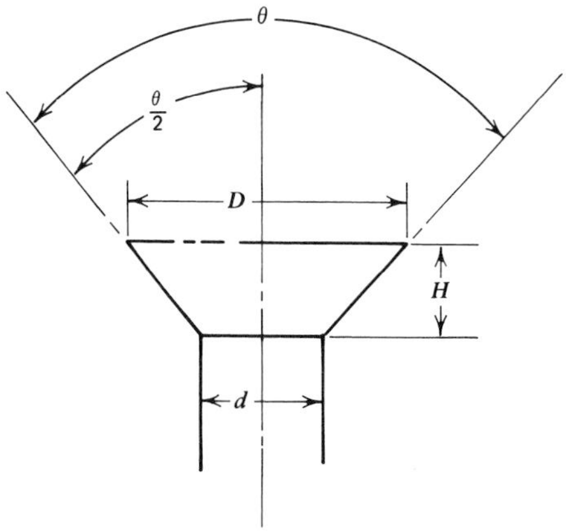

Figure 5.8 A screw head.

are mutually independent. Determine the tolerance for H, the depth of the screw head. The dimensions θ, D, and d and their tolerances are:

$$\theta = 90° \pm 20'$$

$$D = 0.800 \pm 0.002 \text{ in.}$$

$$d = 0.400 \pm 0.001 \text{ in.}$$

8 An electrical circuit is shown in Figure 5.9. V_1 is the input voltage to the transformer. The output from the transformer goes to the amplifier. A phase-shifting synchro is added to the output of the amplifier as shown. Find the design specifications for the output voltage V_0, the expression for which is

$$V_0 = V_2 \cos\theta + V_1 N K \sin\theta$$

where

$$V_1 = 60 \pm 1 \text{ volts}$$

$$N = (3 \text{ to } 1) \pm 2\%$$

$$K = 2 \pm 3\%$$

$$\theta = 60 \pm \frac{1}{2} \text{ degrees}$$

$$V_2 = 80 \pm 0.6 \text{ volts}$$

Assume the random variables V_1, N, K, θ, and V_2 to be independent.

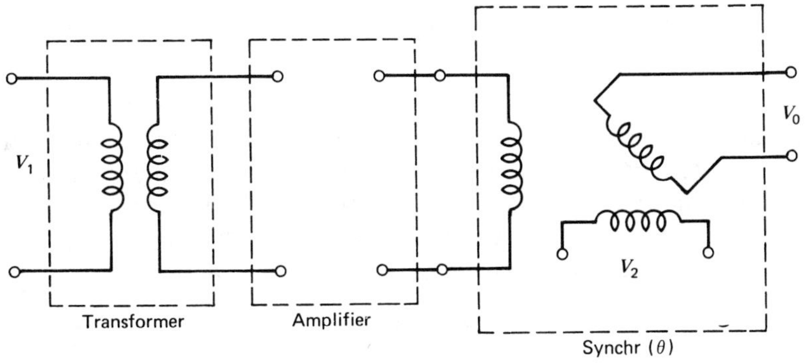

Figure 5.9 An electrical circuit.

9 A random variable $\mathbf{y}$ is a function of three other random variables $\mathbf{x}_1$, $\mathbf{x}_2$, and $\mathbf{x}_3$ and is given by

$$\mathbf{y} = a\mathbf{x}_1/\mathbf{x}_2\mathbf{x}_3$$

The following information is given for the random variables $\mathbf{x}_1$, $\mathbf{x}_2$ and $\mathbf{x}_3$:

$$\mu_{\mathbf{x}_1} = 4.0 \qquad \sigma_{\mathbf{x}_1} = 0.4$$

$$\mu_{\mathbf{x}_2} = 2.0 \qquad \sigma_{\mathbf{x}_2} = 0.2$$

$$\mu_{\mathbf{x}_3} = 1.0 \qquad \sigma_{\mathbf{x}_3} = 0.1$$

$$\rho_{12} = 0.8, \quad \rho_{23} = -0.7, \quad \rho_{13} = -0.5$$

Find the expected value and the standard deviation for the random variable $\mathbf{y}$.

10 The significant endurance limit of a certain kind of steel is given by

$$\mathfrak{S}_n = \mathbf{k}_1\mathbf{k}_2\mathfrak{S}_n'$$

where

$$\mathbf{k}_1 = \text{Surface finish factor}$$

$$\mathbf{k}_2 = \text{Stress concentration factor}$$

$$\mathfrak{S}_n' = \text{Endurance limit}$$

For various reasons, $\mathfrak{S}_n'$, $\mathbf{k}_1$, and $\mathbf{k}_2$ are random variables. The tolerances on these random variables are:

$$\mathbf{k}_1 = 0.8 \pm 0.12$$

$$\mathbf{k}_2 = 0.6 \pm 0.21$$

$$\mathfrak{S}_n' = 200 \pm 60 \text{ kPa}$$

Find the tolerance limits for the significant endurance limit.

11 An automotive component is subjected to a fluctuating stress as shown in Figure 5.10. The maximum value of the stress $\mathbf{s}_{max}$ is a normally distributed random variable with $\mu_{\mathbf{s}_{max}} = 600\,\text{kPa}$ and $\sigma_{\mathbf{s}_{max}} = 40\,\text{kPa}$. The minimum value of the stress $\mathbf{s}_{min}$ is a gamma distributed random variable with parameters $\eta = 17$ and $1/\lambda = 20\,\text{kPa}$. The mean stress is given by

$$\mathbf{s}_{mean} = \frac{\mathbf{s}_{max} + \mathbf{s}_{min}}{2}$$

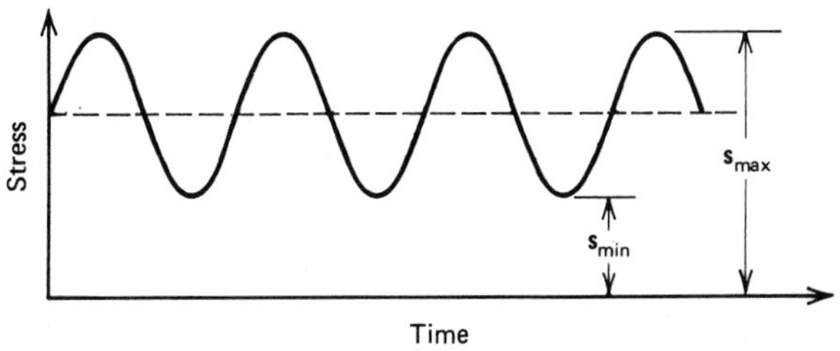

Figure 5.10 Fluctuating stress operating on a component.

and the amplitude of the stress is given by

$$s_{amp} = \frac{s_{max} - s_{min}}{2}$$

a. Approximate s_{min} by a normal distribution and compute the distribution of s_{mean} and s_{amp}.
b. Determine the value which the random variable s_{mean} will exceed only 1.3% of the time.
c. Determine the value which the random variable s_{amp} will not exceed 90% of the time.

12 A beam with a tubular cross-section shown in Figure 5.11 is to be used in an automobile assembly. To compute the stresses, we must compute the moment of inertia of the beam. The moment of inertia (I) of the beam about the neutral axis is

$$I = \pi r^3 t$$

The mean radius r and thickness t of the tubular cross-section have the following dimensions:

$$r = 2.00 \pm 0.06$$

$$t = 0.11 \pm 0.015$$

Find the mean value of the moment of inertia and the standard deviation.

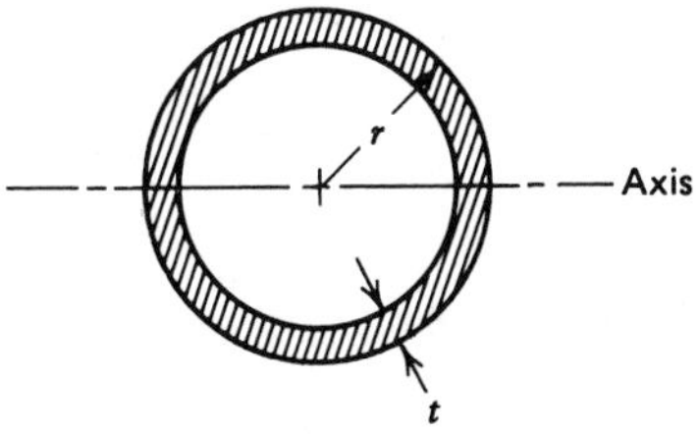

Figure 5.11 Tubular cross-section.

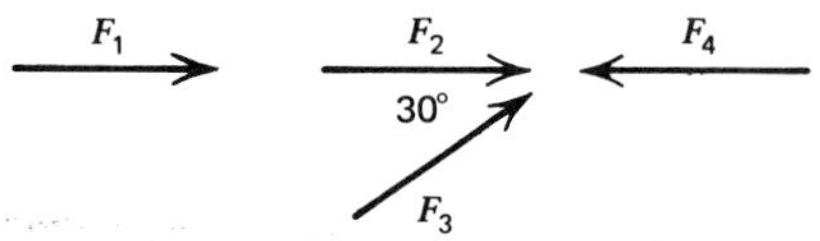

Figure 5.12 Load diagram.

13 The analysis of the loading of a component revealed a load diagram as shown in Figure 5.12. The four forces F_1, F_2, F_3, and F_4 are random variables, the distributions of which are given in the following table.

FORCE	DISTRIBUTION	PARAMETER(S)
F_1	Exponential	$1/\lambda = 2.6$ kN
F_2	Gamma	$1/\lambda = 2.6$ kN, $\eta = 18$
F_3	Normal	$\mu = 37.5$ kN, $\sigma = 3.9$ kN
F_4	Gamma	$1/\lambda = 2.6$ kN, $\eta = 15$

Calculate the mean and the variance of the magnitudes of the resultant load. [*Hint*: The gamma distribution may be approximated by a normal distribution with $\mu = \eta/\lambda$ and $\sigma = \sqrt{\eta}\,/\lambda$.]

14 Find the mean value and the standard deviation for the peak shearing stress r in a helical spring using Taylor's series expression by using the following equation:

$$r = \frac{(1 + d/2D)dGy}{\pi D^2 N}$$

The following information is given for the five random variables in the above equation:

VARIABLE	MEAN	STANDARD DEVIATION
d(in.) wire diameter	0.15	0.008
D(in.) coil diameter	0.80	0.015
G(psi) shearing modulus of elasticity	11.5×10^6	20×10^4
N number of coils	20	0.5
y(in.) deflection	0.95	0.03

BIBLIOGRAPHY

1 Benjamin, J. R., *Statistics for Civil Engineers,* McGraw-Hill, New York, 1964.

2 Bowker, A. H. and Liebermann, G. J., *Engineering Statistics,* Prentice-Hall, Englewood Cliffs, N. J., 1963.

3 Cramer, H., *Mathematical Methods of Statistics,* Princeton University Press, Princeton, N. J., 1946.

4 Haugen, E. B., *Probabilistic Approaches to Design.* John Wiley & Sons, New York, 1968.

5 Parrott, L. G., *Probability and Experimental Error in Science,* John Wiley & Sons, New York, 1961.

6 Parzen, E., *Modern Probability Theory and Its Applications,* John Wiley and Sons, New York, 1960.

7 Shooman, M. L., *Probabilistic Reliability: An Engineering Approach.* McGraw-Hill, New York, 1968.

8 Shooman, M. L., "Reliability Physics Models," *IEEE Transactions on Reliability,* Vol. R-17, No. 1, March 1968, pp. 14–19.

Chapter 6 interference theory and reliability computations

We determine, in this chapter, the probability that a component, a subsystem, or a system fails when the stress, in general, exceeds the strength. In order to compute the reliability we have to know the nature of the stress ($\mathbf{s}$) and strength ($\mathbb{S}$) random variables. The various factors that need to be considered to determine these random variables have been discussed in Chapters 4 and 5. In this chapter, we show how to compute the reliability of a component when the density functions for the stress and the strength random variables are known. The general expressions for reliability are derived in Section 6.1, and expressions for reliability for various density functions, namely, the normal, exponential, log normal, gamma, and Weibull, are presented in Sections 6.2 to 6.8. Section 6.9 examines a graphical method for computing the reliability of a part when only a limited amount of data on the stress and strength is available. Most of the methods are illustrated by appropriate numerical examples. Reliability expressions for extreme value distributions are discussed in Section 6.10.

6.1 GENERAL EXPRESSION FOR RELIABILITY

Let the density function for the stress ($\mathbf{s}$) be denoted by $f_{\mathbf{s}}(\cdot)$, and that for strength ($\mathbb{S}$) by $f_{\mathbb{S}}(\cdot)$, as shown in Figure 6.1. Then, by definition,

$$\text{Reliability} = R = P(\mathbb{S} > \mathbf{s}) = P(\mathbb{S} - \mathbf{s} > 0) \tag{6.1}$$

The shaded portion in Figure 6.1 shows the interference area, which is indicative of the probability of failure. Let us enlarge this interference area, as shown in Figure 6.2, in order to better focus our attention on it. The probability of a stress value lying in a small interval of width ds is equal to the area of the element ds; that is,

$$P\left(s_0 - \frac{ds}{2} \leqslant \mathbf{s} \leqslant s_0 + \frac{ds}{2}\right) = f_{\mathbf{s}}(s_0)\cdot ds$$

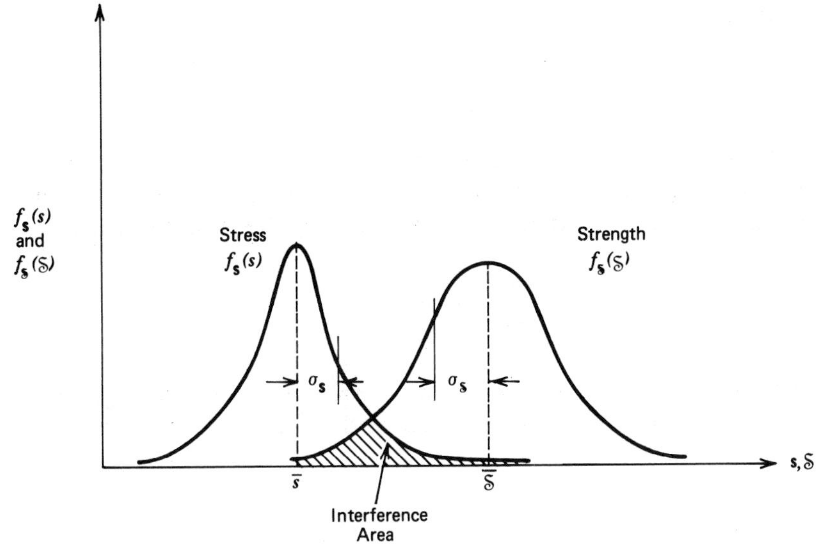

Figure 6.1 Stress-strength interference.

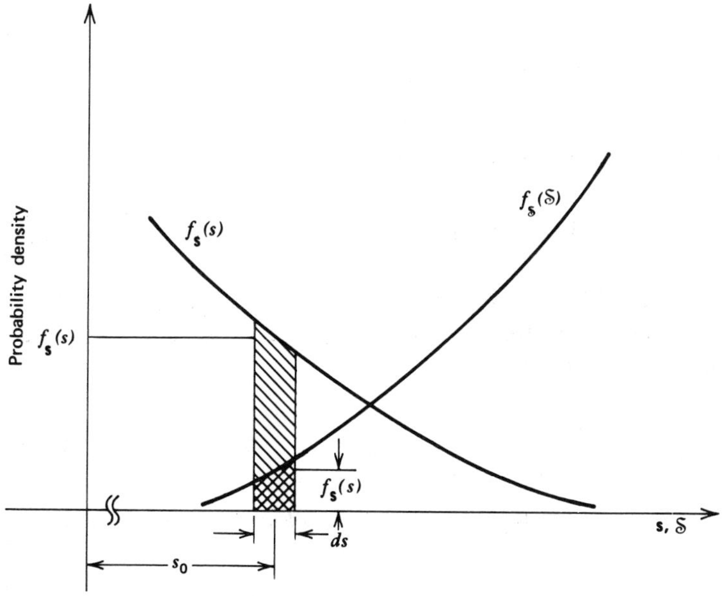

Figure 6.2 Computation of reliability—enlarged portion of the interference diagram for the interference area.

The probability that the strength $\mathbb{S}$ is greater than a certain stress s_0 is given by

$$P(\mathbb{S} > s_0) = \int_{s_0}^{\infty} f_{\mathbb{S}}(\mathbb{S})\, d\mathbb{S}$$

The probability for the stress value lying in the small interval *ds* and the strength $\mathbb{S}$ exceeding the stress given by this small interval *ds* under the assumption that the stress and the strength random variables are independent is given by

$$f_{\mathbf{s}}(s_0)\, ds \cdot \int_{s_0}^{\infty} f_{\mathbb{S}}(\mathbb{S})\, d\mathbb{S} \tag{6.2}$$

Now the reliability of the component is the probability that the strength $\mathbb{S}$ is greater than the stress **s** for all possible values of the stress **s** and hence is given by

$$R = \int_{-\infty}^{\infty} f_{\mathbf{s}}(s) \left[\int_{s}^{\infty} f_{\mathbb{S}}(\mathbb{S})\, d\mathbb{S} \right] ds \tag{6.3}$$

Reliability can also be computed on the basis that the stress remains less than the strength. The probability that the strength $\mathbb{S}$ is within a small interval $d\mathbb{S}$ is

$$P\left(\mathbb{S}_0 - \frac{d\mathbb{S}}{2} \leqslant \mathbb{S} \leqslant \mathbb{S}_0 + \frac{d\mathbb{S}}{2}\right) = f_{\mathbb{S}}(\mathbb{S}_0)\, d\mathbb{S}$$

and the probability of the stress being less than $\mathbb{S}_0$ is given by

$$P(\mathbf{s} \leqslant \mathbb{S}_0) = \int_{-\infty}^{\mathbb{S}_0} f_{\mathbf{s}}(s)\, ds$$

Again assuming that the stress and the strength are independent random variables, the probability of the strength belonging to the small interval $d\mathbb{S}$ and the stress **s** not exceeding $\mathbb{S}_0$ is

$$f_{\mathbb{S}}(\mathbb{S}_0)\, d\mathbb{S} \cdot \int_{-\infty}^{\mathbb{S}_0} f_{\mathbf{s}}(s)\, ds \tag{6.4}$$

Hence the reliability of the component for all the possible values of the strength $\mathbb{S}$ is

$$R = \int_{-\infty}^{\infty} f_{\mathbb{S}}(\mathbb{S}) \left[\int_{-\infty}^{\mathbb{S}} f_{\mathbf{s}}(s)\, ds \right] d\mathbb{S} \tag{6.5}$$

Some other expressions for the reliability and the unreliability are developed next and will be useful later on.

The unreliability, denoted by $\bar{R}$, is defined as

$$\bar{R} = \text{probability of failure} = 1 - R = P(\mathbb{S} \leqslant \mathbf{s})$$

Substituting for R from Equation 6.3 yields

$$\bar{R} = P(\mathbb{S} \leqslant \mathbf{s}) = 1 - \int_{-\infty}^{\infty} f_{\mathbf{s}}(s)\left[\int_{s}^{\infty} f_{\mathbb{S}}(\mathbb{S})\,d\mathbb{S}\right]ds$$

$$= 1 - \int_{-\infty}^{\infty} f_{\mathbf{s}}(s)\left[1 - F_{\mathbb{S}}(s)\right]ds$$

$$= \int_{-\infty}^{\infty} F_{\mathbb{S}}(s)\cdot f_{\mathbf{s}}(s)\,ds \tag{6.6}$$

Alternatively, using Equation 6.5 we have

$$\bar{R} = P(\mathbb{S} \leqslant \mathbf{s}) = 1 - \int_{-\infty}^{\infty} f_{\mathbb{S}}(\mathbb{S})\left[\int_{-\infty}^{\mathbb{S}} f_{\mathbf{s}}(s)\,ds\right]d\mathbb{S}$$

$$= 1 - \int_{-\infty}^{\infty} f_{\mathbb{S}}(\mathbb{S})\cdot F_{\mathbf{s}}(\mathbb{S})\,d\mathbb{S}$$

$$= \int_{-\infty}^{\infty} \left[1 - F_{\mathbf{s}}(\mathbb{S})\right] f_{\mathbb{S}}(\mathbb{S})\,d\mathbb{S} \tag{6.7}$$

Let us define $\mathbf{y} = \mathbb{S} - \mathbf{s}$. Then $\mathbf{y}$ is called the interference random variable. Now we can define the reliability as

$$R = P(\mathbf{y} > 0) \tag{6.8}$$

Now let us assume that $\mathbb{S}$ and $\mathbf{s}$ are independent random variables and greater than or equal to zero. The density function of $\mathbf{y}$ is then given by

$$f_{\mathbf{y}}(y) = \int_{\mathbf{s}} f_{\mathbb{S}}(y+s)\cdot f_{\mathbf{s}}(s)\,ds \tag{6.9}$$

$$= \begin{cases} \displaystyle\int_{0}^{\infty} f_{\mathbb{S}}(y+s)\cdot f_{\mathbf{s}}(s)\,ds, & y \geqslant 0 \\ \displaystyle\int_{-y}^{\infty} f_{\mathbb{S}}(y+s)\cdot f_{\mathbf{s}}(s)\,ds, & y \leqslant 0 \end{cases} \tag{6.10}$$

Hence the probability of failure is given by

$$\bar{R}=\int_{-\infty}^{0} f_{\mathbf{y}}(y)\,dy$$

$$=\int_{-\infty}^{0}\int_{-y}^{\infty} f_{\mathbb{S}}(y+s)\cdot f_{\mathbf{s}}(s)\,ds\,dy \tag{6.11}$$

and the reliability by

$$R=\int_{0}^{\infty} f_{\mathbf{y}}(y)\,dy$$

$$=\int_{0}^{\infty}\int_{0}^{\infty} f_{\mathbb{S}}(y+s)\cdot f_{\mathbf{s}}(s)\,ds\,dy \tag{6.12}$$

6.2 RELIABILITY COMPUTATION FOR NORMALLY DISTRIBUTED STRENGTH AND STRESS

The probability density function for a normally distributed stress **s** is given by

$$f_{\mathbf{s}}(s)=\frac{1}{\sigma_s\sqrt{2\pi}}\exp\left[-\frac{1}{2}\left(\frac{s-\mu_s}{\sigma_s}\right)^2\right], -\infty<s<\infty \tag{6.13}$$

and the probability density function for a normally distributed strength $\mathbb{S}$ is given by

$$f_{\mathbb{S}}(\mathbb{S})=\frac{1}{\sigma_{\mathbb{S}}\sqrt{2\pi}}\exp\left[-\frac{1}{2}\left(\frac{\mathbb{S}-\mu_{\mathbb{S}}}{\sigma_{\mathbb{S}}}\right)^2\right], -\infty<\mathbb{S}<\infty \tag{6.14}$$

where

$\mu_{\mathbf{s}}$ = mean value of the stress

$\sigma_{\mathbf{s}}$ = standard deviation of the stress

$\mu_{\mathbb{S}}$ = mean value for the strength

$\sigma_{\mathbb{S}}$ = standard deviation of the strength

Let us define $\mathbf{y}=\mathbb{S}-\mathbf{s}$. It is well known that the random variable **y** is normally distributed with a mean of

$$\mu_{\mathbf{y}}=\mu_{\mathbb{S}}-\mu_{\mathbf{s}} \tag{6.15}$$

and a standard deviation of

$$\sigma_y = \sqrt{\sigma_{\mathcal{S}}^2 + \sigma_s^2} \tag{6.16}$$

See Figure 6.3.The reliability R can now be expressed in terms of **y** as

$$R = P(y > 0)$$

$$= \int_0^{\infty} \frac{1}{\sigma_y \sqrt{2\pi}} \exp\left[-\frac{1}{2}\left(\frac{y-\mu_y}{\sigma_y} \right)^2 \right] dy$$

If we let $z = (y - \mu_y)/\sigma_y$, then $\sigma_y dz = dy$. When $y = 0$, the lower limit of z is given by

$$z = \frac{0-\mu_y}{\sigma_y} = -\frac{\mu_{\mathcal{S}} - \mu_s}{\sqrt{\sigma_{\mathcal{S}}^2 + \sigma_s^2}} \tag{6.17}$$

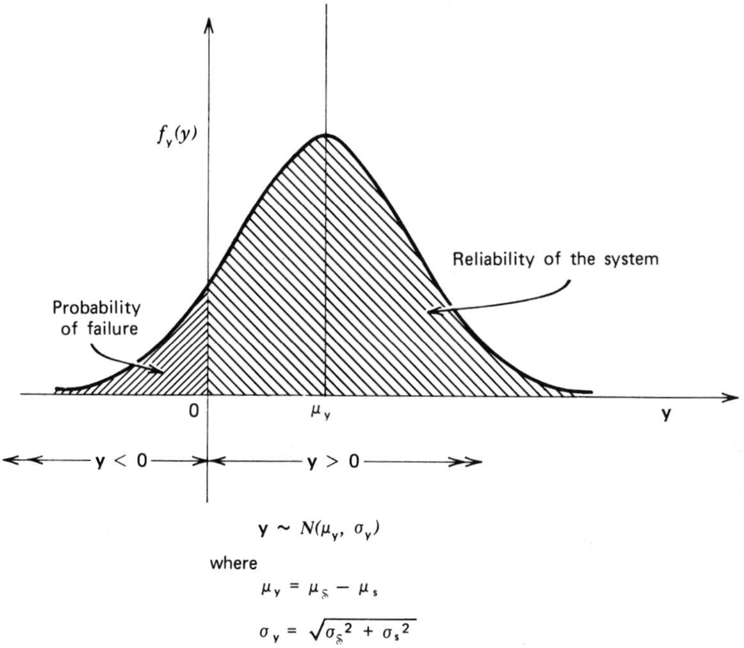

Figure 6.3 Density function of random variable y.

and when $y \to +\infty$, the upper limit of $z \to +\infty$. Therefore,

$$R = \frac{1}{\sqrt{2\pi}} \int_{-\frac{\mu_{\mathfrak{S}} - \mu_s}{\sqrt{\sigma_{\mathfrak{S}}^2 + \sigma_s^2}}}^{\infty} e^{-z^2/2} dz \tag{6.18}$$

Clearly, the random variable $\mathbf{z} = (\mathbf{y} - \mu_y)/\sigma_y$ is the standard normal variable. Hence the reliability can be found by merely referring to the normal tables.

Equation 6.17, which is used to find the lower limit of the standard normal variate $\mathbf{z}$, is commonly known as the coupling equation. Equation 6.18 may be rewritten as

$$R = 1 - \Phi\left(- \frac{\mu_{\mathfrak{S}} - \mu_s}{\sqrt{\sigma_{\mathfrak{S}}^2 + \sigma_s^2}} \right) \tag{6.18a}$$

EXAMPLE 6.1

An automotive component has been designed to withstand certain stresses. It is known from the past experience that, because of variation in loading, the stress on the component is normally distributed with a mean of 30,000 kPa and a standard deviation of 3,000 kPa. The strength of the component is also random because of variations in the material characteristics and the dimensional tolerances. It has been found that the strength is normally distributed with a mean of 40,000 kPa and a standard deviation of 4,000 kPa. Determine the reliability of the component.

We are given that

$$\mathfrak{S} \sim N(40{,}000,\ 4{,}000) \text{ kPa}$$

$$\mathbf{s} \sim N(30{,}000,\ 3{,}000) \text{ kPa}$$

Then the lower limit of the integral for R is given by

$$z = -\frac{40{,}000 - 30{,}000}{\sqrt{(4{,}000)^2 + (3{,}000)^2}} = -\frac{10{,}000}{5{,}000} = -2.0$$

and hence, from the normal tables,

$$R = 0.977$$

EXAMPLE 6.2

The stress developed in an engine component is known to be normally distributed with a mean of 350.00 MPa and a standard deviation of 40.00 MPa. The material strength distribution, based on the expected temperature range and various other factors, is known to be normal with a mean of 820.00 MPa and a standard deviation of 80.00 MPa.

Conventional factor of safety, defined as the ratio of mean strength to mean stress, is given by

$$\text{F.S.} = \frac{\mu_{\mathfrak{S}}}{\mu_s} = \frac{820.00}{350.00} = 2.34$$

To calculate the reliability of the component, we use the coupling equation:

$$z = -\frac{\mu_{\bar{S}} - \mu_{s}}{\sqrt{\sigma_{\bar{S}}^2 + \sigma_{s}^2}} = -\frac{820.00 - 350.00}{\sqrt{(40.00)^2 + (80.00)^2}} = -\frac{470.00}{89.44}$$

$$= -5.25$$

Hence, the reliability of the component is 0.9999999.

Now, suppose that poor heat treatment and larger variations in the environmental temperatures cause the standard deviation for the strength of the component to increase to 150.00 MPa. In that case the factor of safety as defined before remains unchanged, but the reliability is altered. Using the coupling equation,

$$z = -\frac{820.00 - 350.00}{\sqrt{(40.00)^2 + (150.00)^2}} = -\frac{470.00}{155.24} = -3.03$$

the reliability of the component is found to be 0.99877. Thus, we witness a downgrading of reliability resulting from an increased variability in the strength of the component.

EXAMPLE 6.3

A new component is to be designed. A stress analysis revealed that the component is subjected to a tensile stress. But there are variations in the load and the tensile stress is found to be normally distributed with a mean of 35,000 psi and a standard deviation of 4,000 psi. The manufacturing operations create a residual compressive stress that is normally distributed with a mean of 10,000 psi and a standard deviation of 1,500 psi. A strength analysis of the component showed that the mean value of the significant strength is 50,000 psi. The variations introduced by various strength factors are not clear at the present time. The engineer wants to know the maximum value of the standard deviation for the strength that will insure that the component reliability does not drop below 0.999.

We are given that

$$\mathbf{s}_t \sim N(35{,}000, 4{,}000) \text{ psi}$$

$$\mathbf{s}_c \sim N(10{,}000, 1{,}5000) \text{ psi}$$

where $\mathbf{s}_t$ is the tensile stress and $\mathbf{s}_c$ is the residual compressive stress.

The mean effective stress $\bar{\mathbf{s}}$ is obtained by

$$\bar{\mathbf{s}} = \bar{\mathbf{s}}_t - \bar{\mathbf{s}}_c = 35{,}000 - 10{,}000 = 25{,}000 \text{ psi}$$

and its standard deviation by

$$\sigma_{\mathbf{s}} = \sqrt{(\sigma_{s_t})^2 + (\sigma_{s_c})^2}$$

$$= \sqrt{(4{,}000)^2 + (1{,}500)^2}$$

$$= 4{,}272 \text{ psi}$$

From the normal tables, we find the value of z associated with a reliability of 0.999 to be -3.1. Substituting in the coupling equation yields

$$-3.1 = -\frac{50{,}000 - 25{,}000}{\sqrt{(\sigma_{\mathbf{S}})^2 + (4{,}272)^2}}$$

Solving for $\sigma_{\mathbf{S}}$, we get

$$\sigma_{\mathbf{S}} = 6{,}840 \text{ psi}$$

6.3 RELIABILITY COMPUTATION FOR LOG NORMALLY DISTRIBUTED STRENGTH AND STRESS

The standard form of a log normal density function is

$$f_{\mathbf{y}}(y) = \frac{1}{y\sigma\sqrt{2\pi}} \exp\left[-\frac{1}{2\sigma^2}(\ln y - \mu)^2\right], \qquad y > 0 \qquad (6.19)$$

where $\mathbf{y}$ is the random variable. The parameters μ and σ are the mean and the standard deviation, respectively, of the variable $\ln \mathbf{y}$, which is normally distributed. First we develop those relationships for the log normal distribution that are needed later in the analysis.

Let $\mathbf{x} = \ln \mathbf{y}$. Then $dx = (1/y)\,dy$. From Equation 6.19 we have

$$f(x) = \frac{1}{\sigma\sqrt{2\pi}} \exp\left[-\frac{1}{2\sigma^2}(x - \mu)^2\right], \qquad -\infty < x < \infty$$

and hence,

$$E(\mathbf{x}) = E[\ln \mathbf{y}] = \mu$$

and

$$V[\mathbf{x}] = \sigma^2 = V[\ln \mathbf{y}] = \sigma^2_{\ln \mathbf{y}}$$

Now considering the exponent of e in the expression

$$E(\mathbf{y}) = E(e^{\mathbf{x}}) = \int_{-\infty}^{\infty} \frac{1}{\sigma\sqrt{2\pi}} e^x \exp\left\{-\left(\frac{1}{2}\right)\left(\frac{x-\mu}{\sigma}\right)^2\right\} dx$$

we have

$$x-\frac{1}{2}\left(\frac{x-\mu}{\sigma}\right)^2=x-\frac{1}{2\sigma^2}(x^2-2x\mu+\mu^2)$$

$$=-\frac{1}{2\sigma^2}(x^2-2\mu x-2\sigma^2x+\mu^2)$$

$$=-\frac{\mu^2}{2\sigma^2}+\frac{(\mu+\sigma^2)^2}{2\sigma^2}-\frac{1}{2\sigma^2}\left[x^2-2x(\mu+\sigma^2)+(\mu+\sigma^2)^2\right]$$

$$=\frac{1}{2\sigma^2}(2\mu\sigma^2+\sigma^4)-\frac{1}{2\sigma^2}\left[x-(\mu+\sigma^2)\right]^2$$

$$=\mu+\frac{\sigma^2}{2}-\frac{1}{2\sigma^2}\left[x-(\mu+\sigma^2)\right]^2$$

Therefore

$$E(\mathbf{y})=\exp\left(\mu+\frac{\sigma^2}{2}\right)\int_{-\infty}^{\infty}\frac{1}{\sigma\sqrt{2\pi}}\exp\left[-\frac{\{x-(\mu+\sigma^2)\}^2}{2\sigma^2}\right]dx$$

$$=\exp\left(\mu+\frac{\sigma^2}{2}\right) \tag{6.20}$$

To compute the variance of $\mathbf{y}$ we observe that

$$E(\mathbf{y}^2)=\int_{-\infty}^{\infty}\frac{1}{\sigma\sqrt{2\pi}}\exp\left[2x-\frac{1}{2\sigma^2}(x-\mu)^2\right]dx$$

Considering the exponent of e in the expression for $E(\mathbf{y}^2)$, we have

$$2x-\frac{1}{2\sigma^2}(x-\mu)^2$$

$$=-\frac{1}{2\sigma^2}(-4\sigma^2x+x^2-2\mu x+\mu^2)$$

$$=-\frac{1}{2\sigma^2}\left[x^2-2x(\mu+2\sigma^2)+(\mu+2\sigma^2)^2\right]-\frac{\mu^2}{2\sigma^2}+\frac{(\mu+2\sigma^2)^2}{2\sigma^2}$$

$$=-\frac{1}{2\sigma^2}\left[x-(\mu+2\sigma^2)\right]^2+2\mu+2\sigma^2$$

which, when substituted back and simplified as before, yields,

$$E(\mathbf{y}^2) = \exp[2(\mu + \sigma^2)]$$

Hence by the definition of variance, we may write

$$V(\mathbf{y}) = \exp[2(\mu + \sigma^2)] - \{\exp[\mu + \sigma^2/2]\}^2$$
$$= [\exp(2\mu + \sigma^2)][\exp(\sigma^2) - 1] \tag{6.21}$$

We now observe that

$$\frac{V(\mathbf{y})}{[E(\mathbf{y})]^2} = e^{\sigma^2} - 1$$

which, after rearranging, leads to

$$\sigma^2 = \ln\left[\frac{V(\mathbf{y})}{[E(\mathbf{y})]^2} + 1\right] \tag{6.22}$$

We proved in Equation 6.20 that

$$E(\mathbf{y}) = e^{\mu + \sigma^2/2}$$

which implies that

$$\mu = \ln E(\mathbf{y}) - \frac{1}{2}\sigma^2 \tag{6.23}$$

If $\check{\mathbf{y}}$ denotes the median of $\mathbf{y}$, then we may write

$$0.5 = \int_0^{\check{y}} \frac{1}{y\sigma\sqrt{2\pi}} \exp\left[-\frac{1}{2\sigma^2}(\ln y - \mu)^2\right] dy$$

Using the transformation $x = \ln y$, we rewrite it as

$$0.5 = \int_{-\infty}^{\ln \check{y}} \frac{1}{\sigma\sqrt{2\pi}} \exp\left[-\frac{1}{2\sigma^2}(x - \mu)^2\right] dx$$

yielding

$$\mu = \ln \check{\mathbf{y}} \tag{6.24}$$

that is,

$$\check{\mathbf{y}} = e^{\mu}$$

Returning now to the original problem in which $\mathbb{S}$ and $\mathbf{s}$ are log normally distributed, we let $\mathbf{y} = \mathbb{S}/\mathbf{s}$, which means $\ln \mathbf{y} = \ln \mathbb{S} - \ln \mathbf{s}$. We observe that $\ln \mathbf{y}$ is normally distributed since both $\ln \mathbb{S}$ and $\ln \mathbf{s}$ are normally distributed.

The log normal density function is positively skewed and hence the median is a better and more convenient measure of the central tendancy for the log normal distribution than the mean. Clearly, the antilog of the mean of $\ln \mathbb{S}$ is the median of $f_{\mathbb{S}}(\cdot)$, and that of $\ln \mathbf{s}$ the median of $f_{\mathbf{s}}(\cdot)$; that is,

$$\check{\mathbb{S}} = e^{\mu_{\ln \mathbb{S}}}$$

or

$$\mu_{\ln \mathbb{S}} = \ln \check{\mathbb{S}}$$

and

$$\check{\mathbf{s}} = e^{\mu_{\ln \mathbf{s}}}$$

or

$$\mu_{\ln \mathbf{s}} = \ln \check{\mathbf{s}}$$

where $\check{\mathbb{S}}$ and $\check{\mathbf{s}}$ are the medians of $\mathbb{S}$ and $\mathbf{s}$ respectively. By analogy, we add

$$\mu_{\ln \mathbf{y}} = \ln \check{\mathbf{y}}$$

since we know that $\mathbf{y}$ is also log normally distributed. But,

$$\mu_{\ln \mathbf{y}} = \mu_{\ln \mathbb{S}} - \mu_{\ln \mathbf{s}} = \ln \check{\mathbb{S}} - \ln \check{\mathbf{s}} \tag{6.25}$$

Combining the two equations, we get

$$\ln \check{\mathbf{y}} = \ln \check{\mathbb{S}} - \ln \check{\mathbf{s}} = \ln \frac{\check{\mathbb{S}}}{\check{\mathbf{s}}}$$

We also know that

$$\sigma_{\ln \mathbf{y}} = \sqrt{\sigma_{\ln \mathbb{S}}^2 + \sigma_{\ln \mathbf{s}}^2} \tag{6.26}$$

From the definition of reliability, we have

$$R = P\left(\frac{\mathbb{S}}{\mathbf{s}} > 1\right) = P(\mathbf{y} > 1) = \int_1^{\infty} f_{\mathbf{y}}(y)\,dy$$

Let $\mathbf{z} = (\ln \mathbf{y} - \mu_{\ln \mathbf{y}})/\sigma_{\ln \mathbf{y}}$. Then $\mathbf{z}$ is the standard normal variate. We now need to find the new limits of integration. When $y=1$,

$$z=\frac{\ln 1-\mu_{\ln y}}{\sigma_{\ln y}}=-\frac{\ln\check{\mathbb{S}}-\ln\check{\mathbf{s}}}{\sqrt{\sigma_{\ln \mathbb{S}}^{2}+\sigma_{\ln \mathbf{s}}^{2}}}$$

the latter equality following from Equations 6.25 and 6.26, and when $y\to+\infty, z\to+\infty$. The reliability can now be easily computed as

$$R=\int_{-\frac{\ln\check{\mathbb{S}}-\ln\check{\mathbf{s}}}{\sqrt{\sigma_{\ln \mathbb{S}}^{2}+\sigma_{\ln \mathbf{s}}^{2}}}}^{\infty}\phi(z)\,dz \tag{6.27}$$

where $\phi(z)$ is the p.d.f. for the standard normal variate $\mathbf{z}$.

EXAMPLE 6.4

The strength $\mathbb{S}$ and the stress $\mathbf{s}$ are log normally distributed with the following parameters:

$$E(\mathbb{S})=100{,}000 \text{ kPa}$$

$$\text{Standard deviation of } \mathbb{S}=10{,}000 \text{ kPa}$$

$$E(\mathbf{s})=60{,}000 \text{ kPa}$$

$$\text{Standard deviation of } \mathbf{s}=20{,}000 \text{ kPa}$$

We want to compute the reliability.

Let

$$E(\ln\mathbb{S})=\mu_{\mathbb{S}} \quad\text{and}\quad E(\ln\mathbf{s})=\mu_{\mathbf{s}}$$

$$V(\ln\mathbb{S})=\sigma_{\mathbb{S}}^2 \quad\text{and}\quad V(\ln\mathbf{s})=\sigma_{\mathbf{s}}^2$$

From Equation 6.22, we have

$$\sigma_{\mathbb{S}}^2=\ln\left[\frac{V(\mathbb{S})}{(E[\mathbb{S}])^2}+1\right]=\ln\left[\frac{10^8}{10^{10}}+1\right]$$

$$=\ln 1.01=0.00995$$

and from Equation 6.23, we have

$$\mu_{\mathbb{S}}=\ln E[\mathbb{S}]-\frac{1}{2}\sigma_{\mathbb{S}}^2=\ln 100{,}000-\frac{0.00995}{2}$$

$$=11.50795$$

Similarly, for stress $\mathbf{s}$ we have

$$\sigma_{\mathbf{s}}^2=\ln\left[\frac{20{,}000^2}{60{,}000^2}+1\right]=\ln[1.111]$$

$$=0.10535$$

and

$$\mu_s = \ln E[s] - \frac{1}{2}\sigma_s^2 = 11.00209 - \frac{1}{2} \times 0.10535$$

$$= 10.94942$$

Therefore,

$$R = \int_{-z}^{\infty} \phi(u)\,du$$

where z is given by the coupling equation:

$$z = -\frac{\mu_{\mathbb{S}} - \mu_s}{\sqrt{\sigma_{\mathbb{S}}^2 + \sigma_s^2}} = -\frac{11.50795 - 10.94942}{\sqrt{(0.00995) + (0.10535)}}$$

$$= -1.64$$

From the normal tables, we have for $z = -1.64$,

$$R = 0.9495$$

EXAMPLE 6.5

The strength $\mathbb{S}$ and the stress s are log normally distributed with the following parameters:

Mean of $\mathbb{S}$ = 150,000 kPa

Mean of s = 100,000 kPa

Standard deviation of s = 15,000 kPa

We wish to know the maximum allowable standard deviation of strength $\mathbb{S}$ so that the reliability does not fall below 0.990. First we compute

$$\sigma_s^2 = \ln\left[\frac{V(s)}{(E(s))^2} + 1\right] = \ln\left[\frac{15{,}000^2}{100{,}000^2} + 1\right]$$

$$= 0.02225$$

and

$$\mu_s = \ln E[s] - \frac{1}{2}\sigma_s^2$$

$$= 11.50180$$

and

$$\mu_{\mathbb{S}} = \ln E[\mathbb{S}] - \frac{1}{2}\sigma_{\mathbb{S}}^2$$

$$= 11.91839 - \frac{1}{2}\sigma_{\mathbb{S}}^2$$

Now from the normal tables, we find that the value of z associated with a reliability of 0.990 is

$$z = -\frac{\mu_{\mathbb{S}} - \mu_{s}}{\sqrt{\sigma_{\mathbb{S}}^2 + \sigma_{s}^2}} = -2.33$$

On simplification we obtain

$$\mu_{\mathbb{S}}^2 - 2\mu_{\mathbb{S}}\mu_{s} + \mu_{s}^2 = 5.4289(\sigma_{\mathbb{S}}^2 + \sigma_{s}^2)$$

By substituting the values of μ_s, σ_s^2, and $\mu_{\mathbb{S}}$ found earlier and simplifying, we obtain a quadratic equation in $\sigma_{\mathbb{S}}^2$:

$$0.05275 - 5.84549\sigma_{\mathbb{S}}^2 + 0.25(\sigma_{\mathbb{S}}^2)^2 = 0$$

which has the solution:

$$\sigma_{\mathbb{S}}^2 = 0.00903 \quad \text{or} \quad 23.37292$$

Accepting the smaller value of 0.00903, we compute

$$\mu_{\mathbb{S}} = 11.91839 - \frac{1}{2}\sigma_{\mathbb{S}}^2 = 11.91387$$

and hence

$$\begin{aligned} V(\mathbb{S}) &= \left[\exp(2\mu_{\mathbb{S}} + \sigma_{\mathbb{S}}^2)\right]\left(\exp(\sigma_{\mathbb{S}}^2) - 1\right) \\ &= \left[\exp(2(11.91839) + 0.00903)\right]\left[\exp(0.00903) - 1\right] \\ &= 20.405 \times 10^6 \end{aligned}$$

The desired maximum allowable standard deviation of the strength $\mathbb{S}$ is 4.5172×10^3 kPa.

6.4 RELIABILITY COMPUTATION FOR EXPONENTIALLY DISTRIBUTED STRENGTH AND STRESS

In this case we have, for strength $\mathbb{S}$,

$$f_{\mathbb{S}}(\mathbb{S}) = \lambda_{\mathbb{S}} e^{-\lambda_{\mathbb{S}}\mathbb{S}}, \qquad 0 \leqslant \mathbb{S} < \infty$$

and, for stress s,

$$f_s(s) = \lambda_s e^{-\lambda_s s}, \qquad 0 \leqslant s < \infty$$

Using Equation 6.3, we have

$$R=\int_0^{\infty} f_s(s)\left[\int_s^{\infty} f_{\mathbb{S}}(\mathbb{S})\,d\mathbb{S}\right]ds$$

$$=\int_0^{\infty} \lambda_s e^{-\lambda_s s}\left[e^{-\lambda_{\mathbb{S}} s}\right]ds$$

$$=\int_0^{\infty} \lambda_s e^{-(\lambda_s+\lambda_{\mathbb{S}})s}\,ds$$

$$=\frac{\lambda_s}{\lambda_{\mathbb{S}}+\lambda_s}\int_0^{\infty}(\lambda_{\mathbb{S}}+\lambda_s)e^{-(\lambda_{\mathbb{S}}+\lambda_s)s}\,ds$$

$$=\frac{\lambda_s}{\lambda_{\mathbb{S}}+\lambda_s} \tag{6.28}$$

If we denote the mean value of strength by $\bar{\mathbb{S}}=1/\lambda_{\mathbb{S}}$ and the mean value of stress by $\bar{s}=1/\lambda_s$, then $R=\bar{\mathbb{S}}/(\bar{\mathbb{S}}+\bar{s})$.

6.5 RELIABILITY COMPUTATION FOR NORMALLY (EXPONENTIALLY) DISTRIBUTED STRENGTH AND EXPONENTIALLY (NORMALLY) DISTRIBUTED STRESS

For a normally distributed strength the density function is

$$f_{\mathbb{S}}(\mathbb{S})=\frac{1}{\sigma_{\mathbb{S}}\sqrt{2\pi}}\exp\left[-\frac{1}{2}\left(\frac{\mathbb{S}-\mu_{\mathbb{S}}}{\sigma_{\mathbb{S}}}\right)^2\right], \qquad -\infty<\mathbb{S}<\infty$$

and the density function for an exponentially distributed stress is

$$f_s(s)=\lambda e^{-\lambda s}, \qquad s\geqslant 0$$

It is well known that $\mu_s=1/\lambda$ and $\sigma_s=1/\lambda$.

Recall the reliability Equation 6.5:

$$R=\int_0^{\infty} f_{\mathbb{S}}(\mathbb{S})\left[\int_0^{\mathbb{S}} f_s(s)\,ds\right]d\mathbb{S}$$

We realize that

$$\int_0^{\mathbb{S}} f_s(s)\,ds=\int_0^{\mathbb{S}}\lambda e^{-\lambda s}\,ds=1-e^{-\lambda\mathbb{S}}$$

yielding

$$R=\int_0^\infty \frac{1}{\sigma_{\mathbb{S}}\sqrt{2\pi}}\exp\left[-\frac{1}{2}\left(\frac{\mathbb{S}-\mu_{\mathbb{S}}}{\sigma_{\mathbb{S}}}\right)^2\right](1-e^{-\lambda\mathbb{S}})\,d\mathbb{S}$$

$$=\frac{1}{\sigma_{\mathbb{S}}\sqrt{2\pi}}\int_0^\infty \exp\left[-\frac{1}{2}\left(\frac{\mathbb{S}-\mu_{\mathbb{S}}}{\sigma_{\mathbb{S}}}\right)^2\right]d\mathbb{S}$$

$$-\frac{1}{\sigma_{\mathbb{S}}\sqrt{2\pi}}\int_0^\infty \exp\left[-\frac{1}{2}\left(\frac{\mathbb{S}-\mu_{\mathbb{S}}}{\sigma_{\mathbb{S}}}\right)^2\right]e^{-\lambda\mathbb{S}}\,d\mathbb{S}$$

$$=1-\Phi\left(-\frac{\mu_{\mathbb{S}}}{\sigma_{\mathbb{S}}}\right)-\frac{1}{\sigma_s\sqrt{2\pi}}$$

$$\times\int_0^\infty \exp\left[-\frac{1}{2\sigma_{\mathbb{S}}^2}\left((\mathbb{S}-\mu_{\mathbb{S}}+\lambda\sigma_{\mathbb{S}}^2)^2+2\mu_{\mathbb{S}}\sigma_{\mathbb{S}}^2-\lambda^2\sigma_{\mathbb{S}}^4\right)\right]d\mathbb{S} \qquad (6.29)$$

If we let $t=(\mathbb{S}-\mu_{\mathbb{S}}+\lambda\sigma_{\mathbb{S}}^2)/\sigma_{\mathbb{S}}$, then $\sigma_{\mathbb{S}}dt=d\mathbb{S}$. The expression for R now becomes

$$R=1-\Phi\left(-\frac{\mu_{\mathbb{S}}}{\sigma_{\mathbb{S}}}\right)-\frac{1}{\sqrt{2\pi}}\int_{-\frac{\mu_{\mathbb{S}}-\lambda\sigma_{\mathbb{S}}^2}{\sigma_{\mathbb{S}}}}^{\infty}\exp\left[-\frac{t^2}{2}\right]\cdot\exp\left[-\frac{1}{2}(2\mu_{\mathbb{S}}\lambda-\lambda^2\sigma_{\mathbb{S}}^2)\right]dt$$

$$=1-\Phi\left(-\frac{\mu_{\mathbb{S}}}{\sigma_{\mathbb{S}}}\right)-\exp\left[-\frac{1}{2}(2\mu_{\mathbb{S}}\lambda-\lambda^2\sigma_{\mathbb{S}}^2)\right]\left[1-\Phi\left(-\frac{\mu_{\mathbb{S}}-\lambda\sigma_{\mathbb{S}}^2}{\sigma_{\mathbb{S}}}\right)\right] \qquad (6.30)$$

When the distributions for the strength and the stress are interchanged, that is, when the strength has an exponential density function with parameter $\lambda_{\mathbb{S}}$ and the stress is normal with parameters μ_s and σ_s, Equation 6.3 can be used to obtain the following expression for the reliability:

$$R=\int_{-\infty}^{\infty} f_s(s)\left[\int_s^\infty f_{\mathbb{S}}(\mathbb{S})\,d\mathbb{S}\right]ds$$

$$=\int_{-\infty}^{\infty}\frac{1}{\sigma_s\sqrt{2\pi}}\exp\left[-\frac{1}{2}\left(\frac{s-\mu_s}{\sigma_s}\right)^2\right]\cdot\exp[-\lambda_{\mathbb{S}}s]\,ds$$

On simplification, we get

$$R=\Phi\left(-\frac{\mu_{s}}{\sigma_{s}}\right)+\exp\left[-\tfrac{1}{2}\left(2\mu_{s}\lambda_{\mathbb{S}}-\lambda_{\mathbb{S}}^{2}\sigma_{s}^{2}\right)\right]\left[1-\Phi\left(-\frac{\mu_{s}-\lambda_{\mathbb{S}}^{2}\sigma_{s}^{2}}{\sigma_{s}}\right)\right] \tag{6.31}$$

which is a slightly different expression from Equation 6.30.

EXAMPLE 6.6

The strength of a component is normally distributed with $\mu_{\mathbb{S}}=100$ MPa and $\sigma_{\mathbb{S}}=10$ MPa. The stresses acting on the component follow exponential distribution with a mean value of 50 MPa. Compute the reliability of the component.

Using Equation 6.30, we have

$$R=1-\Phi(-10)-\exp\left[-\frac{1}{2}\left(\frac{2\times 100}{50}-\left(\frac{10}{50}\right)^{2}\right)\right]\left[1-\Phi\left[-\frac{100-\frac{10^{2}}{50}}{10}\right]\right]$$

$$=1-0.0-\exp[-1.98][1-0.0]$$

$$=1-0.13806=0.86194$$

6.6 RELIABILITY COMPUTATION FOR GAMMA DISTRIBUTED STRENGTH AND STRESS

The gamma density function for a random variable $\mathbf{x}$ is given by

$$f(x)=\frac{\lambda^{n}x^{n-1}e^{-\lambda x}}{\Gamma(n)};\qquad n>0,\ \lambda>0,\ 0\leqslant x<\infty$$

where λ is called the scale parameter and n the shape parameter.

Let us first consider the case when $\lambda=1$. We have

$$f_{\mathbb{S}}(\mathbb{S})=\frac{1}{\Gamma(m)}\mathbb{S}^{m-1}e^{-\mathbb{S}},\qquad 0\leqslant\mathbb{S}<\infty$$

and

$$f_{s}(s)=\frac{1}{\Gamma(n)}s^{n-1}e^{-s},\qquad 0\leqslant s<\infty$$

Then using Equation 6.9 we have, for $\mathbf{y}=\mathbb{S}-\mathbf{s}$,

$$f_{y}(y)=\frac{1}{\Gamma(m)\Gamma(n)}\int_{0}^{\infty}(y+s)^{m-1}e^{-(y+s)}s^{n-1}e^{-s}\,ds,\qquad y\geqslant 0$$

We let $v = s/y$. Then $dv = (1/y)ds$. We now have

$$f_y(y) = \frac{1}{\Gamma(m)\Gamma(n)} y^{m+n-1} e^{-y} \int_0^\infty v^{n-1}(1+v)^{m-1} e^{-2yv}\, dv$$

Hence,

$$R = \int_0^\infty f_y(y)\, dy$$

$$= \frac{1}{\Gamma(m)\Gamma(n)} \int_0^\infty y^{m+n-1} e^{-y}\, dy \int_0^\infty v^{n-1}(1+v)^{m-1} e^{-2yv}\, dv$$

But

$$\int_0^\infty y^{m+n-1} e^{-(1+2v)y}\, dy = \frac{\Gamma(m+n)}{(1+2v)^{m+n}}$$

Therefore,

$$R = \frac{\Gamma(m+n)}{\Gamma(m)\Gamma(n)} \int_0^\infty \frac{(1+v)^{m-1} v^{n-1}}{(1+2v)^{m+n}}\, dv$$

$$= \frac{\Gamma(m+n)}{\Gamma(m)\Gamma(n)} \int_0^{1/2} (1-u)^{m-1} u^{n-1}\, du$$

where $u = v/(1+2v)$. The above integral is the well-known [10] incomplete beta function $B_{1/2}(m,n)$. Hence,

$$R = \frac{\Gamma(m+n)}{\Gamma(m)\Gamma(n)} B_{1/2}(m,n) \tag{6.32}$$

Next we consider the general case when $\lambda \neq 1$. We have

$$f_{\mathfrak{S}}(\mathfrak{S}) = \frac{\lambda^m}{\Gamma(m)} \mathfrak{S}^{m-1} e^{-\lambda \mathfrak{S}}; \qquad \lambda > 0,\ m > 0,\ 0 \leqslant \mathfrak{S} < \infty$$

and

$$f_s(s) = \frac{\mu^n}{\Gamma(n)} s^{n-1} e^{-\mu s}; \qquad \mu > 0,\ n > 0,\ 0 \leqslant s < \infty$$

Using Equation 6.9 we have, as before,

$$R = \int_0^\infty f_y(y)\,dy$$

$$= \frac{r^n \Gamma(m+n)}{\Gamma(m)\Gamma(n)} \int_0^\infty \frac{(1+v)^{m-1} v^{n-1}}{[1+(1+r)v]^{m+n}}\,dv$$

where $r = \mu/\lambda$. If we let $u = rv/(1+(1+r)v)$, then

$$R = \frac{\Gamma(m+n)}{\Gamma(m)\Gamma(n)} \int_0^{r/(1+r)} (1-u)^{m-1} u^{n-1}\,du$$

which involves r only in the limit of the integration. Hence, the reliability can be expressed in terms of the incomplete beta function whose truncation occurs at $r/(1+r)$ instead of $\frac{1}{2}$ as before; that is,

$$R = \frac{\Gamma(m+n)}{\Gamma(m)\Gamma(n)} B_{r/(1+r)}(m,n) \tag{6.33}$$

Next we briefly discuss three special cases:

1. If $m = n = 1$, then $\mathbb{S}$ and $\mathbf{s}$ are exponentially distributed with

$$R = \frac{\Gamma(2)}{\Gamma(1)\Gamma(1)} \int_0^{r/(1+r)} du = \frac{r}{1+r} = \frac{\mu}{\mu+\lambda}$$

which is the same as the result we obtained in Equation 6.29.

2. If $m = 1$ and $n \neq 1$, then the strength $\mathbb{S}$ is exponentially distributed and the stress $\mathbf{s}$ is gamma distributed. In this case, we have

$$R = \frac{\Gamma(n+1)}{\Gamma(1)\Gamma(n)} \int_0^{r/(1+r)} u^{n-1}\,du = \frac{n\Gamma(n)}{\Gamma(n)} \left(\frac{r}{1+r}\right)^n \frac{1}{n} = \left(\frac{r}{1+r}\right)^n$$

$$= \left(\frac{\mu}{\mu+\lambda}\right)^n \tag{6.34}$$

3. If $m \neq 1$ and $n = 1$, then the strength has a gamma distribution and the stress has an exponential distribution. In this case, we have

$$R = \frac{\Gamma(m+1)}{\Gamma(m)\Gamma(1)} \int_0^{r/(1+r)} (1-u)^{m-1}\,du = 1 - \left(\frac{1}{1+r}\right)^m = 1 - \left(\frac{\lambda}{\mu+\lambda}\right)^m$$

6.7 RELIABILITY COMPUTATION FOR NORMALLY DISTRIBUTED STRESS AND WEIBULL DISTRIBUTED STRENGTH

The probability density function for a Weibull distributed $\mathbb{S}$ is given by

$$f_{\mathbb{S}}(\mathbb{S}) = \left[\frac{\beta}{(\theta - \mathbb{S}_0)^{\beta}}\right](\mathbb{S} - \mathbb{S}_0)^{\beta - 1}\exp\left[-\left(\frac{\mathbb{S} - \mathbb{S}_0}{\theta - \mathbb{S}_0}\right)^{\beta}\right], \qquad \mathbb{S} \geqslant \mathbb{S}_0 \geqslant 0$$

where β is called the slope parameter, and $(\theta - \mathbb{S}_0)$ the scale parameter. $\mathbb{S}_0$ is called the truncation parameter. It is that value of the strength below which the probability of occurrence is zero.

The cumulative probability density function for the strength $\mathbb{S}$ is given by

$$F_{\mathbb{S}}(\mathbb{S}) = 1 - \exp\left[-\left(\frac{\mathbb{S} - \mathbb{S}_0}{\theta - \mathbb{S}_0}\right)^{\beta}\right]$$

and the mean and the variance by

$$\mu_{\mathbb{S}} = \mathbb{S}_0 + (\theta - \mathbb{S}_0)\Gamma\left(\frac{1}{\beta} + 1\right)$$

$$\sigma_{\mathbb{S}}^2 = (\theta - \mathbb{S}_0)^2\left[\Gamma\left(\frac{2}{\beta} + 1\right) - \left\{\Gamma\left(\frac{1}{\beta} + 1\right)\right\}^2\right]$$

The Weibull distribution, being a three-parameter distribution, is extremely flexible, it can assume a wide variety of shapes. For $\beta = 1$, it becomes an exponential distribution.

The probability density function for the normally distributed stress **s** is given by

$$f_{\mathbf{s}}(s) = \frac{1}{\sigma_{\mathbf{s}}\sqrt{2\pi}}\exp\left[-\frac{(s - \mu_{\mathbf{s}})^2}{2\sigma_{\mathbf{s}}^2}\right], \qquad -\infty < s < \infty$$

Substituting in the expression for the probability of failure in Equation 6.6, we get

$$P(\mathfrak{S} \leqslant \mathrm{s}) = \int_{-\infty}^{\infty} f_{\mathrm{s}}(s) F_{\mathfrak{S}}(s)\, ds$$

$$= \int_{\mathfrak{S}_0}^{\infty} \frac{1}{\sigma_{\mathrm{s}}\sqrt{2\pi}} \exp\left[-\frac{(s-\mu_{\mathrm{s}})^2}{2\sigma_{\mathrm{s}}^2} \right] \left[1 - \exp\left[-\left(\frac{s-\mathfrak{S}_0}{\theta-\mathfrak{S}_0} \right)^{\beta} \right] \right] ds$$

$$= \int_{\mathfrak{S}_0}^{\infty} \frac{1}{\sigma_s\sqrt{2\pi}} \exp\left[-\frac{(s-\mu_{\mathrm{s}})^2}{2\sigma_{\mathrm{s}}^2} \right] ds$$

$$- \frac{1}{\sigma_{\mathrm{s}}\sqrt{2\pi}} \int_{\mathfrak{S}_0}^{\infty} \exp\left[-\left\{ \frac{(s-\mu_{\mathrm{s}})^2}{2\sigma_{\mathrm{s}}^2} - \left(\frac{s-\mathfrak{S}_0}{\theta-\mathfrak{S}_0} \right)^{\beta} \right\} \right] ds \qquad (6.35)$$

Consider the transformation $z = (\mathrm{s} - \mu_{\mathrm{s}})/\sigma_{\mathrm{s}}$. Then the first integral is the area under the standard normal density curve from $z = (\mathfrak{S}_0 - \mu_{\mathrm{s}})/\sigma_{\mathrm{s}}$ to $+\infty$. We will denote it by $1 - \Phi((\mathfrak{S}_0 - \mu_{\mathrm{s}})/\sigma_{\mathrm{s}})$. Further let $y = (s - \mathfrak{S}_0)/(\theta - \mathfrak{S}_0)$, then $dy = ds/(\theta - \mathfrak{S}_0)$ and $s = y(\theta - \mathfrak{S}_0) + \mathfrak{S}_0$. Recognizing that

$$\frac{(s-\mu_{\mathrm{s}})^2}{2\sigma_{\mathrm{s}}^2} = \frac{\left[y(\theta-\mathfrak{S}_0) + \mathfrak{S}_0 - \mu_{\mathrm{s}} \right]^2}{2\sigma_{\mathrm{s}}^2} = \frac{1}{2}\left[\left(\frac{\theta-\mathfrak{S}_0}{\sigma_{\mathrm{s}}} \right) y + \frac{\mathfrak{S}_0-\mu_{\mathrm{s}}}{\sigma_{\mathrm{s}}} \right]^2$$

the second term of the right-hand side of Equation 6.35 may be rewritten as

$$\frac{1}{\sqrt{2\pi}} \left(\frac{\theta-\mathfrak{S}_0}{\sigma_{\mathrm{s}}} \right) \int_0^{\infty} \exp\left[-y^{\beta} - \frac{1}{2} \left\{ \left(\frac{\theta-\mathfrak{S}_0}{\sigma_{\mathrm{s}}} \right) y + \frac{\mathfrak{S}_0-\mu_{\mathrm{s}}}{\sigma_{\mathrm{s}}} \right\}^2 \right] dy$$

In the above form, the following three parameters are readily apparent:

$$\beta, \quad \frac{\theta-\mathfrak{S}_0}{\sigma_{\mathrm{s}}}, \quad \text{and} \quad \frac{\mathfrak{S}_0-\mu_{\mathrm{s}}}{\sigma_{\mathrm{s}}}$$

Hence, the probability of failure is given by

$$P(\mathfrak{S} \leqslant \mathrm{s}) = 1 - \Phi\left(\frac{\mathfrak{S}_0-\mu_{\mathrm{s}}}{\sigma_{\mathrm{s}}} \right) - \frac{1}{\sqrt{2\pi}} \left(\frac{\theta-\mathfrak{S}_0}{\sigma_{\mathrm{s}}} \right)$$

$$\times \int_0^{\infty} \exp\left[-y^{\beta} - \frac{1}{2} \left\{ \left(\frac{\theta-\mathfrak{S}_0}{\sigma_{\mathrm{s}}} \right) y + \frac{\mathfrak{S}_0-\mu_{\mathrm{s}}}{\sigma_{\mathrm{s}}} \right\}^2 \right] dy \qquad (6.36)$$

Numerical integration methods have been used to compute the above integral for various values of the parameters. Selected tables are provided in Appendix III.

EXAMPLE 6.7

A spring has to be designed for a failure probability of 10^{-4}. The material out of which the spring is made has the following Weibull parameters:

$$\mathcal{S}_0 = 100{,}000 \text{ psi}$$

$$\beta = 3$$

$$\theta = 130{,}000 \text{ psi}$$

The load acting on the spring is considered to be normally distributed with a coefficient of variation $\sigma_s/\mu_s = 0.02$. We are interested in computing the permissible normal stress parameters that will yield the specified reliability.

Compute

$$C = \frac{\theta - \mathcal{S}_0}{\sigma_s} = \frac{130{,}000 - 100{,}000}{\sigma_s} = \frac{30{,}000}{\sigma_s}$$

$$A = \frac{\mathcal{S}_0 - \mu_s}{\sigma_s} = \frac{100{,}000 - 50\sigma_s}{\sigma_s}$$

$$= \frac{100{,}000}{30{,}000/C} - 50$$

$$= 3.333C - 50$$

or

$$C = 0.3A + 15$$

From the tables in Appendix III, we see that when $A = 0.6$ and $C = 15$ the failure probability is 0.0001. The exact value of C for $A = 0.6$ is

$$C = 0.3 \times 0.6 + 15 = 15.18$$

yielding

$$\sigma_s = \frac{30{,}000}{15.18} = 1{,}970 \text{ psi}$$

and

$$\mu_s = \frac{\sigma_s}{0.02} = 98{,}500 \text{ psi}$$

EXAMPLE 6.8

A component is to be designed for a failure probability of 0.0002. Only the following two parameters of the Weibull distributed strength are known:

$$\beta = 2.0$$

$$\theta = 550{,}000 \text{ kPa}$$

The stress acting on the component is normally distributed with a mean $\mu_s = 100{,}000$ kPa and a standard deviation $\sigma_s = 10{,}000$ kPa. We wish to find the minimum strength parameter $\mathbb{S}_0$ for the component.

We have

$$C = \frac{\theta - \mathbb{S}_0}{\sigma_s} = \frac{550{,}000 - \mathbb{S}_0}{10{,}000}$$

and

$$A = \frac{\mathbb{S}_0 - \mu_s}{\sigma_s} = \frac{\mathbb{S}_0 - 100{,}000}{10{,}000}$$

Eliminating $\mathbb{S}_0$, we get

$$A = 45 - C$$

The tables show that $C = 45$ and $A = 0$ yield the required failure probability of 0.0002. Equating A to zero leads to

$$\mathbb{S}_0 = 100{,}000 \text{ kPa}$$

6.8 RELIABILITY COMPUTATION FOR WEIBULL DISTRIBUTED STRENGTH AND STRESS

The probability density functions for the strength and the stress are

$$f_{\mathbb{S}}(\mathbb{S}) = \frac{\beta_{\mathbb{S}}}{\theta_{\mathbb{S}}}\left(\frac{\mathbb{S} - \mathbb{S}_0}{\theta_{\mathbb{S}}}\right)^{\beta_{\mathbb{S}}-1} \exp\left[-\left(\frac{\mathbb{S} - \mathbb{S}_0}{\theta_{\mathbb{S}}}\right)^{\beta_{\mathbb{S}}}\right], \quad \mathbb{S}_0 \leqslant \mathbb{S} < \infty$$

and

$$f_s(s) = \frac{\beta_s}{\theta_s}\left(\frac{s - s_0}{\theta_s}\right)^{\beta_s-1} \exp\left[-\left(\frac{s - s_0}{\sigma_s}\right)^{\beta_s}\right], \quad s_0 \leqslant s < \infty$$

respectively, where we have replaced $(\theta_{\mathbb{S}} - \mathbb{S}_0)$ by $\theta_{\mathbb{S}}$ and $(\theta_s - s_0)$ by θ_s. The probability of failure given in Equation 6.7 will be

$$\bar{R} = P[\mathbb{S} \leqslant s] = \int_{-\infty}^{\infty} [1 - F_s(\mathbb{S})] f_{\mathbb{S}}(\mathbb{S})\, d\mathbb{S}$$

$$= \int_{\mathbb{S}_0}^{\infty} \exp\left[-\left(\frac{\mathbb{S}-s_0}{\theta_s}\right)^{\beta_s}\right] \frac{\beta_{\mathbb{S}}}{\theta_{\mathbb{S}}}\left(\frac{\mathbb{S}-\mathbb{S}_0}{\theta_{\mathbb{S}}}\right)^{\beta_{\mathbb{S}}-1} \exp\left[-\left(\frac{\mathbb{S}-\mathbb{S}_0}{\theta_{\mathbb{S}}}\right)^{\beta_{\mathbb{S}}}\right] d\mathbb{S}$$

Let

$$y = \left(\frac{\mathbb{S}-\mathbb{S}_0}{\theta_{\mathbb{S}}}\right)^{\beta_{\mathbb{S}}}$$

Then

$$dy = \frac{\beta_{\mathbb{S}}}{\theta_{\mathbb{S}}}\left(\frac{\mathbb{S}-\mathbb{S}_0}{\theta_{\mathbb{S}}}\right)^{\beta_{\mathbb{S}}-1} d\mathbb{S}$$

and

$$\mathbb{S} = y^{1/\beta_{\mathbb{S}}}\, \theta_{\mathbb{S}} + \mathbb{S}_0$$

Therefore,

$$\bar{R} = P[\mathbb{S} \leqslant s] = \int_0^{\infty} e^{-y} \exp\left\{-\left[\frac{\theta_{\mathbb{S}}}{\theta_s} y^{1/\beta_{\mathbb{S}}} + \left(\frac{\mathbb{S}_0 - s_0}{\theta_s}\right)\right]^{\beta_s}\right\} dy \qquad (6.37)$$

The values of the integral in Equation 6.37 have been computed [8,9] by numerical integration methods for different combinations of the parameters for strength and stress.

6.9 GRAPHICAL APPROACH FOR EMPIRICALLY DETERMINED STRESS AND STRENGTH DISTRIBUTIONS

Here we discuss a method of computing the reliability that is useful when no basis exists for assuming any specific distributions for either the stress or the strength but where experimentation has been performed yielding sufficient empirical data.

We define

$$G = \int_s^{\infty} f_{\mathbb{S}}(\mathbb{S})\, d\mathbb{S} = 1 - F_{\mathbb{S}}(s)$$

and

$$H = \int_0^s f_s(u)\, du = F_s(s)$$

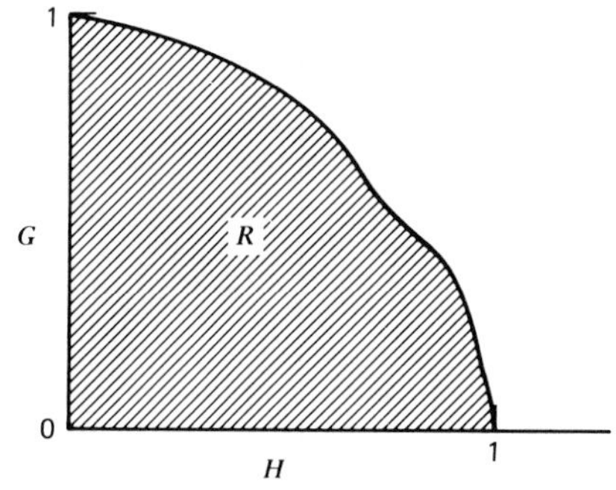

Figure 6.4 Plot of *G* vs. *H*.

Then $dH = f_s(s)\,ds$. The range of H is obviously from 0 to 1. Substituting in Equation 6.3, we get

$$R = \int_0^1 G\,dH \tag{6.38}$$

Equation 6.38 suggests that the area under a G versus H plot would represent the reliability of the component. Based on the strength and the stress data, we can easily determine, for various values of s, the values of $F_S(s)$ and $F_s(s)$, and hence those of G and H. Plotting these values of G and H on a graph paper and as shown in Figure 6.4, measuring the area graphically is all that is needed to determine the reliability. Next we illustrate the use of this graphical method by two examples.

EXAMPLE 6.9

A stress analysis of a component was performed. Ten observations were made of the stress under simulated operating conditions, from which the unknown stress distribution was estimated as shown in Table 6.1.

Table 6.1 Stress data

NUMBER	STRESS s(psi)	$\hat{F}_s(s)$
1	20,750	0.10
2	23,600	0.20
3	24,500	0.30
4	26,250	0.40
5	26,500	0.50
6	27,500	0.60
7	29,250	0.70
8	30,000	0.80
9	33,750	0.90
10	37,500	1.00

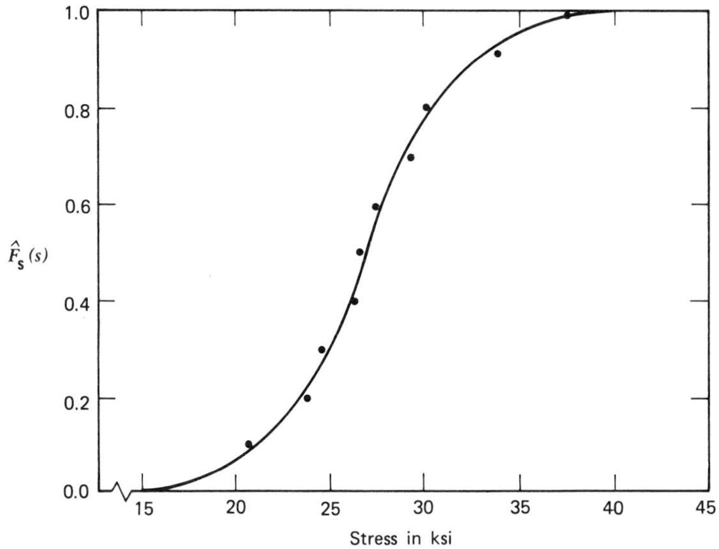

Figure 6.5 Estimated distribution function for stress (s).

A plot of $\hat{F}_s(s)$ versus s is shown in Figure 6.5. The smooth curve shown was used as an estimator of the unknown distribution function $F_s(s)$.

Similarly, a strength analysis of the component yielded 14 values for the strength from which the unknown strength distribution function was estimated as shown in Table 6.2.

Table 6.2 Strength data

NUMBER	STRENGTH $\mathbb{S}$ (psi)	$\hat{F}_{\mathbb{S}}(\mathbb{S})$
1	33,800	0.07
2	34,300	0.14
3	35,400	0.21
4	35,900	0.28
5	36,000	0.35
6	36,000	0.43
7	36,800	0.50
8	37,000	0.57
9	37,100	0.64
10	37,300	0.71
11	38,200	0.78
12	38,500	0,85
13	40,000	0.93
14	42,000	1.00

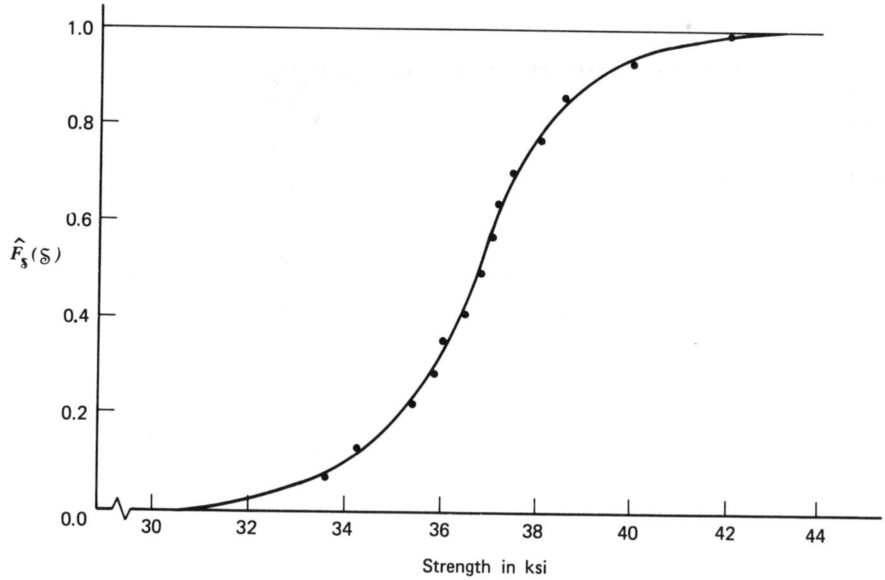

Figure 6.6 Estimated distribution function for strength ($\mathbb{S}$).

A plot of $\hat{F}_{\mathbb{S}}(\mathbb{S})$ versus $\mathbb{S}$ is shown in Figure 6.6. The smooth curve was used as an estimator of the strength distribution. Now that the two distribution functions are established, we can proceed to find the values of G and H for different values of the stress s as shown in Table 6.3.

Table 6.3 *H* and *G* values

STRESS s(psi)	$H=\hat{F}_s(s)$	$G=1-\hat{F}_{\mathbb{S}}(s)$
0	0	1.00
10,000	0	1.00
15,000	0	1.00
20,000	0.07	1.00
25,000	0.31	1.00
30,000	0.77	1.00
32,000	0.87	0.98
33,000	0.90	0.95
34,000	0.94	0.90
35,000	0.96	0.81
36,000	0.98	0.67
37,000	0.99	0.42
38,000	0.995	0.22
39,000	1.00	0.12
40,000	1.00	0.05
41,000	1.00	0.02
42,000	1.00	0.01

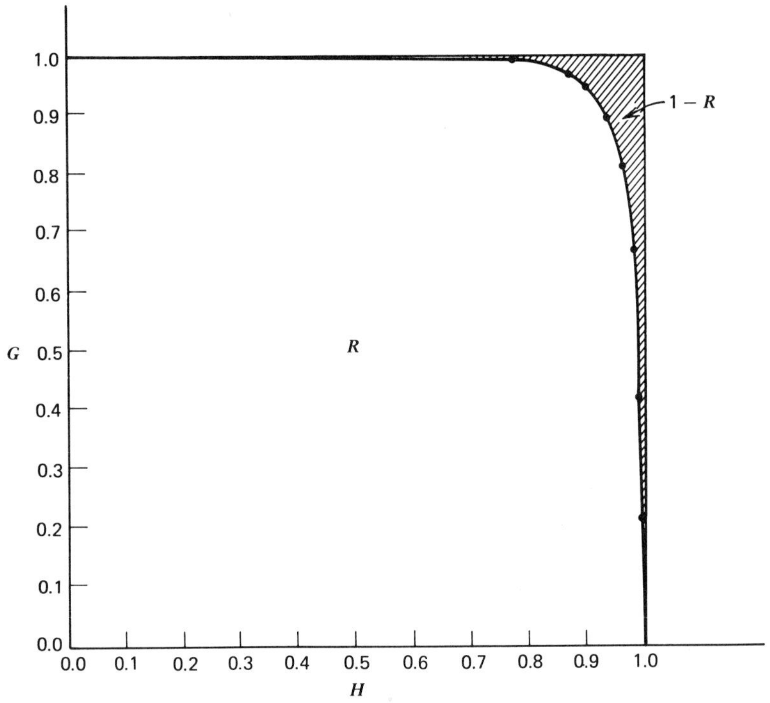

Figure 6.7 Plot of *G* vs. *H*.

A plot of G versus H, shown in Figure 6.7, yields an area under the curve measuring 0.9878, and this value is the reliability of the component.

EXAMPLE 6.10

The stress applied to a component is exponentially distributed. We have known that the stress cannot be less than 10,000 psi. Hence, the density function for the stress may be written as

$$f_s(s) = \begin{cases} 0, & 0 \leqslant s < 10{,}000 \\ \dfrac{1}{10{,}000} \exp\left[-\dfrac{(s-10{,}000)}{10{,}000} \right], & s \geqslant 10{,}000 \end{cases}$$

The strength of the component is known to follow a Weibull distribution. The material used is such that the strength is never less than 15,000 psi. The strength distribution has the parameters:

$$\mathcal{S}_0 = 15{,}000, \quad \theta = 20{,}000, \quad \text{and } \beta = 2$$

Hence the strength p.d.f. is given by

$$f_{\mathbb{S}}(\mathbb{S}) = \frac{2(\mathbb{S} - 15{,}000)}{(20{,}000 - 15{,}000)^2} \exp\left[-\left(\frac{\mathbb{S} - 15{,}000}{20{,}000 - 15{,}000}\right)^2\right]$$

$$= \frac{2(\mathbb{S} - 15{,}000)}{(5{,}000)^2} \exp\left[-\frac{(\mathbb{S} - 15{,}000)^2}{(5{,}000)^2}\right], \qquad \mathbb{S} \geqslant 15{,}000$$

The cumulative distribution functions for the stress and the strength are given by

$$F_s(s) = 1 - \exp\left[-\frac{(s - 10{,}000)}{10{,}000}\right]$$

$$F_{\mathbb{S}}(\mathbb{S}) = 1 - \exp\left[-\frac{(\mathbb{S} - 15{,}000)^2}{(5{,}000)^2}\right]$$

Hence,

$$G = \int_s^\infty f_{\mathbb{S}}(\mathbb{S})\, d\mathbb{S} = 1 - F_{\mathbb{S}}(s) = \exp\left[-\frac{(s - 15{,}000)^2}{(5{,}000)^2}\right]$$

and

$$H = \int_0^s f_s(s)\, ds = F_s(s) = 1 - \exp\left[-\frac{(s - 10{,}000)}{10{,}000}\right]$$

The values of H and G are computed for various values of s as shown in Table 6.4.

Table 6.4 ***H* and *G* values**

STRESS s	H	G
10,000	0	1.0000
12,000	0.1804	1.0000
14,000	0.3288	1.0000
15,000	0.3935	1.0000
16,000	0.4512	0.9600
18,000	0.5496	0.6978
20,000	0.6321	0.3679
22,000	0.6988	0.1408
24,000	0.7534	0.0390
26,000	0.7981	0.0080
28,000	0.8348	0.0011
30,000	0.8647	0.0001
32,000	0.8892	0.0000
34,000	0.9093	0.0000
36,000	0.9259	0.0000
38,000	0.9392	0.0000
40,000	0.9502	0.0000
42,000	0.9593	0.0000
44,000	0.9667	0.0000
46,000	0.9727	0.0000
48,000	0.9777	0.0000
50,000	0.9817	0.0000
52,000	0.9850	0.0000

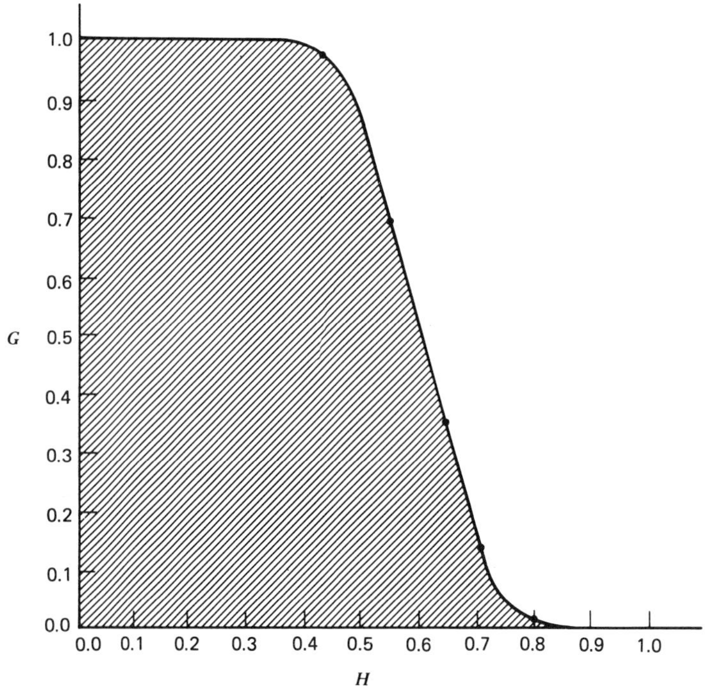

Figure 6.8 Plot of G vs. H.

A plot of G versus H is shown in Figure 6.8. The area under the curve measures 0.6093, which is also the estimated reliability of the component.

6.10 RELIABILITY COMPUTATIONS FOR EXTREME VALUE DISTRIBUTIONS

An explanation of the use of extreme value distributions for failure models was given in Chapter 2 where the failure was attributed to corrosion processes. In general, the extreme value distributions are applicable where the phenomena causing failure depend on the smallest or the largest value from a sequence of random variables. The relevance of this theory to the failure of materials because of fracture or fatigue, to the breakdown of dielectrics, and to the corrosion of metals has been explored by Gumbel [6] and by Epstein [3,4]. Gumbel has described various other applications of extreme value theory to engineering reliability problems.

There are three asymptotic distributions for either the smallest order statistics or the largest order statistics. They have the following form:

Distributions of the Smallest Value

Type I

$$F(x)=1-\exp\left[-\exp\left(\frac{x-\delta}{\theta}\right)\right], \qquad -\infty<x<+\infty,\ \theta>0 \tag{6.39}$$

Type II

$$F(x)=1-\exp\left[-\left(-\frac{x-\delta}{\theta}\right)^{-\beta}\right], \qquad -\infty<x\leqslant\delta,\ \theta>0,\ \beta>0 \tag{6.40}$$

Type III

$$F(x)=1-\exp\left[-\left(\frac{x-\delta}{\theta}\right)^{\beta}\right], \qquad \delta\leqslant x<\infty,\ \theta>0,\ \beta>0 \tag{6.41}$$

Each of these distribution types arises when certain conditions are met. Type I arises when the underlying density function tends to zero exponentially as $x\to-\infty$. On the other hand, if the range of the density function is unbounded from below and if for some $\theta>0$, $\beta>0$ we have

$$\lim_{x\to\infty}(-x)^{\theta}F_{\mathbf{x}}(x)=\beta \tag{6.42}$$

the limiting distribution of the smallest order statistic is of Type II.

Type III distribution arises when the following conditions are met:

1. The range for the underlying density function is bounded from below (i.e., $x\geqslant\delta$)
2. $F_{\mathbf{x}}(x)$ behaves like $(\theta-\delta)^{\beta}$, for some $\theta>0$, $\beta>0$ and as $x\to\delta$.

Type III smallest extreme value distribution is the well-known Weibull distribution.

Distributions of the Largest Value

Type I

$$F(x)=\exp\left[-\exp\left\{-\left(\frac{x-\delta}{\theta}\right)\right\}\right], \qquad -\infty<x<\infty,\ \theta>0 \tag{6.43}$$

Type II

$$F(x)=\exp\left[-\left(\frac{x-\delta}{\theta}\right)^{-\beta}\right], \qquad x\geqslant\delta,\ \theta>0,\ \beta>0 \tag{6.44}$$

Type III

$$F(x)=\exp\left[-\left(-\frac{x-\delta}{\theta}\right)^{\beta}\right], \qquad x\leqslant\delta,\ \theta>0,\ \beta>0 \tag{6.45}$$

Type I distribution arises when the underlying density function $f_x(x)$ approaches zero exponentially as $x \to \infty$. Type II arises if for some $\theta > 0$ and $\beta > 0$, we have

$$\lim_{x\to\infty} x^{\theta}\left[1 - F_x(x)\right] = \beta$$

If the range for the underlying density function is bounded from above (i.e., $x \leqslant \delta$) and if for some finite δ, $1 - F_x(x)$ behaves like $\theta(\delta - x)$ the limiting distribution of the largest order statistic is of Type III.

6.10.1 STRENGTH AND STRESS DISTRIBUTIONS

The failure of a component caused by fracture or fatigue can be described either by the weakest link or by the largest flaw concept. The weakest link concept assumes that the strength of the component is determined by the strength of its weakest link. The largest flaw concept assumes that the strength of the component is determined by the largest flaw present. Flaws are introduced by manufacturing processes, tolerances, material heterogeneity, etc. If the underlying density function of the "links" is normal, the strength distribution function will be Type I smallest value. If we assume that the underlying density function is Weibull, the strength distribution function will be Type III extreme value. Type I largest extreme value distribution, is associated with the largest flaw concept. It was observed by Lipson [8] that the Type I largest extreme value distribution, is a good fit for some fatigue data.

The theory of extreme values has been applied to analyze gust velocities, gust loads, and landing loads for aircrafts. In general, stress distributions are described by the largest extreme value distributions. If the underlying distribution is exponential or normal, the Type I largest extreme value distribution, is applicable. Type II largest extreme value distribution, has been used for the analysis of maximum wind speeds.

6.10.2 RELIABILITY COMPUTATIONS

First consider the case where the strength has the Weibull distribution and the stress has the Type II largest extreme value distribution. We have

$$F_{\mathbb{S}}(\mathbb{S}) = 1 - \exp\left[-\left(\frac{\mathbb{S} - \mathbb{S}_0}{\theta_{\mathbb{S}}}\right)^{\beta_{\mathbb{S}}}\right], \qquad \mathbb{S}_0 \leqslant \mathbb{S} < \infty,\ \theta_{\mathbb{S}} > 0,\ \beta_{\mathbb{S}} > 0 \tag{6.46}$$

and

$$F_s(s) = \exp\left[-\left(\frac{s - s_0}{\theta_s}\right)^{-\beta_s}\right], \qquad s_0 \leqslant s < \infty,\ \theta_s > 0,\ \beta_s > 0 \tag{6.47}$$

Substituting the appropriate values in Equation 6.3, we have

$$R=\int_{\mathbb{S}_0}^{\infty}\exp\left[-\left(\frac{\mathbb{S}-s_0}{\theta_s}\right)^{-\beta_s}\right]\cdot\frac{\beta_{\mathbb{S}}}{\theta_{\mathbb{S}}}\left(\frac{\mathbb{S}-\mathbb{S}_0}{\theta_{\mathbb{S}}}\right)^{\beta_{\mathbb{S}}-1}\exp\left[-\left(\frac{\mathbb{S}-\mathbb{S}_0}{\theta_{\mathbb{S}}}\right)^{\beta_{\mathbb{S}}}\right]d\mathbb{S}$$

Let

$$y=\left(\frac{\mathbb{S}-\mathbb{S}_0}{\theta_{\mathbb{S}}}\right)^{\beta_{\mathbb{S}}}$$

then

$$dy=\frac{\beta_{\mathbb{S}}}{\theta_{\mathbb{S}}}\left(\frac{\mathbb{S}-\mathbb{S}_0}{\theta_{\mathbb{S}}}\right)^{\beta_{\mathbb{S}}-1}d\mathbb{S}$$

and

$$y^{1/\beta_{\mathbb{S}}}.\theta_{\mathbb{S}}+\mathbb{S}_0=\mathbb{S}$$

Hence

$$R=\int_0^{\infty}e^{-y}\exp\left[-\left\{\frac{\theta_{\mathbb{S}}}{\theta_s}y^{1/\beta_{\mathbb{S}}}+\frac{\mathbb{S}_0-s_0}{\theta_s}\right\}^{-\beta_s}\right]dy \tag{6.48}$$

The above integral is evaluated numerically and the values for different values of the parameters are given in Appendix IV.

Similarly, we can compute the reliability in other cases. For example, consider the following:

1. Weibull distributed strength and Type III smallest extreme value distributed stress

$$R=\int_0^{\infty}e^{-y}\exp\left[-\left(\frac{\delta_s-\theta_s\ln y-\mathbb{S}_0}{\theta_{\mathbb{S}}}\right)^{\beta_{\mathbb{S}}}\right]dy \tag{6.49}$$

where

$$y=\exp\left[-\left(\frac{s-\delta_s}{\theta_s}\right)\right]$$

2. Type I smallest extreme value distributed strength and Type I largest extreme value distributed stress

$$R=\int_0^{\infty}e^{-y}\exp\left[-\exp\left\{\frac{\theta_s}{\theta_{\mathbb{S}}}\ln y+\left(\frac{\delta_s-\delta_{\mathbb{S}}}{\theta_{\mathbb{S}}}\right)\right\}\right]dy \tag{6.50}$$

where

$$y=\exp\left[-\left(\frac{s-\delta_s}{\theta_s}\right)\right]$$

3. Type I smallest extreme value distributed strength and Type II largest extreme value distributed stress

$$R=\int_0^{\infty} e^{-y}\exp\left[-\left\{\frac{\theta_{\mathbb{S}}}{\theta_s}\left(\ln y+\frac{\delta_s-s_0}{\theta_{\mathbb{S}}}\right)\right\}^{-\beta_s}\right]dy \tag{6.51}$$

where

$$y=\exp\left(\frac{\mathbb{S}-\delta_{\mathbb{S}}}{\theta_{\mathbb{S}}}\right)$$

4. Type I smallest extreme value strength and Weibull distributed stress

$$R=\int_0^{\infty} e^{-y}\exp\left[-\exp\frac{\theta_s}{\theta_{\mathbb{S}}}\left(y^{1/\beta_s}-\frac{\delta_{\mathbb{S}}-s_0}{\theta_s}\right)\right]dy \tag{6.52}$$

where

$$y=\left(\frac{s-s_0}{\theta_s}\right)^{\beta_s}$$

5. Both stress and strength Type I largest extreme value distributed

$$R=\int_0^{\infty} e^{-y}\exp\left[-\exp\left\{\frac{\theta_{\mathbb{S}}}{\theta_s}\left(\ln y-\frac{\delta_{\mathbb{S}}-\delta_s}{\theta_{\mathbb{S}}}\right)\right\}\right]dy \tag{6.53}$$

where

$$y=\exp\left[-\left(\frac{\mathbb{S}-\delta_{\mathbb{S}}}{\theta_{\mathbb{S}}}\right)\right]$$

6. Strength has Type I largest extreme value distribution and stress has Type II largest extreme value distribution

$$R=\int_0^{\infty} e^{-y}\exp\left[\left(\frac{\theta_{\mathbb{S}}}{\theta_s}\right)^{-\beta_s}\left\{\ln y-\frac{\delta_{\mathbb{S}}-s_0}{\theta_{\mathbb{S}}}\right\}^{-\beta_s}\right]dy \tag{6.54}$$

where

$$y=\exp\left[-\left(\frac{\mathbb{S}-\delta_{\mathbb{S}}}{\theta_{\mathbb{S}}}\right)\right]$$

7. Strength has Type I largest extreme value distribution and stress has Weibull distribution

$$R = 1 - \int_0^{\infty} \exp\left[-y + \left\{ \frac{\theta_{\mathcal{S}}}{\theta_s} \ln y - \left(\frac{\delta_{\mathcal{S}} - s_0}{\theta_s} \right) \right\}^{\beta_s} \right] dy \tag{6.55}$$

where

$$y = \exp\left[-\left(\frac{\mathcal{S} - \delta_{\mathcal{S}}}{\theta_{\mathcal{S}}} \right) \right]$$

8. Weibull distributed strength and Type I largest extreme value distributed stress

$$R = \int_0^{\infty} \exp\left[-y - \exp\left[-\left(\frac{\theta_{\mathcal{S}}}{\theta_s} y^{1/\beta_{\mathcal{S}}} + \frac{\mathcal{S}_0 - \delta_s}{\theta_s} \right) \right] \right] dy$$

where

$$y = \left(\frac{\mathcal{S} - \mathcal{S}_0}{\theta_{\mathcal{S}}} \right)^{\beta_{\mathcal{S}}}$$

The above integral is evaluated numerically and the values of R for different values of the parameters are given in Appendix IV. Details of the numerical integration procedure can be found in Reference 12.

6.11 SUMMARY

Reliability expressions for different stress and strength probability density functions derived in this chapter are given below:

1. If strength and stress are both exponentially distributed, then

$$R = \frac{\lambda_s}{\lambda_{\mathcal{S}} + \lambda_s} = \frac{\bar{\mathcal{S}}}{\bar{s} + \bar{\mathcal{S}}}, \qquad \lambda_s = \frac{1}{\bar{s}}, \qquad \text{and} \qquad \lambda_{\mathcal{S}} = \frac{1}{\bar{\mathcal{S}}}$$

where $\bar{\mathcal{S}}$ and $\bar{s}$ are the mean strength and stress respectively.

2. If strength is normally distributed with mean $\mu_{\mathcal{S}}$ and standard deviation $\sigma_{\mathcal{S}}$ and stress is exponentially distributed with parameter λ_s, ($\lambda_s = 1/\bar{s}$), then

$$R = 1 - \exp\left[-\mu_{\mathcal{S}}\lambda_s + \left(\frac{1}{2} \right) \lambda_s^2 \sigma_{\mathcal{S}}^2 \right]$$

3. If strength and stress are both normally distributed then reliability is given by

$$R = \int_z^{\infty} \phi(u)\,du = 1 - \Phi(z)$$

where $\phi(z)$ is the standardized normal density function, and

$$z = -\frac{\mu_{\mathbb{S}} - \mu_{s}}{\sqrt{\sigma_{\mathbb{S}}^{2} + \sigma_{s}^{2}}}$$

where $(\mu_{\mathbb{S}}, \sigma_{\mathbb{S}})$ are the parameters of the strength, and (μ_{s}, σ_{s}) are the parameters of the stress. However, when the strength and stress are correlated random variables with correlation coefficient ρ, the lower limit z is given by

$$z = -\frac{\mu_{\mathbb{S}} - \mu_{s}}{\sqrt{\sigma_{\mathbb{S}}^{2} + \sigma_{s}^{2} - 2\rho\sigma_{\mathbb{S}}\sigma_{s}}}$$

4. If strength and stress are both log normally distributed, then

$$R = \int_{z}^{\infty} \phi(u)\,du$$

where

$$z = -\frac{\ln \check{\mathbb{S}} - \ln \check{s}}{\sqrt{\sigma_{\ln \mathbb{S}}^{2} + \sigma_{\ln s}^{2}}}$$

where $\check{\mathbb{S}}$ and $\check{s}$ are the medians of the strength and stress distributions, respectively, and $\sigma_{\ln \mathbb{S}}$ and $\sigma_{\ln s}$ are the standard deviations for $\ln \mathbb{S}$ and $\ln s$ respectively.

5. If strength and stress have gamma distributions with parameters (m, λ) and (n, μ) respectively, then

$$R = \frac{\Gamma(m+n)}{\Gamma(m)\Gamma(n)} \int_{0}^{r/(1+r)} (1-u)^{m-1} u^{n-1}\,du$$

where $r = \mu/\lambda$. The integral is the incomplete beta function truncated at $r/(1+r)$.

6. If strength has a Weibull distribution with parameters $(\beta, \theta, \mathbb{S}_0)$ and stress is normally distributed with parameters (μ, σ), then

$$R = 1 - \int_{z_0}^{\infty} \phi(z)\,dz - \frac{1}{\sqrt{2\pi}}\left[\frac{\theta - \mathbb{S}_0}{\sigma}\right]\int_{0}^{\infty} \exp\left\{-y^{\beta} - \frac{1}{2}\left[\left(\frac{\theta - \mathbb{S}_0}{\sigma}\right)y + \frac{\mathbb{S}_0 - \mu}{\sigma}\right]^{2}\right\} dy$$

where

$$z_0 = \frac{\mathbb{S}_0 - \mu}{\sigma}, \quad \text{and} \quad y = \frac{s - \mathbb{S}_0}{\theta - \mathbb{S}_0}$$

7. If strength has Weibull distribution with parameters $(\beta_{\mathbb{S}}, \theta_{\mathbb{S}}, \mathbb{S}_0)$, and stress has a Weibull distribution with parameters (β_s, θ_s, s_0), then

$$R = 1 - \int_{0}^{\infty} e^{-y} \exp\left[-\left[\frac{\theta'_{\mathbb{S}}}{\theta'_{s}}(y)^{1/\beta_{\mathbb{S}}} \quad \frac{\mathbb{S}_0 - s_0}{\theta'_{s}}\right]^{\beta_s}\right] dy$$

where

$$\theta'_{\mathbb{S}} = \theta_{\mathbb{S}} - \mathbb{S}_0, \quad \theta'_s = \theta_s - s_0, \quad \text{and} \quad y = \left[\frac{\mathbb{S} - s_0}{\theta'_{\mathbb{S}}} \right]^{\beta_{\mathbb{S}}}$$

The summary of the results for the extreme value distributions is given in Section 6.10.2. Based on the results given in this chapter, we can compute reliability of any component given the stress and the strength density functions. Expressions for reliability for most of the well-known density functions were developed in this chapter. A graphical approach was given to compute reliability when only limited amount of test data on the stress and strength of a part is available. Many components are subjected to repeated stresses. Reliability expressions when the stress and the strength distributions are functions of time and/or cycles are developed in Chapter 8.

EXERCISES

1 Prove the relationship given by Equation 6.9. Use the method that employs Jacobian transformations.

2 Derive Equation 6.18 by the method of convolution of the stress and the strength normal random variables.

3 Suppose the stress and the strength normal random variables are correlated with a coefficient of correlation r. Derive the reliability expression for this case.

4 Using the relationship developed in Exercise 3 and a coefficient of correlation of 0.5, calculate the reliability for the data given in Example 6.1.

5 The strength $\mathbb{S}$ and the stress $\mathbf{s}$ for the design of a component are log normally distributed with the following information on $\mathbb{S}$ and $\mathbf{s}$:

$$E(\mathbb{S}) = 750.00 \text{ MPa}$$

$$\sigma_{\mathbb{S}} = 50.00 \text{ MPa}$$

$$E(\mathbf{s}) = 500.00 \text{ MPa}$$

$$\sigma_{\mathbf{s}} = 80.00 \text{ MPa}$$

Compute the reliability of the component.

6 A component is to be designed for a specified reliability of 0.990. The stress and the strength random variables are known to be log normally distributed for this component with the following information:

$$E(\mathbb{S})=1100.00 \text{ MPa}$$

$$\sigma_{\mathbb{S}}=100.00 \text{ MPa}$$

$$E(\mathbf{s})=850.00 \text{ MPa}$$

Determine the maximum allowable standard deviation of the stress that can be applied to the component which will give us the desired reliability.

7 Develop Equation 6.31 when the strength of the component has a minimum value $\mathbb{S}_0$. Show what modifications are needed in Equation 6.31.

8 Using the equation developed in Exercise 7 find the reliability of the component with the following data:

$$\mathbb{S}_0=30{,}000 \qquad \mu_s=35{,}000$$

$$\lambda_{\mathbb{S}}=0.001 \qquad \sigma_s=5{,}000$$

9 The strength of a component has a gamma distribution with parameters $\lambda=1$ and $m=4$. The failure inducing stress also is gamma distributed with $\mu=1$ and $n=2$. Compute the reliability of the component.

10 In Exercise 9, assume that $\lambda=4$ and $\mu=2.5$. Compute the reliability of the component for this case.

11 A leaf spring for a truck is to be designed for a reliability of 0.9995 based on the fatigue failure of the leaf spring. The fatigue strength of the material out of which this spring is made is Weibull distributed with the following parameters:

$$\mathbb{S}_0=500.00 \text{ MPa}$$

$$\beta=3.0$$

$$\theta=800.00 \text{ MPa}$$

The random loading of the spring induces stresses that are assumed to be normally distributed with a coefficient of variation of 0.08. Compute the permissible normal stress parameters that would yield the specified reliability.

12 Stress analysis on a component was performed under simulated operating conditions and the following eight observations were obtained:

Stress (kPa) 15.0, 20.5, 22.5, 23.5, 25.0, 26.5, 28.5, 30.5

The strength of the component must be estimated based on the following 10 observations of the strength data:

Strength (kPa) 18.0, 24.5, 28.0, 30.5, 31.5, 33.0, 33.5, 35.5, 40.4, 48.5

Compute the reliability of the component based on these data.

13 The strength of a component is log normally distributed with a mean of 800.00 MPa and standard deviation of 150.00 MPa. The failure governing stresses have normal distribution with a mean of 600.00 MPa and a standard deviation of 110.00 MPa. Compute the reliability of the component using the graphical approach discussed in Section 6.9.

14 The strength and the failure governing stress of a component have a Weibull distribution with the following parameters:

$$\mathbb{S}_0 = 40{,}000 \qquad \beta = 4.0 \qquad \theta = 80{,}000$$

$$s_0 = 20{,}000 \qquad \beta = 3.2 \qquad \theta = 70{,}000$$

Compute the reliability of the component by the graphical method discussed in Section 6.9.

15 The stress acting on a component is uniformly distributed over an interval $[s_{\min}, s_{\max}]$. The strength of the component follows normal distribution $(\mu_{\mathbb{S}}, \sigma_{\mathbb{S}})$. Derive an expression for the reliability of the component. Let

$$s_{\min} = 10, \qquad s_{\max} = 40$$

$$\mu_{\mathbb{S}} = 35, \qquad \mu_{\mathbb{S}} = 5$$

Find R.

16 The stress acting on a component is uniformly distributed over an interval $[s_{\min}, s_{\max}]$. The strength of the component has a three-parameter Weibull distribution with parameters $\mathbb{S}_0$, β, and θ. Derive an expression for the reliability of the component. Let

$$s_{\min} = 10, \qquad s_{\max} = 30$$

$$\mathbb{S}_0 = 20, \qquad \beta = 3.0, \qquad \theta = 30$$

Find R.

17 The stress acting on a component is uniformly distributed over an interval $[s_{\min}, s_{\max}]$. The strength of the component has gamma distribution with parameters η and λ. Derive an expression for the reliability of the component. Let

$$s_{\min} = 10, \qquad s_{\max} = 30$$

$$\eta = 5, \qquad \lambda = 0.2$$

Find R.

18 a. The strength $\mathbb{S}$ and the stress **s** are random variables. We are given histograms for $\mathbb{S}$ and **s**:

$$\text{Stress histogram } (w_i, \Delta s_i), \qquad i=1,\ldots,n$$
$$\text{Strength histogram } (v_j, \Delta \mathbb{S}_j), \qquad j=1,\ldots,m$$

where w_i and v_j are the relative frequencies for the interval Δs_i for stress and the interval $\Delta \mathbb{S}_j$ for strength respectively. Derive an expression for reliability using the histogram data. Let

$$\mathbf{s} \sim \text{Weibull } (s_0 = 10, \beta = 2, \theta = 20)$$

$$\mathbb{S} \sim \text{Weibull } (\mathbb{S}_0 = 15, \beta = 3, \theta = 35)$$

$$n = 20, m = 20$$

Find R.

b. Recompute (a) with the following data:

$$\mathbf{s} \sim \text{gamma } (\eta = 5, \lambda = 0.2)$$

$$\mathbb{S} \sim \text{gamma } (\eta = 7, \lambda = 0.16)$$

In both parts (a) and (b) complete the sensitivity analysis with respect to the errors in the relative frequencies. Make appropriate assumptions.

BIBLIOGRAPHY

1 Colombo, A. G., Reina, G. and Volta, G., "Extreme Value Characteristics of Distributions of Cumulative Processes," *IEEE Transactions on Reliability,* Vol R-23, No. 3, August 1974, pp. 179–193.

2 Disney, R. L., Sheth, N. J. and Lipson, C., "The Determination of the Probability of Failure by Stress/Strength Interference Theory," *Proceedings of Annual Symposium on Reliability,* 1968, pp. 417–422.

3 Epstein, B., "Elements of the Theory of Extreme Values," *Technometrics,* Vol. 2., No. 1, February 1960.

4 Epstein, B. and Brooks, H., "The Theory of Extreme Values and Its Applications in the Study of the Dielectric Strength of Paper Capacitor," *Journal of Applied Physics,* Vol. 19, June 1948, pp. 544–550.

5 Gumbel, E. J., "Statistical Theory of Extreme Values and Some Practical Application," Bureau of Standards, *Applied Mathematics,* Series 33, Washington, D. C., GPO, 1954.

6 Gumbel, E. J., *Statistics of Extremes,* Columbia University Press, New York, 1958.

7 Lipson, C. and Sheth, N. J., *Statistical Design and Analysis of Engineering Experiments,* McGraw-Hill, New York, 1973.

8 Lipson, C., Sheth, N. J. and Disney, R. L., "Reliability Prediction—Mechanical Stress/Strength Interference," Rome Air Development Center, Technical Report RADC-TR-66-710, March 1967.

9 Lipson, C., Sheth, N. J., Disney, R. L. and Altum, M., "Reliability Prediction—Mechanical Stress/Strength Interference," Final Technical Report RADC-TR68-403, Rome Air Development Center, Research and Technology Division, Griffiss Air Force Base, New York, 1969.

10 Pearson, K. (Ed.), *Tables of Incomplete Beta Function,* Biometrika Office, Cambridge University Press, Cambridge, England 1948.

11 Roberts, Norman H., *Mathematical Methods in Reliability Engineering,* McGraw-Hill, New York, 1964.

12 Taraman, S. I., "Design Reliability Models and Determination by Stress–Strength Interference Theory," Unpublished Doctoral Dissertation, Dept. of Industrial Engineering, Wayne State University, Detroit, Michigan, 1975.

Chapter 7 reliability design examples

In this chapter we illustrate the probabilistic approach by discussing the actual design of various components. This approach allows us to get an idea of the reliability of the design. It also helps us perform a sensitivity analysis of the design with respect to the various design parameters.

Functions of the random variables are evaluated approximately by the Taylor's series method discussed in Chapter 5. To keep the illustrations simple, we assume that the stress and the strength random variables have normal density functions.

The problem of designing a tension element is given in Section 7.1. The tensile force, the diameter of the circular cross-section, and the ultimate tensile strength are assumed to be random variables. The tension element is to be designed for a specified reliability. Reliability sensitivity to geometric tolerances and variability in material strength is also analyzed. This gives the designer a quantitative assessment of the impact of variability of the tolerances and material strength. Section 7.2 contains a discussion of the probabilistic design for an I-beam that is simply supported at its ends. We consider a $W \times 8 \times 67$ I-section and compute the depth of the beam for the specified reliability. A shaft with a circular cross-section and subjected to torsion is designed in Section 7.3. The design of a hood torsion bar for a truck is considered in Section 7.4. The probabilistic analysis employs the methods given in Chapter 5, specifically the Taylor's series approximation method. The fatigue failure mode is considered, and reliability for this mode of failure at different cycles is computed.

7.1 DESIGN OF A TENSION ELEMENT

A tension element is to be designed (see Figure 7.1). The load P acting on the element is a random variable. The tension element has a circular cross-section. Its diameter d is a random variable because of the manufacturing tolerance. The ultimate tensile strength of the material used for the element is a random

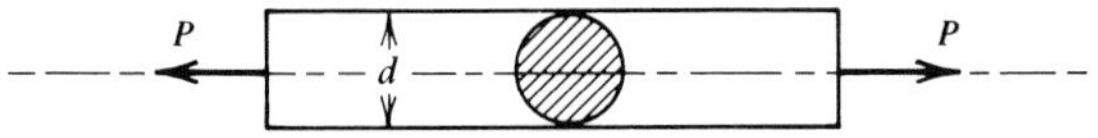

Figure 7.1 Tension element.

variable because the properties of the material vary. The following information is available:

$$\text{Load: } \bar{P} = 4 \times 10^3 \text{ lb}, \sigma_P = 100 \text{ lb}$$

$$\text{Ultimate tensile strength: } \bar{\mathbb{S}} = 100 \times 10^3 \text{ psi}, \sigma_{\mathbb{S}} = 5 \times 10^3 \text{ psi}$$

The specified reliability $R = 0.99990$. Failure is known to occur by tensile fracture.

The tensile stress $\mathbf{s}$ is given by $\mathbf{s} = P/A$ where $A = \pi r^2$. Using the Taylor's series approximation it can be easily established that $\bar{A} = \pi \bar{r}^2$ and $\sigma_A = 2\pi \bar{r} \sigma_r$. If the tolerance on the radius of the circular cross-section is a fraction α of $\bar{r}$, then $3\sigma_r = \alpha \bar{r}$; that is, $\sigma_r = (\alpha/3)\bar{r}$. Now we have

$$\bar{\mathbf{s}} = \frac{\bar{P}}{\bar{A}} = \frac{\bar{P}}{\pi \bar{r}^2} \tag{7.1}$$

and

$$\sigma_{\mathbf{s}}^2 = \sigma_P^2 \left(\frac{1}{\bar{A}} \right)^2 + \sigma_A^2 \left(\frac{\bar{P}}{\bar{A}^2} \right)^2 \tag{7.2}$$

Substituting the values of $\bar{A}$ and σ_A^2 from above in Equation 7.2 yields

$$\sigma_{\mathbf{s}}^2 = \frac{\pi^2 \bar{r}^4 \sigma_P^2 + 4\pi^2 \bar{r}^4 \left(\frac{\alpha}{3} \right)^2 \bar{P}^2}{\pi^4 \bar{r}^8}$$

$$= \frac{\sigma_P^2 + (4/9)\alpha^2 \bar{P}^2}{\pi^2 \bar{r}^4}$$

Substituting the above expressions for $\bar{\mathbf{s}}$ and $\sigma_{\mathbf{s}}^2$ in the coupling equation, under the assumption that the stress and the strength are normal random variables, we get

$$z = -\frac{\bar{\mathbb{S}} - \dfrac{\bar{P}}{\pi \bar{r}^2}}{\sqrt{\sigma_{\mathbb{S}}^2 + \left\{ \dfrac{\sigma_P^2 + (4/9)\alpha^2 \bar{P}^2}{\pi^2 \bar{r}^4} \right\}}} \tag{7.3}$$

But the specified reliability $R=0.99990$ implies $z=-3.72$. If we let the fraction $\alpha=0.015$, then Equation 7.3 becomes

$$-3.72=-\frac{100\times10^3-\dfrac{4\times10^3}{\pi\bar{r}^2}}{\sqrt{(5\times10^3)^2+\dfrac{(100)^2+(4/9)(0.015)^2(4\times10^3)^2}{\pi^2\bar{r}^4}}} \tag{7.4}$$

which can be simplified to

$$95{,}398.38\bar{r}^4-2512\bar{r}^2+15.84804=0 \tag{7.5}$$

The two positive roots are: $\bar{r}_1=0.10235$ in. and $\bar{r}_2=0.12635$ in. The latter gives us the specified reliability of 0.99990. It is interesting to note that the other root $\bar{r}_1=0.10235$ in. would result in a reliability of 0.0001, which is the unreliability of the element.

7.1.1 SENSITIVITY OF RELIABILITY TO GEOMETRIC VARIABILITY

Let $\bar{r}=0.12635$ in Equation 7.3. Table 7.1 shows how the reliability varies according to α, the tolerance expressed as a fraction of the mean radius. The reliability does decrease with an increase in the tolerance as one would logically expect.

7.1.2 SENSITIVITY OF RELIABILITY TO VARIABILITY IN MATERIAL STRENGTH

For $\bar{r}=0.12635$ in Equation 7.3 we compute the reliability for different values of the standard deviation of the strength (σ_S) of the element. As can be verified from Table 7.2, the reliability does decrease with increased variability in the material strength.

Table 7.1 *R* **versus tolerance on** $\bar{r}$

TOLERANCE ON $\bar{r}$ (%)	$-z$	RELIABILITY
0	3.76	0.999914945
0.5	3.75	0.999911481
1.0	3.74	0.999907888
1.5	3.72	0.999900286
3.0	3.61	0.999846797
5.0	3.36	0.999610197
7.0	3.10	0.999032344

Table 7.2 R **versus** σ_S

STANDARD DEVIATION FOR THE STRENGTH (PSI) (σ_S)	RELIABILITY R
2×10^3	1.00000
3×10^3	0.99999
4×10^3	0.99996
5×10^3	0.99990
6×10^3	0.99906
7×10^3	0.99664
8×10^3	0.99157
9×10^3	0.98382
10×10^3	0.97381

7.2 DESIGN OF AN I-BEAM

Consider a beam that is simply supported at A and B as shown in Figure 7.2. The beam is free to rotate at A and B along the longitudinal axis, with roller supports at B. We will neglect the weight of the beam. The load P, the length l of the beam, and the location of the load at a distance a from the end point A are considered to be random variables. The mean values of the random variables are denoted by putting bars at the top of the variables and the standard deviations by σ with the appropriate subscripts.

The maximum moment M is at the point of the application of the load and is given by

$$M=\frac{Pa(l-a)}{l} \tag{7.6}$$

The maximum stress occurs at the top of the upper flange or at the bottom of the lower flange of the beam and is given by

$$s=\frac{Mc}{I} \tag{7.7}$$

where s = fiber stress in psi
M = external bending moment in lb-in.
c = distance from neutral axis to extreme fibers in in.
I = moment of inertia of the beam cross-section about the neutral axis in in.^{4}

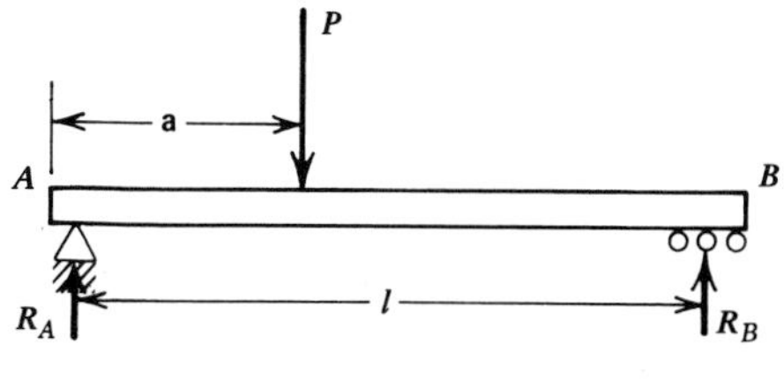

Figure 7.2 Simply supported beam.

The beam is to be designed for a reliability of 0.9990. It is to be made of a steel I-section, the significant strength of which is given by $(\bar{\mathbb{S}}, \sigma_{\mathbb{S}}) = (170{,}000,\ 4{,}760)$ psi. The design parameters and their variabilities are:

$$(\bar{P}, \sigma_P) = (6{,}070,\ 200) \text{ lb}$$

$$l = 120 \pm 1/8 \text{ in., i.e., } \bar{l} = 120 \text{ in., } \sigma_l = 1/24 \text{ in.}$$

$$a = 72 \pm 1/8 \text{ in., i.e., } \bar{a} = 72 \text{ in., } \sigma_a = 1/24 \text{ in.}$$

Using Equation 7.6 and the Taylor's series approximation given in Chapter 5, we obtain

$$\bar{M} = 174{,}816 \text{ lb-in.}$$

and

$$\sigma_M = 5{,}761 \text{ lb-in.}$$

For a $W \times 8 \times 67$ I-section (see Figure 7.3), we have

$$\frac{b_f}{t_f} = 8.88, \qquad \frac{d}{t_w} = 15.7, \qquad \frac{b_f}{d} = 0.92 \tag{7.8}$$

Hence

$$\frac{I}{c} = \frac{b_f d^3 - (b_f - t_w)(d - 2t_f)^3}{6d} = 0.0822 d^3 \tag{7.9}$$

We let $\sigma_d = 0.01\,\bar{d}$, then $(\overline{I/c}) = 0.0822\,\bar{d}^3$ and $\sigma_{(I/c)} = 0.002466\,\bar{d}^3$. From Equation

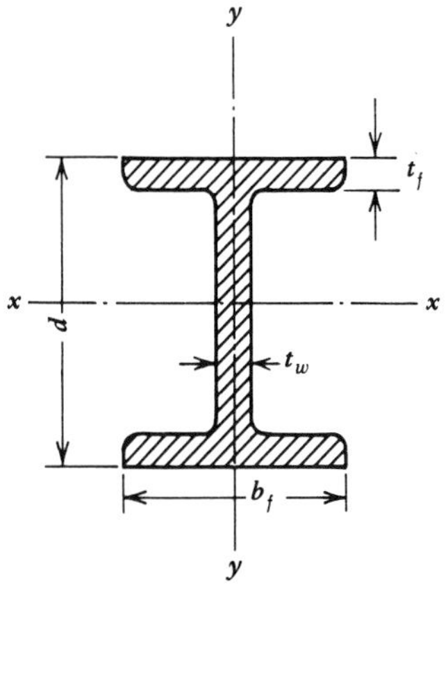

Figure 7.3 Cross-section of an I-beam.

7.7, we have

$$\bar{s} = \frac{\overline{M}}{(\overline{I/c})} = \frac{2{,}126{,}715}{\bar{d}^3}\,\text{psi}$$

and

$$\sigma_s = \left[\left(\frac{1}{(\overline{I/c})}\right)^2 \sigma_M^2 + \left[\frac{-\overline{M}}{(\overline{I/c})^2}\right]^2 \sigma_{(I/c)}^2\right]^{1/2} \tag{7.10}$$

$$= \frac{94{,}776}{\bar{d}^3}$$

For the specified reliability of 0.999, z is equal to -3.09 and, hence, from the coupling equation, we have

$$-3.09 = \frac{-170{,}000 + 2{,}126{,}715/\bar{d}^3}{\left[\left(\dfrac{94{,}776}{\bar{d}^3}\right)^2 + (4{,}760)^2\right]^{1/2}}$$

This simplifies to

$$\bar{d}^6 - 25.2088\bar{d}^3 + 154.6929 = 0$$

Solving the above equation, we find that $\bar{d} = 2.447$ in. gives us the specified reliability of 0.999.

For the above value of $\bar{d} = 2.447$ in., we can study the sensitivity of reliability with respect to variability in material strength by using the following relation:

$$-z = \frac{-170{,}000 + \dfrac{2{,}126{,}715}{14.648}}{\left[(\sigma_{\mathbf{S}})^2 + \left(\dfrac{94{,}776}{14.648}\right)^2\right]^{1/2}}$$

For values of R for different values of $\sigma_{\mathbf{S}}$, see Table 7.3.

Table 7.3 ***R* versus $\sigma_{\mathbf{S}}$**

$\sigma_{\mathbf{S}}$	$-z$	R
5,000	3.035	0.998797
7,000	2.604	0.995393
9,000	2.239	0.987418
11,000	1.945	0.974110
13,000	1.709	0.956276
15,000	1.519	0.935614

7.3 DESIGN OF A SHAFT SUBJECTED TO TORSION

Consider a solid shaft fixed at one end and subjected to a twisting couple at the other end. Let

τ = shear stress, psi
G = modulus of elasticity for shear, psi
d = diameter of the shaft, in.
θ = angle of twist per unit length, rad.

We know that

$$\tau = \left(\tfrac{1}{2}\right) G\theta d \tag{7.11}$$

$$T = G\theta I_p \tag{7.12}$$

where I_p is the polar moment of inertia of the shaft. Combining Equations 7.11 and 7.12, we have

$$\tau = \frac{Td}{2I_p} \tag{7.13}$$

But $I_p = \pi d^4/32$ for a solid circular shaft. Hence,

$$\tau = \frac{16T}{\pi d^3} = \frac{2T}{\pi r^3} \tag{7.14}$$

We wish to design the shaft to have a specified reliability of 0.999. The following information on the design random variables is available:

Applied torque: $\bar{T} = 100{,}000$ lb-in., $\sigma_T = 10{,}000$ lb-in.

Allowable shear stress: $\bar{\mathfrak{S}} = 50{,}000$ psi, $\sigma_{\mathfrak{S}} = 5{,}000$ psi

The variability of the radius of the shaft is given by

$$\sigma_r = (\alpha/3)\bar{r} \tag{7.15}$$

where α is the tolerance fraction.

From Equation 7.14, we have

$$\bar{\tau} = \frac{2\bar{T}}{\pi \bar{r}^3} = \frac{2 \times 100{,}000}{\pi \bar{r}^3}\,\text{psi} \tag{7.16}$$

and

$$\sigma_{\tau} = \sqrt{\frac{4\sigma_T^2}{\pi^2\bar{r}^6} + \frac{36\bar{T}^2\sigma_r^2}{\pi^2\bar{r}^8}}$$

$$= \frac{2\times 10^4}{\pi\bar{r}^3}\sqrt{1+(10\alpha)^2} \tag{7.17}$$

Substituting in the coupling equation, we have

$$-3.09 = -\frac{5\times 10^4 - (2\times 10^5/\pi\bar{r}^3)}{\sqrt{(5\times 10^3)^2 + (2\times 10^4/\pi\bar{r}^3)^2(1+100\alpha^2)}} \tag{7.18}$$

Assuming $\alpha = 0.03$, we can simplify Equation 7.18 to

$$236.9\bar{r}^6 - 667.4\bar{r}^3 + 380.2 = 0$$

which yields $\bar{r} = 1.265$ in. or 0.925 in. It can be verified that $\bar{r} = 1.265$ yields the desired reliability of 0.999.

7.3.1 SENSITIVITY OF RELIABILITY TO GEOMETRIC TOLERANCE

For $\bar{r} = 1.265$ in., we change α in Equation 7.18 and compute the corresponding values of z and hence the reliability R, which are given in Table 7.4.

7.3.2 SENSITIVITY OF RELIABILITY TO VARIABILITY IN MATERIAL STRENGTH

We now hold $\bar{r} = 1.265$ in. and $\alpha = 0.03$ fixed in Equation 7.18, change $\sigma_{\mathbb{S}}$, the standard deviation of the allowable shear strength, and observe the corresponding variability in the reliability. These values are presented in Table 7.5.

Table 7.4 *R* **versus Tolerance on** $\bar{r}$

TOLERANCE ON $\bar{r}$ α	$-z$	RELIABILITY R
0.010	3.136	0.99916
0.020	3.123	0.99910
0.030	3.099	0.99903
0.040	3.072	0.99890
0.050	3.035	0.99880
0.100	2.772	0.99740

Table 7.5 *R* **versus** $\sigma_{\mathfrak{S}}$

STANDARD DEVIATION OF THE SHEAR STRENGTH (PSI) $\sigma_{\mathfrak{S}}$	$-z$	RELIABILITY R
2,000	4.825	0.99999
4,000	3.585	0.99983
5,000	3.090	0.99903
6,000	2.712	0.99664
8,000	2.145	0.98422
10,000	1.763	0.96080

Table 7.6 *R* **versus** $\bar{r}$

MEAN RADIUS $\bar{r}$	$-z$	RELIABILITY R
1.00	−1.640	0.05050
1.20	2.086	0.98169
1.40	4.824	0.99999
1.60	6.555	0.99999
1.80	7.621	0.99999
2.00	8.736	1.00000

7.3.3 RELIABILITY VERSUS BAR RADIUS

Table 7.6 shows the relationship between the mean radius of the torsion bar and the reliability when α and $\sigma_{\mathfrak{S}}$ have their original values.

7.4 DESIGN OF A HOOD TORSION BAR

The purpose of a hood torsion bar is, as the name suggests, to make it easier for a person to open and close the hood of a truck. The bar also lends rigidity to the open and the closed hood positions. Figure 7.4 shows the bar (12) in position with the hood tipped forward to the open position. The bent end of the torsion bar is shown secured to the bracket on the chassis by two bolts (13). The other end of the bar is inserted snugly in the forked bracket (14) fixed to the inside of the hood at the bottom edge as shown in Figure 7.4.

It is given that the closed (horizontal) position and the open (vertical) position of the hood are 90° apart. The two positions will each cause the bar to twist 45° in opposite directions. The various dimensions in the two positions of

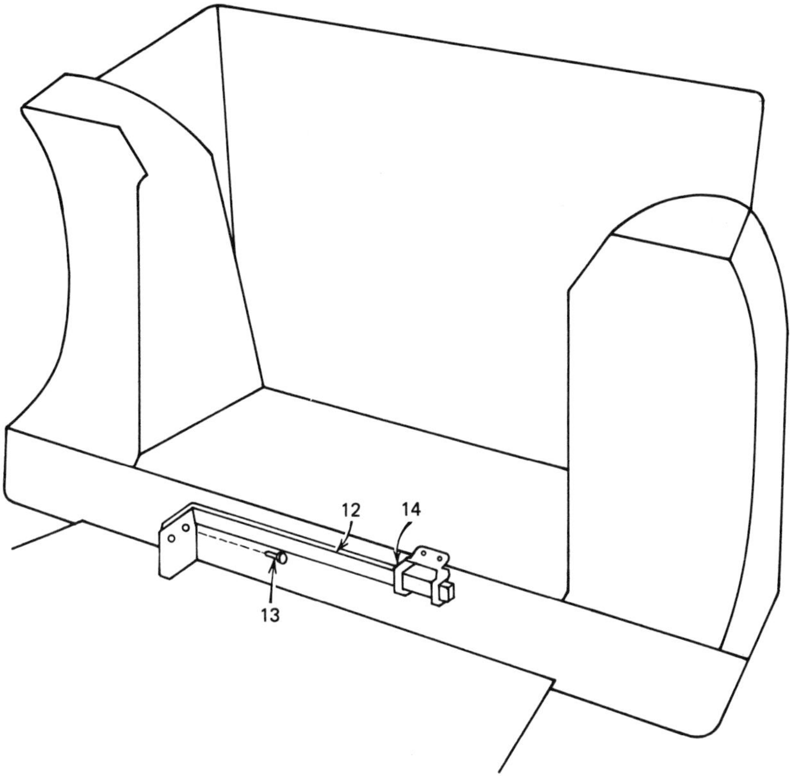

Figure 7.4 Inside view of hood in open position.

the hood are shown in Figure 7.5. The hood is restrained by a cable in its open position.

The rectangular cross-section of the torsion bar is preferred to the more obvious circular section because it is more economical. Figure 7.6 gives the dimensioned sketch of the torsion bar.

7.4.1 THE FAILURE MODE

The bar under consideration will be subjected to alternating shear stresses as the hood is opened and closed. Therefore, a fatigue failure seems to be a logical consequence of these stresses. A complete metallurgical examination of a cracked torsion bar was performed and it was found that the bar was markedly twisted and cracked. The two opposite sides showed similar cracking. It was concluded that the cracks appear to be of a fatigue type.

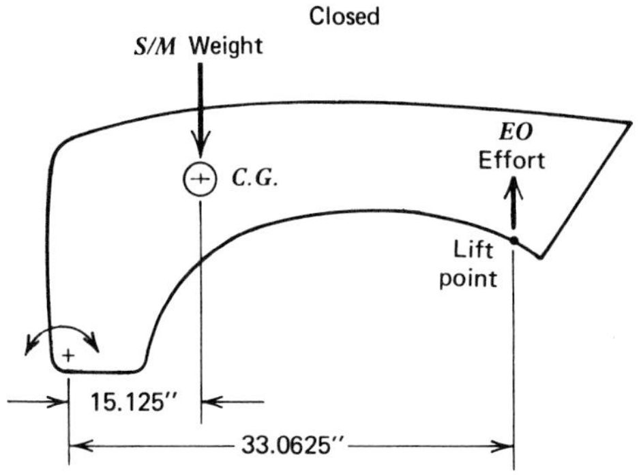

(a) The Closed Position

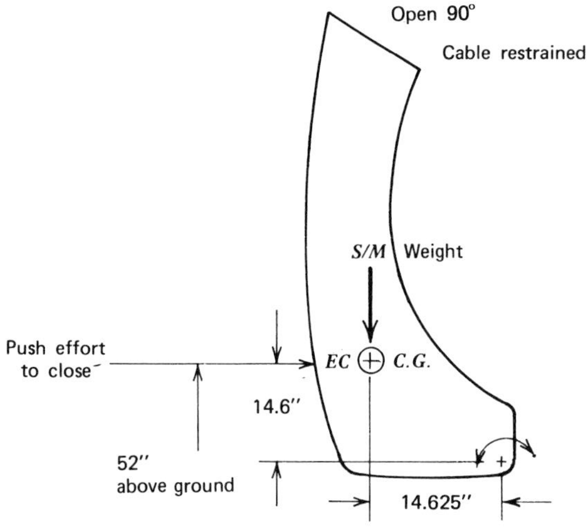

(b) The Open Position

Figure 7.5 The two extreme positions of the hood.

7.4.2 COMPUTATION OF THE MEAN STRESS $\bar{\tau}$ AND σ_τ

The following formulas for a rectangular shaft in pure torsion are found in Seeley and Smith [6], pages 270–271.

$$T=\beta\theta bh^3G/L \tag{7.19}$$

$$\tau=T/\alpha bh^2=(\beta/\alpha)(\theta/L)hG \tag{7.20}$$

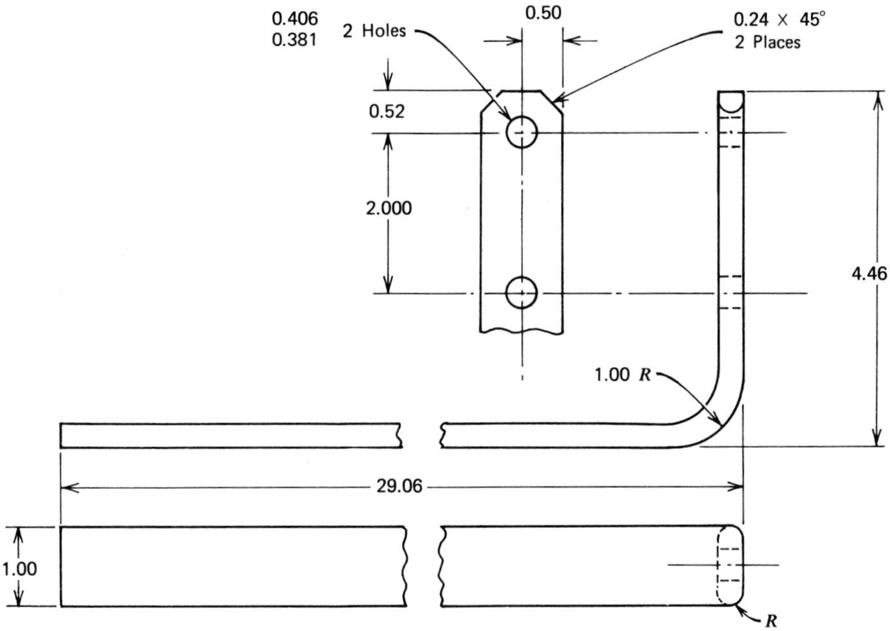

Figure 7.6 Dimensioned sketch of the torsion bar.

where

T = torque in in.-lb

β = coefficient depending on b/h

θ = angle of twist in radian

b = longer side of rectangular bar section in in.

h = shorter side of rectangular bar section in in.

G = modulus of elasticity for material in psi

L = length of bar in in.

τ = shear stress in psi

α = coefficient depending on b/h

The following data are supplied:

$$\bar{b}=1.00 \text{ in.} \qquad \sigma_b=0.01 \text{ in.}$$

$$\bar{h}=0.30 \text{ in.} \qquad \sigma_h=(1/300) \text{ in.}$$

$$\bar{L}=24.625 \text{ in.} \qquad \sigma_L=0.25 \text{ in.}$$

Let $c=\beta/\alpha$. We will now compute $\bar{c}$ and σ_c. Again, referring to Table 13, page 271 of Seely and Smith [6], we observe that $c=\beta/\alpha$ is monotonically increasing in b/h and it is very nearly linear. From these data, we construct the Table 7.7.

Table 7.7 Values of $c=\beta/\alpha$ for Different Values of b/h

	b/h	β	α	$c=\beta/\alpha$
Minimum	$(1-0.03)/(0.3+0.01)=3.129$	0.2655	0.2685	0.98883
Average	$1/0.3 \quad =3.333$	0.2698	0.2720	0.99191
Maximum	$(1+0.03)/(0.3-0.01)=3.552$	0.2739	0.2752	0.99528

The table shows that $\bar{c}=0.99191$ and $\sigma_c=(c_{\max}-c_{\min})/6=(0.99528-0.98883)/6=0.00109$.

The modulus of elasticity G, and the angle of twist θ will be assigned constant values of 12×10^6 psi and $\pi/4$ radians (45°) respectively.

From Equation 7.20, we have

$$\tau=\frac{\theta}{L}chG=f(c,h,L) \tag{7.21}$$

Using the Taylor's series approximation, we have

$$\bar{\tau}\approx f(\bar{c},\bar{h},\bar{L})+\frac{1}{2}\left\{\left.\frac{\partial^2 f}{\partial c^2}\right|_{\bar{c},\bar{h},\bar{L}}\cdot\sigma_c^2+\left.\frac{\partial^2 f}{\partial h^2}\right|_{\bar{c},\bar{h},\bar{L}}\cdot\sigma_h^2+\left.\frac{\partial^2 f}{\partial L^2}\right|_{\bar{c},\bar{h},\bar{L}}\cdot\sigma_L^2\right\}$$

$$=\theta G\left(\bar{c}\bar{h}/\bar{L}\right)+\frac{1}{2}\left\{0+0+2\theta G\left(\bar{c}\bar{h}/\bar{L}^3\right)\sigma_L^2\right\}$$

$$=(\pi/4)(12\times10^6)(0.99191)(0.3)/(24.625)+(1/2)(2)(\pi/4)$$

$$\times(12\times10^6)\times\left((0.99191)(0.3)/(24.625)^3\right)(0.25)^2$$

$$=113{,}891+11.7$$

$$=113{,}903 \text{ psi} \tag{7.22}$$

and

$$\sigma_\tau^2 \approx \left(\frac{\partial f}{\partial c}\bigg|_{\bar{c},\bar{h},\bar{L}} \right)^2 \cdot \sigma_c^2 + \left(\frac{\partial f}{\partial h}\bigg|_{\bar{c},\bar{h},\bar{L}} \right)^2 \cdot \sigma_h^2 + \left(\frac{\partial f}{\partial L}\bigg|_{\bar{c},\bar{h},\bar{L}} \right)^2 \cdot \sigma_L^2$$

$$= \theta^2 G^2 \left\{ \left(\bar{h}/\bar{L}\right)^2 \sigma_c^2 + \left(\bar{c}/\bar{L}\right)^2 \sigma_h^2 + \left(-\bar{c}\bar{h}/\bar{L}^2\right)^2 \sigma_L^2 \right\}$$

$$= (\pi/4)^2 (12 \times 10^6)^2 \left\{ (0.3/24.625)^2 (0.00109)^2 \right.$$

$$+ (0.99191/24.625)^2 (1/300)^2$$

$$\left. + \left((-0.99191)(0.3)/(24.625)^2\right)^2 (0.25)^2 \right\}$$

$$= 2{,}953{,}937 \qquad (7.23)$$

or

$$\sigma_\tau = 1{,}719 \text{ psi}$$

7.4.3 COMPUTATION OF THE MEAN TORQUE $\bar{T}$ AND σ_T

From Table 7.7, we first estimate $\bar{\alpha} = 0.2720$ and $\sigma_\alpha = (\alpha_{\max} - \alpha_{\min})/6 = (0.2752 - 0.2685)/6 = 0.001117$. Proceeding exactly as before, we apply the Taylor's series approximation to

$$T = \alpha b h^2 \tau = f(\alpha, b, h, \tau) \qquad (7.24)$$

to obtain

$$\bar{T} \approx \bar{\alpha}\bar{b}\bar{h}^2\bar{\tau} + \frac{1}{2}\left\{0 + 0 + (2\bar{\alpha}\bar{b}\bar{\tau})\sigma_h^2 + 0\right\}$$

$$= (0.2720)(1)(0.3)^2(113{,}903) + (1/2)(2)(0.2720)(1)$$

$$\times (113{,}903) \times (1/300)^2$$

$$= 2{,}788.3 + 0.3$$

$$= 2{,}789 \text{ lb-in.}$$

and

$$\sigma_T^2 \approx (\bar{b}\bar{h}^2\bar{\tau})^2\sigma_\alpha^2 + (\bar{\alpha}\bar{h}^2\bar{\tau})^2\sigma_b^2 + (2\bar{\alpha}\bar{b}\bar{h}\bar{\tau})^2\sigma_h^2 + (\bar{\alpha}\bar{b}\bar{h}^2)^2\sigma_\tau^2$$

$$= \left[(1)(0.3)^2(113{,}903)\right]^2(0.00117)^2$$

$$+ \left[(0.272)(0.3)^2(113{,}903)\right]^2 \times (0.01)^2$$

$$+ \left[2(0.272)(1)(0.3)(113{,}903)\right]^2(1/300)^2$$

$$+ \left[(0.272)(1)(0.3)^2\right]^2 \times (1{,}719)^2$$

$$= 131.72 + 777.49 + 3{,}839.44 + 1{,}770.82$$

$$= 6{,}518.87$$

or

$$\sigma_T = 80.7 \text{ lb-in.}$$

7.4.4 COMPUTATION OF THE MEAN OPENING EFFORT $\overline{EO}$ AND σ_{EO}

Referring to the sketch of the closed position in Figure 7.5 and taking clockwise moments about the hinge, we have $Wa - T - EO \cdot b = 0$, that is,

$$EO = (Wa - T)/b \tag{7.25}$$

where

W = Weight of hood in lb

a = Lever arm of weight in in.

T = Torque provided by torsion bar in lb-in.

EO = Opening effort in lb

b = Lever arm of opening effort in in.

We will assume W to be 215 ± 7.5 lb, and a and b to be 15.125 ± 0.375 in. and 33.0625 ± 1.5 in., respectively. Note that the tolerance for the effort arm is rather generous because of the human element involved. We find σ's by dividing the tolerance by 3. Hence,

$\bar{W} = 215$ lb $\qquad \sigma_W = 2.5$ lb

$\bar{a} = 15.125$ in. $\qquad \sigma_a = 0.125$ in.

$\bar{b} = 33.0625$ in. $\qquad \sigma_b = 0.5$ in.

We will now compute $\overline{EO}$ and σ_{EO} using the same methodology as before.

$$\overline{EO} \approx (\overline{W}\overline{a} - \overline{T})/\overline{b} + \frac{1}{2}\left\{0+0+0+\left(2(\overline{W}\overline{a} - \overline{T})/\overline{b}^3\right)\sigma_b^2\right\}$$

$$= ((215)(15.125) - 2{,}789)/33.0625$$

$$+\left(\tfrac{1}{2}\right)\left\{(2)\left[((215)(15.125) - 2{,}789)/33.0625^3\right](0.5)^2\right\}$$

$$= 11.2552 + 0.0026$$

$$= 11.26 \text{ lb}$$

$$\sigma_{EO}^2 \approx \left(\overline{a}/\overline{b}\right)^2 \sigma_W^2 + \left(\overline{W}/\overline{b}\right)^2 \sigma_a^2 + \left(-1/\overline{b}\right)^2 \sigma_T^2 + \left(-(\overline{W}\overline{a} - \overline{T})/\overline{b}^2\right)^2 \sigma_b^2$$

$$= (15.125/33.0625)^2(2.5)^2 + (215/33.0625)^2(0.125)^2$$

$$+(-1/33.0625)^2(80.7)^2 + \left[-((215)(15.125) - 2{,}789)/33.0625^2\right]^2(0.5)^2$$

$$= 1.308 + 0.661 + 5.958 + 0.045$$

$$= 7.971$$

or

$$\sigma_{EO} = 2.82 \text{ lb}$$

7.4.5 COMPUTATION OF THE MEAN CLOSING EFFORT $\overline{EC}$ AND σ_{EC}

Referring to the open position in Figure 7.5 and proceeding exactly as in Section 7.4.4, we have

$$EC = (Wd - T)/e \tag{7.26}$$

where

EC = Closing effort in lb

d = Lever arm of weight in the open position in in.

e = Lever arm of closing effort in in.

Let d and e be 14.625 ± 0.375 in. and 14.6 ± 3.0 in. respectively, that is,

$$\bar{d} = 14.625 \text{ in.} \qquad \sigma_d = 0.125 \text{ in.}$$

$$\bar{e} = 14.6 \text{ in.} \qquad \sigma_e = 1.0 \text{ in.}$$

Now we compute

$$EC \approx (\bar{W}\bar{d} - \bar{T})/\bar{e} + \left((\bar{W}\bar{d} - \bar{T})/\bar{e}^3\right)\sigma_e^2$$

$$= \left[(2.5)(14.625) - 2{,}789\right]/14.6 + \left[\left[(215)(14.625) - 2{,}789\right]/14.6^3\right](1)^2$$

$$= 24.34 + 0.11$$

$$= 24.45 \text{ lb}$$

and

$$\sigma_{EC}^2 \approx \left(\bar{d}/\bar{e}\right)^2 \sigma_W^2 + \left(\bar{W}/\bar{e}\right)^2 \sigma_d^2 + (-1/\bar{e})^2 \sigma_T^2 + \left(-(\bar{W}\bar{d} - \bar{T})/\bar{e}^2\right)^2 \sigma_e^2$$

$$= (14.625/14.6)^2(2.5)^2 + (215/14.6)^2(0.125)^2 + (-1/14.6)^2\,(80.7)^2$$

$$+ \left[-((215)(14.625) - 2{,}789)/14.6^2\right]^2(1)^2$$

$$= 6.271 + 3.388 + 30.552 + 2.779$$

$$= 42.991$$

or

$$\sigma_{EC} = 6.56 \text{ lb}$$

7.4.6 DEVELOPMENT OF 3σ LIMITS ON S-N DIAGRAM

The specified material for the torsion bar is AISI 5160 steel with a specified BHN range of 415–477. Referring to Table 3-6 on page 44 of the *Fatigue Design Handbook* [10], we find that the corresponding range for ultimate tensile strength $\mathfrak{S}_u$ to be 210–247 ksi. We write the following two formulas that will enable us to establish the desired limits.

At 10^3 cycles,

$$\mathfrak{S}_a = k_1 k_2 \mathfrak{S}_u \tag{7.27}$$

and at 10^6 cycles,

$$\mathfrak{S}_a = k_3 k_4 k_5 \mathfrak{S}_n' \tag{7.28}$$

where

$\mathbb{S}_a$ = Applied alternating stress

$\mathbb{S}_u$ = Ultimate tensile stress

$\mathbb{S}_n'$ = Endurance limit

k_1 = Factor to account for 10^3 cycles

k_2 = Factor to convert tensile limit to shear limit

k_3 = Factor for torsion

k_4 = Factor for size

k_5 = Factor for surface

Most of the average values and variation of the various factors defined above were either directly obtained or estimated from Chapter 12 of Lipson and Juvinall [5]. They are assembled in Table 7.8. Note all the stress values are in ksi units.

Using Equation 7.27, at 10^3 cycles we have

$$\bar{\mathbb{S}}_a = \bar{k}_1 \bar{k}_2 \bar{\mathbb{S}}_u = (0.92)(0.82)(228.5) = 172.38 \text{ ksi}$$

and

$$\sigma^2_{\mathbb{S}a} = \left(\bar{k}_2 \bar{\mathbb{S}}_u\right)^2 \sigma^2_{k_1} + \left(\bar{k}_1 \bar{\mathbb{S}}_u\right)^2 \sigma^2_{k_2} + \left(\bar{k}_1 \bar{k}_2\right)^2 \sigma^2_{\mathbb{S}_u}$$

$$= \left[(0.82)(228.5)\right]^2 0.01^2 + \left[(0.92)(228.5)\right]^2 0.01^2 + \left[(0.92)(0.82)\right]^2 6.197^2$$

$$= 3.51 + 4.42 + 21.64$$

$$= 29.57$$

Table 7.8 Values of various quantities

QUANTITY	MEAN	TOLERANCE	STANDARD DEVIATION (σ = tolerance/3)	LIMITS
$\mathbb{S}_u$	228.5	±18.5	6.167	210.00–247
$\mathbb{S}_n'(=0.5\mathbb{S}_u)$	114.25	± 9.25	3.089	105.0 –123.5
k_1	0.92	± 0.03	0.01	0.89– 0.95
k_2	0.82	± 0.03	0.01	0.79– 0.85
k_3	0.60	± 0.03	0.01	0.57– 0.63
k_4	0.94	± 0.06	0.02	0.88– 1.00
k_5	0.44	± 0.14	0.0467	0.30– 0.58

or

$$\sigma_{\mathbb{S}_a} = 5.44 \text{ ksi}$$

Hence the 3σ limits on $\mathbb{S}_a$ at 10^3 cycles are: $172.30 \pm 3(5.44)$, or (156.06, 188.70 ksi).

Similarly, using Equation 7.28, at 10^6 cycles,

$$\bar{\mathbb{S}}_a = \bar{k}_3 \bar{k}_4 \bar{k}_5 \bar{\mathbb{S}}_n' = (0.6)(0.94)(0.44)(114.25) = 28.35 \text{ ksi}$$

and

$$\begin{aligned}
\sigma_{\mathbb{S}_a}^2 &= \left(\bar{k}_4 \bar{k}_5 \bar{\mathbb{S}}_n'\right)^2 \sigma_{k_3}^2 + \left(\bar{k}_3 \bar{k}_5 \bar{\mathbb{S}}_n'\right)^2 \sigma_{k_4}^2 + \left(\bar{k}_3 \bar{k}_4 \bar{\mathbb{S}}_n'\right)^2 \sigma_{k_5}^2 + \left(\bar{k}_3 \bar{k}_4 \bar{k}_5\right)^2 \sigma_{\mathbb{S}_n'}^2 \\
&= \left[(0.94)(0.44)(114.25)\right]^2 (0.0467)^2 \\
&\quad + \left[(0.6)(0.44)(114.25)\right]^2 (0.02)^2 + \left[(0.6)(0.94)(114.25)\right]^2 (0.0467)^2 \\
&\quad + \left[(0.6)(0.94)(0.44)\right]^2 (3.089)^2 \\
&= 0.22 + 0.36 + 9.04 + 0.59 \\
&= 10.22
\end{aligned}$$

or

$$\sigma_{\mathbb{S}_a} = 3.20 \text{ ksi}$$

Hence, at 10^6 cycles, the 3σ limits become:

$$28.35 \pm 3(3.20), \text{ or } (18.76, 37.94 \text{ ksi})$$

Now if we plot $\log_{10} \mathbb{S}_a$ versus $\log_{10} N_f$ (the number of cycles to failure) on a graph paper, or equivalently, $\mathbb{S}_a$ versus N_f on a log-log graph paper, the straight line joining the two points (10^3 cycles, 188.70 ksi; and 10^6 cycles, 37.94 ksi) can be treated as an upper 3σ limit of the S-N diagram. Similarly, the straight line joining the two points (10^3 cycles, 156.06 ksi; and 10^6 cycles, 18.76 ksi) can be assumed to represent the lower 3σ limit on the S-N line. Observe the envelope generated by these limits intersected with the limits on $\mathbb{S}_a$ of $\tau \pm 3\sigma_\tau$ as shown in Figure 7.7. The extreme points A and B of the envelope represent the minimum and maximum values of the cycles to failure, N_f.

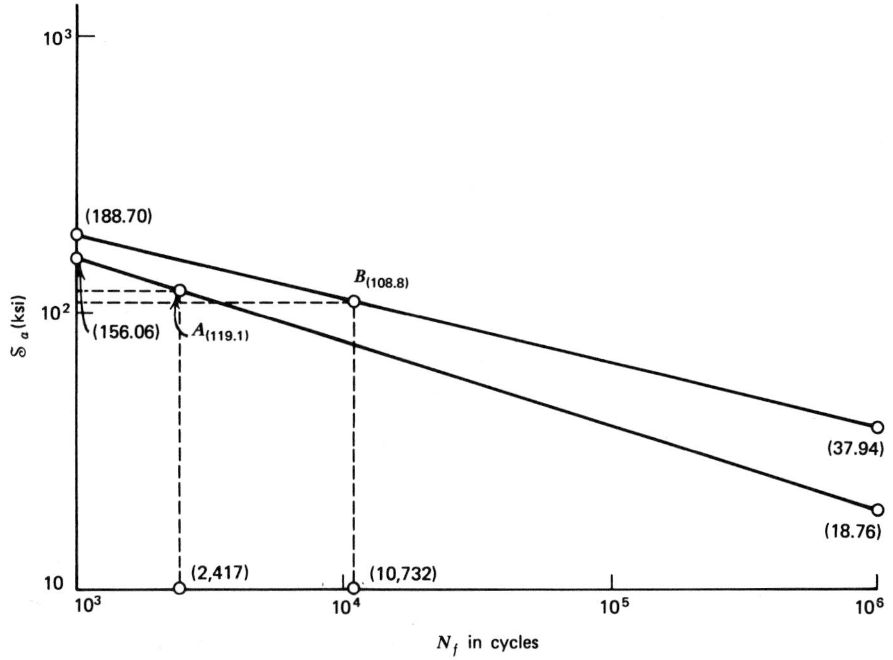

Figure 7.7 The S-N diagram.

Assuming N_f to be distributed log normally, we get

$$\bar{n} = (n_A + n_B)/2 \tag{7.29}$$

$$\sigma_n = (n_B - n_A)/6 \tag{7.30}$$

where $n = \log N_f$. From the graph in Figure 7.7, $N_A = 2{,}417$ and $N_B = 10{,}732$ cycles. Application of Equations 7.29 and 7.30 yields

$$\bar{n} = 3.7070, \qquad \sigma_n = 0.1246$$

Hence the median life of the bar is given by

$$10^{\bar{n}} = 5{,}093 \text{ cycles}$$

7.4.7 COMPUTATION OF RELIABILITY

Suppose we want to compute the reliability of the torsion bar under discussion at (1) 2,000 cycles, and (2) 10,000 cycles.

$$
\begin{aligned}
(1)\ R_{2,000} &= P\left[N_f > 2{,}000\right] \\
&= P\left[n > 3.301\right] \\
&= 1-\Phi\left[(3.301-\bar{n})/\sigma_n\right] \\
&= 1-\Phi\left[(3.301-3.707)/0.1246\right] \\
&= 1-\Phi\left[-3.258\right] \\
&= 0.99944 \\
(2)\ R_{10,000} &= 1-\Phi\left[(4-3.707)/0.1246\right] \\
&= 1-\Phi\left[2.352\right] \\
&= 1-0.99071 \\
&= 0.00929
\end{aligned}
$$

Thus the torsion bar reliability can be estimated at any desired life in cycles as shown above. The decision to accept this design may have to be based on the desired life for the bar and the seriousness of its failure.

The S-N diagram with its 3σ limits was constructed from the known properties of the specified material after applying various factors such as surface factor and size factor. A better method would have been actual life testing of the bar. But life testing is often prohibitively time-consuming and is almost always expensive.

For a quick comparison of the results of our approach to the conventional one, we refer the reader to Table 7.9.

Table 7.9 Comparison between the reliability approach and the conventional approach

QUANTITY		RELIABILITY APPROACH: MEAN	RELIABILITY APPROACH: $\pm 3\sigma$ RANGE	CONVENTIONAL APPROACH: VALUE
Shear stress	τ	113.9 ksi	(108.7, 119.1 ksi)	117.3 ksi
Torque	T	2,789 lb-in.	(2,584, 3,030 lb-in.)	(2,539, 3,050)
Opening effort	EO	11.3 lb	(2.8, 19.7 lb)	(6.1, 21.6 lb)
Closing effort	EC	24.4 lb	(4.7, 44.1 lb)	(6.5, 41.5 lb)
Life to failure	N_f	5,093[a] cycles	(2,417, 10,732 cycles)	2,800 cycles

[a]This value represents the median and not the mean.

7.5 SUMMARY

We have illustrated the probabilistic design methodology with the help of four design problems. Elements that are subjected to tension, bending and torsion were considered. In all of these design problems, the engineering variables such as loads, dimensions, and strength were considered as random variables. Only the first two moments of the random variables were used in the analysis. In the analysis, the random variables were combined and the first two moments of the new variable, which is a function of the other random variables, were computed by Taylor's series approximation method. The first two moments of the stress and the strength random variables were computed. Reliability was found by assuming that the stress and the strength are normal random variables with the first two moments computed in the analysis. It should be realized that in the actual design problems, other distributions may be used based on the available data. In this case, the results given in Chapter 6 can be used. Many components are subjected to repeated stresses and they suffer strength degradation. The reliability analysis for these components is given in Chapter 8 and this can be used for the design of the components subjected to repeated stresses.

EXERCISES

1 An automotive linkage bar is to be designed for tension (after the failure mode analysis, it was found that the bar is most likely to fail in the tension mode i.e., failure by tensile fracture). Ultimate tensile strength of the material for the bar has an average value of 10,000 MPa and a standard deviation of 500 MPa. The bar should have a circular cross-section and the manufacturing tolerances for the bar are $\pm 2\%$ of the diameter, based on 4σ limits (i.e., $\sigma_D = 0.5\%$ of diameter). The load acting on the rod has an average value of 10,000 N and a standard deviation of 1,000 N. The rod should be designed in such a way that the probability of failure is no more than one in ten thousand (i.e., reliability goal is 0.99990). Design the rod for the above reliability objective and with the data as given above. Study the sensitivity of the reliability of the above design with respect to the standard deviation (varying from 20 to 1,000 MPa) of the material strength and with respect to the tolerances (varying from $\pm 0.5\%$ to 10% of the diameter, based on 4σ limits).

2 A simply supported beam of rectangular cross-section is to be designed to carry a concentrated load. The load, the length, and the location of the load are random variables.

Load **P**: $(\bar{P}, \sigma_p) = (30{,}000, \; 2{,}000)$ N

Length of the beam **l**: $(\bar{l}, \sigma_l) = (3.0, 0.01)$ m

Location of the load at a distance '**a**' from the one end of the beam: $(\bar{a}, \sigma_a) = (2.0, 0.01)$ m

The significant strength $\mathbb{S}$ of the beam: $(\bar{\mathbb{S}}, \sigma_{\mathbb{S}}) = (400, 20)$ MPa
The width of the rectangular cross-section is assumed to be one-half of the depth of the beam. Assume that the tolerances on the dimensions of the beam are 3% of the dimensions of the beam. Design the beam for a reliability of 0.9990. With the dimensions of the beam obtained by the above design, study the sensitivity of the reliability of the above design with respect to the following design variables:

(a) Standard deviation of the strength (varying from 10 MPa to 80 MPa)
(b) Standard deviation of the load (varying from 500 N to 4,000 N)
(c) Tolerances on the rectangular cross-section (varying from 1% to 10%)

3 An eccentrically loaded hollow cylindrical column is to be designed for a specified reliability of 0.999. The maximum stress, σ_{max}, is given by

$$\sigma_{max} = \frac{P}{A}\left[1 + \frac{ec}{r^2}\sec\left(\frac{1}{r}\sqrt{P/AE}\right)\right]$$

where

P = load, lb

A = area, in.^{2}

e = eccentricity, in.

c = distance to extreme fibers, in.

r = radius of gyration, in.

E = modulus of elasticity, psi

Let the outside radius R_0 be 1.25 times the inside radius R_i. The standard deviation for the radius is assumed as 5% of the value of the radius. The column is made of AISI-1018 cold drawn steel with the mean value of yield point as 78,000 psi and standard deviation of 5,000 psi. Design the column with the following data:

VARIABLE	MEAN VALUE	STANDARD DEVIATION
Load, **P**, lb	80,000	4,000
Eccentricity, **e**, in.	0.5	0.03
Modulus of elasticity, **E**, psi	30×10^6	10^6
Length, **l**, in.	70	0.6

4 The solid transmission shaft discussed in Section 7.3 is subjected to a bending moment in addition to torsion. The expected value of the bending moment is 30,000 lb-in. and the standard deviation of the bending moment is 2,000 lb-in. Determine the radius of the shaft for a required reliability of 0.9990 with the additional information given in Section 7.3.

5 A propped cantilever beam of rectangular cross-section is to be designed for a specified reliability of 0.990. The loading of the beam is as given in Figure 7.8.

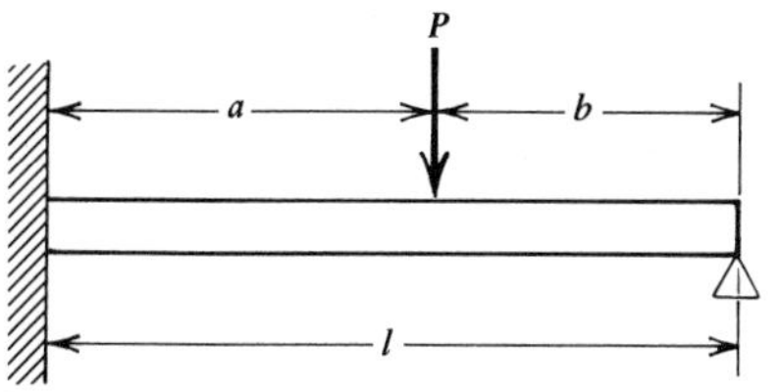

Figure 7.8 Propped cantilever.

We are given the following data:

$$\bar{P}=20{,}000 \text{ N} \qquad \sigma_P=2000 \text{ N}$$

$$\bar{l}=5.0 \text{ m} \qquad \sigma_l=0.2 \text{ m}$$

$$\bar{a}=3.0 \text{ m} \qquad \sigma_a=0.3 \text{ m}$$

Depth d of the beam is twice the width w. Maximum allowable stress has an expected value of 560 MPa and a standard deviation of 30 MPa. The tolerances on the dimensions of the beam are 3% of the nominal values of the dimensions.

6 Design the I-beam given in Section 7.2 when the beam has a uniformly distributed load with an expected value of 40 lb/ft and a standard deviation of 2 lb/ft. All the other information about the beam is given in Section 7.2.

BIBLIOGRAPHY

1 Haugen, E. B., *Analysis of Electromagnetic Devices by Probabilistic Methods*, NAA/S & ID, January 1966.

2 Haugen, E. B., *Probabilistic Approaches to Design*, John Wiley & Sons, 1968.

3 Haugen, E. B., *Probabilistic Mechanical Design*, The University of Arizona, Tucson, Arizona, 1974.

4 Kivenson, Gilbert, *Durability and Reliability in Engineering Design*, Hayden Book Company, New York, 1971.

5 Lipson, C. and Juvinall, R. C., *Handbook of Stress and Strength of Manufactured Parts*, Macmillan, New York, 1963.

6 Seeley, F. B. and Smith, J. O., *Advanced Mechanics of Materials*, John Wiley & Sons, 1952.

7 Smith, C. O., "Design of Pressure Vessels to Probabilistic Criteria," First International Conference on Structural Mechanics in Reactor Technology, Berlin, Germany, September 1971.

8 Smith, C. O., "Structural Designs Based on Probability," *Design News*, December 8, 1969, pp. 98–108.

9 Smith, C. O. and Haugen, E. B., "Design of Circular Members in Torsion to Probabilistic Criteria," *Design News*, November 8, 1968, pp. 108–115.

10 Society of Automotive Engineers, *Fatigue Design Handbook*, Vol. 4, 1968.

Chapter 8 time dependent stress-strength models

Stress-strength interference (SSI) models for the static case were examined in Chapter 6. These models are good for a single stress application. In this chapter we develop models that consider the repeated application of stresses and also consider the change of the distribution of strength with time.

Such reliability models are generally known as stress-strength-time (SST) models [1,2,4–6]. Some examples of the practical motivation behind considering time-dependent reliability models may be cited as weakening caused by aging or cumulative damage. Also, as better estimates of distributions become available from a history of nonfailures, these models provide a means for reassessing reliability.

The use of words such as "stress" or "strength" is unfortunate. The imputed meaning is unnecessarily restrictive (to mechanical loading) in the minds of many individuals, especially engineers. We use it in a broader sense, applicable in a variety of situations well beyond the traditional mechanical or structural systems. "Stress" is used to indicate any agency that tends to induce "failure," while "strength" indicates any agency resisting "failure." "Failure" itself is taken to mean failure to function as intended; it is defined to have occurred when the actual stress exceeds the actual strength for the first time.

Figure 8.1 shows various patterns of load variation with respect to time. Cyclic load with constant amplitude is shown in Figure 8.1*b*. This pattern generally occurs in laboratory testing. In general, the components are subjected to a complicated pattern of randomly varying load amplitudes and frequencies. Two examples of this are the load in an aircraft created by atmospheric gusts, and the loads produced in a surface vehicle's suspension components by random irregularity of road surfaces. Figure 8.1*c* shows such a random pattern.

Even when different components are subjected to the same fluctuating stress, they fail at different cycles because of nonhomogeneity in material properties, variation in surface conditions, etc. To predict the average life of a component, a number of test specimens are tested at various stress levels until

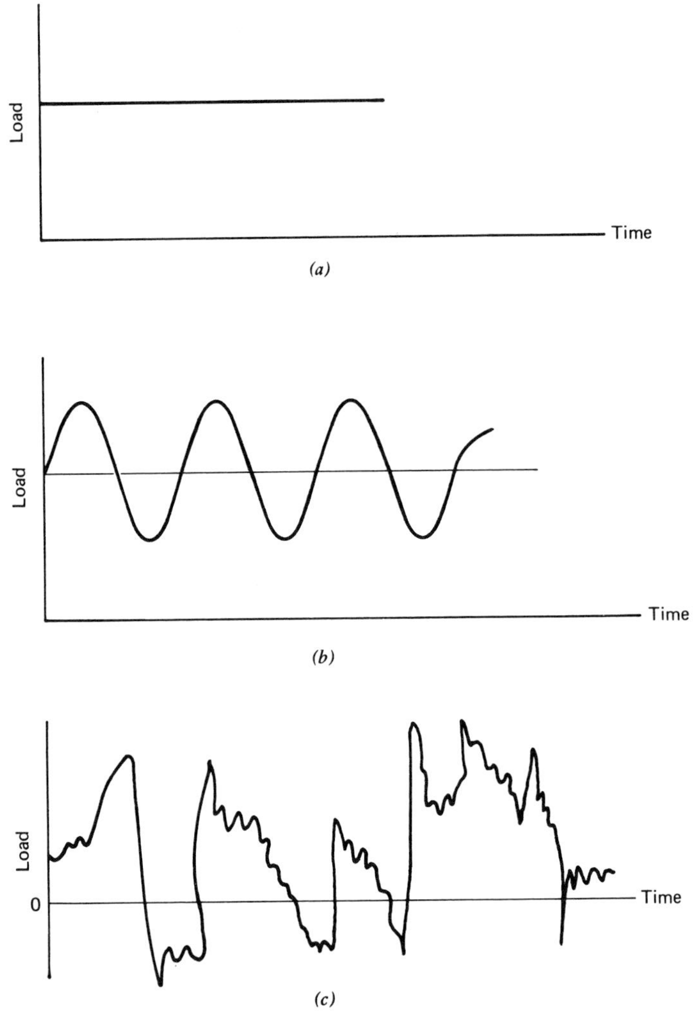

Figure 8.1 Patterns of load (stress) variation. (*a*) Constant load; (*b*) cyclic load with constant amplitude; (*c*) random-load spectrum.

failure. The results are plotted on log-log paper with the stresses on the ordinate and the corresponding lives on the abscissa, as shown in Figure 8.2. A line representing the average life is fitted, and the horizontal portion of this line indicates that the component will have infinite life if subjected to stresses below this line, and this stress corresponding to this line is called the endurance limit. In practice, several components are tested at a given stress level to estimate the fatigue life. Figure 8.3 shows the scatter in fatigue life at a given stress level.

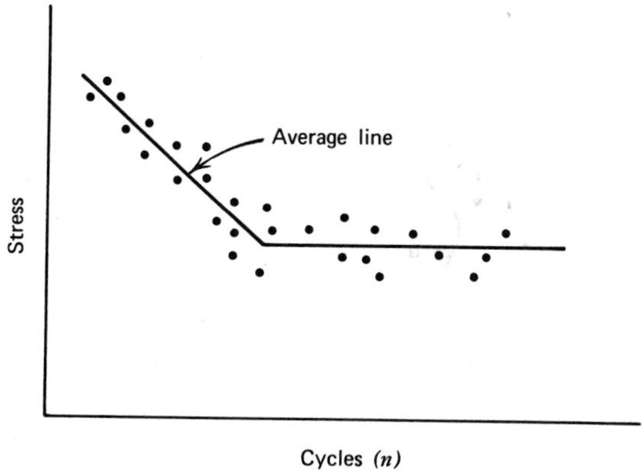

Figure 8.2 Conventional S-N diagram (plotted on a log-log scale).

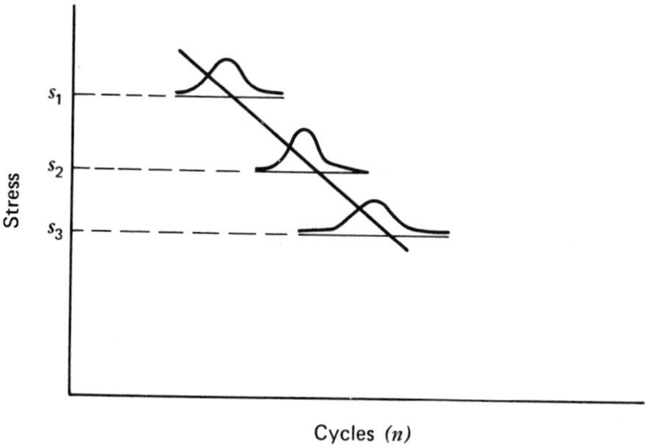

Figure 8.3 Scatter in fatigue life at a given stress (plotted on a log-log scale).

There is uncertainty about the stress and the strength random variables at any instant of time and also about the behavior of the random variables with respect to time and/or cycles. The two terms "random-fixed" and "random-independent" are used to describe these two uncertainties. Random refers to the behavior of the variable at the beginning of time and the word "fixed" means that the behavior of the variable with respect to time is fixed, or the variable varies in time in a known manner. For example, when one selects a component from a population of components, the value of strength that one obtains is random before the selection is made. However, once the component is selected and placed in the system the strength is fixed until failure occurs, or the strength is assumed to vary in a predetermined (nonrandom) fashion. The word "independent" means that the successive values assumed by the variable are statistically independent, and thus one value gives us no information about the size of the subsequent values. But in the case of "random-fixed" the subsequent values of the variable are dependent because they are given by a known equation. This classification is explained in detail in Section 8.1. Basically there are three classifications for both the stress and the strength variables, and they are: deterministic, random-fixed, and random-independent. In general, we can develop other classifications but then the models become very complicated [7, 8, 9]. Section 8.2 develops the expressions for reliability after n applications of loading for all the nine possible combinations of the stress and strength variables. Some of these cases are considered in References 3, 4, 5, and 6. In Section 8.3, the expressions for reliability R_n are changed to the value of reliability at any instant of time t, denoted by $R(t)$, under the assumption that the number of cycles occurring in a given time interval follow the Poisson distribution. Section 8.4 develops the general models for reliability with aging, cyclic damage, and cumulative damage. This introduces a great deal of complexity in the analytical models. Simulation models [1, 8] have been developed to study the reliability of a component for these models.

8.1 STRESS-STRENGTH CLASSIFICATIONS

The failure of components under repeated stresses has been investigated primarily with respect to the fatigue behavior of materials. The design of these components involves the use of stress versus fatigue life, or S-N diagrams. The maximum allowable reversed stress is defined by multiplying the endurance limit by a suitable safety factor. The strength of the component may be defined as the maximum stress or load that causes failure under specified environmental conditions. Strength may also be specified in terms of the life of the system under a particular load history or spectrum of stresses. The lifetime may be measured in terms of the number of stress cycles. The strength may also depend on the previous stress history in various ways.

Repeated stresses are characterized by the time each load is applied and the behavior of time intervals between the application of loads. The load occurrences may be classified as:

1. The loads are applied at known times $t_1, t_2, \ldots, t_n$. In this case, the changes may follow a cyclic pattern. These may be seasonal changes, day/night, on-off, up-down cycles, etc. The cycle times are known exactly beforehand. The load occurrences may be constant or may vary from cycle to cycle.

2. The loads are applied at random times. Thus, in this case, the cycle times are random and independent rather than known. The randomness of the cycle times may be described by exponential or gamma probability density functions. The loading may be deterministic or stochastic in nature.

The uncertainty about the stress and strength variables may be classified in the following three categories [4]:

Deterministic

The variable assumes values that are exactly known *à priori*. This may suggest that the design and the manufacturing process is fully controlled. This is rarely the case in the real world, but, in some cases, may serve as an acceptable approximation.

Random-Fixed

We are interested in the behavior of the strength variable with respect to time or cycles. The strength is a random variable at any particular instant of time. The word "fixed" in this classification refers to the behavior of the random variable with respect to time and/or cycles. This means that the random variable changes or varies in time in a known manner. Let $g_0(y_0)$ be the p.d.f. of the strength random variable $\mathbf{y}_0$ at the initial time. Then the strength $y(t)$ at any instant of time t is given by

$$y(t) = y_0 \phi(t)$$

where y_0 is the initial strength and $\phi(t)$ is a known or given function that describes exactly how strength decreases with time. For example, let

$$\phi(t) = (1 - 0.001t)$$

which means that with the passage of each unit of time, strength decreases by 0.1% of the initial strength y_0. Thus, the initial strength is a random variable, but once we know its value, the value of the strength in the future is exactly known. Let $\mathbf{y}_k$ denote the strength for the kth application of load, and let us assume that $\mathbf{y}_k$ is a function of $\mathbf{y}_0$ and the occurrence number k. Thus, we have

$$\mathbf{y}_k = g(\mathbf{y}_0, k)$$

where $\mathbf{y}_0$ is a random variable with a known probability density function. Thus, the behavior of $\mathbf{y}_k$ is fixed by the behavior of $\mathbf{y}_0$. Some examples are

$$\mathbf{y}_k = \mathbf{y}_0 k^{-\alpha}$$

or

$$\mathbf{y}_k = \mathbf{y}_0 + bk$$

In the above equations, the strength decreases with the number of load occurrences for $\alpha > 0$ and $b < 0$, respectively. If increase in strength occurs because of work hardening, we have $\alpha < 0$ and $b > 0$.

Random-Independent

The variable is not only random but, unlike the random-fixed case, the successive values assumed by the variable are statistically independent. Observation of one stress or strength value gives us no information about the size of the subsequent values.

Successive stresses are generally independent. Strength will vary randomly and will be independent from cycle to cycle only if it is being affected by other environmental factors, such as temperature and vibrations, which are independent of the process.

Strength depends on the number of load applications, their magnitudes and time durations. If the strength varies only with time, the effect is called aging. Corrosion is an example of aging. When aging occurs the parameters of the variable change with respect to time. If the strength is a function of the number of load occurrences, the effect is called cyclic damage. If the value of the strength depends on the number of load occurrences as well as their magnitudes, the effect is called cumulative damage. The strength may also depend on the sequence of the loading. Cumulative damage adds considerable complexity to the reliability models. In addition, we must consider the stress application frequency factor or the usage rate, and this is done in Section 8.3 by converting R_n to $R(t)$ under the Poisson assumption. Similarly, other models can be developed with other assumptions about the stress application frequency factor.

8.2 RELIABILITY COMPUTATIONS FOR DETERMINISTIC CYCLE TIMES

We now develop individual expressions for reliability in different cases. We obtain these expressions by combining the three levels of uncertainty classifications for both stress and strength and two types of cycle occurrences. We prefer R_n, the reliability after n cycles (the probability of not having a failure on any one of the n cycles), to $R(t)$, the reliability at time t, the argument t being continuous. We do so because the former can be converted to the latter very simply when cycle times are deterministically known. For example,

$$R(t) = R_n, \qquad t_n < t \leqslant t_{n+1}, \quad n = 1, 2, \ldots \tag{8.1}$$

where t_i is the instant in time at which the ith cycle occurs.

case 1 deterministic stress and deterministic strength

Let x_i and y_i, $i = 1, 2, \ldots, n$, be the stress and strength, respectively, on the ith cycle. Then

$$R_n = P[E_1, E_2, \ldots, E_n] \tag{8.2}$$

where

$$R_n = \text{reliability after } n \text{ cycles}$$

$$E_i = \text{event no failure occurs on the } i\text{th cycle}$$

$$\text{Hence, } R_n = \begin{cases} 0 & \text{if } x_i > y_i \quad \text{for some } i, \quad 1 \leqslant i \leqslant n \\ 1 & \text{if } x_i \leqslant y_i \quad \text{for all } i, \quad 1 \leqslant i \leqslant n \end{cases} \tag{8.3}$$

Consider a special case where the stress is nondecreasing and the strength is nonincreasing in time. Then R_n can be determined by only one comparison (between x_n and y_n).

case 2 deterministic stress and random-fixed strength

Let the stress be x_0, a constant, and the strength on the ith cycle be $\mathbf{y}_i$ given by

$$\mathbf{y}_i = \mathbf{y}_0 - a_i, \qquad i = 1, 2, \ldots \tag{8.4}$$

where $a_i \geqslant 0$ are known constants. Further, the a_i's are assumed nondecreasing in time. The p.d.f. of $\mathbf{y}_0$, $g_0(y_0)$, is assumed known. Then

$$\begin{aligned} P[E_n] &= P(x_n \leqslant y_n) \\ &= P(x_0 \leqslant y_0 - a_n) \\ &= P(y_0 \geqslant x_0 + a_n) \\ &= \int_{x_0 + a_n}^{\infty} g_0(y_0)\, dy_0 \end{aligned} \tag{8.5}$$

But

$$\begin{aligned} R_n &= P[E_1, E_2, \ldots, E_n] \\ &= P[E_1 | E_2, E_3, \ldots, E_n] \cdot P[E_2, E_3, \ldots, E_n] \\ &= P[E_1 | E_2, E_3, \ldots, E_n] \cdot P[E_2 | E_3, E_4, \ldots, E_n] \cdot P[E_3, E_4, \ldots, E_n] \\ &= P[E_1 | E_2, E_3, \ldots, E_n] \times P[E_2 | E_3, E_4, \ldots, E_n] \times \ldots P[E_{n-1} | E_n] \times P[E_n] \end{aligned} \tag{8.6}$$

All but the last term ($P[E_n]$) in the R.H.S. of Equation 8.6 are 1's because of the restrictions on the a_i's which cause the strength $\mathbf{y}_i$ to decrease in time. Hence

$$R_n = P[E_n] = \int_{x_0 + a_n}^{\infty} g_0(y_0)\, dy_0 \tag{8.7}$$

Let $R_{n,n-1}$ be the conditional reliability for the nth cycle given that the device has survived the previous $(n-1)$ cycles. Then

$$R_{n,n-1} = P[E_n | E_1, E_2, \ldots, E_{n-1}] = \frac{P[E_1, \ldots, E_n]}{P[E_1, \ldots, E_{n-1}]} = \frac{R_n}{R_{n-1}} \tag{8.8}$$

case 3 deterministic stress and random-independent strength

Let the stress be constant at x_0 as in case 2. Let $g_i(y)$ be the p.d.f. of the r.v. strength $\mathbf{y}_i$ during cycle i. Since successive values of $\mathbf{y}_i$ are independent, we get

$$R_n = P[E_1, E_2, \ldots, E_n] = P[E_1] \times P[E_2] \times \ldots \times P[E_n] \tag{8.9}$$

where

$$P[E_i] = P(x_0 \leqslant \mathbf{y}_i) = \int_{x_0}^{\infty} g_i(y)\,dy, \qquad i = 1, 2, \ldots, n \tag{8.10}$$

In particular, if the p.d.f. remains unchanged over time, that is, if

$$g_1(y) = g_2(y) = \ldots = g_n(y) = g(y)$$

then

$$R_n = (P[E_i])^n = \left\{ \int_{x_0}^{\infty} g(y)\,dy \right\}^n \tag{8.11}$$

case 4 random-fixed stress and deterministic strength

This case is similar to case 2, except that the roles of stress and strength are reversed. Let

$$\mathbf{x}_i = \mathbf{x}_0 + b_i, \qquad i = 1, 2, \ldots \tag{8.12}$$

denote the stress in cycle i, where b_i's are known nonnegative constants, nondecreasing in time. Further, let the strength be held constant at y_0. The p.d.f. of $\mathbf{x}_0$, $f_0(x_0)$ is assumed known.

The restrictions on the b_i's guarantee nondecreasing stress, which in turn ensure, as in case 2, that

$$R_n = P[E_n] \tag{8.13}$$

$$= P(\mathbf{x}_n \leqslant y_n)$$

$$= P(\mathbf{x}_0 + b_n \leqslant y_0)$$

$$= (\mathbf{x}_0 \leqslant y_0 - b_n)$$

$$= \int_0^{y_0 - b_n} f_0(x_0)\,dx_0 \tag{8.14}$$

case 5 random-fixed stress and random-fixed strength

Let the stress be given by

$$\mathbf{x}_i = \mathbf{x}_0 + b_i, \qquad i = 1, 2, \ldots \tag{8.15}$$

where the b_i's have the same restrictions as in case 4, *viz.*, the b_i's are known nonnegative constants and nondecreasing in time. Further, let the strength be given by

$$\mathbf{y}_i = \mathbf{y}_0 - a_i, \qquad i = 1, 2, \ldots \tag{8.16}$$

where the a_i's have the same restrictions as in case 2; that is, the a_i's are nonnegative known constants and are nondecreasing in time. The p.d.f.'s $f_0(x_0)$ and $g_0(y_0)$ are assumed known. We have required the stress to be nondecreasing and the strength to be nonincreasing. Hence arguments identical to those in case 2 would yield

$$R_n = P[E_n] \tag{8.17}$$

$$= P(\mathbf{x}_n \leqslant \mathbf{y}_n)$$

$$= P(\mathbf{x}_0 + b_n \leqslant \mathbf{y}_0 - a_n)$$

$$= P(\mathbf{x}_0 \leqslant \mathbf{y}_0 - a_n - b_n)$$

$$= \int_0^{\infty} g_0(y_0)\left(\int_0^{y_0 - a_n - b_n} f_0(x_0)\,dx_0\right)dy_0 \tag{8.18}$$

In particular, if the stress and strength do not vary with time, that is, $a_i = b_i = 0$, $i = 1, 2, \ldots,$ the R.H.S. of Equation 8.18 can be easily recognized as the standard expression for reliability based on stress-strength interference given in Chapter 6.

case 6 random-fixed stress and random-independent strength

In this case the stress $\mathbf{x}$ is a random variable with a known p.d.f. $f(x)$. The successive random strengths $\mathbf{y}_1, \mathbf{y}_2, \ldots, \mathbf{y}_n$ are independent and identically distributed with p.d.f. $g(y)$. Thus

$$R_n = P[E_1, E_2, \ldots, E_n]$$

The E_i event means that

$$E_i \sim (\mathbf{y}_i > \mathbf{x})$$

Hence

$$R_n = P\big[(\mathbf{y}_1 > \mathbf{x}) \cap (\mathbf{y}_2 > \mathbf{x}) \cap \ldots \cap (\mathbf{y}_n > \mathbf{x})\big]$$

$$= P\big[\min(\mathbf{y}_1, \mathbf{y}_2, \ldots, \mathbf{y}_n) > \mathbf{x}\big] \tag{8.19}$$

The distribution function of the random variable

$$\mathbf{y}_{\min} = \min(\mathbf{y}_1, \mathbf{y}_2, \ldots, \mathbf{y}_n)$$

was derived in Chapter 2 and is given by

$$G_n(y) = 1 - [1 - G(y)]^n \tag{8.20}$$

Now, Equation 8.19 can be rewritten as

$$R_n = P[\mathbf{y}_{\min} > \mathbf{x}]$$

Hence, using Equation 6.3, we have

$$R_n = \int_0^\infty f(x)[1 - G(x)]^n \, dx \tag{8.21}$$

case 7 random-independent stress and deterministic strength

This is an exact reversal of case 3. By reciprocity between stress x and strength y, we observe

$$R_n = P(E_1) \times P(E_2) \times \ldots \times P(E_n) \tag{8.22}$$

where

$$\begin{aligned} P(E_i) &= P(\mathbf{x}_i \leqslant y_0) \\ &= \int_0^{y_0} f_i(x)\, dx \end{aligned} \tag{8.23}$$

where y_0 is the deterministic strength and $f_i(x)$ represents the p.d.f. of the r.v. stress $\mathbf{x}_i$ during the ith cycle. In particular, if $f_1(x) = f_2(x) = \ldots = f_n(x) = f(x)$, then

$$R_n = (P(E_1))^n = \left\{ \int_0^{y_0} f(x)\, dx \right\}^n \tag{8.24}$$

case 8 random-independent stress and random-fixed strength

This case is similar to case 6 and we have

$$\begin{aligned} R_n &= P\left[(\mathbf{x}_1 < \mathbf{y}) \cap (\mathbf{x}_2 < \mathbf{y}) \cap \ldots \cap (\mathbf{x}_n < \mathbf{y})\right] \\ &= P\left[\max(\mathbf{x}_1, \ldots, \mathbf{x}_n) < \mathbf{y}\right] \end{aligned}$$

Let $\mathbf{x}_{\max} = \max(\mathbf{x}_1, \mathbf{x}_2, \ldots, \mathbf{x}_n)$. Then the distribution $F_n(x)$ of $\mathbf{x}_{\max}$ is given by

$$F_n(x) = [F(x)]^n$$

and hence

$$R_n = \int_0^\infty g(y)[F(y)]^n \, dy \tag{8.25}$$

If the stresses for each load application are independent but not identically distributed, then Equation 8.25 becomes

$$R_n = \int_0^\infty g(y) \left[\prod_{k=1}^{n} F_{\mathbf{x}_k}(y) \right] dy \tag{8.26}$$

where $\prod_{k=1}^{n} F_{\mathbf{x}_k}(y)$ is the distribution function for the maximum in a series of n independent stresses.

Another way to derive Equation 8.25 is based on the notion of conditional probabilities and conditional densities. Using the compound probability law, we have

$$R_n = P[E_1,\ldots,E_n] = P(E_1)P(E_2|E_1)\ldots P(E_n|E_1,\ldots,E_{n-1}) \tag{8.27}$$

or this can be expressed as a recurrence relation

$$R_n = R_{n-1}P(E_n|E_1,\ldots,E_{n-1}) \tag{8.28}$$

Now, let us evaluate the conditional probabilities in Equation 8.27. First we have

$$P(E_1) = P(\mathbf{x}_1 < \mathbf{y}) = \int_0^\infty g(y)\left\{\int_0^y f(x)\,dx\right\}dy \tag{8.29}$$

The nonfailure of a component on each successive application of stress reduces the uncertainty about the strength random variable. Thus, even though the actual system strength is unknown, it is evident that it must be greater than the unknown magnitude of stress that was applied. Thus the conditional strength density depends on the previous load history. In order to find $P(E_n|E_1,\ldots,E_{n-1})$, we replace $g(y)$ in Equation 8.29 by appropriate conditional strength density. Thus, we have

$$\begin{aligned} P(E_n|E_1,\ldots,E_{n-1}) &= P(\mathbf{x}_n < \mathbf{y}|E_1,\ldots,E_{n-1}) \\ &= \int_0^\infty g_n(y|E_1,\ldots,E_{n-1})\left\{\int_0^y f(x)\,dx\right\}dy \end{aligned} \tag{8.30}$$

where $g_n(y|E_1,\ldots,E_{n-1})$ is the conditional p.d.f. of random fixed strength $\mathbf{y}$ given nonfailures were observed on each of the first $(n-1)$ cycles. The conditional probability element is written as

$$\begin{aligned} g_n(y^*|E_1,\ldots,E_{n-1})\,dy &= P\left[y^* - \frac{dy}{2} \leqslant \mathbf{y} \leqslant y^* + \frac{dy}{2}\,|E_1,\ldots,E_{n-1}\right] \\ &= \frac{P\left[y^* - \frac{dy}{2} \leqslant \mathbf{y} \leqslant y^* + \frac{dy}{2}\,; E_1,\ldots,E_{n-1}\right]}{P[E_1,\ldots,E_{n-1}]} \end{aligned} \tag{8.31}$$

Let event A be defined as $y^* - (dy/2) \leqslant \mathbf{y} \leqslant y^* + (dy/2)$ and event E_i as $\mathbf{x}_i < y^*$, and further, event B_n as $E_1 \cap E_2 \cap \ldots \cap E_{n-1}$. Then, Equation 8.31 can be written as

$$g_n(y^*|E_1,\ldots,E_{n-1})\,dy = \frac{P(A \cap B_n)}{P(B_n)}$$

Again, we have

$$\begin{aligned} P(A \cap B_n) &= P(B_n|A)\cdot P(A) \\ &= P(A)P(E_1,\ldots,E_{n-1}|A) \\ &= P(A)P(E_1|A)P(E_2|A \cap E_1)\ldots P(E_{n-1}|A \cap E_1 \cap \ldots \cap E_{n-2}) \end{aligned} \tag{8.32}$$

Since the stresses are mutually independent, each of the last $(n-1)$ terms in Equation 8.32 is nothing but $\int_0^{y^*} f(x)dx$. Thus

$$P(A \cap B_n) = P(y^* + dy/2 \leqslant \mathbf{y} \leqslant y^* - dy/2)\left[\int_0^{y^*} f(x)\,dx\right]^{n-1}$$

$$= g(y^*)\,dy\left[\int_0^{y^*} f(x)\,dx\right]^{n-1}$$

Thus, in terms of the density function Equation 8.31 may be written as

$$g_n(y|E_1,\ldots,E_{n-1}) = \frac{g(y)\left[\int_0^{y} f(x)\,dx\right]^{n-1}}{R_{n-1}} \tag{8.33}$$

We note that this is the well-known hazard function. Thus, substituting Equation 8.33 in Equation 8.30, we get the conditional probability of success for the nth occurrence as

$$P[E_n|E_1,\ldots,E_{n-1}] = \frac{\int_0^{\infty} g(y)\left[\int_0^{y} f(x)\,dx\right]^{n} dy}{R_{n-1}} \tag{8.34}$$

Comparing Equation 8.34 with Equation 8.28, we have

$$R_n = \int_0^{\infty} g(y)\left[\int_0^{y} f(x)\,dx\right]^{n} dy$$

which is the same as Equation 8.25.

case 9 random-independent stress and random-independent strength

Let $f_i(x)$ and $g_i(y)$ be p.d.f.'s of stress $\mathbf{x}_i$ and strength $\mathbf{y}_i$ respectively in cycle $i = 1, 2, \ldots$. Then, since $\mathbf{x}_i$'s and $\mathbf{y}_i$'s are independent,

$$R_n = P[E_n, E_{n-1}, \ldots, E_1]$$

$$= P(E_n)\cdot P(E_{n-1})\cdot \ldots \cdot P(E_1)$$

$$= \prod_{i=1}^{n} P(E_i)$$

where

$$P(E_i) = P(\mathbf{x}_i \leqslant \mathbf{y}_i)$$

$$= \int_0^{\infty} f_i(x) \int_x^{\infty} g_i(y)\,dy\,dx$$

In particular, if f and g do not change with time, then

$$R_n = \prod_{i=1}^{n} P(E_i) = (P(E_1))^n \tag{8.35}$$

where $P(E_1)$ is nothing but the reliability for the stress-strength interference models discussed in Chapter 6.

8.3 RELIABILITY COMPUTATIONS FOR RANDOM CYCLE TIMES

In the previous section, all the reliabilities were computed in terms of cycles of stresses and/or strengths. To connect these to reliabilities in terms of time is very simple if cycle times, constant or varying, are deterministically known. However, if the cycles occur at random times, then

$$R(t) = \sum_{i=0}^{\infty} \pi_i(t) R_i \tag{8.36}$$

where $\pi_i(t)$ is the probability of i cycles occurring in the time interval $[0, t]$, and R_i is, as before, the probability of all i successes. Clearly, the case of deterministic cycle times becomes a special case of the Equation 8.36. In some cases it is appropriate to assume that the number of cycles occurring in a given time interval are Poisson distributed. Hence,

$$\pi_k(t) = P(N_t = k) = \frac{e^{-\alpha t}(\alpha t)^k}{k!} \tag{8.37}$$

where α is a parameter equal to the mean occurrences per unit time. It should be pointed out that other distributions may also be applicable but they lead to more complicated results.

In this section, we will derive expressions for reliability $R(t)$ for all the nine cases using Equations 8.36 and 8.37. We shall assume a Poisson distribution for $\pi_k(t)$ and borrow the appropriate expressions for R_i from the previous section.

case 1 deterministic stress and deterministic strength

Let the stress x_i be known and nondecreasing and the strength y_i be known and nonincreasing. Let n^* be such that $R_{n^*} = 1$ and $R_{n^*+1} = 0$. Then we have $R_i = 1$ for $i = 0, 1, 2, \ldots, n^*$ and $R_i = 0$ for $i = n^* + 1, n^* + 2, \ldots$. Hence

$$R(t) = \sum_{i=0}^{\infty} \pi_i(t) \cdot R_i = \sum_{i=0}^{n^*} \pi_i(t) = \sum_{i=0}^{n^*} \frac{e^{-\alpha t}(\alpha t)^i}{i!} \tag{8.38}$$

In particular, if $x_i = x_0$ and $y_i = y_0$, $i = 1, 2, \ldots$, then n^* is either zero $(x_0 > y_0)$ or infinity

($x_0 \leq y_0$), yielding, respectively,

$$R(t) = \pi_0(t)R_0 = e^{-\alpha t} \cdot 1 = e^{-\alpha t} \tag{8.39}$$

or

$$R(t) = \sum_{i=0}^{\infty} \pi_i(t)R_i = 1 \cdot \sum_{i=0}^{\infty} \pi_i(t) = 1 \tag{8.40}$$

case 2 deterministic stress and random-fixed strength

Let the stress be a known constant x_0. Let $\mathbf{y}_i = \mathbf{y}_0$, $i = 1, 2, \ldots$, be the r.v. strength with a known p.d.f. $f_0(y_0)$. Then

$$R_i = P[E_i] = \int_{x_0}^{\infty} f_0(y_0)\, dy_0 \triangleq R, \qquad i = 1, 2, \ldots \tag{8.41}$$

Clearly, the expression for R_i is independent of the cycle number i. Hence

$$\begin{aligned} R(t) &= \sum_{i=0}^{\infty} \pi_i(t) \cdot R_i \\ &= \pi_0(t)R_0 + \sum_{i=1}^{\infty} \pi_i(t)R_i \\ &= \frac{e^{-\alpha t}(\alpha t)^0}{0!} \cdot (1) + R \sum_{i=1}^{\infty} \pi_i(t) \\ &= e^{-\alpha t} + R(1 - \pi_0(t)) \\ &= e^{-\alpha t} + R(1 - e^{-\alpha t}) \\ &= R + (1 - R)e^{-\alpha t} \end{aligned} \tag{8.42}$$

where R is given by Equation 8.41. Note that R happens to be the reliability under static conditions given in Chapter 6. Also note that we have forced all a_i's in the previous section to be zero. We would fail to get a close form expression such as Equation 8.42 if the a_i's were not assumed to be zero. If the strength was allowed to decrease with time, the reliability R_i would depend on the cycle number i, and the resulting reliability expression would be more involved.

case 3 deterministic stress and random-independent strength

Let the stress be a known constant x_0. Let $g(y)$ be the p.d.f. of random independent strength $\mathbf{y}$. Then

$$R_n = R^n, \qquad n = 1, 2, \ldots,$$

where

$$R = \int_{x_0}^{\infty} g(y)\, dy \tag{8.43}$$

Note that R is the reliability for one stress cycle.

$R_n = R^n$ is also true for $n=0$ as $R_0 \equiv 1$. Hence,

$$
\begin{aligned}
R(t) &= \sum_{i=0}^{\infty} \pi_i(t)\cdot R_i \\
&= \sum_{i=0}^{\infty} \frac{e^{-\alpha t}(\alpha t)^i}{i!} R^i \\
&= \frac{e^{-\alpha t}}{e^{-R\alpha t}} \cdot \sum_{i=0}^{\infty} \frac{e^{-R\alpha t}(R\alpha t)^i}{i!} \\
&= e^{-\alpha t + R\alpha t}\cdot 1 \\
&= e^{-\alpha t(1-R)}
\end{aligned}
\tag{8.44}
$$

where

$$R = \int_{x_0}^{\infty} g(y)\,dy$$

case 4 random-fixed stress and deterministic strength

By reciprocity to case 2, we write

$$R(t) = R + (1-R)e^{-\alpha t} \tag{8.45}$$

where $R = \int_0^{y_0} f_0(x_0)\,dx_0$, is the reliability for one stress cycle. Here $\mathbf{x}_0$ is the random fixed stress with a known p.d.f. $f_0(x_0)$ that does not vary with time and y_0 is the deterministically known strength that is constant in time.

case 5 random-fixed stress and random-fixed strength

In this case, we will let $\mathbf{x}_0$ and $\mathbf{y}_0$ be the random fixed stress and strength with known p.d.f.'s $f_0(x_0)$ and $g_0(y_0)$ respectively. $\mathbf{x}_0$ and $\mathbf{y}_0$ will be assumed not to vary with time; that is, $a_i = b_i = 0$, $i = 1, 2, \ldots$. Hence,

$$R_i = \int_0^{\infty} g_0(y_0) \int_0^{y_0} f_0(x_0)\,dx_0\,dy_0 = R, \qquad i = 1, 2, \ldots \tag{8.46}$$

Once again R_i is independent of i. Hence

$$
\begin{aligned}
R(t) &= \sum_{i=0}^{\infty} \pi_i(t)\cdot R_i \\
&= \pi_0(t)\cdot R_0 + \sum_{i=1}^{\infty} \pi_i(t)\cdot R_i \\
&= e^{-\alpha t}\cdot 1 + R(1-e^{-\alpha t}) \\
&= R + (1-R)e^{-\alpha t}
\end{aligned}
\tag{8.47}
$$

where R is given by Equation 8.46.

case 6 random-fixed stress and random-independent strength

We have

$$R_n = \int_0^\infty f(x)\left(\int_x^\infty g(y)\,dy\right)^n dx, \qquad n=0,1,2,\ldots, \tag{8.48}$$

where $f(x)$ is the p.d.f. of random fixed stress $\mathbf{x}$ and $g(y)$ is the p.d.f. of random independent strength $\mathbf{y}$. Hence,

$$R(t) = \sum_{i=0}^{\infty} \pi_i(t)\cdot R_i$$

$$= \sum_{i=0}^{\infty} \frac{e^{-\alpha t}(\alpha t)^i}{i!}\cdot \int_0^\infty f(x)\left(\int_x^\infty g(y)\,dy\right)^i dx \tag{8.49}$$

Interchanging the order of the first integration and summation yields

$$R(t) = e^{-\alpha t}\int_0^\infty f(x)\left\{\sum_{i=0}^{\infty} \frac{\left(\alpha t\int_x^\infty g(y)\,dy\right)^i}{i!}\right\} dx$$

$$= \int_0^\infty f(x)\cdot e^{-\alpha t}\exp\left\{\alpha t\int_x^\infty g(y)\,dy\right\} dx$$

Defining $G(x) = \int_0^x g(y)dy$ and substituting, we get

$$R(t) = \int_0^\infty f(x)e^{-\alpha t}\cdot e^{\alpha t(1-G(x))}\,dx$$

$$= \int_0^\infty f(x)e^{-\alpha t G(x)}\,dx \tag{8.50}$$

For small (αt), the first few terms of the infinite series in Equation 8.49 may serve as a good approximation for $R(t)$.

case 7 random-independent stress and deterministic strength

By reciprocity with case 3, we get

$$R(t) = e^{-\alpha t(1-R)} \tag{8.51}$$

where $R = \int_0^{y_0} f(x)dx$, is the reliability for one stress cycle. The random independent stress is denoted by $\mathbf{x}$ and the deterministic strength by y_0 which is constant in time.

case 8 random-independent stress and random-fixed strength

By reciprocity with case 6, we write

$$R(t)=\sum_{i=0}^{\infty}\frac{e^{-\alpha t}(\alpha t)^i}{i!}\int_0^{\infty}g(y)\left(\int_0^y f(x)\,dx\right)^i dy \tag{8.52}$$

$$=\int_0^{\infty}g(y)\cdot e^{-\alpha t}\cdot\sum_{i=0}^{\infty}\frac{\left(\alpha t\int_0^y f(x)\,dx\right)^i}{i!}\,dy$$

$$=\int_0^{\infty}g(y)\cdot e^{-\alpha t}\cdot e^{\alpha t F(y)}\cdot dy$$

where $F(y)=\int_0^y f(x)\,dx$. Hence,

$$R(t)=\int_0^{\infty}g(y)e^{-\alpha t(1-F(y))}\,dy \tag{8.53}$$

For small values of αt, the first few terms of the infinite series in Equation 8.52 may adequately approximate the exact answer given by Equation 8.53.

case 9 random-independent stress and random-independent strength

Let $f(x)$ and $g(y)$ represent the p.d.f.'s for stress **x** and strength **y** respectively. Further let the r.v.'s be independent on each cycle. Then

$$R_n=R^n,\qquad n=1,2,\ldots,\qquad\text{and}\qquad R_0\equiv R^0=1,$$

where $R=\int_0^{\infty}f(x)\int_x^{\infty}g(y)\,dy\,dx$, is the reliability for one stress cycle. Hence,

$$R(t)=\sum_{i=0}^{\infty}\pi_i(t)\cdot R_i$$

$$=\sum_{i=0}^{\infty}\frac{e^{-\alpha t}(\alpha t)^i}{i!}\cdot R^i$$

$$=e^{-\alpha t}\cdot e^{R\alpha t}\sum_{i=0}^{\infty}\frac{e^{-R\alpha t}(R\alpha t)^i}{i!}$$

$$=e^{-\alpha t+R\alpha t}\cdot 1$$

$$=e^{-\alpha t(1-R)} \tag{8.54}$$

8.4 RELIABILITY IN CASE OF AGING, CYCLIC DAMAGE, AND CUMULATIVE DAMAGE

In many situations the probability density functions for the stress and the strength change with time and/or number of previous stress applications. If they change with the passage of time, the effect is called aging and results in a decrease in strength. If the changes in the strength are functions of number of stress occurrences, the effect is called cyclic damage. In the case of cumulative damage, the strength is affected by both the number of stress occurrences and their magnitude.

Consider the case of independent stresses and random-fixed strength with a known cycle dependence. The strength may be considered as a normal random variable with a known variance but a variable mean given by

$$E(\mathbf{y}_n) = a + be^{-cn}, \quad b > 0, \quad c > 0 \tag{8.55}$$

Here the expected value decreases from an initial value of $(a+b)$ to a final value of a. Thus, the conditional density function $g_n(y,n)$ is given by

$$g_n(y,n) = P\left[\, y \leqslant \mathbf{y}_n \leqslant y + dy \,|\, (n-1)\text{successes}\right] \tag{8.56}$$

This should be substituted in case 6 to get the corresponding reliability expressions. Another way to introduce cycle dependence is to change the stresses with n. Thus, we replace $f(x)$ by $f(x,n)$. Thus, in case 8, we have, instead of Equation 8.33, the following equation:

$$g_n(y,n) = g_1(y)\left[\int_0^y f(x,1)\,dx\right]\left[\int_0^y f(x,2)\,dx\right]\cdots\left[\int_0^y f(x,n)\,dx\right] \Big/ R_{n-1} \tag{8.57}$$

In the case of aging the strength is a function of time t instead of just the number of stress occurrences. The conditional probability density function, like that defined by Equation 8.31, will take the following form:

$$g_n(y;t_1,t_2,\ldots,t_n)\,dy$$

$$= P\left[\, y - \frac{dy}{2} \leqslant \mathbf{y} \leqslant y + \frac{dy}{2} \,|\, \text{successes at times } t_1,t_2,\ldots,t_n\right] \tag{8.58}$$

Using Equation 8.58, reliability functions defined previously can be generalized. For the cumulative damage models, if we assume that the cumulative damage weakens the strength of the part at each occurrence by an amount proportional to the applied stress, we get a cumulative damage law based on the sum of the applied stresses given by

$$\mathbf{y}_n = \mathbf{y}_1 - \sum_{i=1}^{n} c_i \mathbf{x}_i \tag{8.59}$$

where the c_i's are proportionality constants.

Let us now model the situation where the strength changes with time or the number of load occurrences. Let $\mathbf{y}_k$ be the strength after k load occurrences, and $\mathbf{y}_k$ be modeled as a deterministic function of the initial (random) strength $\mathbf{y}_0$ and k. Thus, we have

$$\mathbf{y}_k = h(\mathbf{y}_0, k) \tag{8.60}$$

One example of the above function may be

$$\mathbf{y}_k = \mathbf{y}_0 \phi(k) \tag{8.61}$$

where $\phi(k)$ is a monotonically decreasing function of k, $k = 1, 2, \ldots$. Let us assume that the stresses $\mathbf{x}_0, \mathbf{x}_1, \mathbf{x}_2, \ldots, \mathbf{x}_n$ are independent with distribution functions $F_0, F_1, F_2, \ldots, F_n$, respectively. Then

$$R_n = P\left[(\mathbf{x}_0 < \mathbf{y}_0) \cap (\mathbf{x}_1 < \mathbf{y}_1) \cap \ldots \cap (\mathbf{x}_n < \mathbf{y}_n)\right] \tag{8.62}$$

Since the strength $\mathbf{y}_k$ is a known function of the initial strength $\mathbf{y}_0$, we may rewrite Equation 8.62 as

$$R_n = P\left[(\mathbf{x}_0 < \mathbf{y}_0) \cap (\mathbf{x}_1 < h(\mathbf{y}_0, 1)) \cap \ldots \cap (\mathbf{x}_n < h(\mathbf{y}_0, n))\right] \tag{8.63}$$

If the initial strength $\mathbf{y}_0$ is between y and $y + dy$, we may express the probability of success (defined by event E) as given by Equation 8.64 because of the independence of the stresses

$$P\left[E \mid y < \mathbf{y}_0 < y + dy\right] = P\left[\mathbf{x}_1 < y\right] P\left[\mathbf{x}_2 < h(y, 1)\right] \ldots P\left[\mathbf{x}_n < h(y, n)\right] \tag{8.64}$$

Hence, the reliability R_n is

$$R_n = \int_0^{\infty} F_0(y) F_1(h(y,1)) F_2(h(y,2)) \ldots F_n(h(y,n))\, g(y)\, dy \tag{8.65}$$

where $g(y)$ is the p.d.f. of the initial strength. If $\mathbf{y}_k$ is given by Equation 8.61 and all the stresses are independent and identically distributed, we have

$$R_n = \int_0^{\infty} g(y) \left\{ \prod_{i=0}^{n} F(y\phi(i)) \right\} dy \tag{8.66}$$

R_n given by Equation 8.65 may be approximated as follows. Let $\bar{F} = 1 - F$, and $\mathbf{y}_0 = h(\mathbf{y}_0, 0)$. Then

$$R_n = \int_0^\infty g(y)\left\{ \prod_{i=0}^{n} (1 - \bar{F}_i(h(y,i)) \right\} dy \tag{8.67}$$

Now, we have the following inequality

$$\prod_{i=0}^{n} (1 - \bar{F}_i(h(y,i)) > 1 - \sum_{i=0}^{n} \bar{F}_i(h(y,i)) \tag{8.68}$$

and thus the lower bound on R_n is given by

$$R_n > 1 - \sum_{i=1}^{n} \int_0^\infty g(y)\bar{F}_i(h(y,i))\, dy \tag{8.69}$$

If $\Sigma_{i=0}^{n} \bar{F}_i(h(y,i)) \ll 1$, then the lower bound given by Equation 8.69 is fairly close to R_n. Now, the probability of failure on the ith stress application is

$$P_f(i) = \int_0^\infty g(y)\bar{F}_i(h(y,i))\, dy \tag{8.70}$$

Hence, R_n can be expressed as

$$R_n > 1 - \sum_{i=0}^{n} P_f(i) \approx \exp\left[- \sum_{i=0}^{n} P_f(i) \right] \tag{8.71}$$

The relations given by Equation 8.71 are close approximations to R_n when

$$\sum_{i=0}^{n} P_f(i) \ll 1$$

Actual computations of $R(t)$ are complex, but it may be possible to establish upper and lower bounds by sliding to adjacent easier cases. How tight these bounds would be depends on the specific situation.

Next we solve a few examples that may help clarify any points that may not have been obvious to the reader of the preceding theoretical derivations.

EXAMPLE 8.1

A machine tool component is known to be stressed cyclically with uniformly increasing values on each successive cycle. The actual value of the initial stress is a normally distributed random variable with mean $\mu_{\mathbf{x}} = 70$ MPa and standard deviation $\sigma_{\mathbf{x}} = 3$ MPa. It has been observed that the stress values obey the following linear law:

$$\mathbf{x}_i = \mathbf{x}_0 + ai, \qquad i = 1, 2, \ldots$$

where

$$\mathbf{x}_i = \text{the stress on the } i\text{th cycle}$$
$$\mathbf{x}_0 = \text{the initial stress}$$
$$a = \text{a constant}$$

The strength of the component material is also a normally distributed random variable. The initial strength has $\mu_{\mathbf{y}} = 95$ MPa and $\sigma_{\mathbf{y}} = 4$ MPa. But the strength can be assumed to deteriorate according to the law

$$\mathbf{y}_i = y_0 - bi, \qquad i = 1, 2, \ldots$$

where

$$\mathbf{y}_i = \text{the strength on the } i\text{th cycle}$$
$$\mathbf{y}_0 = \text{the initial strength}$$
$$b = \text{a constant}$$

If $a = 0.001$ MPa/cycle and $b = 0.0001$ MPa/cycle, we wish to find the reliability after 10,000 cycles.

In this case, we have both stress and strength as random-fixed. Substituting in the standard coupling equation we get, at 10,000 cycles,

$$z = -\frac{\mu_{\mathbf{y}} - \mu_{\mathbf{x}}}{\sqrt{\sigma_{\mathbf{y}}^2 + \sigma_{\mathbf{x}}^2}}$$

$$= -\frac{[95 - (10^{-4})(10^4)] - [70 + (10^{-3})(10^4)]}{\sqrt{3^2 + 4^2}}$$

$$= -2.8$$

Hence

$$R_{10^4} = 1 - \Phi[z]$$
$$= 1 - 0.00256 = 0.99744$$

Note that working out Equation 8.18 would yield the same reliability.

EXAMPLE 8.2

If, in Example 8.1, no deterioration of strength with cycles is assumed, how much longer will it be before the reliability drops to the original value at 10,000 cycles?

If we wish the same reliability as before, we need to have the same z; that is, we need to have the same $\mu_{\mathbf{y}} - \mu_{\mathbf{x}}$ as before, the denominator remaining unchanged. Hence

$$[95] - [70 + (10^{-3})(n)] = [95 - (10^{-4})(10^4)] - [70 + (10^{-3})(10^4)]$$

or

$$n = 11{,}000 \text{ cycles}$$

Thus the component will attain the same reliability after 1,000 extra cycles, if the deterioration is prevented.

EXAMPLE 8.3

The reliability of an airplane component found by SSI methods is 0.99, which is constant for all cycles. If the cycles are Poisson distributed with a mean of 0.5 cycle per hour, find the reliability of the component at time = 200 hours.

Using Equation 8.47, we get

$$
\begin{aligned}
R(200) &= 0.99 + (1-0.99)e^{-(0.5)(200)} \\
&= 0.99 + 0.01e^{-100} \\
&= 0.9900045
\end{aligned}
$$

8.5 SUMMARY

The single stress-strength models considered in Chapter 6 are good only for single stress situations and must be modified to describe the commonly occurring situation of repeated stresses. This chapter developed models that consider the repeated application of stresses as well as the change in the distribution of the strength with time, which may be caused by aging and/or cumulative damage. All the possible stress and strength combinations were considered for completeness of the analysis. Two stress occurrence models were considered for repeated stresses, namely: deterministic occurrence and random occurrence governed by the Poisson law. In many practical situations, stress and strength vary with time as a result of varying loads or degradation of materials. This introduces a great deal of complexity in analytical models. Simulation models can be developed [8] to study the failure phenomena for a component subjected to random loading and strength degradation due to aging, cyclic damage, and/or cumulative damage.

EXERCISES

1 The stresses acting on a component on various cycles are independent and identically distributed as normal random variables with $\mu = 20.0$ and $\sigma = 3.0$. The strength of the component is a normal random variable at time zero with $\mu = 45.0$ and $\sigma = 4.0$. The strength of the component changes with each application of the load and follows the following law:

$$\mathbf{y}(k) = \mathbf{y}_0 \phi(k)$$

where $\phi(k) = 1 - 0.0001k$.

The time between the occurrences of stresses is a random variable with exponential density function with $\lambda = 10$. Find an expression for the reliability of the component as a function of time t. Show how you would compute the reliability of the component at $t = 50$.

2 Solve Problem 1 when the stress and the strength random variables are exponentially distributed with mean values of 20.0 and 45.0, respectively. All the other data is the same as given in Problem 1.

3 The strength of the component in Problems 1 and 2 changes with time according to the following law:

$$y(t) = y(0)\phi(t)$$

where $\phi(t)$ is a linear function such that

$$\phi(0) = 1 \qquad \text{and} \qquad \phi(100) = 0.3$$

How will you find the reliability of the component at time t given that n stress applications have occurred in time zero to t for both Problems 1 and 2?

4 Equation 8.75 gives an expression for R_n for the case of cyclic damage. Develop an expression for R_n for the case of cumulative damage under suitable assumptions.

5 Discuss the role of simulation models for time dependent stress-strength models.

BIBLIOGRAPHY

1 Bratt, M. J., Reethoff, G., and Wieber, G. W., "A Model for Time Varying and Interfering Stress/Strength Probability Density Distributions with Consideration for Failure Incidence and Property Degradation," Proceedings Third Annual Aerospace Reliability and Maintainability Conference, 1969, pp. 566–575.

2 Freudenthal, A. M. et al, "The Analysis of Structural Safety," *Journal of the Structural Division, ASCE*, Vol. 92, No. 572, February 1966, pp. 267–325.

3 Parzen, E., *Stochastic Processes*, Holden-Day, San Francisco, 1962.

4 Schatz, R., Shooman, M., and Shaw, L., "Application of Time Dependent Stress-Strength Models of Non-Electrical and Electrical Systems," Proceedings Reliability and Maintainability Symposium, January 1974, pp. 540–547.

5 Schatz, R., "Time/Cycle Reliability Modeling Techniques Utilizing the Stress-Strength Interference Method," Westinghouse Astronuclear Laboratory Memo No. 54071.

6 Shaw, L., Shooman, M., and Schatz, R., "Time Dependent Stress-Strength Models for Non-Electrical and Electrical Systems," Proceedings Reliability and Maintainability Symposium, January, 1973, pp. 186–197.

7 Sweet, A. L. and Kozin, F., "Investigation of a Random Cumulative Damage Theory," *Journal of Materials*, Vol. 3, No. 4, 1968.

8 Taraman, S. I., "Design Reliability Models and Determination by Stress-Strength Interference Theory," unpublished Ph.D. Dissertation, Department of Industrial Engineering & Operations Research, Wayne State University, Detroit, Michigan, 1975.

9 Tumolillo, T. A., "Methods for Calculating the Reliability Function for Systems Subjected to Random Stresses," IEEE Transactions on Reliability, October 1974.

Chapter 9 dynamic reliability models

Static models utilizing a constant reliability level were considered earlier. These models are now extended to dynamic models where the reliability level is time dependent. Time dependent models are more difficult to develop and evaluate than static models. Sections 9.1 through 9.4 consider various dynamic reliability models, but the emphasis changes in Section 9.5, which introduces measures of system effectiveness that are more encompassing than the strict definition of reliability.

The dynamic reliability models covered in Sections 9.1 through 9.4 encompass the basic series model, the chain model, and various parallel models including the standby redundant and load sharing parallel models.

In Section 9.5 such measures as serviceability, availability, and maintainability are introduced. These measures add dimensions that should also be considered when assessing product reliability.

9.1 THE SERIES SYSTEM AND RELATED MODELS

The series system is one in which every component must function if the system is to function. Complex systems are sometimes subdivided into a series arrangement for analysis by properly grouping components into a unit such that the units are in series.

If we let $\mathbf{t}_i$ be the random variable of the time to failure for the ith component, then for an n component series system the system reliability is

$$R_S(t) = P[\mathbf{t}_1 > t \cap \mathbf{t}_2 > t \cap \cdots \cap \mathbf{t}_n > t] \tag{9.1}$$

If we assume independence then Equation 9.1 can be written as

$$R_S(t) = P(\mathbf{t}_1 > t)P(\mathbf{t}_2 > t) \cdots P(\mathbf{t}_n > t)$$

but by definition $P(\mathbf{t}_i > t) = R_i(t)$ and

$$R_S(t) = \prod_{i=1}^{n} R_i(t) \tag{9.2}$$

where $R_i(t)$ is the reliability function for the ith component.

The hazard function for a series system also is a rather convenient expression. Taking the natural logarithm of the series system reliability (Equation 9.2) we have

$$\ln R_S(t) = \sum_{i=1}^{n} \ln R_i(t) \tag{9.3}$$

Now recall that

$$R(t) = \exp\left[-\int_0^t h(\tau)\,d\tau\right]$$

which means that

$$\int_0^t h(\tau)\,d\tau = -\ln R(t)$$

or

$$h(t) = -\frac{d}{dt}\ln R(t) \tag{9.4}$$

Applying this to Equation 9.3 we first say that

$$-\frac{d}{dt}\ln R_S(t) = \sum_{i=1}^{n} -\frac{d}{dt}\ln R_i(t)$$

and then using Equation 9.4

$$h_S(t) = \sum_{i=1}^{n} h_i(t) \tag{9.5}$$

Thus, the system hazard function is the sum of the component hazard functions under the assumption of independence, regardless of the form of the component p.d.f.'s.

EXAMPLE 9.1

Let us consider an n component series system where the ith component has a constant failure rate λ_i. Here each component has an exponentially distributed time to failure. We will determine the system hazard rate and reliability function.

Assuming independence and applying Equation 9.5 the system hazard rate is

$$h_S(t) = \sum_{i=1}^{n} \lambda_i \tag{9.6}$$

and of course, this is a constant. Thus, the reliability function is

$$R_S(t) = \exp\left[-t \sum_{i=1}^{n} \lambda_i \right] \tag{9.7}$$

Here we can find the expected time to failure for the system by recalling that

$$E(\mathrm{t}) = \int_0^{\infty} R_S(t)\, dt$$

which in this case is

$$\mathrm{MTBF}_S = 1 / \sum_{i=1}^{n} \lambda_i$$

and one can make the intuitively obvious conclusion that the system MTBF will decrease as series components are added.

Unfortunately, if many p.d.f.'s other than the exponential are used for the component failure distribution, the results will not be as easy to obtain as in the previous example.

9.1.1 THE SERIES CHAIN MODEL

The series chain model is really not a time dependent model. It is presented here because the techniques used in its solution are consistent with others in this chapter.

The series chain model is a series system in that if any one component fails the system will fail; however, the concept of how a component fails is different. As an example of this concept of failure, consider a circuit composed of n identical components, and this circuit is subjected to thermal stresses. Let us assume for simplicity that the thermal stresses are the main cause of failure. In this situation the one component having the least resistance to the thermal stresses will be the first to fail. Then, in this case, the system reliability will be

$$R_S = \min_i R_i \tag{9.8}$$

where R_i is the reliability of the ith component and describes the component's resistance to failure from thermal stresses.

This model can be compared to a chain composed of n links where the chain will break if the applied stress exceeds the strength of any one link. Hence,

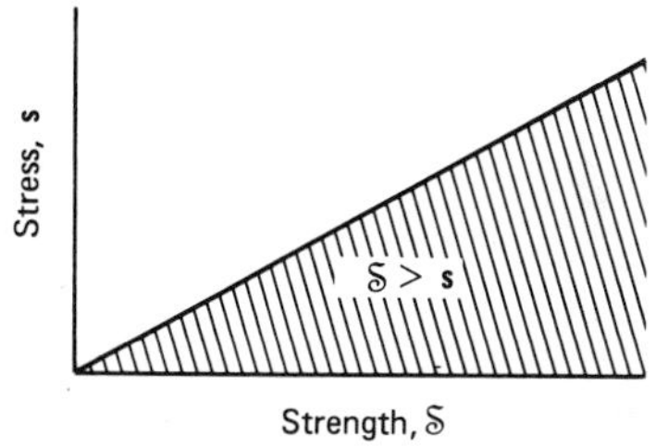

Figure 9.1 Area representing ($\mathbb{S} > s$).

the name "chain model" or "weakest link model" is used. Let

$$f_s(s) = \text{the p.d.f. for the stress random variable, } \mathbf{s}$$

and

$$f_{\mathbb{S}}(\mathbb{S}) = \text{the p.d.f. for the strength random variable, } \mathbb{S}$$

then the reliability for any one link is

$$R_i = P(\mathbb{S} > \mathbf{s})$$

and considering Figure 9.1 (see also Equation 6.3), this reliability is

$$R_i = \int_0^{\infty} \int_s^{\infty} f_s(s) f_{\mathbb{S}}(\mathbb{S})\, d\mathbb{S}\, ds \tag{9.9}$$

Or this can be written as

$$R_i = \int_0^{\infty} f_s(s)\left[1 - F_{\mathbb{S}}(s)\right] ds \tag{9.10}$$

Now if the chain is composed of n randomly selected links then this is equivalent to randomly selecting a sample of size n from the strength distribution, $f_{\mathbb{S}}(\mathbb{S})$. Let $\mathbb{S}_n$ be the random variable representing the strength of the n-link chain. This will be

$$\mathbb{S}_n = \min_i (\mathbb{S}_i) \tag{9.11}$$

where $\mathbb{S}_i$ is the strength of the ith link. Applying the extreme value distribution concept (Section 2.7.4)

$$G(\mathbb{S}_n) = 1 - \left[1 - F_{\mathbb{S}}(\mathbb{S}_n)\right]^n \tag{9.12}$$

where $G(\mathbb{S}_n)$ is the cumulative distribution representing chain strength.

The system reliability for the chain is

$$R_n = P[\mathbb{S}_n > \mathbf{s}]$$

and, using the concept of Equation 9.10, this is

$$R_n = \int_0^{\infty} f_{\mathbf{s}}(s)\left[1 - F_{\mathbb{S}}(s)\right]^n ds \tag{9.13}$$

This gives the system reliability in terms of the number of components (n), the stress p.d.f. $f_{\mathbf{s}}(s)$ acting on the system, and the strength distribution $F_{\mathbb{S}}(\mathbb{S})$ for an individual component. Also, observe the similarity of Equation 9.13 with Equation 8.21.

9.2 PARALLEL SYSTEM MODELS

Duplicate components or alternate modes for sustaining system operation are forms of parallel models. The pure parallel system is one in which all components are initially activated, and any one component can sustain system operation. For an n component, pure parallel arrangement the system unreliability is

$$Q_S(t) = P[\mathbf{t}_1 < t \cap \mathbf{t}_2 < t \cap \cdots \cap \mathbf{t}_n < t] \tag{9.14}$$

and assuming independence this becomes

$$Q_S(t) = P(\mathbf{t}_1 < t)P(\mathbf{t}_2 < t)\cdots P(\mathbf{t}_n < t) \tag{9.15}$$

Here independence implies that the probability that the surviving components fail remains unchanged as components fail.

Substituting the reliability function into Equation 9.15 we have

$$Q_S(t) = \prod_{i=1}^{n} \left[1 - R_i(t)\right] \tag{9.16}$$

or

$$R_S(t) = 1 - \prod_{i=1}^{n} \left[1 - R_i(t)\right] \tag{9.17}$$

This is the system reliability for a pure parallel arrangement where all subsystems are energized at time zero. Next, some specific application of Equation 9.17 will be considered.

9.2.1 PARALLEL REDUNDANT SYSTEMS WITH EXPONENTIAL COMPONENT FAILURE DISTRIBUTIONS

Consider a special case of the parallel redundant system where each subsystem has an exponentially distributed time to failure. For the ith subsystem, the reliability function is

$$R_i(t) = e^{-t\lambda_i}, \qquad t \geqslant 0$$

where λ_i is the failure rate. Then by Equation 9.17 the system reliability is

$$R_S(t) = 1 - \prod_{i=1}^{n} (1 - e^{-t\lambda_i}) \tag{9.18}$$

Here it can be noted that the system time to failure is not exponentially distributed.

Consider now the two-subsystem case where $n = 2$. Expanding Equation 9.18, we have

$$R_S(t) = e^{-t\lambda_1} + e^{-t\lambda_2} - e^{-t(\lambda_1+\lambda_2)}, \qquad t \geqslant 0 \tag{9.19}$$

Then the expected time to system failure, found by integrating Equation 9.19 over the range of t, is

$$E_S(\mathbf{t}) = \frac{1}{\lambda_1} + \frac{1}{\lambda_2} - \frac{1}{\lambda_1 + \lambda_2} \tag{9.20}$$

The pattern for $R_S(t)$ and $E_S(\mathbf{t})$ when the subsystem failure distribution is exponential is easily deduced. For, when $n = 3$, we have

$$R_S(t) = e^{-t\lambda_1} + e^{-t\lambda_2} + e^{-t\lambda_3} - e^{-t(\lambda_1+\lambda_2)} - e^{-t(\lambda_1+\lambda_3)}$$
$$- e^{-t(\lambda_2+\lambda_3)} + e^{-t(\lambda_1+\lambda_2+\lambda_3)} \tag{9.21}$$

and

$$E_S(\mathbf{t}) = \frac{1}{\lambda_1} + \frac{1}{\lambda_2} + \frac{1}{\lambda_3} - \frac{1}{\lambda_1+\lambda_2} - \frac{1}{\lambda_1+\lambda_3} - \frac{1}{\lambda_2+\lambda_3} + \frac{1}{\lambda_1+\lambda_2+\lambda_3} \tag{9.22}$$

Thus the system reliability function and mean time to failure can be written down directly.

Now consider the case where all subsystems have the same failure rate λ. Then Equation 9.20 gives the mean time to failure for a two-component system as

$$E_S(\mathbf{t}) = \frac{2}{\lambda} - \frac{1}{2\lambda} = \frac{3}{2\lambda}$$

and, by Equation 9.22 for a three-component system, we have

$$E_S(\mathbf{t}) = \frac{3}{\lambda} - \frac{3}{2\lambda} + \frac{1}{3\lambda} = \frac{11}{6\lambda}$$

Or, in general for an n component system, the mean time to failure is

$$E_S(\mathbf{t}) = \sum_{i=1}^{n} \frac{\theta}{i} \tag{9.23}$$

where $\theta = 1/\lambda$. Here it can be seen that the marginal gain in the mean time to system failure decreases with each subsystem added

We now consider other forms of redundancy that are more appropriate to practical systems.

9.3 STANDBY REDUNDANT SYSTEMS

The standby redundant system configuration is a form of paralleling where only one subsystem is in operation; if the operating subsystem fails then another subsystem is brought into operation. For example, providing spare parts is a form of standby redundancy.

The standby redundant configuration is depicted in Figure 9.2. The switch (S) may represent an operator replacing a failed component or an elaborate piece of sensing and switching equipment. The switch can also have various modes of failure, and the inactive subsystem might be such that it can fail in standby. Thus, there are several different possibilities with standby redundant systems.

In the following analysis the subsystems will be designated $1, 2, \ldots, n$ in the order that they would be called into operation. The event E_i will be defined as the event that the ith subsystem is operating and $\mathbf{t}_i$ will be the random variable representing the life of the ith subsystem with p.d.f. $f_i(t_i)$.

9.3.1 THE PERFECT SWITCHING CASE

The assumption will now be made that the switch is failure free. Let us first consider a two-unit standby system. The possible modes for sustaining operation to time $\mathbf{t}$ are depicted in Figure 9.3.

The reliability function for a standby system that has n subsystems will be denoted by $R_S^n(t)$. Then considering the two-unit system, the success modes as

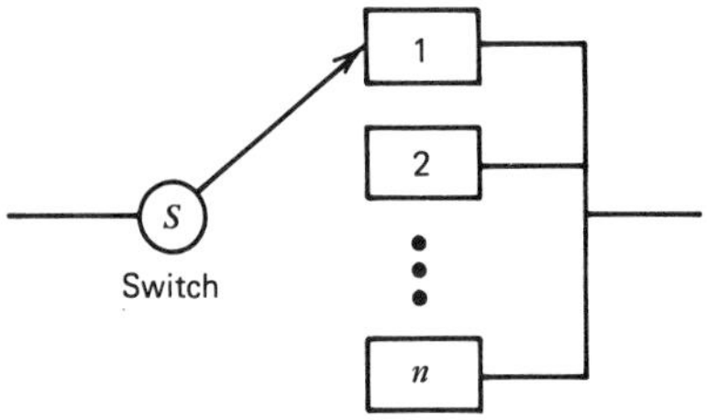

Figure 9.2 Standby redundant system.

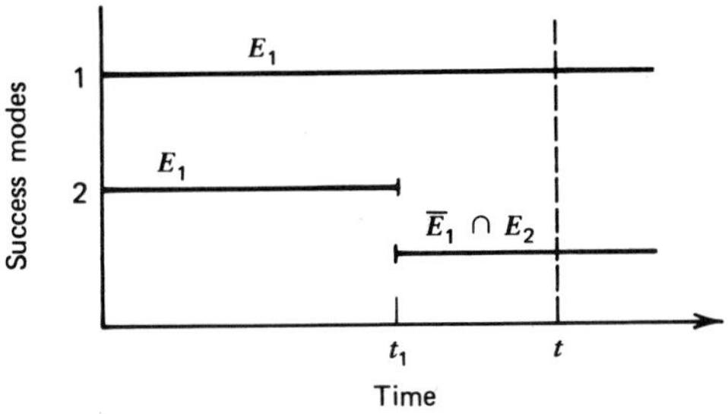

Figure 9.3 Success modes for a two-unit standby system.

depicted in Figure 9.3 lead to the following

$$R_S^2(t) = P[(\mathbf{t}_1 > t) \cup (\mathbf{t}_1 \leq t \cap \mathbf{t}_2 > t - \mathbf{t}_1)] \tag{9.24}$$

Because of the mutually exclusive success modes

$$R_S^2(t) = P[\mathbf{t}_1 > t] + P[\mathbf{t}_1 \leq t \cap \mathbf{t}_2 > t - \mathbf{t}_1]$$

This becomes

$$R_S^2(t) = R_1(t) + \int_0^t f_1(t_1) R_2(t - t_1)\, dt_1 \tag{9.25}$$

So for specified subsystem p.d.f.'s, Equation 9.25 gives the system reliability function for a two-unit standby redundant system.

If we consider the special case where all subsystems have a constant failure rate of λ, then Equation 9.25 reduces to

$$R_S^2(t) = e^{-\lambda t}(1 + \lambda t), \qquad t \geq 0 \tag{9.26}$$

Now we will proceed to the analysis of a three-unit standby system. The possible success modes are depicted in Figure 9.4. Here it can be seen that the

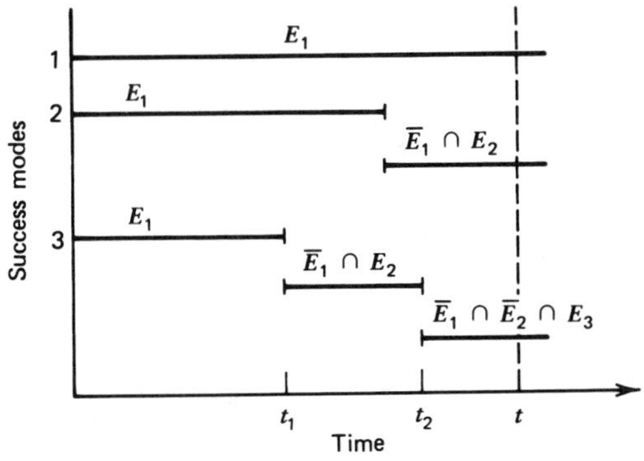

Figure 9.4 Success modes for a three-unit standby system.

first two modes are identical to those encountered in the two-unit standby system, and hence we only need consider the third success mode. The probability that this mode will occur is

$$P_3 = P[(\mathbf{t}_1 < t) \cap (\mathbf{t}_2 < t - \mathbf{t}_1) \cap (\mathbf{t}_3 > t - \mathbf{t}_1 - \mathbf{t}_2)]$$

This is found by

$$P_3 = \int_0^t f_1(t_1) \int_0^{t-t_1} f_2(t_2) R_3(t - t_1 - t_2)\, dt_2\, dt_1 \tag{9.27}$$

Then the system reliability is

$$R_S^3(t) = R_S^2(t) + P_3 \tag{9.28}$$

Here again if we consider the special case where each subsystem has a constant failure rate λ, Equation 9.28 reduces to

$$R_S^3(t) = e^{-\lambda t}[1 + \lambda t + (\lambda t)^2/2], \qquad t \geqslant 0 \tag{9.29}$$

It should now be obvious that if we want to consider a four-unit standby system we need only one additional success mode. This additional success mode would have the probability

$$P_4 = P[(\mathbf{t}_1 < t) \cap (\mathbf{t}_2 < t - \mathbf{t}_1) \cap (\mathbf{t}_3 < t - \mathbf{t}_1 - \mathbf{t}_2) \cap (\mathbf{t}_4 > t - \mathbf{t}_1 - \mathbf{t}_2 - \mathbf{t}_3)] \tag{9.30}$$

which is found by

$$P_4 = \int_0^t f_1(t_1) \int_0^{t-t_1} f_2(t_2) \int_0^{t-t_1-t_2} f_3(t_3) R_4(t - t_1 - t_2 - t_3)\, dt_3\, dt_2\, dt_1 \tag{9.31}$$

Then the reliability function for the four-unit standby system is

$$R_S^4(t) = R_S^3(t) + P_4 \tag{9.32}$$

Again if we consider the special case of a constant failure rate λ, then

$$P_4 = \frac{(\lambda t)^3}{6} e^{-\lambda t}$$

and

$$R_S^4(t) = e^{-\lambda t}[1 + (\lambda t) + (\lambda t)^2/2 + (\lambda t)^3/6], \qquad t \geqslant 0 \tag{9.33}$$

The general case is now easily recognized from Equation 9.33 as

$$R_S^n(t) = e^{-\lambda t} \sum_{i=0}^{n-1} (\lambda t)^i / i! \tag{9.34}$$

where λ is the failure rate and is identical for each standby unit.

We have now illustrated the general approach for obtaining the reliability function for standby systems. Problems in obtaining the reliability function can arise as a result of integration difficulties.

9.3.2 IMPERFECT SWITCHING

There are several ways that switch failure can occur in standby redundant systems. The possible modes of switch failure depend on the particular switching mechanism and system. Two possibilities are considered here.

Let us first look at a situation where the switch simply fails to operate when called upon. The probability that the switch performs when required is p_S. For the two-unit standby system it is easy to see that Equation 9.25 is modified as

$$R_S^2(t)' = R_1(t) + p_S \int_0^t f_1(t_1) R_2(t - t_1)\, dt_1 \tag{9.35}$$

Or for a three-unit system Equation 9.28 becomes

$$R_S^3(t)' = R_S^2(t)' + p_S^2 P_3 \tag{9.36}$$

Thus, the previous developments are easily modified to handle this form of static switch failure.

It will be assumed that the switch is a complex piece of equipment and has a constant failure rate of λ_s. Thus the reliability function for the switch is

$$R_S(t) = e^{-\lambda_s t}, \qquad t \geqslant 0$$

and the switch can fail before it is needed.

Now let us reconsider the two-unit standby system. The reliability at time t is

$$R_S^2(t)'' = P\left[(\mathbf{t}_1 > t) \cup (\mathbf{t}_1 \leqslant t \cap \mathbf{t}_s > \mathbf{t}_1 \cap \mathbf{t}_2 > t - \mathbf{t}_1)\right] \tag{9.37}$$

where $\mathbf{t}_s$ is the random variable representing time to switch failure. Equation 9.37 becomes

$$R_S^2(t)'' = R_1(t) + \int_0^t f_1(t_1) R_S(t_1) R_2(t - t_1)\, dt_1 \tag{9.38}$$

Or substituting in for $R_S(t_1)$

$$R_S^2(t)'' = R_1(t) + \int_0^t f_1(t_1) e^{-\lambda_s t_1} R_2(t-t_1)\, dt_1 \tag{9.39}$$

If we consider the special case where all subsystems have a constant failure rate λ, then Equation 9.39 reduces to

$$R_S^2(t)'' = e^{-\lambda t}\left[1 + \frac{\lambda}{\lambda_s}(1 - e^{-\lambda_s t})\right], \qquad t \geqslant 0 \tag{9.40}$$

We now examine the three-unit standby situation, and this will be sufficient to illustrate the necessary modifications to the previous developments.

For the three-unit standby system we need only add a third success mode to Equation 9.38. The additional success mode will have the probability

$$P_3'' = P\left[(\mathbf{t}_1 < t) \cap (\mathbf{t}_2 < t - \mathbf{t}_1) \cap (\mathbf{t}_3 > t - \mathbf{t}_1 - \mathbf{t}_2) \cap (\mathbf{t}_s > \mathbf{t}_1 + \mathbf{t}_2)\right] \tag{9.41}$$

and this becomes

$$P_3'' = \int_0^t f_1(t_1) \int_0^{t-t_1} f_2(t_2) R_3(t - t_1 - t_2) R_S(t_1 + t_2)\, dt_2\, dt_1 \tag{9.42}$$

Then the system reliability function is

$$R_S^3(t)'' = R_S^2(t)'' + P_3'' \tag{9.43}$$

Again considering the situation where each subsystem has a constant failure rate λ, Equation 9.43 yields

$$R_S^3(t)'' = e^{-\lambda t}\left[1 + \frac{\lambda}{\lambda_s}(1 - e^{-\lambda_s t})\right] + e^{-\lambda t}(\lambda/\lambda_s)^2\left[1 - e^{-\lambda_s t} - \lambda_s t e^{-\lambda_s t}\right], \qquad t \geqslant 0 \tag{9.44}$$

Many other types of switch failure may be encountered in practical situations. For example, a switch may fail to hold a subsystem on line, or the switch may inadvertently sense a failure. Each situation must be analyzed on an individual basis, and great simplicity is provided if the use of a constant failure rate is appropriate.

9.4 SHARED LOAD PARALLEL MODELS

In this configuration the parallel subsystems equally share the load, and as a subsystem fails, the surviving subsystems must sustain an increased load. Thus as successive subsystems fail, the failure rate of the surviving components

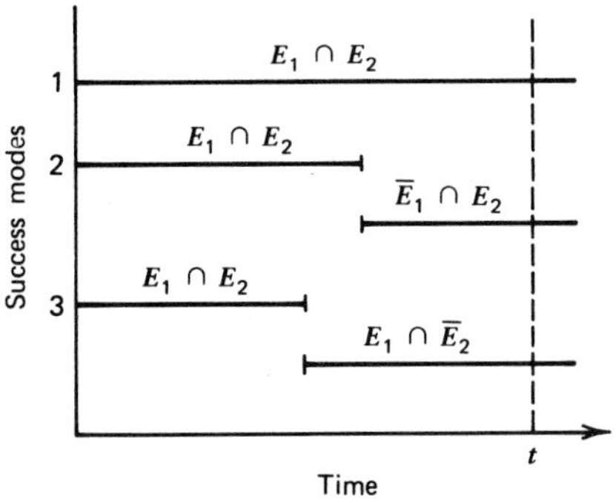

Figure 9.5 Success modes.

increases. An example of a shared parallel configuration would be when bolts are used to hold a machine member; if one bolt breaks the remainder must support the load.

Here analysis will be limited to the case of two subsystems. We define

$$h(t) = \text{p.d.f. for time to failure under half load}$$

$$f(t) = \text{p.d.f. for time to failure under full load}$$

It will be assumed that when a failure occurs the survivor then follows the p.d.f. $f(t)$, and that this p.d.f. does not depend on the interval of elapsed time.

For this situation the possible modes of survival are depicted in Figure 9.5. We consider the probability of each mode separately then add the probabilities because the events represented by each mode are mutually exclusive.

Consider the first mode where both components survive. The probability of this occurring is simply

$$P[\mathbf{t}_1 > t \cap \mathbf{t}_2 > t] = [R_h(t)]^2$$

where

$$R_h(t) = \int_t^{\infty} h(\tau)\, d\tau$$

Now consider the second mode. Here we see that

$$P[(\mathbf{t}_1 \leqslant t, \text{ under half load}) \cap (\mathbf{t}_2 > \mathbf{t}_1, \text{ under half load}) \cap (\mathbf{t}_2 > t - \mathbf{t}_1, \text{ under full load})]$$
$$= \int_0^t h(t_1) R_h(t_1) R_f(t - t_1)\, dt_1$$

where

$$R_f(t) = \int_t^{\infty} f(\tau)\, d\tau$$

The third mode is identical to the second if we assume that the components are identical. Thus the system reliability becomes

$$R_S(t) = [R_h(t)]^2 + 2\int_0^t h(t_1) R_h(t_1) R_f(t - t_1)\, dt_1 \tag{9.45}$$

For many p.d.f.'s the calculation of $R_S(t)$ is difficult because of the integration problem. However, as an easy example let us consider the special case where the failure rates are constant. If we let

$$\lambda_h = \text{half load failure rate}$$

and

$$\lambda_f = \text{full load failure rate}$$

then Equation 9.45 reduces to

$$R_S(t) = e^{-2\lambda_h t} + 2\int_0^t \lambda_h e^{-\lambda_h \tau} e^{-\lambda_h \tau} e^{-\lambda_f (t-\tau)}\, d\tau \tag{9.46}$$

or

$$R_S(t) = e^{-2\lambda_h t} + \frac{2\lambda_h}{(2\lambda_h - \lambda_f)} \left[e^{-\lambda_f t} - e^{-2\lambda_h t} \right], \qquad t \geqslant 0 \tag{9.47}$$

which is the system reliability for a two-unit shared parallel system.

The approach to dynamic models has been illustrated for various design configurations. There are many other possibilities, depending on the particular application. Also, if a constant failure rate is appropriate, another approach utilizing Markov processes can be employed. The Markov approach was not considered in this chapter, and Reference 4 is recommended as a source of further information.

9.5 SYSTEM EFFECTIVENESS MEASURES

So far this book has been primarily concerned with the evaluation of the reliability of a system. However, to a customer, reliability usually has a more encompassing meaning than that provided by the reliability function. For example, suppose a consumer must choose between a highly reliable product, that is very difficult to repair and a product that is slightly less reliable but easier to repair. In this situation it might be more economical to purchase the less reliable product. Clearly, other descriptors may be necessary to appropriately describe the reliability of a product from the customer's viewpoint. We now will proceed to define these various measures.

The ease of servicing a product is desirable from a cost point of view. A measure termed *serviceability* is defined as the ease with which a system can be repaired. Serviceability is a characteristic of the system's design and must be planned at the design stage. For example, consider the family automobile where the ease of maintenance of the engine not only depends on the basic engine design but also on whether all parts of the engine are easily accessible. If important parts of the engine are not accessible, engine repair can be both difficult and costly.

Serviceability is difficult to measure on a numerical scale. It is usually measured by ranking, or by a specifically developed rating procedure. This requires that systems be compared and ranked according to the ease of servicing.

Maintainability is a more widely known term. Specifically, *maintainability* is defined as the probability that a failed system can be made operable in a specified interval of downtime. Here the downtime includes the total time that the system is out of service. Downtime is a function of the failure detection time, repair time, administrative time, and the logistics time connected with the repair cycle. Theoretically, for a product there exists a maintainability density function just as there exists a reliability function. The maintainability function describes probabilistically how long a system remains in a failed state.

The term repairability is more restrictive than maintainability and is concerned with only that segment of the maintenance cycle during which repairmen are actively repairing the system. Specifically, *repairability* is defined as the probability that a failed system will be restored to a satisfactory operating condition in a specified interval of active repair time. This measure is probably more valuable to the administration of the repair facility since it helps quantify workload for the repairmen.

The term *operational readiness* is defined as the probability that either a system is operating or can operate satisfactorily when the system is used under stated conditions. Operational readiness is more encompassing than the term availability. *Availability* is defined as the probability that a system is operating satisfactorily at any point in time and considers only operating time and downtime, thus excluding idle time. Availability is a measure of the ratio of the operating time of the system to the operating time plus the downtime. Thus it includes both reliability and maintainability. Operating satisfactorily implies operating under the stated environmental and load conditions just as it does in the definition of reliability.

Intrinsic availability is defined as the probability that a system is operating in a satisfactory manner at any point in time when used under stated conditions. In this context time is limited to operating and active repair time. Intrinsic availability is a more restrictive measure than availability; hence the availability will always be less than or equal to the intrinsic availability. The decrease will be caused by the administrative and logistics time connected with the repair cycle, which is included in the availability measure. For instance, the unavailability of spare parts will decrease the availability measure but not the intrinsic availability measure.

The various measures are illustrated in Figure 9.6, where the displayed time horizon is taken as the basis for the analysis. Each increment on the time scale will be referred to as one time unit. Based on this graphic description of the performance of the system over time we now estimate the various measures.

The total downtime in one complete cycle is six time units, and the active repair time is four time units. The six time units represent one sample from the maintainability function for this system while the four time units represent one sample from the repairability function. An appropriate model for these functions would have to be just as for the reliability function.

The operational readiness (O.R.) would be estimated by

$$\text{O.R.} = \frac{\text{Operating time} + \text{idle time}}{\text{Operating time} + \text{idle time} + \text{downtime}}$$

$$= \frac{14}{20} = 0.70 \qquad (9.48)$$

which means that the system is ready to perform its function 70% of the time.

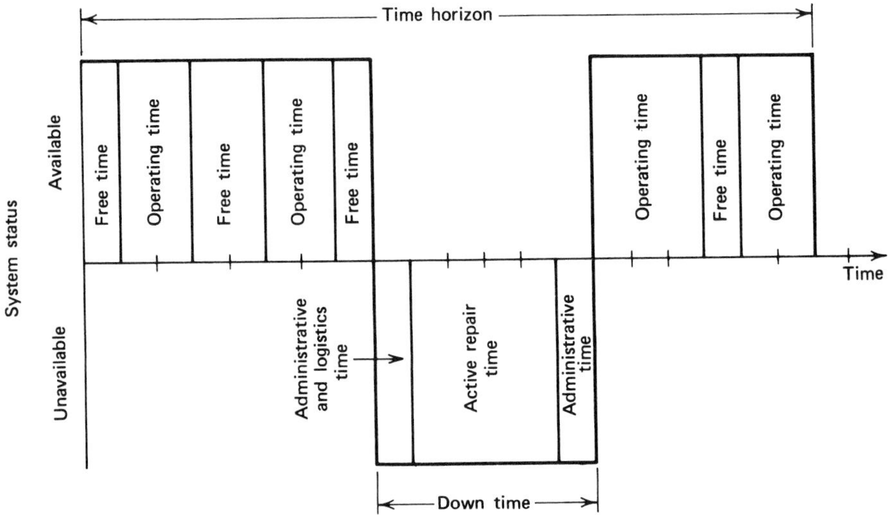

Figure 9.6 A graphical description of system status over a time horizon.

The availability function excludes free time and would be estimated as

$$A = \frac{\text{Operating time}}{\text{Operating time} + \text{downtime}}$$

$$= \frac{9}{15} = 0.60 \tag{9.49}$$

Now we will evaluate the intrinsic availability function which excludes the administrative and logistic time from the repair cycle. The intrinsic availability is estimated as

$$A_I = \frac{\text{Operating time}}{\text{Operating time} + \text{active repair time}}$$

$$= \frac{9}{13} = 0.69 \tag{9.50}$$

So by eliminating the administrative and logistics time in the repair cycle, the current availability of 0.60 can be increased in the limit to the intrinsic availability of 0.69. Or, there is a potential for a 9% improvement in availability.

The availability and maintainability functions, like the reliability function, are time dependent. The previous example estimated these functions in a static fashion. We now assume specific models for both the failure and downtime distributions and proceed to derive the maintainability and availability functions. The simplest possible case (using the exponential distribution) will be considered. However, the reader is warned, once more, that the exponential may not be a good model for downtime for certain practical applications.

Let us assume that we have a system for which the time to failure is exponential with failure rate λ. The downtime is also assumed to have an exponential distribution with a repair rate of μ. Since we have specified the total downtime distribution, the maintainability function $M(t)$ is

$$P(\mathbf{t} \leqslant t) = M(t) = 1 - e^{-\mu t}, \qquad t \geqslant 0 \tag{9.51}$$

where $\mathbf{t}$ = total downtime.

For the exponential distribution and for some small interval of time Δt we have

$$P[\text{system failure during } \Delta t] = \lambda \Delta t \tag{9.52}$$

and

$$P[\text{repair during } \Delta t | \text{system failure}] = \mu \Delta t \tag{9.53}$$

Let the availability function be denoted by $A(t)$. Using Equations 9.52 and 9.53 we see that

$$\begin{aligned} A(t+\Delta t) &= A(t)(1-\lambda\Delta t)+\left[1-A(t)\right]\mu\Delta t \\ &= A(t)-\lambda A(t)\Delta t+\mu\Delta t-\mu A(t)\Delta t \end{aligned}$$

or

$$\frac{A(t+\Delta t)-A(t)}{\Delta t} = -(\lambda+\mu)A(t)+\mu$$

Taking the limit as $\Delta t\to 0$, we have

$$\frac{d}{dt}A(t) = -(\lambda+\mu)A(t)+\mu$$

which is readily recognizable as a differential equation whose solution is

$$A(t)=\frac{\mu}{\mu+\lambda}+\frac{\lambda}{\mu+\lambda}e^{-(\mu+\lambda)t} \tag{9.54}$$

As t becomes large, the availability $A(t)$ clearly approaches a constant value. The steady state value for the availability is given by

$$\begin{aligned} A = \lim_{t\to\infty} A(t) &= \frac{\mu}{\lambda+\mu} \\ &= \frac{\text{Mean repair rate}}{\text{Mean failure rate} + \text{mean repair rate}} \end{aligned} \tag{9.55}$$

or equivalently,

$$A = \frac{\text{Mean time to failure}}{\text{Mean time to repair} + \text{mean time to failure}}$$

Fortunately, it so happens that if the downtime p.d.f. is taken as something other than the exponential, the same steady state solution results. In practice, frequently the log normal is used as the downtime p.d.f.

The intrinsic availability can be approached in the same way as the availability. The only difference will be that the mean repair rate in the availability function will be replaced by the mean active repair rate. Thus the steady state intrinsic availability is given by

$$A_I = \frac{\text{Mean time to failure}}{\text{Mean active repair time} + \text{mean time to failure}} \tag{9.56}$$

The steady state solutions A and A_I can also be interpreted as the long-run averages.

The system measures as presented depend on the rules and definitions that are agreed upon for any particular application. Unless these definitions are carefully and clearly established the resulting measures will be built on an unstable foundation and will be subject to controversy. Further, there is no set of definitions that universally applies to all consumer products.

9.6 SUMMARY

The dynamic or time dependent models are a natural extension of the static models previously considered in Chapter 3. To use dynamic models the p.d.f. of time to failure for each subsystem must be known, which means that one needs considerable knowledge about subsystem failure. Frequently, in order to apply dynamic models, it is convenient to assume a constant failure rate. The constant failure rate assumption is probably reasonable if the system is not subdivided too far because a large subsystem comprised of many components will tend to have a constant failure rate during its useful life.

System effectiveness measures such as operational readiness, serviceability, maintainability, and availability are other important descriptors that help to measure the total effectiveness of a system. Measures like these are affected by factors external to the system such as the abundance of spare parts, the accessibility of repair facilities, and the ease of repair. Unfortunately, in many consumer products things such as serviceability are considered long after the design has been completed—and often only when it is time to write the service manual.

EXERCISES

1 Two subsystems operate functionally in series and have a failure p.d.f. given by

$$f_i(t) = (t/\theta_i)\exp[-t^2/2\theta_i], \qquad t \geq 0$$

where θ_i is the parameter for the ith subsystem. (a) Find the system hazard function, $h_S(t)$. (b) Find the system reliability function, $R_S(t)$. (c) Find the system p.d.f. (d) If $\theta_1 = 300$ hr and $\theta_2 = 400$ hr, find $R_S(200\text{ hr})$. (e) For the values in (d) find t^* such that $R_S(t^*) = 0.90$.

2 Consider a series system composed of two subsystems where the first subsystem has a Weibull failure time distribution with parameters $\beta = 2$ and $\theta = 200$ hr. The second subsystem has an exponential time to failure distribution with $\theta = 300$ hr. For the system (a) find the hazard function; (b) find the reliability function; (c) find the expectea tıme to failure.

3 Three critical transistors in an electronic control circuit are subject to thermal stresses. The magnitude of the thermal stress **x** is described by an exponential distribution with

$$f(x)=\frac{1}{50}e^{-x/50}, \qquad x \geqslant 0$$

The ability of transistors to withstand thermal stresses is described by

$$f(y)=\frac{1}{60}e^{-y/60}, \qquad y \geqslant 0$$

where **y** = strength. When any one transistor fails, the circuit fails. Find the circuit reliability.

4 Consider a parallel system composed of two identical subsystems where the subsystem failure rate is λ, a constant. (a) Assume a pure parallel arrangement and plot the reliability function using a normalized time scale for the abscissa as $t'=t/\lambda$. (b) Assume a standby system with perfect switching and plot this reliability function on the same graph. (c) Assume that the standby system has a switch with a probability of failure of 0.20, and plot this reliability function on the same graph. (d) Compare the three systems.

5 Consider a two-unit pure parallel arrangement where each subsystem has a constant failure rate of λ, and compare this to a standby redundant arrangement that has a constant switch failure rate of λ_S. Specifically, what is the maximum permissable value of λ_S such that the pure parallel arrangement is superior to the standby arrangement?

6 Consider a two-unit standby redundant system with perfect switching where the time to failure for a unit is given by

$$f(t)=\theta t e^{-\theta t^2/2}, \qquad t \geqslant 0$$

where θ is a constant. (a) Find the system reliability function. (b) If $\theta=10$, find the system reliability for $t=20$.

7 Consider a two-unit standby redundant system that has a constant switch failure rate of λ_S. If the switch fails, the system fails. In this system both units have identical time to failure p.d.f.'s given by $f(t)$. (a) Find the reliability function for the system. (b) If $\lambda_S=0.01/\text{hr}$ and the subsystems both have a constant failure rate of 0.02/hr, find $R(50\text{ hr})$.

8 Consider a two-unit standby redundant system where the failure rate for the operating unit is a constant λ. The standby unit can fail while in standby and has a constant failure rate of μ in the standby mode. (a) Develop the reliability function for this situation assuming perfect switching. (b) Find the mean time to failure.

9 For a system the maintenance density function is

$$m(t)=\frac{1}{t\sqrt{2\pi}}\exp\left[-\frac{1}{2}(\ln t-4)^2\right], \qquad t\geqslant 0$$

where t = downtime in hours, and the reliability function is given by

$$R(t)=\exp[-t/20], \qquad t\geqslant 0$$

(a) Find the maintainability function. (b) Find the steady state availability.

10 Consider a two-unit standby redundant system where the on-line unit has a constant failure rate of λ_1 and the off-line unit has a failure rate of λ_2 when in operation. Assume perfect switching and no failures in the off-line position. (a) Find $R_S^2(t)$, the reliability function for the system. (b) Show that when $\lambda_1=\lambda_2=\lambda$ the result reduces to Equation 9.26.

11 Consider a two-unit shared parallel system where

$$h(t)=\lambda e^{-\lambda t}, \qquad t\geqslant 0$$

and

$$f(t)=5\lambda e^{-5\lambda t}, \qquad t\geqslant 0$$

(a) Find the system reliability function. (b) Find the mean time to failure.

12 Consider the reliability block diagram of a simple transmitter subsystem shown in Figure 9.7. The transmitters are identical and have a MTBF of 40 hours (exponential failure distribution). A failed transmitter can be serviced while the survivor carries the load. The maintainability function is given by $M(t)=1-\exp(-t/3)$ where t = the repair time in hours.

What is the probability that this transmitter subsystem fails during a 28 hour mission?

13 Figure 9.8 illustrates the circuit diagram for a power supply that is used to power transducers which are part of various fatigue testing machines. The components in this power supply act as series elements with the exception of the two diodes and two capacitors. The power supply will still function if one of the diodes and/or one of the capacitors fails.

Using the component failure rates given below develop the dynamic model for this power supply and find the mean time to failure.

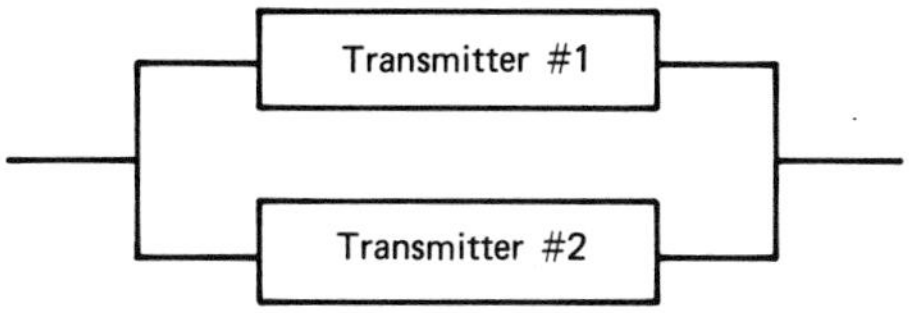

Figure 9.7 Reliability block diagram of a transmitter subsystem.

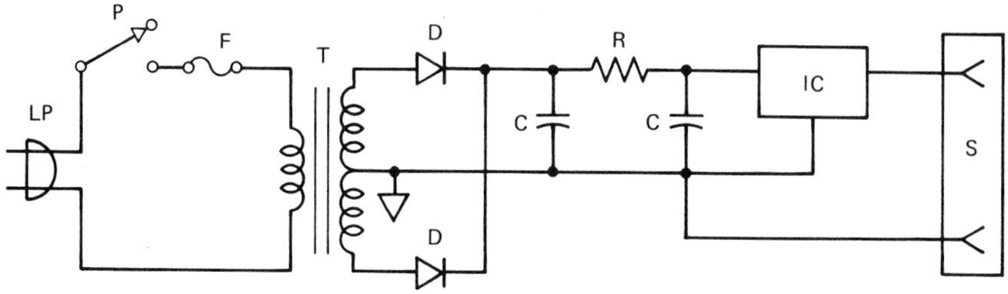

Figure 9.8 Power transducer circuit diagram.

Component	Name	Failure Rate × 10^6
LP&S	Line plug	0.25
F	Fuse	0.10
T	Transformer	0.34
D	Diode	0.17
C	Capacitor, filter	0.44
R	Resistor	0.0035
IC	Integrated circuit regulator	0.40
P	Power switch	0.003

BIBLIOGRAPHY

1 Balaban, H. S., "Some Effects of Redundancy on System Reliability," *Proceedings of the Sixth National Symposium on Reliability and Quality Control*, Washington, D. C., January 11–13, 1960.

2 Bazovsky, I., *Reliability Theory and Practice*, Prentice-Hall, Englewood Cliffs, N. J., 1961.

3 *Maintainability Engineering Handbook*, NAVORD OD 39223, Naval Ordnance Systems Command, Louisville, Kentucky, February, 1970.

4 Sandler, G. H., *System Reliability Engineering*, Prentice-Hall, Englewood Cliffs, N. J., 1963.

5 Shooman, M. L., *Probabilistic Reliability*, McGraw-Hill, New York, 1968.

6 Von Alven, W. H. (Ed.); ARINC Research Corporation, *Reliability Engineering*, Prentice-Hall, Englewood Cliffs, N. J., 1964.

Chapter 10 reliability estimation: exponential distribution

The exponential distribution is undoubtedly the most commonly used distribution in life testing applications. Unfortunately, in many cases, it is used because it is easy to apply rather than because it is a choice based on a thorough understanding of the fundamentals. This chapter is concerned with the theory and application of the exponential distribution.

The basic properties of the exponential distribution are presented in Section 10.1. This section provides familiarity with the exponential distribution and its related hazard and reliability functions. Familiarity with the exponential distribution will reveal its limitations.

Once life testing data is obtained the applicability of the exponential as a failure model can be assessed. Section 10.2 considers making this assessment in various life testing situations.

In statistical inference the concern is with parameter estimation, confidence intervals, or hypothesis testing. These are precisely the topics considered in Sections 10.3, 10.4, and 10.7, respectively. Section 10.5 is concerned with reliability and percentile estimates while Section 10.6 briefly considers an exponential distribution with a nonzero minimum life.

Section 10.8 is concerned with the expected time to complete a life test and shows the trade-offs that one can make to accelerate testing. Appendixes at the end of this chapter summarize the theoretical background of the material presented.

10.1 STATISTICAL PROPERTIES OF THE EXPONENTIAL DISTRIBUTION

The p.d.f. for an exponentially distributed random variable **x** is given by

$$f(x;\theta)=\frac{1}{\theta}e^{-x/\theta}, \qquad x\geqslant 0 \tag{10.1}$$

where **x** represents such quantities as time between failures or kilometers between failures. The reliability function is given by

$$R(x) = e^{-x/\theta}, \qquad x \geqslant 0 \tag{10.2}$$

The distribution has one parameter θ where $\theta > 0$. The parameter θ is the mean of the distribution (i.e., $E(\mathbf{x}) = \theta$). The quantity θ is frequently referred to as the mean time between failures (MTBF). A second quantity $\lambda = 1/\theta$ is termed the failure rate.

Frequently the MTBF is quoted as implying high reliability up to this time. However, $R(x=\theta) = \exp(-1) = 0.368$; so the system actually has a rather low chance of surviving beyond the MTBF, if the reliability function is exponential.

The exponential and Poisson distributions are directly related, and much of the theoretical development utilizes this relationship. Let us illustrate the underlying process and the relationship by a physical example. Consider an automotive front suspension system. Assume that over a large population of cars and customers, the average rate of severe shocks to the front suspension system is given by λ. Here a severe shock will be taken as an impact of sufficient magnitude to cause catastrophic failure. The shocks might be caused by deep pot holes, curb impacts, etc.

We will apply the usual Poisson assumptions, which are

1. The number of shocks during any given time interval Δt is independent of the number of shocks prior to the beginning of the interval.
2. The probability of exactly one shock in any interval of length Δt is proportional to the length of the interval with a constant of proportionality λ.

Define

$$P_t(r) = \text{probability of } r \text{ road shocks prior to time } t$$

Then

$$P_{t+\Delta t}(0) = [P_t(0)][1 - \lambda\Delta t] \tag{10.3}$$

and

$$P_{t+\Delta t}(r) = P_t(r)[1 - \lambda\Delta t] + P_t(r-1)[\lambda\Delta t], \qquad r > 0 \tag{10.4}$$

from Equations 10.3 and 10.4 as $\Delta t \to 0$, we obtain the set of differential equations

$$\frac{d}{dt}P_t(0) = -\lambda P_t(0)$$

$$\frac{d}{dt}P_t(r) = \lambda[P_t(r-1) - P_t(r)], \qquad r > 0$$

The initial conditions are $P_0(0) = 1$ and $P_r(0) = 0$ for $r > 0$.

The unique solution to this set of differential equations is

$$P_t(r) = \frac{(\lambda t)^r e^{-\lambda t}}{r!}, \qquad r = 0, 1, 2, \ldots$$

This is the Poisson distribution. Thus, under the assumptions, the number of failures over an interval of time t is Poisson distributed with parameter λt.

Now reliability, by definition, is

$$R(t) = P[\mathbf{t} > t]$$

This means that no failure can occur prior to t. Hence

$$R(t) = P_t(r=0) = e^{-\lambda t}$$

This gives the p.d.f.

$$f(t) = \lambda e^{-\lambda t}, \qquad t \geqslant 0$$

which is precisely the exponential distribution. So if the number of failures occurring over an interval of time is Poisson distributed, then the time between failures is exponentially distributed. A converse proof may be found in Appendix 10.A.

Assumption 1 of independence in the previous example is sometimes referred to as the "no memory" property of the Poisson or exponential distribution. Illustrating this property another way, consider $P[\mathbf{x} > t + a | \mathbf{x} > t]$. Here, a device has lasted a length of time t, and we wish to know the probability that it will survive an additional length of time a. Note that $(\mathbf{x} > t + a) \cap (\mathbf{x} > t) = (\mathbf{x} > t + a)$ and using the conditional probability concept

$$P[\mathbf{x} > t + a | \mathbf{x} > t] = \frac{e^{-(t+a)/\theta}}{e^{-t/\theta}} = e^{-a/\theta}$$

It is clear that the desired probability is independent of t, the length of time that the device has survived prior to the interval a. Physically this means that the system experiences no effects of wearout and this is a property of the exponential distribution.

In the automotive suspension system example, if certain components were to experience work hardening as a result of previous deformations below the elastic limit, the system failure rate may change. In this case the exponential distribution would not be an appropriate model. Thus particular care must be exercised when considering components. However, for system level reliability calculations, the exponential is usually a good model. This is further illustrated in the development to follow.

Consider a total system composed of many components that probably have different failure distributions. The time to system failure distribution will approach the exponential even though component wearout, which may not follow the exponential, is the cause of failure. As an illustration of this phenomenon, let us consider a highly simplified example.

Table 10.1 Wearout distribution

TIME PERIOD	PROBABILITY OF FAILURE
1	0.05
2	0.25
3	0.40
4	0.25
5	0.05

Let us assume that we have a system made up of 1,000 components in series where the time-to-failure distribution for each component is the same and is as given in Table 10.1. This distribution is certainly not exponential.

The components will be replaced as they fail. We assume that replaced components do not start accumulating time until the end of the period in which they were put into operation. Now we can calculate the number of failures during successive time periods as follows:

TIME PERIOD	CALCULATION		NUMBER FAILING
1.	$.05 \times 1000$	=	50
2.	$.25 \times 1000 + .05 \times 50$	=	252
3.	$.40 \times 1000 + .05 \times 252 + .25 \times 50$	=	425
4.	$.25 \times 1000 + .05 \times 425 + .25 \times 252 + .40 \times 50$	=	354
5.	$.05 \times 1000 + .05 \times 354 + .25 \times 425 + .4 \times 252 + .25 \times 50$	=	287
6.	$.05 \times 287 + .25 \times 354 + .4 \times 425 + .25 \times 252 + .05 \times 50$	=	338
7.	$.05 \times 338 + .25 \times 287 + .4 \times 354 + .25 \times 425 + .05 \times 252$	=	349
8.	$.05 \times 349 + .25 \times 338 + .4 \times 287 + .25 \times 354 + .05 \times 425$	=	326
9.	$.05 \times 326 + .25 \times 349 + .4 \times 338 + .25 \times 287 + .05 \times 354$	=	328
10.	$.05 \times 328 + .25 \times 326 + .4 \times 349 + .25 \times 338 + .05 \times 287$	=	336
11.	$.05 \times 336 + .25 \times 328 + .4 \times 326 + .25 \times 349 + .05 \times 338$	=	333
12.	$.05 \times 333 + .25 \times 336 + .4 \times 328 + .25 \times 326 + .05 \times 349$	=	331
13.	$.05 \times 331 + .25 \times 333 + .4 \times 336 + .25 \times 328 + .05 \times 326$	=	333
14.	$.05 \times 333 + .25 \times 331 + .4 \times 333 + .25 \times 336 + .05 \times 328$	=	333
15.	$.05 \times 333 + .25 \times 333 + .4 \times 331 + .25 \times 333 + .05 \times 336$	=	333

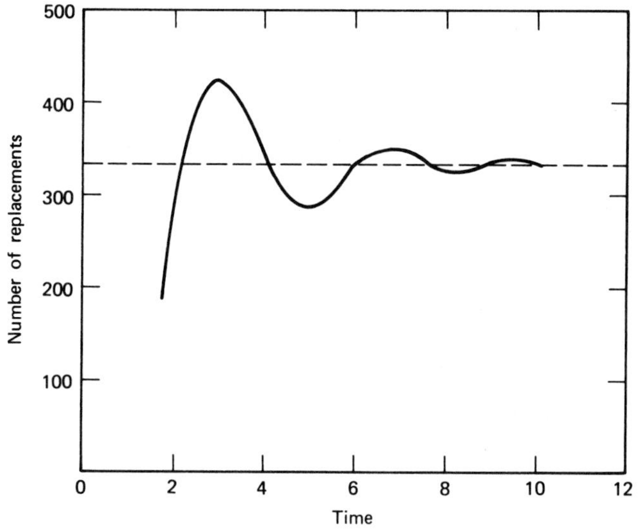

Figure 10.1 System failure rate over time.

From the calculations, the system failure rate seems stabilized at 333. Theoretically the mean of the time-to-failure distribution is 3 and the failure rate should stabilize at $1{,}000/3=333$. That is, after a certain length of time the system failure rate will stabilize to a constant value. If a continuous version of this problem was developed the failure rate would exhibit behavior as shown in Figure 10.1

The mixing of various components, the replacement of some components prior to failure (preventive maintenance), and the different failure patterns of dissimilar components all tend to contribute to a system failure pattern that might be reasonably well approximated by an exponential distribution. This is particularly true for many consumer products that receive practically no preventive maintenance.

10.2 DATA ANALYSIS

Data should be analyzed carefully both from an engineering and a statistical standpoint; for, sometimes, results that prove insignificant from a statistical standpoint may still provide insight from an engineering standpoint. This section illustrates some very basic statistical tests that can be conducted to assess the validity of using the exponential distribution as a failure model. There exist several tests for this purpose, and the one employed here is not the most common one encountered in practice. However, based on recent studies, this test

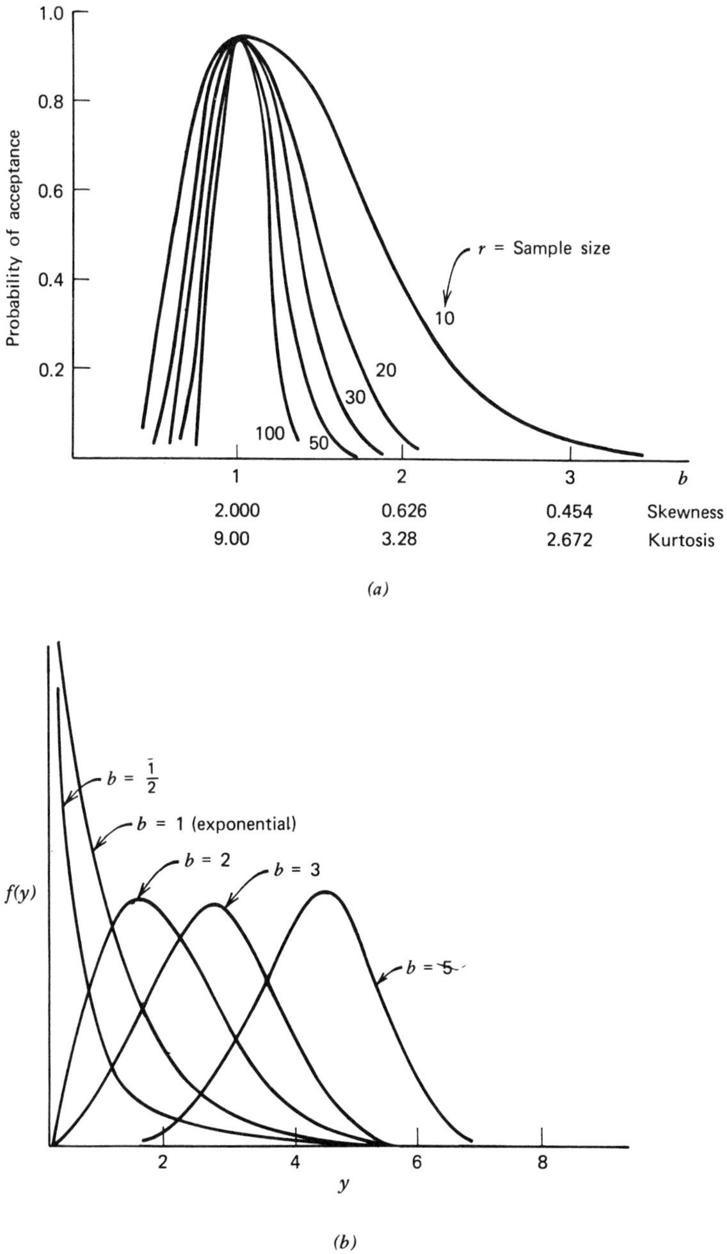

Figure 10.2 Operating characteristic curves for Bartlett's test ($\alpha = 0.05$); (*a*) O.C. curves; (*b*) shape of the probability density functions as used to generate O.C. curves.

Table 10.2 Vibration simulation: failure time data on a half-ton truck
(Data is in cumulative hours; total test time = 245 hr)

21.2	74.7	108.6	157.4
47.9	76.8	112.9	164.7
59.2	84.3	127.0	196.8
62.0	91.0	143.9	214.4
74.6	93.3	151.6	218.9

is more powerful than the other available tests in detecting either an increasing or a decreasing failure rate.

The basic test is termed Bartlett's test, and the test statistic is given by

$$B_r = \frac{2r\left[\ln\left(\frac{t_r}{r}\right) - \frac{1}{r}\left(\sum_{i=1}^{r} \ln x_i\right)\right]}{1+(r+1)/6r}$$

where $\mathbf{x}_i$ is the random variable representing time to failure, r is the number of failures, and $t_r = \Sigma_{i=1}^{r} x_i$ [see References 12 and 22].

Under the hypothesis of an exponential distribution, the statistic B_r is chi-square distributed with $r-1$ degrees of freedom, and a two-tailed chi-square test is in order.

The relative power of Bartlett's test is illustrated in Figure 10.2. Specifically, Figure 10.2*a* shows operating characteristic curves for Bartlett's test. These curves were developed by simulation from the p.d.f.'s shown in Figure 10.2*b*. The p.d.f.'s are all Weibull with a variance of unity and a shape parameter as indicated by the value of "*b*" (The Weibull distribution will be covered in detail in Chapter 11). For the various shapes of the p.d.f.'s given in Figure 10.2*b*, one can observe the probability of accepting the hypothesis that the p.d.f. is exponential by using Figure 10.2*a* and finding the appropriate *b* value on the abscissa. From Figure 10.2*a* it appears that a sample size of at least 20 failures is necessary for the test to have a reasonable power of discrimination.

10.2.1 VALIDATING THE EXPONENTIAL FAILURE MODEL

Here we examine the validity of the hypothesis that the data is representative of an exponential distribution. If the data contradicts this hypothesis, then we must search for another distribution model to represent time to failure.

The Bartlett's test will now be applied to several different situations representative of those encountered in practice.

EXAMPLE 10.1

The data in Table 10.2 represents a 245 hour vibration simulation test on a half-ton truck. That is, the truck was shaken on a simulator for a total time of 245 hours. The time into testing at which a failure occurred was recorded. Let us examine the data to determine if the time between failures can be reasonably well approximated by the exponential distribution.

In this case the $\mathbf{x}_i$ values represent time between failures and must be calculated from the original data. The failure times as obtained are given in Table 10.3.

Table 10.3 Time between failures for half-ton truck data

21.2	0.1	15.3	5.8
26.7	2.1	4.3	7.3
11.3	7.5	14.1	32.1
2.8	6.7	16.9	17.6
12.6	2.3	7.7	4.5

Thus we have

$$\Sigma \ln x_i = 38.80$$

$$t_r = 218.9$$

and

$$B_{20} = \frac{2(20)\left[\ln\left(\frac{218.9}{20}\right) - \frac{1}{20}(38.80)\right]}{1 + \frac{21}{120}} = 15.42$$

The critical values for a two-tailed test with $\alpha = 0.10$ are

$$\chi^2_{0.95,\,19} = 10.12 \qquad \text{and} \qquad \chi^2_{0.05,\,19} = 30.14$$

Therefore, the test does not contradict the hypothesis that the exponential distribution can be used to model the time to failure for this situation.

EXAMPLE 10.2

In Table 10.4, data is displayed as obtained from life testing of 20 heater switches. In this test each switch was cycled to failure. An overload voltage was applied to accelerate the test.

In this situation each value obtained represents a time to failure $\mathbf{x}_i$. Naturally the failure data occurred in an ordered fashion.

Table 10.4 Cycles to failure for 20 heater switches

100	7,120	24,110	36,860
340	12,910	28,570	38,540
1,940	13,670	31,620	42,110
5,670	19,490	32,800	43,970
6,010	23,700	34,910	64,730

Since all switches were run to failure, this set of data is analogous to a random sample of 20 where each data point represents an $\mathbf{x}_i$ value.

The intermediate values for calculating the test statistic are:

$$\sum_{i=1}^{20} \ln x_i = 188.22$$

$$t_r = \sum_{i=1}^{20} x_i = 469{,}170$$

Hence

$$B_{20} = 22.19$$

Using the same critical values as in the previous example we can conclude that the assumption of an exponential time to failure distribution is not contradicted.

EXAMPLE 10.3

Table 10.5, contains data from a slightly different test situation than in the previous example. In this test nine testing stands were used, and as switches failed, they were replaced. Each stand was cycled 20,000 times, and a counter recorded the cycle number at which failures occurred.

Table 10.5 Heater switch cycle test

STAND NO.	CYCLES AT WHICH FAILURES OCCURRED
1	6,700
2	4,600
3	4,100; 18,100; 18,950[a]
4	5,400
5	3,100; 8,100
6	2,600
7	No failure
8	4,700
9	No failure

[a] Counters were not reset when a new switch was placed on test. Thus counts are continuous from zero. Total test is 20,000 cycles per stand with replacement.

One may be tempted to treat the failure time data from the nine test stands as if they came from a single "equivalent" test stand with nine times the failure rate of a single stand. However, this procedure will tend to produce a phenomenon which appears exponentially distributed even if the individual test stand failure times are not exponentially distributed [4]. This is why the failure times should be taken directly as generated by each stand. These are given in Table 10.6.

Table 10.6 Heater switch cycle test data: cycles to failure

STAND NO.	CYCLES TO FAILURE
1	6,700
2	4,600
3	4,100; 14,000; 850
4	5,400
5	3,100; 5,000
6	2,600
7	—
8	4,700
9	—

Now we calculate

$$t_r = 51{,}050$$
$$\Sigma \ln x_i = 83.32$$
$$B_{10} = 4.11$$

and the critical values are

$$\chi^2_{0.95,9} = 3.32 \qquad \text{and} \qquad \chi^2_{0.05,9} = 16.92$$

Thus the hypothesis of exponentiality cannot be disputed.

EXAMPLE 10.4

The data in Table 10.7 comes from the maintenance records of a small fleet of cars. Odometer readings were recorded when certain types of unscheduled maintenance had to be performed. At the time of this survey, the total kilometers were recorded, and none of the automobiles were in a failed state.

Table 10.7 Automobile fleet data

CAR NO.	ODOMETER AT FAILURE (km)	TOTAL ODOMETER (km)
01	2,467; 3,128; 3,283; 7,988	8,012
02	None	6,147
03	1,870; 6,121; 6,175	9,002
04	3,721; 4,393; 5,848; 6,425; 6,535	11,000
05	498	4,651
06	184; 216; 561; 2,804	5,012
07	2,342; 4,213	12,718

In this example, the total kilometers for each automobile are different, and we will apply the principle of "total time to obtain a failure." In order to illustrate this principle

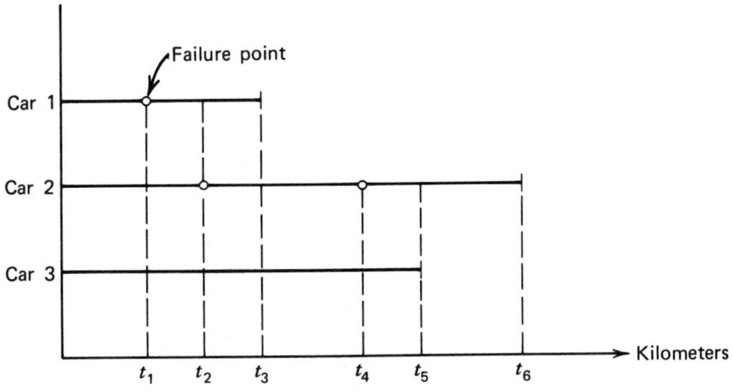

Figure 10.3 Suspension at different kilometers.

we refer to Figure 10.3. In the figure, the t_i's are the kilometers at which either a failure or a suspension occurred. Let τ_i be the total kilometers up to the ith failure. The basic reasoning here is that τ_i total kilometers occurred to produce a failure. Or, in other words, τ_i represents the total kilometers prior to the ith failure. Then

$$\tau_1 = 3t_1$$

$$\tau_2 = 3t_1 + 3(t_2 - t_1) = \tau_1 + 3(t_2 - t_1)$$

$$\tau_3 = 2t_4 + t_3$$

and the total kilometers between failures is τ_1, $(\tau_2 - \tau_1)$, and $(\tau_3 - \tau_2)$. These quantities are then used to compute the test statistic. It should be noted that this method could have been applied in the previous example.

The automotive fleet data is summarized in Table 10.8. Then

$$t_r = 47{,}762$$

$$\Sigma \ln x_i = 139.42$$

and

$$B_{19} = 15.89$$

The χ^2 values are

$$\chi^2_{0.95,18} = 9.39 \qquad \text{and} \qquad \chi^2_{0.05,18} = 28.87$$

Thus we cannot reject the hypothesis of exponentiality.

Table 10.8 Automotive fleet data calculations

KILOMETER POINTS	NO. CARS OPERATING	TOTAL KILOMETERS	KILOMETERS BETWEEN FAILURES (x_i)
184	7	1,288	1,288
216	7	1,512	224
498	7	3,486	1,974
561	7	3,927	441
1,870	7	13,090	9,163
2,342	7	16,394	3,304
2,467	7	17,269	875
2,804	7	19,628	2,359
3,128	7	21,896	2,268
3,283	7	22,981	1,085
3,721	7	26,047	3,066
4,213	7	29,491	3,444
4,393	7	30,751	1,260
4,651 Suspend	7	32,557	—
5,012 Suspend	6	34,723[a]	—
5,848	5	38,903[b]	8,152
6,121	5	40,268	1,365
6,147 Suspend	5	40,398	—
6,175	4	40,510	242
6,425	4	41,510	1,000
6,535	4	41,950	440
7,988	4	47,762	5,812
8,012 Suspend	3		
9,002 Suspend	2		
11,000 Suspend	1		

[a] $6 \times 5{,}012 + 1 \times 4{,}651 = 34{,}723$

[b] $5 \times 5{,}848 + 5{,}012 + 4{,}651 = 38{,}903$

EXAMPLE 10.5

Twenty rubber seals designed to prevent dirt from collecting on the wearing surfaces of ball joints were subjected to an accelerated life test. The test was discontinued when the tenth failure occurred. The data is given in the first column of Table 10.9.

In this situation the same principle can be applied as on the previous test. The last three columns of the table show the steps in the calculations. From the data we obtain

$$t_r = 2{,}495{,}050$$

$$\sum_{i=1}^{r} \ln x_i = 118.22$$

and

$$B_{10} = 10.22$$

Table 10.9 Ball joint seal test data

CYCLES TO FAILURE	CYCLES BETWEEN FAILURE	NUMBER ON TEST	ACCUMULATED TEST TIME BETWEEN FAILURES
20,400	20,400	20	408,000
30,000	9,600	19	182,400
50,700	20,700	18	372,600
57,750	7,050	17	119,850
60,300	2,550	16	40,800
74,100	13,800	15	207,000
78,300	4,200	14	58,800
144,000	65,700	13	854,100
153,500	9,500	12	114,000
166,000	12,500	11	137,500

The critical values are

$$\chi^2_{0.95,9} = 3.32 \qquad \text{and} \qquad \chi^2_{0.05,9} = 16.92$$

Thus the calculated value is not significant. We therefore fail to reject the hypothesis of exponentiality.

EXAMPLE 10.6

Table 10.10 displays failure data from the test track records of three automobiles. The test consisted of a 100,000 kilometer scheduled run.

The nature of the reporting system made it impossible to obtain the time between failures on a vehicle. It does no good to argue that this was a poor reporting system, because the data has already been collected. If we can use an exponential distribution to model this data, then we can still salvage much information from it. If it is not reasonable to assume an exponential distribution, then we need the time to second failure, etc., since the failure pattern may be changing.

This situation can be approached by utilizing a basic property of the exponential distribution; namely, that intervals of the same length should theoretically produce the same number of failures. In this case the theoretical number of failures per interval can be estimated by

$$f_{ti} = \frac{\text{total number of failures}}{\text{total number of intervals}} = \frac{75}{25} = 3.0 \tag{10.5}$$

Now this must be compared to the observed frequency f_o for each interval. This can be easily done by utilizing the chi-square test given by

$$\chi^2_c = \sum_{i=1}^{k} \frac{(f_{oi} - f_{ti})^2}{f_{ti}} \tag{10.6}$$

Table 10.10 Test track failure data reported by kilometer intervals

KILOMETER INTERVAL ($\times 10^{-3}$)	FAILURES	$(f_0 - f_t)$	$(f_0 - f_t)^2$
$\leqslant 4$	0	−3	9
8	3	0	0
12	8	5	25
16	4	1	1
20	8	5	25
24	1	−2	4
28	1	−2	4
32	6	3	9
36	5	2	4
40	0	−3	9
44	8	5	25
48	5	2	4
52	0	−3	9
56	2	−1	1
60	0	−3	9
64	1	−2	4
68	7	4	16
72	0	0	0
76	4	1	1
80	2	−1	1
84	3	0	0
88	2	−1	1
92	2	−1	1
96	2	−1	1
100	1	−2	4
	$\Sigma = 75$		167

where

f_{oi} = observed number of failures in the ith interval
f_{ti} = theoretical number of failures in the ith interval
k = number of intervals

The calculated value χ_c^2 is compared to $\chi_{\alpha,k-2}^2$ to determine significance. That is, a significant departure from the exponential is assumed when $\chi_c^2 > \chi_{\alpha,k-2}^2$.

Returning to our example

$$\chi_c^2 = 55.67$$

and the critical value is

$$\chi_{0.05,23}^2 = 35.2$$

Hence it is concluded that the exponential is not a good failure model for this situation.

In product testing one premature or short failure time can significantly dominate the estimation process to produce poor results. One would like a procedure to judge whether or not a presumed early failure is really representative of the product's failure pattern. Procedures for testing for an abnormal (nonrepresentative) early failure will now be considered.

10.2.2 TESTING FOR ABNORMALLY SHORT FAILURE TIMES

In product testing, short failure times can be caused by such things as manufacturing defects or substandard material. Such defects result in components that are not representative of the population as a whole, and thus the failures can be eliminated from further analysis. Undoubtedly the best way to judge whether or not a particular failure should be included in a reliability estimate is by an engineering analysis to determine the cause of the failure. If this cannot be done, the test to follow can be applied to support one's intuition in identifying short failure times. The method for developing this test is easy to follow, and by understanding the development one can apply the test to other situations.

Let $(\mathbf{x}_1, \mathbf{x}_2, \ldots, \mathbf{x}_r)$ be a sequence of r independent and identically distributed exponential random variables. For example, in vehicle testing these variables would represent the kilometers between failures for the first r failures. Then the quantity $2\mathbf{x}_i/\theta$ is chi-square distributed with two degrees of freedom (see Appendix 10.B). It is well known that when two or more independent chi-square distributed random variables are summed, the new variable is also chi-square distributed with degrees of freedom equal to the sum of the degrees of freedom for the individual variables. So the quantity $(2/\theta)\ \Sigma_{i=2}^{r}\mathbf{x}_i$ is chi-square distributed with $(2r-2)$ degrees of freedom. Thus an F-distributed random variable can be formed by

$$F_{2,2r-2} = \frac{(r-1)\mathbf{x}_1}{\sum_{i=2}^{r} \mathbf{x}_i} \tag{10.7}$$

Here we have assumed that $\mathbf{x}_1$ is the short failure time.

If the failure time $(\mathbf{x}_1)$ is significantly small, this ratio will be disproportionately small. This means that if

$$F_{1-\alpha,2,2r-2} > \frac{(r-1)\mathbf{x}_1}{\sum_{i=2}^{r} \mathbf{x}_i} \tag{10.8}$$

then there is evidence that $\mathbf{x}_1$ represents an abnormally early failure.

Since $F_{1-\alpha,2,(2r-2)}$ cannot be located directly in the F table, the reciprocal relationship for F variables can be utilized which gives the rejection criterion

$$F_{\alpha,2r-2,2} < \frac{\sum_{i=2}^{r} \mathbf{x}_i}{(r-1)\mathbf{x}_1} \tag{10.9}$$

EXAMPLE 10.7

Consider the failure data in Table 10.11, which represents cycles to failure for 20 turbine blades.

Table 10.11 Kilocycles to failure for turbine blades

193	1,793	3,479	5,310
1,582	2,028	4,235	6,809
1,637	2,260	4,264	8,317
1,658	2,272	4,635	9,728
1,786	2,700	4,919	10,700

The sample F value is calculated as follows:

$$\sum_{i=2}^{20} x_i = 80{,}112$$

$$x_1 = 193$$

$$F_c = \frac{80{,}112}{(19)193} = 21.8$$

The critical F value is

$$F_{0.05,38,2} = 19.47$$

indicating that the first failure time of 193 kilocycles is not representative of the rest of the data.

EXAMPLE 10.8

Reconsider the heater switch test data as given in Table 10.4. The first two failures appear to have occurred significantly earlier than the rest of the failures. Let us run a test to determine the validity of the hypothesis that the first two failures are from the same population as the remaining failures.

Reworking the previous theory we know that $(2/\theta)(\mathbf{x}_1 + \mathbf{x}_2)$ is chi-square distributed with four degrees of freedom and $(2/\theta)\Sigma_{i=3}^{20}\mathbf{x}_i$ is chi-square distributed with 36 degrees of freedom. Thus

$$F_{4,36} = \frac{(\mathbf{x}_1 + \mathbf{x}_2)/4}{\left(\sum_{i=3}^{20} \mathbf{x}_i\right)/36} = \frac{9(\mathbf{x}_1 + \mathbf{x}_2)}{\sum_{i=3}^{20} \mathbf{x}_i}$$

This F ratio will be small if the first two failure times are abnormally short. Converting this to an upper-tailed F test, we reject the hypothesis that the first two failures are representative of the remainder if

$$F_{0.05,36,4} < \frac{\sum_{i=3}^{20} \mathbf{x}_i}{9(\mathbf{x}_1 + \mathbf{x}_2)}$$

In this example

$$\frac{\sum_{i=3}^{20} \mathbf{x}_i}{9(\mathbf{x}_1 + \mathbf{x}_2)} = \frac{468,730}{9(440)} = 118.37$$

and the critical F value is

$$F_{0.05,36,4} = 5.74$$

Thus there is evidence that the first two failures occurred abnormally early.

10.2.3 TESTING FOR ABNORMALLY LONG FAILURE TIMES

The same type of reasoning as used in the previous test can be applied to determine if a failure time is abnormally long. Redeveloping the previous test procedure, $\mathbf{x}_1$ represents an abnormally long failure time if

$$F_{0.05,2,2r-2} < \frac{(r-1)\mathbf{x}_1}{\sum_{i=2}^{r} \mathbf{x}_i} \tag{10.10}$$

where $\mathbf{x}_1$ could be any failure time, not necessarily the first failure time.

EXAMPLE 10.9

The data in Table 10.12 represents a sample of the kilometers at which the original equipment mufflers were replaced on a certain make of automobile. Here we will consider the first failure time of 43,850 kilometers to determine if it is abnormally long.

Table 10.12 Muffler replacement data (in km)

43,850	65,324	83,541	89,950
47,737	67,105	84,543	100,791
49,111	67,549	84,899	102,431
61,900	69,291	88,191	104,343
64,511	81,154	88,901	105,062

We compute

$$\sum_{i=2}^{20} x_i = 1{,}550{,}184$$

$$F_c = \frac{(19)(43{,}850)}{1{,}550{,}184} = 0.54$$

Since F_c is less than unity it will obviously not be significant. Thus it cannot be concluded that the first failure time is abnormally long.

10.2.4 DETECTING CHANGES IN THE FAILURE RATE

It should be obvious by now that the previous test can be applied in a number of different ways to detect changes in the failure rate. This is usually done when one has à priori belief that a change has taken place.

EXAMPLE 10.10

Reconsider the data in Table 10.2, and assume that we suspect that the first five failures do not represent the remainder. Here we will not make any judgment as to the direction of the change, so a two-tailed test is appropriate. The F ratio becomes

$$\frac{\sum_{i=1}^{5} x_i/10}{\sum_{i=6}^{20} x_i/30} = \frac{3(74.6)}{(218.9 - 74.6)} = 1.55$$

Using a significance level of 0.05 the critical F values for 10 and 30 degrees of freedom are (0.30, 2.51). Hence we do not detect a significant change.

Until now we have attempted to assess the validity of the assumption that the time to failure is exponential. The remainder of this chapter is concerned with statistical inference when the underlying life distribution is assumed to be exponential.

10.3 ESTIMATION OF MEAN LIFE

Estimating the mean life in the case of the exponential distribution is a straightforward computation. This fact has undoubtedly contributed to the popularity of the exponential distribution.

The estimation of the mean life parameter θ is described by

$$\hat{\theta} = \frac{T}{r} \tag{10.11}$$

where

T = total test time accumulated on all items including those that failed and those that did not fail
r = total number of failures

The estimator $\hat{\theta}$ is termed a maximum likelihood estimator. It has the properties of unbiasedness, minimum variance, efficiency, and sufficiency (see Appendix 10.C for the development of $\hat{\theta}$ as a maximum likelihood estimator).

Now we will reexamine some of the previous examples and estimate θ, the mean life.

EXAMPLE 10.11

Reconsider Example 10.1, in which a half-ton truck was tested on a vibration simulator. Here we have

$$T = 245 \text{ hr and } r = 20 \text{ failures}$$

yielding

$$\hat{\theta} = 12.25 \text{ hr}$$

EXAMPLE 10.12

In Example 10.3, nine test stands were used to cycle heater switches, and the failed switches were replaced. Each stand ran for 20,000 cycles. In this case, we have

$$T = (9 \text{ stands})\ (20{,}000 \text{ cycles/stand}) = 180{,}000 \text{ cycles}$$

$$r = 10$$

yielding

$$\hat{\theta} = \frac{180{,}000 \text{ cycles}}{10} = 18{,}000 \text{ cycles}$$

EXAMPLE 10.13

Reconsider Example 10.4, in which fleet data was maintained on seven automobiles. Here we have

$$T = 56{,}542 \text{ km} \quad \text{and} \quad r = 19$$

Thus

$$\hat{\theta} = \frac{56{,}542 \text{ km}}{19} = 2{,}976 \text{ km between maintenance actions}$$

The above examples illustrate various applications. However, to be complete, we will discuss two commonly quoted formulas for estimating θ. First consider a life testing situation where n items are simultaneously placed on test, and the test is terminated when the first r out of n failures occur ($r \leqslant n$). Failed items are not replaced. This is termed Type II censoring. Then, by our previous reasoning

$$\hat{\theta} = \frac{\sum_{i=1}^{r} x_i + (n-r)x_r}{r} \tag{10.12}$$

where x_i is the time to failure for the ith item.

Consider a second situation where we have n test stands, and we cycle each test stand for τ cycles. As items fail they are replaced. Where a truncation time is specified this is called Type I censoring. Here we have

$$\hat{\theta} = \frac{n\tau}{r} \tag{10.13}$$

where r is the number of failures.

As can be seen, a test can be censored at a particular time, or when a particular number of failures occur, or, in fact, all items can be run to failure. In test planning it is important to remember that the accuracy of the resulting estimation will be determined by the number of failures obtained from testing when using the exponential distribution. This will be clarified in Section 10.4 on confidence intervals. Also, the more items one places on test the quicker one will obtain a preselected number of failures (Section 10.8). However, the items obtained for testing and the physical test facilities cost money. Thus one must balance the economic advantage of a shorter test duration with the economic penalty of placing more items on test.

10.4 CONFIDENCE INTERVALS FOR THE MEAN TIME TO FAILURE ASSUMING A ZERO MINIMUM LIFE

Confidence intervals for the mean life are now developed where the time-to-failure distribution is exponential. Two-sided confidence intervals are considered, leaving it to the reader to make the necessary changes for one-sided limits.

Confidence limits for the exponential distribution are based on the following two different test situations, depending on how the data are recorded (see Appendix 10.D for the development of these limits).

10.4.1 RECORDING FAILURE TIMES

First consider the test situation where failure times $\mathbf{x}_1, \mathbf{x}_2, \ldots, \mathbf{x}_r$ are observed ($r \leqslant n$). In this case the quantity $2r\hat{\theta}/\theta$ is chi-square distributed with $2r$ degrees of freedom. Thus, an appropriate probability statement is

$$P\left[\chi^2_{1-\alpha/2,2r} \leqslant \frac{2r\hat{\theta}}{\theta} \leqslant \chi^2_{\alpha/2,2r}\right] = 1-\alpha \tag{10.14}$$

which can be algebraically changed to

$$\frac{2T}{\chi^2_{\alpha/2,2r}} \leqslant \theta \leqslant \frac{2T}{\chi^2_{1-\alpha/2,2r}} \tag{10.15}$$

where $T = r\hat{\theta}$. Note that this T is consistent with the previous definition.

The $100(1-\alpha)\%$ two-sided confidence interval on the mean life will have a width of

$$w = 2r\left[\frac{1}{\chi^2_{1-\alpha/2,2r}} - \frac{1}{\chi^2_{\alpha/2,2r}}\right]\hat{\theta} \tag{10.16}$$

Since $\hat{\theta}$ is a random variable, then the width w is a random variable, and the expected width would be given by Equation 10.16 with $\hat{\theta}$ replaced by θ. Of course, the variance of the width can also be found if needed.

EXAMPLE 10.14

Eight leaf springs were cycle tested to failure on an accelerated life test. The results follow:

8,712	39,400	79,000	151,208
21,915	54,613	110,200	204,312

The mean life is estimated by

$$\hat{\theta} = \frac{\sum_{i=1}^{8} x_i}{8} = \frac{669{,}360 \text{ cycles}}{8}$$

$$= 83{,}670 \text{ cycles}$$

Now let us set a 95% two-sided confidence interval on the mean life. Appropriate chi-square test values are

$$\chi^2_{0.975,16} = 6.91 \qquad \text{and} \qquad \chi^2_{0.025,16} = 28.84$$

The limits on θ are given by

$$\frac{(2)669{,}360}{28.84} \leqslant \theta \leqslant \frac{(2)669{,}360}{6.91}$$

or

$$46{,}419 \text{ cycles} \leqslant \theta \leqslant 193{,}736 \text{ cycles}$$

EXAMPLE 10.15

Fifteen automotive a/c switches were cycled and observed for failure. The test was suspended when the fifth failure occurred. Failed switches were not replaced. The failures occurred at the following cycles:

1,410	3,138	6,971
1,872	4,218	

We have

$$T = 87{,}319 \text{ cycles}$$

yielding

$$\hat{\theta} \doteq 17{,}464 \text{ cycles}$$

For a 95% two-sided confidence interval on the mean life, the critical chi-square values are

$$\chi^2_{0.975,\,10} = 3.25 \qquad \text{and} \qquad \chi^2_{0.025,\,10} = 20.48$$

Hence the desired interval is given by

$$\frac{(2)87{,}319}{20.48} \leqslant \theta \leqslant \frac{(2)87{,}319}{3.25}$$

or

$$8{,}527 \text{ cycles} \leqslant \theta \leqslant 53{,}735 \text{ cycles}$$

10.4.2 COUNTING FAILURES OVER A TIME INTERVAL

Let us assume that in a test situation we count the number of failures that occur over an interval of test time T. This situation could arise in practice in different ways. For example, we might have n test stands where we replace items as they fail and discontinue the test at a predetermined time. Or we might drive vehicles over a 40,000 km test schedule and elect to count failures rather than failure intervals.

In the above situation where we have observed r failures over an interval of test time T, the $100(1-\alpha)\%$ two-sided confidence interval is

$$\frac{2T}{\chi^2_{\frac{\alpha}{2},2(r+1)}} \leqslant \theta \leqslant \frac{2T}{\chi^2_{1-\frac{\alpha}{2},2r}} \tag{10.17}$$

The limits resemble the previous ones, and in fact, the upper limit is the same. However, the derivation is different. Essentially, this derivation proceeds by recognizing that a Poisson process is observed for a period of time T, which means that the confidence limits are on the Poisson parameter $1/\theta$. Then there exists a well-known relationship between the Poisson and the chi-square distribution that is used to make the limits easy to find in standard tables. The derivation is given in Appendix 10.D.2.

The next example illustrates not only an application of the above confidence limits but also the use of an approximation of the χ^2 distribution for large degrees of freedom. This approximation is necessary because these values are usually not tabled.

EXAMPLE 10.16

Each of six, half-ton, heavy-duty military vehicles was run over a 100,000 km test track schedule. The test produced 84 failures of a certain predefined category. The 90% confidence limits for the MTBF are desired.

In order to calculate the confidence limits we need $\chi^2_{0.05,170}$ and $\chi^2_{0.95,168}$. Chi-square values for such high degrees of freedom are usually not tabled; however, the quantity $\sqrt{2\chi^2}$ is approximately normally distributed with mean $\mu=\sqrt{2\nu-1}$ and variance $\sigma^2=1$ where ν represents the degrees of freedom.

Using this approximation, we first find $\chi^2_{0.95,168}$. The mean of the approximate normal distribution is

$$\mu=\sqrt{2(168)-1}=18.30$$

Then

$$z_{0.95}=-1.645=\frac{\sqrt{2\chi^2}-18.30}{1.0}$$

Solving for χ^2, we get

$$\chi^2_{0.95,168}=138.69$$

The chi-square approximation can be reduced to

$$\chi^2_{\alpha,\nu}\approx\left[\mathbf{z}_\alpha+\sqrt{2\nu-1}\right]^2/2$$

where $\mathbf{z}$ is the standard normal deviate. Applying this to find $\chi^2_{0.35,170}$ we obtain

$$\chi^2_{0.05,170}\approx\left[1.645+\sqrt{2(170)-1}\right]^2/2=201.10$$

Thus the confidence limits are

$$\frac{2(600{,}000)}{201.10}\leqslant \text{MTBF}\leqslant\frac{2(600{,}000)}{138.69}$$

or

$$5{,}967\text{ km}\leqslant\text{MTBF}\leqslant 8{,}652\text{ km}$$

In many test situations, one will have a choice of either setting confidence limits in the above fashion or of using the previous method where failure times were employed. If failure intervals are disregarded, the width of the confidence interval will obviously be larger than in the previous method, which uses each failure interval. Essentially, in this method by disregarding the failure times we discard some of the information and this naturally results in a wider confidence interval.

This method becomes rather useful when no failures occur over an interval of length T. Then a lower confidence limit can still be obtained by this method. In particular, if no failures are observed, the $100(1-\alpha)\%$ one-sided lower confidence limit on the mean life is

$$\frac{2T}{\chi^2_{\alpha,2}} \leqslant \theta \tag{10.18}$$

EXAMPLE 10.17

A truck-tractor was rebuilt for testing a turbine engine. The truck was run over a test schedule with a 50,000 kg gross load on the trailer. The test was suspended at 14,400 km because of a starter failure. No other failures were observed. The starter failure was not credited as a failure of the core turbine engine. Thus, the turbine lasted for 14,400 km with no failures credited to it. What can we say about the mean life for this engine?

Well, since $\chi^2_{0.10,2}=4.605$, we can say with 90% confidence that the mean life

$$\theta \geqslant \frac{2(14{,}400)}{\chi^2_{0.10,2}} = 6{,}254 \text{ km}$$

In the previous example, if the truck had traveled farther, obviously the lower bound on mean life would have increased. Here we must be very careful to distinguish between the statistical and the engineering interpretation of the results. The use of the lower bound to compare designs or as a decision criteria can be misleading. That is, suppose we have a second truck engine that we wish to compare to this one, and the second engine was run until a failure was observed. The 90% lower confidence limit on the mean life for the second engine was found to be 75,000 km. One frequently finds decision makers using this information to conclude that the second engine has a better mean life. This may not be true, and, in any event, one is penalizing the first engine since it was taken off the test for reasons other than failure. If the first engine was allowed to run to failure it may have demonstrated a superior mean life.

10.5 RELIABILITY AND PERCENTILE ESTIMATION WITH CONFIDENCE LIMITS

The reliability function for the exponential distribution is estimated by

$$\hat{R}(x) = e^{-x/\hat{\theta}}, \qquad x \geqslant 0 \tag{10.19}$$

where $\hat{\theta}$ is the estimator for θ as previously developed.

Confidence intervals can also be obtained from previous developments. If the upper and lower confidence limits on θ are denoted by U and L, respectively, then the confidence limits on the reliability function are

$$e^{-x/L} \leqslant R(x) \leqslant e^{-x/U} \tag{10.20}$$

Sometimes the focus is on the time or the number of kilometers a certain percentage of the population will survive. We now determine an expression for such an event. If p represents the fraction failing at time x_p, we write

$$R(x_p) = 1 - p$$

But for the exponential

$$R(x_p) = e^{-x_p/\theta}$$

Combining the two and solving for x_p, we get

$$x_p = \theta \ln\left(\frac{1}{1-p}\right) \tag{10.21}$$

which is estimated by

$$\hat{x}_p = \hat{\theta} \ln\left(\frac{1}{1-p}\right) \tag{10.22}$$

Confidence limits on x_p can be obtained by merely substituting L and U for θ.

EXAMPLE 10.18

For a prototype test vehicle, electrical failures for the engine operation occurred at the following kilometers: 28,820; 36,707; 46,128; and 68,345. The total test schedule was 72,000 kilometers.

First let us estimate the reliability function and determine a 90% lower confidence limit for the 12,000 kilometer reliability.

Here we have

$$\hat{\theta} = \frac{72{,}000}{4} = 18{,}000 \text{ km}$$

Hence the reliability function is

$$\hat{R}(x) = e^{-x/18{,}000}$$

Now since we have recorded the failure times, a chi-square value with eight degrees of freedom is appropriate. Hence

$$L = \frac{(2)72{,}000}{\chi^2_{0.10,\,8}} = 10{,}777 \text{ km}$$

Therefore

$$R(12{,}000) \geqslant e^{-12{,}000/10{,}777} = 0.33$$

That is, we are 90% confident that the 12,000 km reliability is at least 0.33.

Now let us consider estimating the kilometers at which 10% of the vehicles will have failed. Using the formula

$$\hat{x}_p = \hat{\theta} \ln \frac{1}{1-p}$$

we have

$$\hat{x}_{0.1} = 18{,}000 \ln \frac{1}{0.9} = 1896 \text{ km}$$

The 10% point is sometimes referred to as the B_{10} life—a term that is a carryover from ball bearing testing.

10.6 THE NONZERO MINIMUM LIFE SITUATION

In some life testing situations the two-parameter exponential distribution is the appropriate model. For example, a warranty period might be viewed as a period of no failures from the customer's standpoint. Also in the study of fatigue of metals, it is well known that there is an initial period of no failures, and, for example, bearing failures in reciprocating engines, which primarily fail in fatigue, exhibit a nonzero minimum life. There undoubtedly exist many other examples.

The probability density function for the two-parameter exponential distribution is given by

$$f(x;\theta,\delta) = \frac{1}{\theta} \exp[-(x-\delta)/\theta], \qquad x \geqslant \delta > 0, \quad \theta > 0 \tag{10.23}$$

and the reliability function is

$$R(x) = \exp[-(x-\delta)/\theta], \qquad x \geqslant \delta > 0 \tag{10.24}$$

The parameter δ is frequently referred to as the minimum life. This p.d.f. has a mean of $(\theta + \delta)$.

10.6.1 PARAMETER ESTIMATION

Consider the situation where n items are placed on life test, and the test is terminated at the time of the rth failure (i.e., Type II censoring). In this case the estimators for θ and δ are $\hat{\theta}'$ and $\hat{\delta}$ respectively and are defined by

$$\hat{\theta}' = \frac{\sum_{i=2}^{r} (x_i - x_1) + (n-r)(x_r - x_1)}{r-1} \tag{10.25}$$

and

$$\hat{\delta} = x_1 - \frac{\hat{\theta}'}{n} \tag{10.26}$$

Here the prime used with $\hat{\theta}$ distinguishes it from the estimator used in the zero minimum life case. These estimators are the best choice in the sense that they are minimum variance unbiased estimators (see Appendix 10.E for a detailed theoretical development).

The reliability function is estimated by

$$\hat{R}(x) = \exp\left[-(x-\hat{\delta})/\hat{\theta}'\right], \qquad x \geqslant \hat{\delta} \tag{10.27}$$

EXAMPLE 10.19

Consider the data in Table 10.13, which represents cycles to failure for throttle return springs. Twenty springs were tested under conditions similar to those encountered in actual use. The test was truncated at the time of the tenth failure.

Table 10.13 Throttle return spring data

CYCLES TO FAILURE (x_i)	($x_i - x_1$)
190,437	0
245,593	55,156
277,761	87,324
432,298	241,861
530,100	339,663
626,300	435,863
1,043,307	852,870
1,055,528	865,091
1,221,393	1,030,956
2,099,199	1,908,762
	$\Sigma = 5{,}817{,}546$

Then we have

$$\hat{\theta}' = \frac{5{,}817{,}546 + (10)(1{,}908{,}762)}{9}$$

$$= 2{,}767{,}241 \text{ cycles}$$

and

$$\hat{\delta} = 190{,}437 - \frac{2{,}767{,}241}{20}$$

$$= 52{,}075 \text{ cycles}$$

The reliability function is given by

$$\hat{R}(x)=\exp\left[-\frac{(x-52{,}075)}{2{,}767{,}241}\right], \qquad x \geqslant 52{,}075$$

So, for example, the reliability estimate for two million cycles would be found by letting $x=2M$ cycles and would be $\hat{R}(2M)=0.49$.

10.6.2 CONFIDENCE INTERVALS FOR MEAN LIFE, MINIMUM LIFE, AND RELIABILITY

The confidence interval for θ, the mean life is developed in a fashion similar to the one-parameter case. The quantity $2(r-1)\hat{\theta}'/\theta$ is distributed as $\chi^2(2r-2)$ in this case. Thus an appropriate probability statement is

$$P\left[\chi^2_{1-(\alpha/2),2r-2} \leqslant 2(r-1)\hat{\theta}'/\theta \leqslant \chi^2_{(\alpha/2),2r-2}\right]=1-\alpha \qquad (10.28)$$

Rearranging the inequality gives the $100(1-\alpha)\%$ two-sided confidence interval as

$$\frac{2(r-1)\hat{\theta}'}{\chi^2_{(\alpha/2),2r-2}} \leqslant \theta \leqslant \frac{2(r-1)\hat{\theta}'}{\chi^2_{1-(\alpha/2),2r-2}} \qquad (10.29)$$

The confidence interval for the minimum life (δ) is based on $2n(\mathbf{x}_1-\delta)/\theta$, where $\mathbf{x}_1$ is the minimum failure time, and on $2(r-1)\hat{\theta}'/\theta$. These two quantities are distributed as $\chi^2(2)$ and $\chi^2(2r-2)$, respectively, and they are independent. This suggests forming an F random variable.

Now since δ is the true minimum life, the upper limit on δ is the minimum observed failure time. We call this minimum time $\mathbf{x}_1$ and can construct the random variable

$$\mathbf{F}_{2,2r-2}=\frac{n(\mathbf{x}_1-\delta)}{\hat{\theta}'} \qquad (10.30)$$

The probability statement necessary to obtain the shortest confidence interval is

$$P\left[0 \leqslant \frac{n(\mathbf{x}_1-\delta)}{\hat{\theta}'} \leqslant F_{\beta,2,2r-2}\right]=1-\beta \qquad (10.31)$$

and by rearranging the inequality, the $100(1-\beta)\%$ confidence limits on δ are

$$x_1-\frac{\hat{\theta}'}{n}F_{\beta,2,2r-2} \leqslant \delta \leqslant x_1 \qquad (10.32)$$

The expression for the reliability function contains both δ and θ. The confidence limits for the reliability function can be constructed in the following fashion.

If the lower and the upper $100(1-\alpha)\%$ confidence limits for θ are designated L' and U', respectively, and the lower and upper $100(1-\beta)\%$ confidence limits on δ are designated L and U, respectively, then a confidence interval on $R(x)$ is given by

$$\exp\left[-(x-L)/L'\right] \leqslant R(x) \leqslant \exp\left[-(x-U)/U'\right] \tag{10.33}$$

at a level of significance $\alpha' = \alpha + \beta - \alpha\beta$. Thus we have a $100(1-\alpha')\%$ confidence interval on the reliability.

EXAMPLE 10.20

Refer again to the throttle spring data in Table 10.13. The confidence limits on the mean life would be found as follows: Assume $\alpha = 0.10$. Then, for $r = 10$,

$$\chi^2_{0.95,18} = 9.390 \qquad \text{and} \qquad \chi^2_{0.05,18} = 28.869$$

Hence

$$\frac{2(9)(2{,}767{,}241)}{28.869} \leqslant \theta \leqslant \frac{2(9)(2{,}767{,}241)}{9.390}$$

or

$$1{,}725{,}392 \leqslant \theta \leqslant 5{,}304{,}615$$

The confidence limits on the minimum life for $\alpha = 0.10$ would be found as follows:

$$F_{0.10,2,18} = 2.62$$

So

$$190{,}437 - \frac{(2{,}767{,}241)(2.62)}{20} \leqslant \delta \leqslant 190{,}437$$

In this case since the lower limit is negative, we must modify the limits to

$$0 \leqslant \delta \leqslant 190{,}437$$

Since this interval includes zero, it is not a very informative interval.

10.7 HYPOTHESIS TESTING

In life testing, situations frequently arise where it is important to determine if a new system meets a design goal or an established standard. This leads to the area of statistical inference called hypothesis testing.

Strictly speaking, a hypothesis is an assumption of either the value of the population's parameters or the distributional form. For example, some of the tests discussed in the section on data analysis were, in reality, simply the assessment of the hypothesis that the distributional form was exponential.

Here we will be concerned with the mean and the minimum life parameters. Using the mean life parameter as an example, the null hypothesis might be H_0: $\theta \geqslant \theta_0$ where θ_0 is some specific value. The alternate hypothesis is then, H_1: $\theta < \theta_0$. Here we have what is termed a composite null and alternate hypothesis. A hypothesis such as H: $\theta = \theta_0$ is termed a simple hypothesis and is rarely encountered in practice.

The basic scheme of hypothesis testing is to take a random sample (say, $x_1, x_2, \ldots, x_n$) from the population of interest and calculate a test statistic $\zeta = h(x_1, x_2, \ldots, x_n)$. If the test statistic falls in some predetermined critical region C, we reject the hypothesis H_0. If $\theta = \theta_0$, then the probability of accepting H_0 is given by

$$Pa = P[\zeta \notin C \mid \theta = \theta_0] = 1 - \alpha$$

The quantity α is termed the level of significance. Also, α is the probability of committing a Type I error, which is defined as rejecting H_0 when it is true.

Now let us assume that $\theta = \theta_1$, some value such that $\theta_1 \neq \theta_0$. Then the probability of accepting H_0 is given by

$$Pa = P[\zeta \notin C \mid \theta = \theta_1] = \beta$$

The quantity β is the probability of committing a Type II error, which is defined as accepting H_0 when it is false. It is essential to recognize that in hypothesis testing, one is never completely certain that a correct decision has been made; however, the error probabilities can be controlled.

The quantity Pa is a function of the true value of the parameter θ and of the number of failures r. Pa can be calculated for different values of θ and the results plotted as a graph termed the operating characteristic (O.C.) curve for the test.

In life testing, the statistical problems associated with determining the critical region are relatively minor, and the O.C. curve for the test is not difficult to develop. However, the assurance of a representative sample for a particular population is frequently a problem. Prototypes or preproduction models form the usual sample. They are tested against a test schedule that was developed to be representative of some typical usage. All of these factors add a further dimension to the total problem, and possibly the only way to consider all the facets is to include a degree of engineering judgment along with the statistical test to follow.

10.7.1 HYPOTHESIS CONCERNING THE MINIMUM LIFE

We assume that n items are placed on test and the test ends at the time of the rth failure ($r \leqslant n$). Thus we obtain the ordered failure times $x_1, x_2, \ldots, x_r$. The hypotheses of interest are

$$H_0\colon \delta = 0$$

$$H_1\colon \delta > 0$$

The previously developed statistic (Equation 10.30) can be used for this situation where under the null hypothesis $\delta = 0$. The basic procedure is

1. Calculate

$$F_c = \frac{nx_1}{\hat{\theta}'}$$

where $\hat{\theta}'$ is as defined in Equation 10.25.

2. Reject H_0 if $F_c > F_{\alpha,2,2r-2}$

where α is the level of significance.

EXAMPLE 10.21

In our calculations for Example 10.19, concerning the throttle return springs, we found

$$\hat{\theta}' = 2{,}767{,}241$$

Then

$$F_c = \frac{nx_1}{\hat{\theta}'} = \frac{15(190{,}437)}{2{,}767{,}421} = 1.03$$

The critical F value for $\alpha = 0.05$ is

$$F_{0.05,2,18} = 3.55$$

Thus we cannot conclude that the minimum life is other than zero. Of course, this conclusion was obvious from the confidence limits on the minimum life set earlier. The equivalent between setting confidence limits and hypothesis testing should be recognized.

10.7.2 HYPOTHESES CONCERNING THE MEAN LIFE

The development of this hypothesis test is based on the fact that $2r\hat{\theta}/\theta$ is distributed as $\chi^2(2r)$ in the case where n items are placed on test and the test is truncated at the rth failure.

Consider the hypotheses

$$H_0\colon \theta \leqslant \theta_0$$

$$H_1\colon \theta > \theta_0$$

Then, for a significance level of α, the probability of accepting H_0 is

$$Pa = P\left[\frac{2r\hat{\theta}}{\theta_0} \leqslant \chi^2_{\alpha,2r} \middle| \theta = \theta_0\right] = 1 - \alpha \tag{10.34}$$

Thus the procedure is:

1. Calculate $\chi_c^2 = 2r\hat{\theta}/\theta_0$
2. Reject H_0 if $\chi_c^2 > \chi_{\alpha,2r}^2$

The O.C. curve for the test is given by

$$Pa = P\left[\chi_{2r}^2 \leqslant \frac{\theta_0}{\theta_1}\chi_{\alpha,2r}^2\right] \tag{10.35}$$

The O.C. curves have been plotted for different values of r versus the discrimination ratio θ_0/θ_1 and can be found in Appendix VII.

EXAMPLE 10.22

Reconsider Example 10.19, concerning the throttle return springs, and assume a zero mean life.

Suppose we are interested in the hypotheses

$$H_0: \theta \leqslant 1{,}000{,}000 \text{ cycles}$$

$$H_1: \theta > 1{,}000{,}000 \text{ cycles}$$

Then $\hat{\theta}$ for the zero minimum life case is found from Equation 10.12 as

$$\hat{\theta} = 2{,}871{,}391 \text{ cycles}$$

$$\chi_c^2 = \frac{2(10)2{,}871{,}391}{1{,}000{,}000} = 57$$

The critical value for $\alpha = 0.05$ is

$$\chi_{0.05,20}^2 = 31.41$$

Thus there is sufficient evidence to reject H_0 and conclude that $\theta > 1{,}000{,}000$ cycles.

By referring to Appendix VII and using $\alpha = 0.05$, the O.C. curves for this test can be found. From these O.C. curves, we can see, for example, that if the true value of θ was 1,500,000 cycles the Type II error is approximately 0.59.

Another possible hypothesis situation is

$$H_0: \theta \geqslant \theta_0$$

$$H_1: \theta < \theta_0$$

It takes a simple reworking of the previous theory to get

$$Pa = P\left[\frac{2r\hat{\theta}}{\theta_0} \geqslant \chi_{1-\alpha,2r}^2 \middle| \theta = \theta_0\right] = 1 - \alpha \tag{10.36}$$

Thus the procedure is

1. Calculate $\chi_c^2 = 2r\hat{\theta}/\theta_0$
2. Reject H_0 if $\chi_c^2 < \chi_{1-\alpha,2r}^2$

O.C. curves for this situation are also given in Appendix VII.

10.7.3 COMPARISON OF TWO DESIGNS

Here we consider the situation where two designs are being tested. The two-parameter exponential is taken as the population p.d.f. The samples will be designated S_1 and S_2 where $S_1 = (x_{11}, x_{12}, \ldots, x_{1r_1})$ and $S_2 = (x_{21}, x_{22}, \ldots, x_{2r_2})$, where n_1, n_2 are the respective sample sizes and r_1, r_2 are the respective truncation points ($r_1 \leqslant n_1, r_2 \leqslant n_2$). For simplicity, S_1 will be assigned such that $x_{11} \leqslant x_{21}$.

Consider the hypotheses

$$H_0: \theta_1 = \theta_2$$

$$H_1: \theta_1 \neq \theta_2$$

The test procedure is

1. Calculate

$$C = \frac{\sum_{j=1}^{r_2} (x_{2j} - x_{21}) + (n_2 - r_2)(x_{2r_2} - x_{21})}{\sum_{j=1}^{r_1} (x_{1j} - x_{11}) + (n_1 - r_1)(x_{1r_1} - x_{11})} \tag{10.37}$$

2. Calculate

$$F_c = \frac{(r_1 - 1)}{(r_2 - 1)} C \tag{10.38}$$

3. Reject H_0 if either $F_c > F_{\frac{\alpha}{2}, 2r_2 - 2, 2r_1 - 2}$ or $F_c < \dfrac{1}{F_{\frac{\alpha}{2}, 2r_1 - 2, 2r_2 - 2}}$

EXAMPLE 10.23

Ten electronic tubes were life tested until five failures had occurred. At a later date, 10 more tubes of the same type were again tested to determine if a different method of processing affected the average life. The failure times, in hours, were recorded as

S_1	S_2
102	160
137	161
161	205
195	241
230	270
825	1,037

In this case the hypotheses are

$$H_0\colon \theta_1 = \theta_2$$

and

$$H_1\colon \theta_1 \neq \theta_2$$

We compute

$$C = \frac{1{,}367}{955} = 1.43$$

and

$$F_c = C = 1.43$$

The critical values are

$$\frac{1}{F_{0.025,8.8}} = \frac{1}{4.43} = 0.23$$

and

$$F_{0.025,8,8} = 4.43$$

Thus we cannot reject H_0.

A second set of hypotheses of interest concerning the minimum life is

$$H_0\colon \delta_1 = \delta_2$$

$$H_1\colon \delta_1 \neq \delta_2$$

The test procedure is

1. Calculate

$$d = \frac{n_2(x_{21} - x_{11})}{\sum_{j=1}^{r_2} (x_{2j} - x_{21}) + (n_2 - r_2)(x_{2r_2} - x_{21})} \tag{10.39}$$

2. Calculate

$$F_c = \frac{(2r_2 - 2)}{2} d \tag{10.40}$$

3. Reject H_0 if $F_c > F_{\alpha,2,2r_2-2}$.

10.8 EXPECTED TIME TO COMPLETE TESTING

In the previous work in estimating the mean life and reliability, the precision of the estimator was a function of r, the truncation point of the test, and not of n, the number of items on test. That is, the same precision is provided whether only r items are tested to failure, or n items were placed on test and the test is truncated at the time of the rth failure.

Now a logical question is: Why place more than r items on test? The answer is in the savings in test time. The test will be terminated at time t_r, and the expected time to test completion (see Appendix 10.F) is

$$E(\mathbf{t}_r) = \theta \sum_{i=1}^{r} \frac{1}{n-i+1} \tag{10.41}$$

The variance is given by

$$V(\mathbf{t}_r) = \theta^2 \sum_{i=1}^{r} \frac{1}{(n-i+1)^2} \tag{10.42}$$

The savings in test time can be compared by considering the ratio $E(\mathbf{t}_{r,n}) / E(\mathbf{t}_{r,r})$, where $E(\mathbf{t}_{r,n})$ is the expected test time when n items are placed on test and the test is truncated at the rth failure. The quantity $E(t_{r,r})$ is the expected test time when r items are on test and all run to failure. Table 10.14, summarizes the results. For example, if 10 items are placed on test and the test is truncated at the fifth failure, the test will be only 28% as long as one in which five items are placed on test and all are allowed to fail.

10.9 SUMMARY

In this chapter we have tried to provide a comprehensive coverage of statistical inference when using the exponential distribution as a failure model. The exponential distribution is particularly applicable to modeling system failures if the system has many components. The mixing of the component failure distributions tends to make the system failure times follow the exponential distribution.

The one parameter of the exponential distribution θ, which is termed the MTBF, is estimated by $\hat{\theta} = T/r$, and the quantity $2r\hat{\theta}/\theta$ is chi-square distributed with $2r$ degrees of freedom. Remembering these two relationships allows one to estimate reliability or develop confidence intervals for the exponential distribution as used in life testing.

Table 10.14 Fraction of test time for truncation at r out of n

$n =$ NO. ON TEST	$r =$ TEST TERMINATION POINT																			
	1	2	3	4	5	6	7	8	9	10	11	12	13	14	15	16	17	18	19	20
1	1.000																			
2	0.500	1.000																		
3	0.333	0.556	1.000																	
4	0.250	0.389	0.591	1.000																
5	0.200	0.300	0.427	0.616	1.000															
6	0.167	0.244	0.336	0.456	0.635	1.000														
7	0.143	0.206	0.278	0.365	0.479	0.650	1.000													
8	0.125	0.179	0.237	0.305	0.387	0.497	0.663	1.000												
9	0.111	0.157	0.207	0.262	0.327	0.406	0.513	0.673	1.000											
10	0.100	0.141	0.183	0.230	0.283	0.345	0.423	0.526	0.682	1.000										
11	0.091	0.127	0.165	0.205	0.250	0.301	0.361	0.437	0.537	0.690	1.000									
12	0.083	0.116	0.150	0.185	0.224	0.267	0.316	0.375	0.449	0.547	0.696	1.000								
13	0.077	0.107	0.137	0.169	0.202	0.240	0.282	0.330	0.388	0.460	0.556	0.703	1.000							
14	0.071	0.099	0.126	0.155	0.185	0.218	0.254	0.295	0.342	0.399	0.470	0.564	0.708	1.000						
15	0.067	0.092	0.117	0.143	0.170	0.200	0.232	0.267	0.307	0.353	0.409	0.479	0.572	0.713	1.000					
16	0.063	0.086	0.109	0.133	0.158	0.184	0.213	0.244	0.279	0.318	0.363	0.418	0.487	0.578	0.717	1.000				
17	0.059	0.081	0.103	0.125	0.147	0.171	0.197	0.225	0.255	0.289	0.328	0.373	0.426	0.494	0.585	0.722	1.000			
18	0.056	0.076	0.096	0.117	0.138	0.160	0.183	0.208	0.235	0.265	0.299	0.337	0.381	0.434	0.501	0.590	0.725	1.000		
19	0.053	0.072	0.091	0.110	0.130	0.150	0.171	0.194	0.219	0.245	0.275	0.308	0.345	0.389	0.441	0.507	0.595	0.729	1.000	
20	0.050	0.068	0.086	0.104	0.122	0.141	0.161	0.182	0.204	0.228	0.255	0.284	0.316	0.353	0.396	0.448	0.513	0.600	0.732	1.000

EXERCISES

1 The weather radar system on a commercial aircraft has a MTBF of 1,140 hours. Assume an exponential time to failure distribution and answer the following questions: (a) What is the probability of failure during a four-hour flight? (b) What is the maximum length of flight such that the reliability will not be less than 0.99? (Assume that the system is in continuous operation during flight.)

2 An XM-1 tank has a MTBF of 810 kilometers. Assuming an exponential distribution: (a) What is the maximum mission length such that there will be a 0.98 chance of the tank returning? (b) What is the probability of the tank returning from a 160 kilometer mission? (c) How many tanks should be sent out on the 160 kilometer mission to obtain a probability of 0.99 that at least five tanks will arrive at the target area (assume 80 kilometers to target).

3 Ten engines of a new design were each driven the equivalent of 50,000 kilometers. Odometer readings were recorded whenever an unscheduled maintenance action occurred. The odometer readings for each vehicle follow:

ENGINE NUMBER	ODOMETER READINGS
01	220; 11,970; 21,397; 27,766
02	45,270; 48,836
03	25,695; 25,989; 30,980; 32,769; 47,459
04	4,200; 14,672; 21,831; 29,187; 31,964; 36,535; 44,094
05	3,900; 29,147; 31,613; 37,524; 43,601; 45,208
06	12,750; 21,183; 23,649; 33,348; 40,907
07	3,730; 6,300; 11,840
08	22,565; 22,710; 28,301; 31,628; 45,784; 47,213
09	12,759; 14,548; 19,539; 41,108; 44,550
10	12,212; 18,727; 41,854; 42,169; 47,996

(a) Can the data be represented by the exponential distribution (use $\alpha = 0.10$)?
(b) If answer to part (a) is yes, estimate the MTBF.

4 The following data represent kilometers to failure:

43,000	27,200	10,600	12,400
27,000	4,100	200,000	18,200
68,000	40,500	109,000	14,200
46,000	2,600	2,400	24,500

(a) Assess the feasibility of using the exponential distribution to model this situation. Assuming that the exponential is applicable: (b) estimate the MTBF; (c) set a 90% lower confidence limit on the 10% failure kilometer. (d) With 90% confidence, quote the 2,400 km reliability.

5 Additional testing for the situation in Problem 4 produced the following data:

1,650	215,000	65,000
14,300	1,400	121,400
20,900	21,700	182,400
61,000	184,300	28,700
61,700	1,670	

Reassess part (a) of Problem 4 using $\alpha = 0.10$.

6 An automobile was driven over a 120,000 kilometer test course. The following represent odometer readings at which a particular type of failure occurred.

4,123	27,720	63,582
4,497	28,496	66,057
10,506	40,887	100,763
12,317	48,323	

Assuming an exponential distribution as representative, is there any evidence that the failure rate in the first 40,000 kilometers is different than in the last 80,000 kilometers (use $\alpha = 0.10$)?

7 The following data represent time between failures for air conditioning systems on a fleet of Boeing 720 jet airplanes:

Intervals between failures

PLANE												
7907	7908	7909	7910	7911	7912	7913	7914	7915	7916	7917	8044	8045
194	413	90	74	55	23	97	50	359	50	130	487	102
15	14	10	57	320	261	51	44	9	254	493	18	209
41	58	60	48	56	87	11	102	12	5		100	14
29	37	186	29	104	7	4	72	270	283		7	57
33	100	61	502	220	120	141	22	603	35		98	54
181	65	49	12	239	14	18	39	3	12		5	32
	9	14	70	47	62	142	3	104			85	67
	169	24	21	246	47	68	15	2			91	59
	447	56	29	176	225	77	197	438			43	134
	184	20	386	182	71	80	188				230	152
	36	79	59	33	246	1	79				3	27
	201	84	27	***	21	16	88				130	14
	118	44	***	15	42	106	46					230
	***	59	153	104	20	206	5					66
	34	29	26	35	5	82	5					61

Intervals between failures

						PLANE						
7907	7908	7909	7910	7911	7912	7913	7914	7915	7916	7917	8044	8045
	31	118	326		12	54	36					34
	18	25			120	31	22					
	67	156			11	216	139					
	57	310			3	46	210					
	62	76			14	111	97					
	7	26			71	39	30					
	22	44			11	63	23					
	34	23			14	18	13					
		62			11	191	14					
		***			16	18						
		130			90	163						
		208			1	24						
		70			16							
		101			52							
		208			95							

The asterisks (***) represent a major overhaul of the air conditioning unit; however, the failure interval containing the major overhaul was omitted.
Adapted with permission from F. Prochan, "Theoretical Explanation of Observed Decreasing Failure Rate," *Technometrics*, Vol. 5, No. 3, August, 1963.

(a) Using the procedures from Chapter 2, plot the failure rate for the air conditioning system of airplane number 7909, and plot the failure rate for all air conditioning systems combined. (b) Run Bartlett's test on the air conditioning system of airplane number 7909 and for all air conditioning systems combined using $\alpha = 0.10$. (c) Estimate the MTBF. (d) Plot the estimated cumulative distribution using the calculated MTBF and the cumulative fraction surviving as calculated from the data on the same graph and compare.

8 For a test vehicle, major electrical failures occurred at the following kilometers:

63	17,393	23,128
114	18,707	24,145
14,820	19,179	33,832
16,105	22,642	34,345

The vehicle was driven a total of 36,000 kilometers. (a) Estimate the MTBF. (b) Determine the 90% two-sided confidence interval for the MTBF. (c) Estimate the reliability function. (d) Determine the 95% lower confidence limit for the 1,200 kilometer reliability. (e) With 90% confidence estimate the kilometer at which 10% of the population will fail.

9 In 600,000 test kilometers accumulated on six vehicles, a total of 69 failures occurred. Assuming an exponential failure distribution: (a) Estimate the MTBF. (b) Find the 90% lower confidence limit on the MTBF. (c) Find the 90% lower confidence limit on the reliability function.

10 A transmission valve operated for 9,276 cycles before the test was discontinued. The test was stopped because an oil pump failure caused the valve to overheat and burn out. (a) Set a 90% lower confidence limit on the MTBF. (b) A second valve of a different design failed at 19,460 cycles. Management would like your recommendation as to which valve is best. What would be your answer?

11 A new design is to be tested and if it is judged to be superior to the old design then the old design will be replaced. From extensive testing over the years the old design is known to have an MTBF of 1,250 hours. If the new design has an MTBF that is 50% better than the old design, we desire an 80% chance of selecting the new design. If the new design is no better than the old one, we desire a 95% chance of not selecting the new design. (a) How many failures will we have to observe? (b) Assuming that the new design has an MTBF of 1,565 hours and that all items placed on test are allowed to fail, what is the expected time of testing? (c) How many items must be placed on test to obtain a 30% reduction in test time?

12 The following data represent kilometers to failure:

14,240	18,656	41,810	16,432
25,988	16,870	22,128	8,170
14,749	16,152	14,749	8,169
27,950	10,170	21,089	12,195
22,819	21,909		

Assume that the minimum life is not zero and (a) Estimate the parameters of the two-parameter exponential. (b) Set 95% confidence limits on the parameters. (c) Estimate the kilometers at which 10% of the population will fail and also set a 90% lower confidence limit on this quantity.

13 The following represents the time to failure for two designs for laser aiming devices as tested under severe usage. (a) Is there a significant difference in the two designs? (b) Is the minimum life of either design other than zero?

1	2
726 hr	411
189	203
209	306
8	535
113	417
260	394
1,604	246
430	279
25	554
39	570
177	1,259
660	346
1,880	708
260	519
905	
751	

14 For the following life testing data assume a nonzero minimum life and find the 90% lower confidence limit on the reliability ($n = 20$).

508 hr, 525, 539, 613, 677, 689, 709, 760, 930,
1,160, 1,226, 1,251, 1,405, 1,683.

15 From taxi fleet maintenance records it is estimated that on the average a wheel bearing fails at 200,000 kilometers with normal maintenance. Assuming an exponentially distributed time to failure, what is the probability that one or more bearings fail on a taxi before 100,000 kilometers?

16 The cylinder components of a diesel engine consist of a piston assembly, cylinder liner, and connecting rod. This system has a failure rate of 0.03000 failures per 200,000 kilometers. Failure creates a significant increase in oil consumption, and, as a result, customers complain.

It is estimated that the failure rate can be decreased by 0.00004 failures per 200,000 kilometers with an expenditure of $125,000 for engineering redesign and minor tooling changes.

The diesel engine under consideration has six cylinders and approximately 150,000 engines are sold per year for use in long-haul trucks.

From the company's standpoint is this expenditure worthwhile. Analyze the situation and write a brief engineering report for the company manager.

17 A military vehicle has a required 200 kilometer mission reliability of 97.5%. (a) What is the vehicle's required MTBF? (b) How many missions can be run before exceeding a 10% chance for mission failure?

18 A military vehicle contract requires a contractor to demonstrate with 90% confidence that a vehicle can achieve a 150 kilometer mission with a reliability of 98%. The prescribed method of testing is to run six vehicles over a 60,000 km test track. (a) What is the contractual MTBF? (b) What is the maximum number of failures that the contractor can experience on the demonstration test and still satisfy contractual requirements? (c) Construct O.C. curves for the test (use $\alpha = 0.05$).

BIBLIOGRAPHY

1 Altman, O. L. and C. G. Goor, "Actuarial Analysis of the Operating Life of B-29 Aircraft Engines," *Journal of The American Statistical Association*, Vol. 41, 1946.

2 Bazovsky, I., *Reliability Theory and Practice*; Prentice-Hall, Englewood Cliffs, N. J., 1961.

3 Bulgren, W. G., and J. E. Hewett, "Double Sample Tests for Hypotheses about the Mean of an Exponential Distribution," *Technometrics*, Vol. 15, No. 1, February 1973.

4 Cox, D. R. and W. L. Smith, "On the Superposition of Renewal Processes," *Biometrika*, Vol. 41, 1954.

5 Davis, D. J.; "An Analysis of Some Failure Data," *Journal of The American Statistical Association*, Vol. 47, No. 258, June 1952.

6 Drenick, R. F.; "Mathematical Aspects of the Reliability Problem," *Journal of the Society of Industrial and Applied Mathematics*, Vol. 8, No. 1, March 1960.

7 Epstein, B.; "Truncated Life Tests in the Exponential Case," *Annals of Mathematical Statistics*, Vol. 25, 1954.

8 Epstein, B.; "The Exponential Distribution and its Role in Life Testing," *Industrial Quality Control*, December 1958.

9 Epstein, B.; "Estimation from Life Test Data," *Technometrics*, Vol. 2, No. 4, November 1960.

10 Epstein, B.; "Life Test Acceptance Sampling Plans When the Underlying Distribution of Life is Exponential," *Proceedings of the Sixth National Symposium on Reliability and Quality Control*, Washington, D. C., January 11–13, 1960.

11 Epstein, B.; "Statistical Life Test Acceptance Procedures," *Technometrics*, Vol. 2, No. 4, November 1960.

12 Epstein, B.; "Tests for the Validity of the Assumption that the Underlying Distribution of Life is Exponential," *Technometrics*, Vol. 2, No. 1, February 1960.

13 Epstein, B, "Tests for the Validity of the Assumption That the Underlying Distribution of Life is Exponential: Part II," *Technometrics*, Vol. 2, No. 2, May 1960.

14 Epstein, B., *Statistical Techniques in Life Testing*, PB-171 580 U. S. Department of Commerce, Office of Technical Services, Washington, D. C., 1959.

15 Epstein, B. and M. Sobel, "Life Testing," *Journal of the American Statistical Association*, Vol. 48, 1953.

16 Epstein, B. and M. Sobel, "Some Theorems Relevant to Life Testing from an Exponential Distribution", *Annals of Mathematical Statistics*, Vol. 25, 1954.

17 Epstein, B. and C. K. Tsao, "Some Tests Based on Ordered Observations from Two Exponential Populations," *Annals of Mathematical Statistics*, Vol. 24, 1953.

18 Grubbs, F. E., "Determination of Number of Failures to Control Risks of Erroneous Judgements in Exponential Life Testing," *Technometrics*, Vol. 15, No. 1, February 1973.

19 Halperin, M., "Maximum Likelihood Estimation in Truncated Samples," *Annals of Mathematical Statistics*, Vol. 23, 1952.

20 Hogg, R. V. and A. T. Craig, *Introduction to Mathematical Statistics*, Macmillan, New York, Second Edition, 1965.

21 Kumar, S. and H. I. Patel, "A Test for the Comparison of Two Exponential Distributions," *Technometrics*, Vol. 13, No. 1, February 1971.

22 Lamberson, L. R., "An Evaluation and Comparison of Some Test for the Validity of the Assumption that the Underlying Distribution of Life is Exponential," *AIIE Transactions*, Vol. 12, December 1974.

23 Military Standard 781B, *Reliability Tests: Exponential Distribution*, Department of Defense, November 1967.

24 Mood, A. M. and F. G. Graybill, *Introduction to the Theory of Statistics*, McGraw-Hill, New York, Second Edition, 1963.

25 Paulson, E., "On Certain Likelihood-Ratio Test Associated with the Exponential Distribution," *Annals of Mathematical Statistics*, Vol. 12, 1941.

26 Prochan, F., "Theoretical Explanation of Observed Decreasing Failure Rate," *Technometrics*, Vol. 5, No. 2, August 1963.

27 Sasieni, M., A. Yaspan, and L. Friedman, *Operations Research: Methods and Problems*, John Wiley & Sons, New York, 1959.

APPENDIX

RELATED TECHNICAL MATERIAL

Appendixes 10.A to 10.F contain some of the motivation and proofs for the topics of the exponential distribution. We believe the material to be sufficient to provide a basis for understanding the theory.

10.A EXPONENTIAL AND POISSON RELATIONSHIP

The exponential and Poisson distributions are related by the following theorem.

theorem

If the mean time between failures is an exponential random variable with parameter θ, then the number of failures in an interval of length t time units is a Poisson distributed random variable with parameter t/θ; that is, if

$$f(x)=\frac{1}{\theta}\exp(-x/\theta), \qquad x\geqslant 0$$

where $\mathbf{x}$ = the random variable for time between failures. Then for an interval of length t time units, we have

$$P(r)=\frac{\left(\frac{t}{\theta}\right)^r \exp(-t/\theta)}{r!}$$

where r = number of failures observed over the interval of length t time units.

proof

Let $\mathbf{N}(t)$ = number of failures in an interval $[0,t]$

$\mathbf{T}_r$ = time of rth failure

Note that

$$\mathbf{N}(t)<r$$

if and only if

$$\mathbf{T}_r>t$$

Or

$$P(\mathbf{N}(t)<r)=P(\mathbf{T}_r>t)$$

and

$$P[\mathbf{N}(t)<r+1]=P(\mathbf{T}_{r+1}>t)$$

Hence

$$P[\mathbf{N}(t)=r]=P(\mathbf{T}_{r+1}>t)-P(\mathbf{T}_r>t)$$
$$=(1-P(\mathbf{T}_{r+1}\leqslant t))-(1-P(\mathbf{T}_r\leqslant t))$$
$$=P(\mathbf{T}_r\leqslant t)-P(\mathbf{T}_{r+1}\leqslant t)$$
$$=F_{\mathbf{T}_r}(t)-F_{\mathbf{T}_{r+1}}(t)$$

But $\mathbf{T}_r=\mathbf{x}_1+\mathbf{x}_2+\cdots+\mathbf{x}_r$ where all $\mathbf{x}_i$'s are independent and exponentially distributed. The moment generating function for $\mathbf{T}_r$ is given by

$$M_{\mathbf{T}_r}(s)=[M_{\mathbf{x}}(s)]^r$$

But

$$M_{\mathbf{x}}(s)=\frac{1}{1-\theta s}$$

for the exponential. Therefore

$$M_{\mathbf{T}_r}(s)=\left[\frac{1}{1-\theta s}\right]^r$$

which yields

$$f(T_r)=\frac{\frac{1}{\theta}e^{-T_r/\theta}\left(\frac{1}{\theta}T_r\right)^{r-1}}{(r-1)!},\qquad T_r\geqslant 0$$

Integrating

$$F_{\mathbf{T}_r}(t)=\int_0^t\frac{\frac{1}{\theta}e^{-T_r/\theta}\left(\frac{1}{\theta}T_r\right)^{r-1}}{(r-1)!}\,dT_r$$
$$=\frac{1}{\theta^r}\frac{1}{(r-1)!}\int_0^t T_r^{r-1}e^{-T_r/\theta}\,dT_r$$

Applying integration by parts yields a pattern that is recognizable as

$$F_{\mathbf{T}_r}(t)=-e^{-t/\theta}\left[\frac{1}{(r-1)!}\left(\frac{t}{\theta}\right)^{r-1}+\frac{1}{(r-2)!}\left(\frac{t}{\theta}\right)^{r-2}+\cdots+1\right]+1$$

and

$$F_{\mathbf{T}_{r+1}}(t) = -e^{-t/\theta}\left[\frac{1}{r!}\left(\frac{t}{\theta}\right)^{r} + \frac{1}{(r-1)!}\left(\frac{t}{\theta}\right)^{r-1} + \cdots + 1\right] + 1$$

yielding

$$P(\mathbf{N}(t) = r) = F_{\mathbf{T}_r}(t) - F_{\mathbf{T}_{r+1}}(t) = -e^{-t/\theta}\left[-\frac{1}{r!}\left(\frac{t}{\theta}\right)^{r}\right]$$

$$= \frac{(t/\theta)^r e^{-t/\theta}}{r!}$$

which is precisely a Poisson distribution with parameter t/θ.

Therefore, for any interval $[0, t]$, if the time to failure is exponential with parameter θ, the arrivals are Poisson with parameter (t/θ). It is interesting to note that only t, the length of the interval, matters and not the time the interval begins.

10.B RELATIONSHIP BETWEEN EXPONENTIAL AND CHI-SQUARE DISTRIBUTION

Consider an interval of time **t** to the first failure where

$$f(t) = \frac{1}{\theta}e^{-t/\theta}, \qquad t \geqslant 0$$

theorem

The random variable $\mathbf{y} = 2\mathbf{t}/\theta$ is chi-square distributed with two degrees of freedom.

proof

We have $dy = (2/\theta)\,dt$ and $t = (\theta/2)y$. A simple application of the fact that

$$g(y)\,dy = f(h^{-1}(y))\left|\frac{dh^{-1}(y)}{dy}\right|$$

yields (see Section 5.1)

$$g(y)\,dy = \frac{1}{2}e^{-y/2}\,dy, \qquad y \geqslant 0$$

which is precisely a chi-square distribution with two degrees of freedom.

10.C THE MAXIMUM LIKELIHOOD ESTIMATOR FOR MEAN LIFE

First let us consider the problem of finding the maximum likelihood function for certain life testing situations. Suppose we place two items on life test and observe the failure times t_1 and t_2. Now $t_1 \leqslant t_2$; that is, the failure times are ordered, and hence are not independent. Figure 10.C.1 illustrates this dependence.

So the likelihood function $L(\cdot)$ for the situation of Figure 10.C.1 is

$$L(t_1, t_2; \theta) = 2f(t_1; \theta) \cdot f(t_2; \theta), \qquad 0 \leqslant t_1 \leqslant t_2$$

In general, for n ordered failures, the likelihood function will be

$$L(t_1, t_2, \ldots, t_n; \theta) = n! \prod_{i=1}^{n} f(t_i; \theta)$$

which for the exponential distribution is

$$n! \frac{1}{\theta^n} \exp\left(- \sum_{i=1}^{n} t_i / \theta \right)$$

Now, it can be rightly argued that this ordering makes no difference as long as all n items are allowed to reach failure. However, if we stop the test after r failures where $r < n$, then we will have observed the weakest members of the sample, and then this ordering must be taken into account.

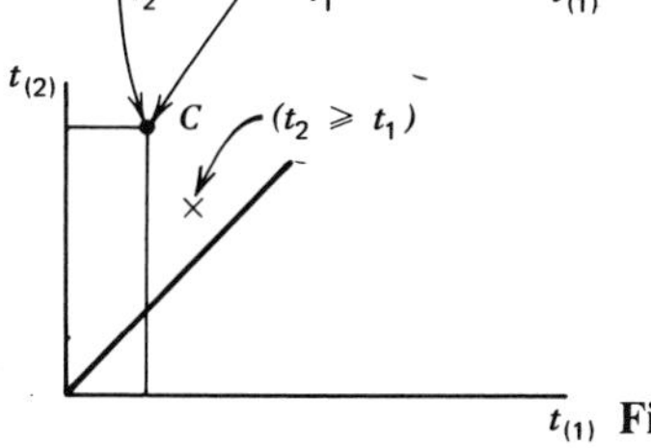

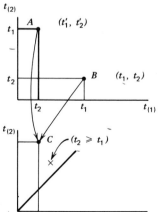

Figure 10.C.1 Two items on life test.

Maximizing the likelihood function for n items leads to the estimate of the parameter θ,

$$\hat{\theta} = \sum_{i=1}^{n} \frac{t_i}{n} = \bar{t}$$

10.C.1 THE NONREPLACEMENT CASE

Now we will consider the case where n items are placed on test and the test is truncated at the rth failure $(r < n)$. The likelihood function is

$$L = \frac{n!}{(n-r)!} \frac{1}{\theta} e^{-t_1/\theta} \frac{1}{\theta} e^{-t_2/\theta} \cdots \frac{1}{\theta} e^{-t_r/\theta} \cdot e^{-(n-r)t_r/\theta}$$

$$= \frac{n!}{(n-r)!} \frac{1}{\theta^r} \exp\left[-\frac{1}{\theta} \left[\sum_{i=1}^{r} t_i + (n-r)t_r \right] \right] \qquad (10.C.1)$$

Taking the natural log

$$\ln L = \ln \frac{n!}{(n-r)!} - r \ln \theta - \frac{1}{\theta} \left[\sum_{i=1}^{r} t_i + (n-r)t_r \right]$$

Differentiating w.r.t. θ and equating to zero,

$$-\frac{r}{\theta} + \frac{1}{\theta^2} \left[\sum_{i=1}^{r} t_i + (n-r)t_r \right] = 0$$

Solving for θ, gives the estimator $\hat{\theta}$ that maximizes $\ln L$, and hence the likelihood function L

$$\hat{\theta} = \frac{\sum_{i=1}^{r} t_i + (n-r)t_r}{r}$$

Since the joint density of $\mathbf{t}_1, \mathbf{t}_2, \ldots, \mathbf{t}_r$, which appears in Equation 10.C.1, can be factored as $g(\hat{\theta}, \theta) h(t_1, t_2, \ldots, t_r)$, the estimator $\hat{\theta}$ is sufficient.

10.C.2 THE REPLACEMENT CASE

Consider n items, which are replaced as soon as they fail, placed on a test that is truncated after a specified time interval τ. Let us visualize this by saying that we have test stands numbered $1, 2, \ldots, n$. Let

$\mathbf{x}_i$ = number of failures on ith stand

$\mathbf{y}$ = total number of failures observed during τ

Or

$$\mathbf{y} = \sum_{i=1}^{n} x_i$$

Therefore, if $\mathbf{x}_i$ is Poisson with parameter τ/θ, then $\mathbf{y}$ is also Poisson with parameter $n\tau/\theta$. That is,

$$P(y) = \frac{\left(\frac{n\tau}{\theta}\right)^y \exp(-n\tau/\theta)}{y!}, \qquad y = 0, 1, 2, \ldots$$

For $y = r$, a constant

$$L(r;\theta) = \frac{\left(\frac{n\tau}{\theta}\right)^r \exp(-n\tau/\theta)}{r!}$$

and

$$\ln L = r\left[-\ln\theta + \ln(n\tau)\right] - \ln r! - \frac{n\tau}{\theta}$$

For $\ln L$ to be maximum, we must have

$$-r\frac{1}{\theta} + \frac{n\tau}{\theta^2} = 0$$

or

$$\hat{\theta} = \frac{n\tau}{r}$$

Since maximum likelihood estimators are invariant, we may write the maximum likelihood estimator for reliability in all cases as

$$\hat{R}(x) = \exp(-x/\hat{\theta}), \qquad x \geqslant 0$$

An interesting problem left for the reader is to find the distribution of $\hat{R}(x)$, and then find its mean and variance.

10.D CONFIDENCE LIMIT DEVELOPMENT

10.D.1 FAILURE TIME OBSERVED

Let us consider the problem of setting confidence limits on θ (or R). First consider the case where n items are placed on a test that is truncated after r

failures ($r \leqslant n$). We find that

$$\hat{\theta} = \frac{\sum_{i=1}^{r} t_i + (n-r)t_r}{r}$$

We will introduce new variables

$$\mathbf{y}_i = \mathbf{t}_i - \mathbf{t}_{i-1}, \qquad i = 1, 2, \ldots, r; \quad t_0 = 0 \tag{10.D.1}$$

theorem
The random variables $\mathbf{y}_i$ are mutually independent, and each $\mathbf{y}_i$ has the p.d.f. $((n-i+1)/\theta)\exp[-[(n-i+1)y_i/\theta]]$.

proof
Substituting into the joint distribution of the t_i's given in Equation 10.C.1,

$$f(y_1, y_2, \ldots, y_r; \theta) = \frac{n!}{(n-r)!} \frac{1}{\theta^r} e^{-\frac{k}{\theta}} |J| \tag{10.D.2}$$

where $\mathbf{y}_i$ must be substituted into

$$k = \sum_{i=1}^{r} \mathbf{t}_i + (n-r)\mathbf{t}_r \tag{10.D.3}$$

A simple algebraic manipulation of the definition in Equation 10.D.1 leads to

$$\mathbf{t}_i = \sum_{j=1}^{i} \mathbf{y}_j, \qquad i = 1, 2, \ldots, r$$

which, when substituted into Equation 10.D.3, yields

$$k = \sum_{i=1}^{r} (n-i+1)\mathbf{y}_i$$

Now consider the Jacobian for this transformation (note that this is a simple one-to-one transformation). We have, for $i = 1, 2, \ldots, r$

$$\frac{\partial t_i}{\partial y_j} = \begin{cases} 1 & \text{if} \quad j \leqslant i \\ 0 & \text{if} \quad j > i \end{cases}$$

Hence

$$|J| = 1$$

Substituting in Equation 10.D.2

$$f(y_1,y_2,\ldots,y_r;\theta)=\frac{n!}{(n-r)!}\frac{1}{\theta^r}\exp\left[-\sum_{i=1}^{r}(n-i+1)y_i/\theta\right],$$

$$y_i\geqslant 0$$

The marginal distribution for $\mathbf{y}_k$ may be obtained by integrating the joint density function $r-1$ times

$$h(y_k)=\int\int\cdots\int\frac{n!}{(n-r)!}\exp\left[-\sum_{i\neq p}(n-i+1)y_i/\theta\right]$$

$$\times\exp\left[-(n-p+1)y_p/\theta\right]dy_p\prod_{\substack{i=1\\ i\neq k,p}}^{r}dy_i$$

Let $u=-(1/\theta)(n-p+1)y_p$ then $du=-(1/\theta)(n-p+1)dy_p$. Hence after one integration, w.r.t. $\mathbf{y}_p$, the marginal distribution can be expressed in terms of $r-2$ integrations,

$$h(y_k)=\frac{n!}{(n-r)!}\frac{1}{(n-p+1)}\frac{1}{\theta^{r-1}}\int\int\cdots\int\exp\left[-\sum_{i\neq p}(n-i+1)y_i/\theta\right]$$

$$\times\left[\exp\left[-(n-p+1)y_p/\theta\right]\right]_{\infty}^{0}\prod_{\substack{i=1\\ i\neq k,p}}^{r}dy_i$$

Note that $(n!/(n-r)!)(1/(n-p+1))$ reduces to $(n-k+1)$ when all but $\mathbf{y}_k$ are integrated. Hence the marginal for $\mathbf{y}_k$ becomes, after integrating $r-1$ times,

$$h(y_k)=(n-k+1)\frac{1}{\theta}\exp\left[-(n-k+1)y_k/\theta\right],\qquad y_k\geqslant 0$$

In general mutual independence exists if

$$f(y_1,y_2,\ldots,y_r)=h(y_1)h(y_2)\cdots h(y_r)$$

and in this case

$$\prod_{i=1}^{r}\left\{(n-i+1)\frac{1}{\theta}\exp\left[-(n-i+1)y_i/\theta\right]\right\}$$

$$=\frac{n!}{(n-r)!}\frac{1}{\theta^r}\exp\left[-\sum_{i=1}^{r}(n-i+1)y_i/\theta\right]$$

Therefore, $\mathbf{y}_i$ and $\mathbf{y}_j$, $j \neq i$, are mutually independent and are distributed as stated.

Now let us find the distribution of $\hat{\theta}$. Consider the estimator expressed in terms of the $\mathbf{y}_i$'s as

$$\hat{\theta} = \frac{1}{r} \sum_{i=1}^{r} (n-i+1)\mathbf{y}_i$$

Since the $\mathbf{y}_i$'s are mutually independent, we can use moment generating functions to find the distribution of $\hat{\theta}$. For exponential $\mathbf{y}_i$,

$$M_{\mathbf{y}_i}(s) = \left[\frac{1}{1 - \dfrac{\theta}{n-i+1} s} \right]$$

Then

$$M_{\hat{\theta}}(s) = M_{\mathbf{y}_1}\left(\frac{n}{r}s\right) M_{\mathbf{y}_2}\left(\frac{n-1}{r}s\right) \cdots M_{\mathbf{y}_r}\left(\frac{n-r+1}{r}s\right)$$

$$= \left[\frac{1}{1 - \dfrac{\theta}{r}s} \right]^r$$

which implies that

$$f(\hat{\theta}) = \frac{1}{(r-1)!} \left(\frac{r}{\theta}\right)^r \hat{\theta}^{r-1} \exp\left[-(\hat{\theta} r/\theta)\right] \tag{10.D.5}$$

Note that $E(\hat{\theta}) = \theta$; therefore $\hat{\theta}$ is an unbiased estimator for θ. Also $\operatorname{var}(\hat{\theta}) = \theta^2/r$, which is identical to the Cramer-Rao lower bound. Thus $\hat{\theta}$ has minimum variance and is efficient.

Obviously we can find confidence limits for $\hat{\theta}$ directly from Equation 10.D.5; however, there is great advantage in being able to use a distribution that has previously tabled percentage points. Recall that if $\mathbf{z}$ is a chi-square distributed random variable with k degrees of freedom, then

$$f(z) = \frac{1}{2^{k/2}\Gamma\left(\dfrac{k}{2}\right)} z^{(k/2)-1} e^{-z/2}, \qquad z \geqslant 0$$

So if we make the simple transformation in Equation 10.D.5

$$\frac{z}{2} = \frac{r\hat{\theta}}{\theta}$$

or

$$z = \frac{2r\hat{\theta}}{\theta}$$

then

$$dz = \frac{2r}{\theta} d\hat{\theta}$$

which after substitution yields

$$f(z) = \frac{1}{2^r(r-1)!} z^{r-1} e^{-z/2}$$

It is obvious that $2r\hat{\theta}/\theta$ is distributed as $\chi^2(2r)$. We may now set the $(1-\alpha)$ confidence limits by

$$P\left[\chi^2_{1-(\alpha/2),2r} \leqslant \frac{2\hat{\theta}r}{\theta} < \chi^2_{\alpha/2,2r}\right] = 1-\alpha$$

or

$$\frac{2\hat{\theta}r}{\chi^2_{\alpha/2,2r}} \leqslant \theta \leqslant \frac{2\hat{\theta}r}{\chi^2_{1-\alpha/2,2r}} \qquad (10.D.6)$$

If the lower and the upper limits on θ in Equation 10.D.6 are denoted by L and U, respectively, the confidence interval for the reliability may be written as

$$e^{-t/L} \leqslant R(t) \leqslant e^{-t/U} \qquad (10.D.7)$$

10.D.2 NUMBER OF FAILURES OBSERVED OVER AN INTERVAL

In some test situations the number of failures over an interval of time are recorded. To set confidence intervals in this case, the well-known relationship between the sum of Poisson terms and the integral of the chi-square distribution will be used. Basically we would be setting confidence limits on the mean of a Poisson distribution.

We will develop $100(1-\alpha)\%$ confidence limits for the case where r, the observed number of failures over an interval of time T, is known. For this situation the Poisson distribution will be written as

$$P(x) = \frac{\lambda^x e^{-\lambda}}{x!}, \qquad x = 0, 1, 2, \ldots$$

where $\lambda = T/\theta$ and **x** is the number of failures occurring during the interval of time T. The confidence limits are found by solving for λ in the equations

$$\sum_{x=0}^{r} \frac{\lambda^x e^{-\lambda}}{x!} = \frac{\alpha}{2} \tag{10.D.8}$$

and

$$\sum_{x=r}^{\infty} \frac{\lambda^x e^{-\lambda}}{x!} = \frac{\alpha}{2} \tag{10.D.9}$$

In Equation 10.D.8, $\alpha/2$ represents the probability of observing r or fewer failures for a given λ. This probability decreases as λ is increased for a given r. Thus this equation will give the upper limit on λ. The opposite is true for Equation 10.D.9. Equations 10.D.8 and 10.D.9 can be rewritten as

$$P(x \leqslant r|\lambda_u) = \frac{\alpha}{2}$$

$$P(x \geqslant r|\lambda_l) = \frac{\alpha}{2}$$

Now consider the incomplete gamma function

$$\Gamma(w; \sigma, \beta) = \int_0^w \frac{x^\sigma e^{-x/\beta}}{\sigma! \beta^{\sigma+1}} dx$$

Integrating successively by parts creates a pattern that is easily recognizable as

$$\Gamma(y; \sigma, \beta) = 1 - \sum_{x=0}^{\sigma} \frac{y^x e^{-y/\beta}}{x! \beta^x} \tag{10.D.10}$$

In order to match this with the desired Poisson sum we let $y = \lambda$, $\beta = 1$, and $\sigma = r$. Then Equation 10.D.10 becomes

$$\Gamma(\lambda; r, 1) = 1 - P(x \leqslant r|\lambda)$$

That is,

$$\Gamma(\lambda_u; r, 1) = 1 - \frac{\alpha}{2}$$

and

$$\Gamma(\lambda_l; r-1, 1) = \frac{\alpha}{2}$$

The above expressions could be used to find λ_u and λ_l by referring to incomplete gamma tables; however, it is well known that the gamma function can be readily related to the chi-square distribution.

Consider the incomplete gamma function

$$\Gamma(\lambda; r, 1) = \int_0^{\lambda} \frac{x^r e^{-x}}{r!} dx$$

Let $x = z/2$ and $r = (v/2) - 1$, then $2\,dx = dz$. Substituting,

$$\Gamma(\lambda; r, 1) = \int_0^{2\lambda_u} \frac{z^{(v/2)-1}}{\Gamma\left(\frac{v}{2}\right) 2^{v/2}} e^{-z/2} dz$$

which is a chi-square distribution with degrees of freedom $v = 2(r+1)$. Thus

$$2\lambda_u = \chi^2_{\frac{\alpha}{2}, 2(r+1)}$$

The lower limit would be found in a similar manner as

$$2\lambda_l = \chi^2_{1-\alpha/2, 2r}$$

Combining the two, the $100(1-\alpha)\%$ confidence limits are

$$\frac{\chi^2_{1-\frac{\alpha}{2}, 2r}}{2} \leqslant \lambda \leqslant \frac{\chi^2_{\frac{\alpha}{2}, 2(r+1)}}{2}$$

or

$$\frac{2T}{\chi^2_{\frac{\alpha}{2}, 2(r+1)}} \leqslant \theta \leqslant \frac{2T}{\chi^2_{1-\alpha/2, 2r}} \qquad (10.D.11)$$

Note that if zero failures are observed in T miles, a lower confidence limit can still be obtained.

10.E NONZERO MINIMUM LIFE

The p.d.f. in this case is given by

$$f(x; \theta, \delta) = \frac{1}{\theta} e^{-(1/\theta)(x-\delta)}, \qquad x \geqslant \delta > 0 \qquad (10.E.1)$$

For a life testing experiment truncated at the rth failure, the likelihood function is

$$L = \frac{n!}{(n-r)!} \frac{1}{\theta^r} \exp\left[-\left[\sum_{i=1}^{r} (x_i - \delta) + (n-r)(x_r - \delta) \right] \Big/ \theta \right] \qquad (10.E.2)$$

Taking the natural logarithm,

$$\ln L = \ln\frac{n!}{(n-r)!} - r\ln\theta - \frac{1}{\theta}\left[\sum_{i=1}^{r}(x_i-\delta)+(n-r)(x_r-\delta)\right]$$

Differentiating with respect to θ, setting $(d/d\theta)\ln L=0$, and solving for θ, we get

$$\hat{\theta}=\frac{\sum_{i=1}^{r}(x_i-\delta)+(n-r)(x_r-\delta)}{r} \tag{10.E.3}$$

Observing that L is a monotone increasing function in δ and $\delta \leqslant x_1$ leads to

$$\hat{\delta}=x_1$$

Since $\mathbf{x}_1$ is the smallest order statistic in a sample of size n, the p.d.f. for $\mathbf{x}_1$ is

$$\begin{aligned} g(x_1) &= n\left[1-F(x_1)\right]^{n-1} f(x_1) \\ &= \frac{n}{\theta}\exp\left[-n(x_1-\delta)/\theta\right], \qquad x_1 \geqslant \delta \end{aligned} \tag{10.E.4}$$

and the expected value of $\mathbf{x}_1$ is

$$E(\mathbf{x}_1)=\delta+\frac{\theta}{n}$$

This suggests that δ should be estimated by

$$\hat{\delta}=x_1-\frac{\hat{\theta}'}{n} \tag{10.E.5}$$

where $\hat{\theta}'$ is an unbiased estimator of θ.

Now consider the numerator of Equation 10.E.3 with x_1 substituted for δ. We have

$$\nu=\sum_{i=2}^{r}(x_i-x_1)+(n-r)(x_r-x_1) \tag{10.E.6}$$

Using a procedure similar to the zero minimum life case (Equation 10.D.1), we will define

$$w_i=(n-i+1)(x_i-x_{i-1}), \qquad x_0=\delta$$

Then changing variables in Equation 10.E.2 gives

$$f(w_1, w_2, \ldots, w_r) = \frac{1}{\theta^r} \exp\left[-\sum_{i=1}^{r} w_i/\theta \right]$$

The Jacobian for the transformation this time would be $(n-r)!/n!$. Note that the marginal distribution for $i = 1, 2, \ldots, r$ is

$$g(w_i) = \frac{1}{\theta} \exp[-w_i/\theta], \qquad w_i \geqslant 0 \tag{10.E.7}$$

Thus the $\mathbf{w}_i$'s are independent and identically distributed.

Now Equation 10.E.6 in terms of the $\mathbf{w}_i$'s becomes

$$\boldsymbol{\nu} = \sum_{i=2}^{r} \mathbf{w}_i \tag{10.E.8}$$

Where $2\boldsymbol{\nu}/\theta$ is chi-square distributed with $(2r-2)$ degrees of freedom. Hence

$$E\left(\frac{2\boldsymbol{\nu}}{\theta}\right) = 2r - 2$$

or

$$E(\boldsymbol{\nu}) = \theta(r-1)$$

Therefore an unbiased estimator for θ is

$$\hat{\theta}' = \frac{\sum_{i=2}^{r} (x_i - x_1) + (n-r)(x_r - x_1)}{r-1} \tag{10.E.9}$$

10.E.1 CONFIDENCE LIMITS ON THE MEAN LIFE (θ)

We know that $2(r-1)\hat{\theta}'/\theta$ is distributed as $\chi^2(2r-2)$. Hence the development of confidence limits is very similar to the previous ones. The $100(1-\alpha)\%$ two-sided limits are given by

$$\frac{2(r-1)\hat{\theta}'}{\chi^2_{\alpha/2, 2r-2}} \leqslant \theta \leqslant \frac{2(r-1)\hat{\theta}'}{\chi^2_{1-\alpha/2, 2r-2}}$$

10.E.2 CONFIDENCE LIMITS FOR THE MINIMUM LIFE (δ)

From Equation 10.E.4 and the definition of $\mathbf{w}_1$ we have

$$\frac{2n(\mathbf{x}_1 - \delta)}{\theta} = \frac{2}{\theta} \mathbf{w}_1$$

which is distributed as $\chi^2(2)$. From Equations 10.E.6 and 10.E.8 we have

$$\frac{2\mathbf{v}}{\theta}=\frac{2}{\theta}\sum_{i=2}^{r}\mathbf{w}_i$$

which is distributed as $\chi^2(2r-2)$. From Equation 10.E.7 we can conclude that $\mathbf{w}_1$ and $\mathbf{v}$ are independent.

Thus an F-distributed random variable can be constructed as

$$F_{2,2r-2}=\frac{n(r-1)(\mathbf{x}_1-\delta)}{\mathbf{v}}$$

We know that $\delta\leqslant\mathbf{x}_1$, and the $100(1-\beta)\%$ confidence interval on δ is given by

$$P\left[0\leqslant\frac{n(r-1)(\mathbf{x}_1-\delta)}{\mathbf{v}}\leqslant F_{\beta,2,2r-2}\right]=1-\beta$$

Transposing, we can write the $100(1-\beta)\%$ confidence limits on δ as

$$x_1-\frac{\hat{\theta}'F_{\beta,2,2r-2}}{n}\leqslant\delta\leqslant x_1$$

10.F EXPECTED TEST TIME

In the situation where n items are placed on test and the test is truncated at the rth failure, the length of the test is found by

$$\mathbf{t}_r=\sum_{i=1}^{r}\mathbf{y}_i$$

where $\mathbf{y}_i$ is as defined in Equation 10.D.1. Therefore

$$E(\mathbf{t}_r)=\sum_{i=1}^{r}E(\mathbf{y}_i)$$

We have earlier shown that $\mathbf{y}_i$ is exponential with parameter $\theta/(n-i+1)$. Thus

$$E(\mathbf{t}_r)=\theta\sum_{i=1}^{r}\frac{1}{(n-i+1)}\qquad(10.\text{F}.1)$$

Note that $E(\mathbf{t}_r)$ decreases as n increases.

Chapter 11 reliability estimation: weibull distribution

After the exponential distribution, the Weibull distribution is probably the most widely used distribution for life testing applications. Before it was used in life testing the Weibull distribution was known to statisticians as the Fisher-Tippett Type III distribution of smallest values or as the third asymptotic distribution of smallest extreme values. This extreme value distribution was also derived by a Swedish scientist-engineer named Waloddi Weibull as a probabilistic characterization for the breaking strength of materials. The derivation was based on practical assumptions about the failure of materials. Weibull's subsequent publications in popular engineering journals promoted the use of his developments. Today the Weibull distribution is well known and can be found in many introductory statistics textbooks.

The statistical properties of the Weibull distribution are considered first in Section 11.1. The purpose of this section is to provide familiarity with the Weibull p.d.f., the different distributional forms available by changing parameter values, and the related hazard and reliability functions.

Graphical estimation can be used with the Weibull distribution to estimate the reliability or the Weibull parameters. Section 11.2 provides a detailed treatment of the graphical procedures available for the Weibull distribution and includes procedures for suspended items. Confidence limits are also considered.

Section 11.3 presents statistical estimation procedures that have recently been developed for the Weibull distribution. With these procedures the subjectivity of graphical estimation can be alleviated. However, tables must be provided to use the statistical estimation procedures, and these procedures are somewhat more cumbersome to use than the graphical procedures.

Section 11.4 considers a goodness-of-fit test specifically developed for the Weibull distribution, while Section 11.5 considers the problem of calculating savings in test time by using Type II censoring.

11.1 STATISTICAL PROPERTIES OF THE WEIBULL DISTRIBUTION

The Weibull distribution is easiest to remember in its cumulative form. The cumulative distribution for a random variable **x** distributed as the three-parameter Weibull is given by

$$F(x;\theta,\beta,\delta)=1-e^{-\left(\frac{x-\delta}{\theta-\delta}\right)^{\beta}}, \qquad x\geqslant\delta \tag{11.1}$$

where $\beta>0$, $\theta>0$ and $\delta\geqslant 0$. The parameter β is called the shape parameter or the Weibull slope; the parameter θ is called the scale parameter or the characteristic life; and the parameter δ is called the location parameter or the minimum life.

The two-parameter Weibull has a minimum life of zero, and the cumulative distribution is given by

$$F(x;\theta,\beta)=1-e^{-(x/\theta)^{\beta}}, \qquad x\geqslant 0 \tag{11.2}$$

Since the three-parameter distribution can always be converted to the two-parameter distribution by a simple linear transformation, the two-parameter Weibull will be used to illustrate the properties of the distribution.

The p.d.f. is obtained by differentiating Equation 11.2 as

$$f(x;\theta,\beta)=\frac{\beta}{\theta}\left(\frac{x}{\theta}\right)^{\beta-1}e^{-(x/\theta)^{\beta}}, \qquad x\geqslant 0 \tag{11.3}$$

The hazard function given by

$$h(x)=\frac{\beta}{\theta}\left(\frac{x}{\theta}\right)^{\beta-1}, \qquad x\geqslant 0 \tag{11.4}$$

will decrease in time if $\beta<1$, will increase if $\beta>1$, or will be a constant if $\beta=1$.

The kth moment of the Weibull distribution is found as follows:

$$\mu_k'=E(\mathbf{x}^k)=\int_0^{\infty}x^k\frac{\beta}{\theta}\left(\frac{x}{\theta}\right)^{\beta-1}e^{-(x/\theta)^{\beta}}dx \tag{11.5}$$

Using the transformation

$$u=\left(\frac{x}{\theta}\right)^{\beta}$$

then

$$du=\frac{\beta}{\theta}\left(\frac{x}{\theta}\right)^{\beta-1}dx$$

and we have

$$\mu_k' = \theta^k \int_0^\infty u^{k/\beta} e^{-u}\,du \tag{11.6}$$

which is readily recognizable as a gamma function and yields

$$\mu_k' = \theta^k \Gamma\left(1 + \frac{k}{\beta}\right) \tag{11.7}$$

Hence the mean of the Weibull distribution is

$$\mu = \theta \Gamma\left(1 + \frac{1}{\beta}\right) \tag{11.8}$$

and the variance is

$$\sigma^2 = \theta^2 \left[\Gamma\left(1 + \frac{2}{\beta}\right) - \Gamma^2\left(1 + \frac{1}{\beta}\right)\right] \tag{11.9}$$

The shape parameter β, as the name implies, determines the shape of the distribution. As β increases, the mean of the distribution approaches the characteristic life θ, and the variance approaches zero as illustrated by the numerical values in Table 11.1.

Table 11.1 Mean and variance of Weibull distribution as a function of the shape parameter β

β	MEAN/θ	VARIANCE/θ^2
0.5	2.0000	20.000
1.0	1.0000	1.000
2.0	0.8862	0.215
3.0	0.8934	0.105
3.5	0.8998	0.081
4.0	0.9064	0.065
5.0	0.9182	0.044
6.0	0.9275	0.033
10.0	0.9514	0.013
20.0	0.9730	0.004

Table 11.2 Skewness and kurtosis for the Weibull distribution

β	SKEWNESS	KURTOSIS
0.5	6.619	87.720
1.0	2.000	9.000
2.0	0.631	3.245
3.0	0.168	2.729
3.5	0.025	2.713
4.0	−0.087	2.748
5.0	−0.254	2.881
6.0	−0.373	3.035
10.0	−0.637	3.562
20.0	−0.860	4.247

The skewness and kurtosis for the Weibull distribution are also a function of the shape parameter β. The skewness is given by

$$\alpha_3 = \frac{\Gamma\left(1+\frac{3}{\beta}\right) - 3\Gamma\left(1+\frac{1}{\beta}\right)\Gamma\left(1+\frac{2}{\beta}\right) + 2\Gamma^3\left(1+\frac{1}{\beta}\right)}{\left[\Gamma\left(1+\frac{2}{\beta}\right) - \Gamma^2\left(1+\frac{1}{\beta}\right)\right]^{3/2}} \tag{11.10}$$

By using Equations 11.7 and 11.9, the kurtosis can also be found. Table 11.2 shows various numerical values for the skewness and kurtosis. Recall that a normal distribution has a skewness of zero and a kurtosis of 3. The Weibull can be seen to approximate these values when the shape parameter (β) is between 3.5 and 4.0. This is the basis for the claim that the Weibull approximates the normal distribution. It should be noted that this approximation is not a mathematical approximation, and values of β other than 3.5 to 4.0 can be obtained from seemingly normal data. This will be clear in an example to be discussed later.

Figure 11.1 shows various forms of the distribution where the scale parameter (θ) is taken as unity. The scale parameter helps locate the distribution along the x axis. This can be seen by considering the cumulative distribution. Substituting $x=\theta$ into the cumulative distribution given in Equation 11.2 we get

$$F(x=\theta)=1-e^{-1}=0.632 \tag{11.11}$$

So for any Weibull distribution the probability of failure prior to θ is equal to 0.632. Thus θ will always divide the area under the p.d.f. into 0.632 and 0.368 for all values of β. This is why θ is called the characteristic life.

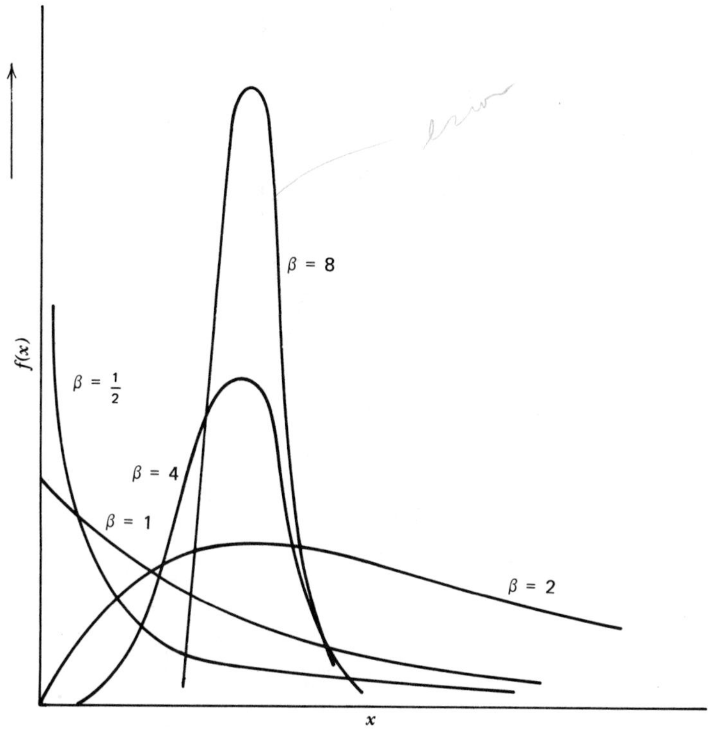

Figure 11.1 The Weibull distribution for different values of β.

11.2 GRAPHICAL ESTIMATION

Graphical estimation of parameters and graphical prediction have great appeal in practice. Basically, in order to use a graphical estimation procedure, a convenient transformation of the cumulative distribution must be available that changes it into a linear form.

Consider the cumulative distribution function for the Weibull

$$F(t) = 1 - e^{-(t/\theta)^{\beta}}$$

which, after it is rearranged and the natural logarithm is taken twice, yields

$$\ln\left(\ln\frac{1}{1-F(t)}\right) = \beta\ln t - \beta\ln\theta \tag{11.12}$$

Rearranging this into the standard form for the dependent and the independent

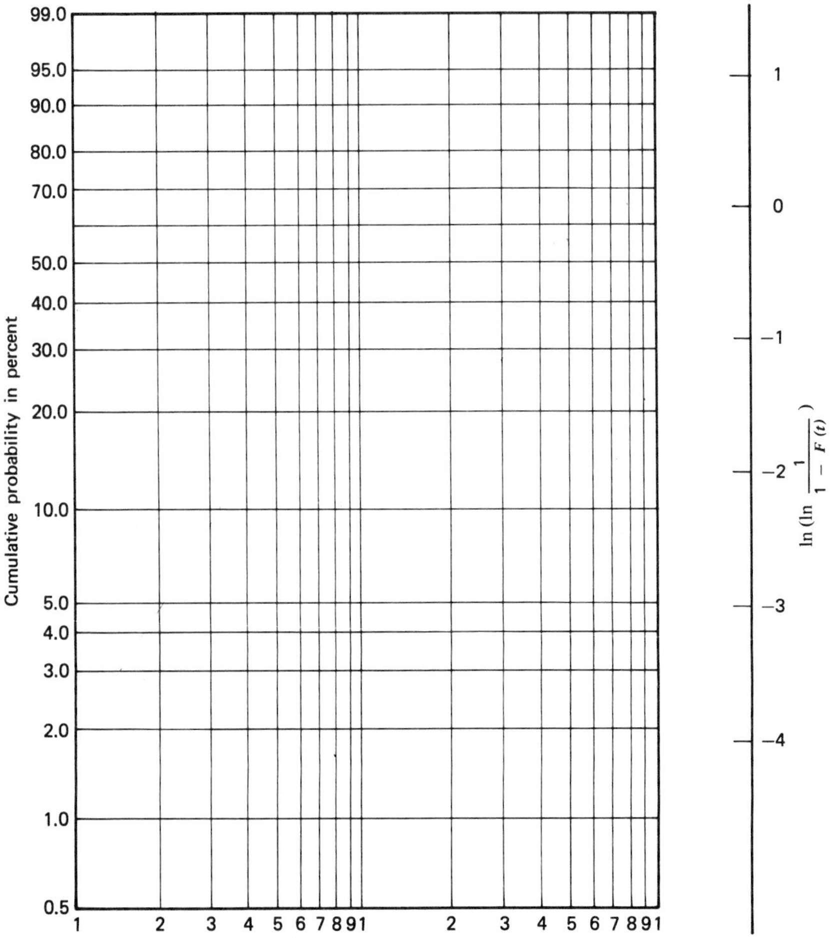

Figure 11.2 Weibull probability paper.

variable gives

$$\ln t = \frac{1}{\beta} \ln\left(\ln \frac{1}{1 - F(t)}\right) + \ln \theta$$

which is clearly of the form $Y = (1/\beta)X + A$, and will plot as a straight line on rectangular (X, Y) graph paper.

Weibull graph paper can be constructed by relabeling the grid lines on the rectangular paper as $Y = \ln t$ and $X = \ln[\ln(1/(1 - F(t)))]$. In addition, the axes are usually reversed and then β is the slope of the straight-line plot. Figure 11.2 shows such a Weibull paper.

Now suppose that five components are placed on test, and the observed failure times are 67, 120, 130, 220, and 290 hours. These failure times establish the abscissa values for plotting on the Weibull paper as given in Figure 11.2. We also need the corresponding values of the cumulative distribution function $F(x)$ to establish the ordinate values. Or we must know the fraction of the population failing prior to each sample value.

Let us designate the jth ordered observation ${}_o\mathbf{x}_j$ and the fraction of the population below this observation as $\mathbf{p}_j$, or $\mathbf{p}_j = F({}_o\mathbf{x}_j)$. Since the value of ${}_o\mathbf{x}_j$ varies from sample to sample, $\mathbf{p}_j$ is a random variable. Thus we must consider the distribution of $\mathbf{p}_j$. At first glance one may be tempted to estimate $\mathbf{p}_j$ by the fraction j/n, where n is the sample size. However, j/n is the fraction of the sample failing, and surely we would not believe, based on observing a sample of size 5, that 100% of the population would fail prior to the largest observed failure time. We will now proceed with the development of the distribution for $\mathbf{p}_j$.

11.2.1 DEVELOPMENT OF THE RANK DISTRIBUTION

Given an ordered random sample ${}_o\mathbf{x}_1, {}_o\mathbf{x}_2, \ldots, {}_o\mathbf{x}_n$ of size n from a population having a cumulative distribution function $F(x)$, where $\mathbf{x}$ is continuous, we wish to determine estimators for $F({}_o\mathbf{x}_1), F({}_o\mathbf{x}_2), \ldots, F({}_o\mathbf{x}_n)$. Let us define

$$ {}_n\mathbf{p}_j = F({}_o\mathbf{x}_j) \tag{11.13} $$

that is, ${}_n\mathbf{p}_j$ is the fraction of the population failing prior to the jth ordered observation in a sample of size n.

Next we partition the population into three regions as shown in Table 11.3. For the jth outcome $({}_o\mathbf{x}_j)$ to occur in region 2, $j-1$ observations must occur in region 1 and $n-j$ in region 3. The regions are considered mutually exclusive, and the probability that each observation falls in a particular region is constant. Clearly, the multinomial distribution is applicable in this situation.

Recall that the multinomial distribution is given by

$$ P(y_1, y_2, \ldots, y_k) = \frac{m!}{y_1! y_2! \ldots y_k!} (\Theta_1)^{y_1} (\Theta_2)^{y_2} \ldots (\Theta_k)^{y_k} $$

where $\Sigma_{i=1}^k \Theta_i = 1$ and $\Sigma_{i=1}^k y_i = m$. Note that Θ_i is the probability of obtaining

Table 11.3 Partitioning of population

Region	1	2	3
Limits of Region	0 to ${}_ox_j - \frac{1}{2}d_ox_j$	${}_ox_j - \frac{1}{2}d_ox_j$ to ${}_ox_j + \frac{1}{2}d_ox_j$	${}_ox_j + \frac{1}{2}d_ox_j$ to ∞
Probability of Obtaining an Outcome in Region	$F({}_ox_j - \frac{1}{2}d_ox_j)$	$f({}_ox_j)\,d_ox_j$	$1 - F({}_ox_j + \frac{1}{2}d_ox_j)$

the ith outcome and is constant from trial to trial, and $\mathbf{y}_i$ is the random variable representing the number of outcomes of the ith type.

If we ignore terms of the order of $\frac{1}{2}d_o x_j$ then we have

Number of Outcomes	Probability
$(j-1)$	$F(_o x_j)$
1	$f(_o x_j)\,dx$
$(n-j)$	$1-F(_o x_j)$

So the probability of $j-1$ outcomes falling in region 1, one falling in region 2, and $n-j$ falling in region 3 is given by

$$\frac{n!}{(j-1)!\,1!\,(n-j)!}\left[F(_o x_j)\right]^{j-1} f(_o x_j)\,dx\left[1-F(_o x_j)\right]^{n-j} \qquad (11.14)$$

This is precisely the distribution of $_o\mathbf{x}_j$, the value of the jth ordered observation.

Now we know that

$$F(_o x_j) = {}_n p_j$$

which by differentiating yields

$$dF(_o x_j) = f(_o x_j)\,dx = d_n p_j$$

Hence the probability element becomes

$$g(_n p_j)\,d_n p_j = \frac{n!}{(j-1)!(n-j)!}\,{}_n p_j^{\,j-1}(1-{}_n p_j)^{n-j}\,d_n p_j \qquad (11.15)$$

where

$$0 \leqslant {}_n p_j \leqslant 1$$

This distribution is commonly termed the rank distribution when used in this context in life testing. Actually the p.d.f. of the random variable $_n\mathbf{p}_j$ is the well-known beta distribution. Figure 11.3 shows the p.d.f. for different values of j for a sample of size $n=10$.

It is important to note that in the derivation of the rank distribution the only assumption made was that $F(x)$ was differentiable. Thus the rank distribution does not apply exclusively to the Weibull distribution but would be equally valid for the normal or the exponential, or for a host of other continuous distributions.

Since $_n\mathbf{p}_j$ is a random variable we face the problem of selecting a single value representative of $_n\mathbf{p}_j$ to use in plotting on Weibull paper. In this effort we will consider various alternatives.

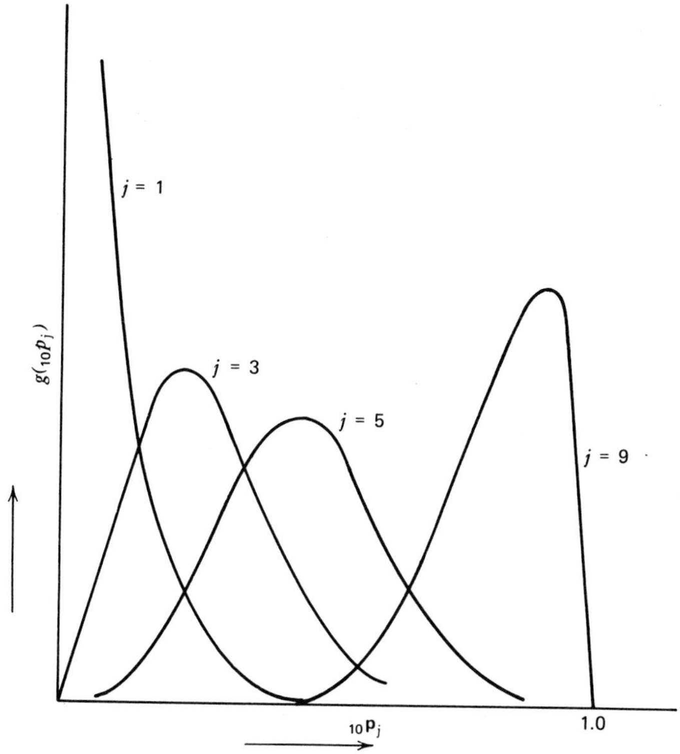

Figure 11.3 The rank distribution for a sample of size 10.

The mean value of $_n\mathbf{p}_j$ is given by

$$E\,(_n\mathbf{p}_j) = \int_0^1 {}_np_j \frac{n!}{(j-1)!(n-j)!}\, {}_np_j^{j-1}\,(1 - {}_np_j)^{n-j}\, d_np_j$$

$$= \frac{n!}{(j-1)!(n-j)!} \int_0^1 {}_np_j^j\,(1 - {}_np_j)^{n-j}\, d_np_j \tag{11.16}$$

Now a beta function in the standard form is given by

$$B\,(m,k) = \int_0^1 x^{m-1}\,(1-x)^{k-1}\,dx = \frac{\Gamma(m)\Gamma(k)}{\Gamma(m+k)} \tag{11.17}$$

which, when applied to Equation 11.16, yields

$$E\,(_n\mathbf{p}_j) = \frac{n!}{(j-1)!(n-j)!}\,\frac{\Gamma(j+1)\Gamma(n-j+1)}{\Gamma(n-j+1+j+1)} = \frac{j}{n+1} \tag{11.18}$$

This means that the mean value of the fraction failing prior to the jth ranked observation in a sample of size n equals $j/(n+1)$, which differs significantly from j/n for small values of n, the sample size.

The median value, commonly called the median rank, is somewhat more difficult to find. Consider the integral

$$\mathbf{P} = \int_0^{\tilde{x}} g({}_np_j)\, d\,{}_np_j \tag{11.19}$$

where $g({}_np_j)$ is as defined in Equation 11.15. The value of $\tilde{x}$ for which $\mathbf{P}=0.50$ would be the desired median value.

An incomplete beta function is given by

$$B_x(m,n) = \int_0^x y^{m-1}(1-y)^{n-1}\,dy \tag{11.20}$$

Recalling the definition of a beta function as given in Equation 11.17, the integral in Equation 11.19 becomes

$$I_{\tilde{x}}(j,n-j+1) = \frac{B_{\tilde{x}}(j,n-j+1)}{B(j,n-j+1)} \tag{11.21}$$

The incomplete beta ratios $I_x(m,k)$ are extensively tabled (e.g., see Karl Pearson, *Tables of the Incomplete Beta Function*, Cambridge University Press, Cambridge, England, 1932). For various values of $\mathbf{P}=I_{\tilde{x}}(j,n-j+1)$ values of $\tilde{x}$ can be found easily. However, for convenience, the median rank values have been summarized in Appendix VIII.

An approximation to the median rank value is given by

$$\tilde{x} = \frac{j-0.3}{n+0.4} \tag{11.22}$$

The use of this approximation will be necessary in some later applications.

At this point we have two possible candidates for the value of $F({}_ox_j)$. They are (1) the mean value given by $E[F({}_ox_j)]=j/(n+1)$ and (2) the median value $\tilde{F}({}_ox_j)\approx(j-0.3)/(n+0.4)$. Both values are found in use. The mean value is used because the mean is commonly taken as the representative value of a sample from a distribution. However, in highly skewed distributions, as are most rank distributions, the median may be a better descriptor. The rank distributions are systematically skewed from right to left as one proceeds from the lowest to the highest sample value. Thus the probability of a sample value falling below the mean value is highest for the first ordered sample value and lowest for the last ordered sample value. This implies that if mean rank plotting is used, the rank assigned to the first observation will probably be too high and the rank assigned to the last observation too low with successive changes in error for intermediate observations. When a straight line is fitted to these observations using mean ranks, the line will most likely be rotated clockwise, and the slope will be underestimated. This is the basic argument advanced in favor of plotting positions other than the mean ranks.

Other plotting positions have been proposed. For example, White [37] considers ${}_n\mathbf{y}_j = \ln {}_n\mathbf{t}_j$ where $\mathbf{t}$ is Weibull distributed. Recall that this is the transformation made on the Weibull paper. White then finds $E[{}_ny_j] = E[\ln {}_nt_j]$ and provides these values in tabular form for use as plotting positions.

Kimball [21] has done a limited study comparing some of the possible contenders. His results are summarized in Table 11.4. These results are based on complete samples of size 6 and consider only the estimation of the shape parameter β. Based on Kimball's results for estimating β, White's plotting positions appear to be the best. The plotting positions $(j-\frac{3}{8})/(n+\frac{1}{4})$ give results close to White's method and are simple to use. Note that the popular median rank convention appears to be inferior to both of these, but better than mean ranks.

It should be emphasized that Kimball's results are very limited and a more comprehensive comparison study still remains to be done. Hence we feel justified in using the popular median rank procedure in the subsequent examples.

Table 11.5 shows the median and the mean ranks for the previous sample of five. Although the differences between the mean and median rank values are numerically small, the plotted lines can differ significantly as a result of the scale used on the Weibull paper. Figure 11.4 shows graphical estimates using both the mean and median ranks. Figure 11.4 also illustrates another problem with graphical estimation. The points obtained from a sample rarely fall on an exact straight line because of sampling error, and the best straight line must be

Table 11.4 Kimball's comparison of plotting positions for estimating the shape parameter β

PLOTTING CONVENTIONS	BIAS/β	MEAN SQUARE ERROR/β^2
Mean $= j/(n+1)$	0.2284	0.3011
Median	0.0632	0.1928
$(j-\frac{3}{8})/(n+\frac{1}{4})$	0.0220	0.1758
$(j-\frac{1}{2})/n$	−0.0547	0.1543
Mode	0.1916	0.2692
$E[{}_ny_j]^a$	0	0.1686

^{a}Equivalent to White's method.

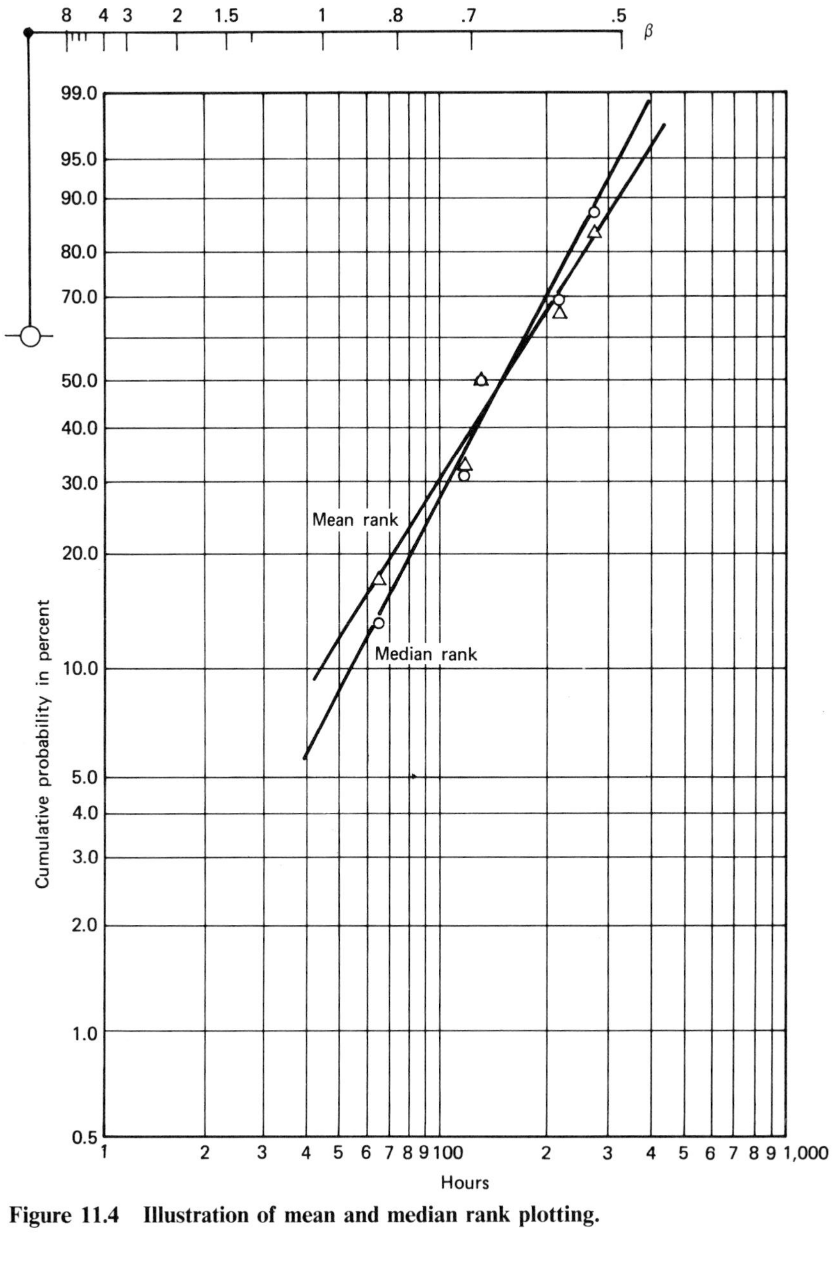

Figure 11.4 Illustration of mean and median rank plotting.

Table 11.5 Mean and median rank values for a sample

FAILURE TIME (HRS)	MEAN RANK	MEDIAN RANK
67	0.167	0.129
120	0.333	0.314
130	0.500	0.500
220	0.667	0.686
290	0.833	0.871

visually fitted to the data. A least-square fitting procedure could be used; however, this somewhat defeats the purpose, namely, the ease of graphical estimation.

11.2.2 GRAPHICAL ESTIMATION OF PARAMETERS

From the basic straight line as plotted on Weibull paper we have

$$\ln\left(\ln\frac{1}{1-F(x)}\right)=\beta\ln x-\beta\ln\theta$$

Clearly β can be estimated by the slope of the Weibull plot. Figure 11.5 shows the estimated slope, $\hat{\beta}=1.6$. In commercial Weibull paper a printed scale is usually provided for estimating the slope. In Figure 11.5, the scale is shown at the top.

The characteristic life parameter θ can be estimated by recalling that $F(x=\theta)=0.632$. Thus projecting the 63.2% point from the ordinate to the corresponding value on the abscissa gives an estimate of θ. In Figure 11.5, we find that $\hat{\theta}=190$ hr.

The mean can also be estimated graphically by substituting $\mu=\theta\Gamma(1+1/\beta)$ to yield

$$F(x=\mu)=1-e^{-[(\Gamma(1+1/\beta))]^{\beta}} \tag{11.23}$$

which is a function of β only. This function is presented in Table 11.6. So by locating $F(\mu)$ on the ordinate for a given slope, the mean is found as the corresponding value on the abscissa. In Figure 11.5, a slope of 1.6 leads to an estimated mean of 170 hours. The mean computed from the original data is 165.4 hours. Because graphical errors are inevitable, it is preferable to calculate the mean directly from the original data.

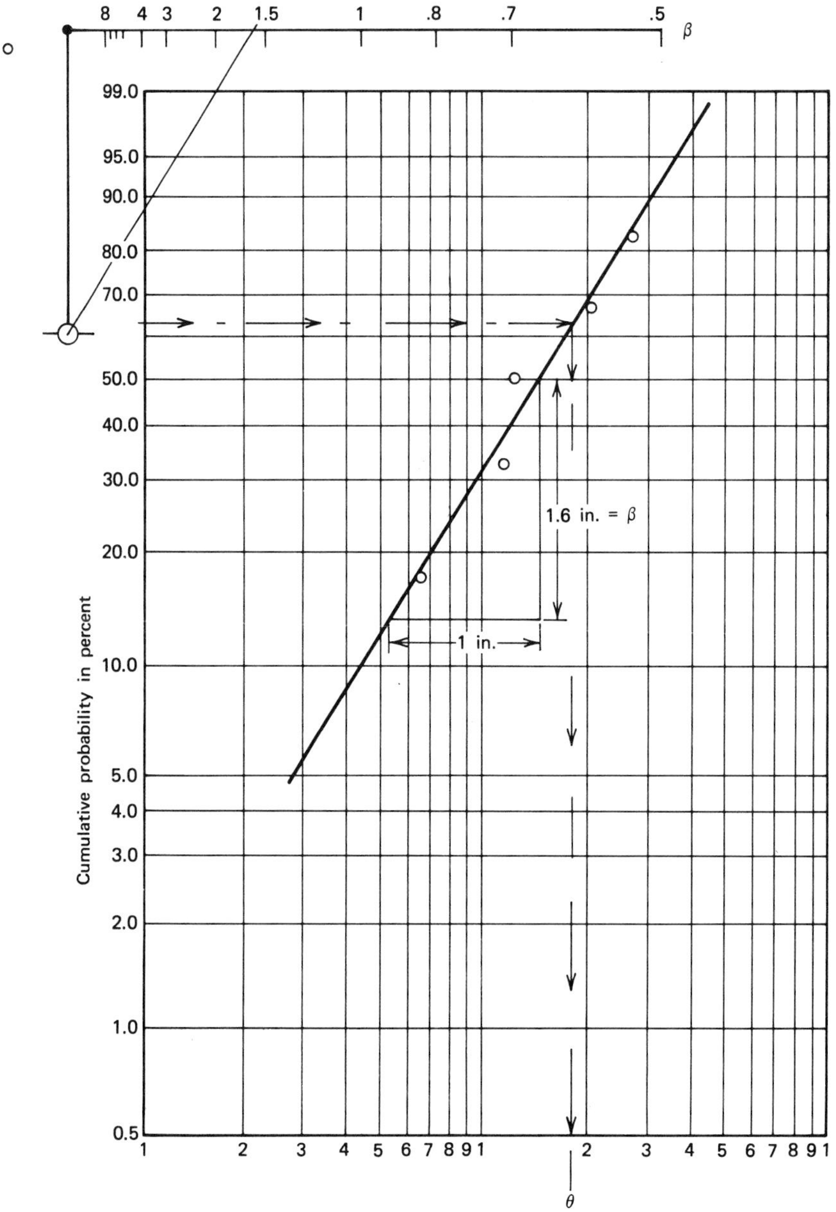

Figure 11.5 Graphical parameter estimation for the Weibull distribution.

Table 11.6 Theoretical relationship between cumulative distribution and slope

SLOPE β	$F(\mu)$
0.5	0.757
1.0	0.632
1.1	0.618
1.2	0.605
1.3	0.594
1.4	0.584
1.5	0.576
1.6	0.568
1.8	0.555
2.0	0.544
2.2	0.535
2.5	0.524
2.7	0.517
3.0	0.509
3.2	0.505
4.0	0.491

The following example will demonstrate that normally distributed data does not necessarily have a slope of approximately 3.6 when plotted on Weibull paper.

EXAMPLE 11.1

Consider the data in Table 11.7, which appears to be normally distributed (in fact, these data plot very well on normal probability paper). Let us see how it looks when plotted on Weibull paper.

Table 11.7 Pull strength of spot welds

MIDPOINT (N)	FREQUENCY
142	3
146	5
150	20
154	36
158	18
162	13
166	5
Total =	100

We calculate the median ranks by assuming that one-half of the observations in each cell fall below the midpoint value. Thus for the first midpoint of 142 Newtons, we have 3/2 observations, and the median rank is

$$\left(\frac{1.5-0.3}{100+0.4}\right)=1.20\%$$

Proceeding in a similar fashion we obtain the remaining ranks, as shown in Table 11.8.

Table 11.8 Spot weld data for plotting on Weibull paper

STRENGTH (N)	MEDIAN RANK (%)
142	1.20
146	5.18
150	17.63
154	45.52
158	72.41
162	87.85
166	96.81

Now if these data were to be plotted on the Weibull paper of the type shown earlier the result would be an almost vertical line. Figure 11.6 shows a plot of the data with the abscissa scale expanded by a factor of 10. The data appear to plot reasonably well as a straight line, and the parameters are estimated as

$$\hat{\beta}=29$$

$$\hat{\theta}=157 \text{ N}$$

Note the high value of the slope for this seemingly normal data, which is due to the large displacement from zero.

The mean and the variance can be estimated from Equations 11.8 and 11.9 as

$$\hat{\mu}=157\Gamma\left(1+\frac{1}{29}\right)=154 \text{ N}$$

$$\hat{\sigma}^2=(157)^2\left[\Gamma\left(1+\frac{2}{29}\right)-\Gamma^2\left(1+\frac{1}{29}\right)\right]=20.4 \text{ N}^2$$

For a sample as large as this, we can obtain better estimates of μ and σ^2 directly from the data. The calculated values are

$$\text{sample mean}=\bar{x}=154.9 \text{ N}$$

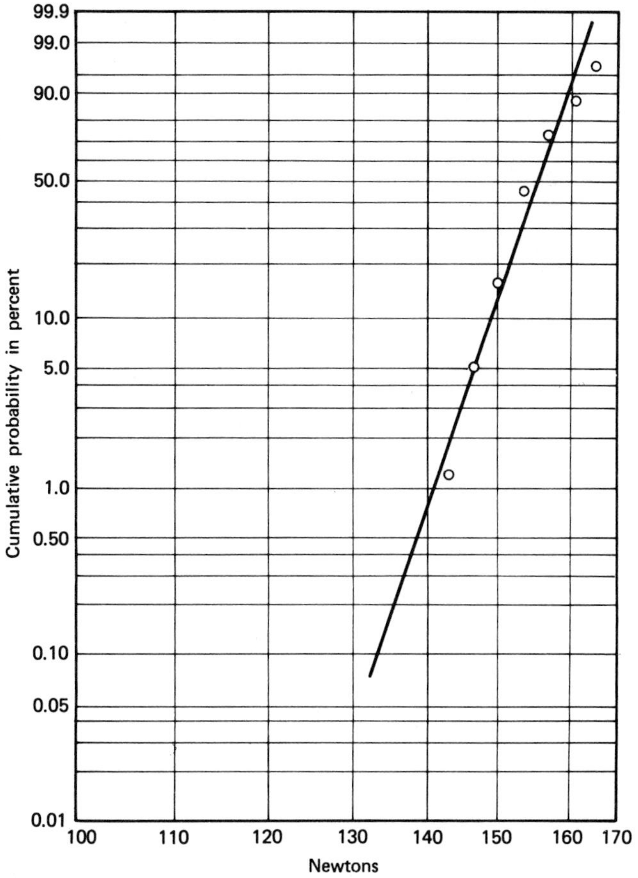

Figure 11.6 Spot weld test data.

and

$$\text{sample variance} = s^2 = 21.4\ \text{N}^2$$

Thus the graphical procedure did provide fair estimates when compared to the values as calculated from the sample.

The Weibull distribution with $\theta = 157$, $\beta = 29$ and the normal distribution with mean $\mu = 154.9$ and variance $\sigma^2 = 21.4$ are graphically illustrated in Figure 11.7. The histogram is based on the original data. Visual comparison among the three plots can be made because they all share the same scale.

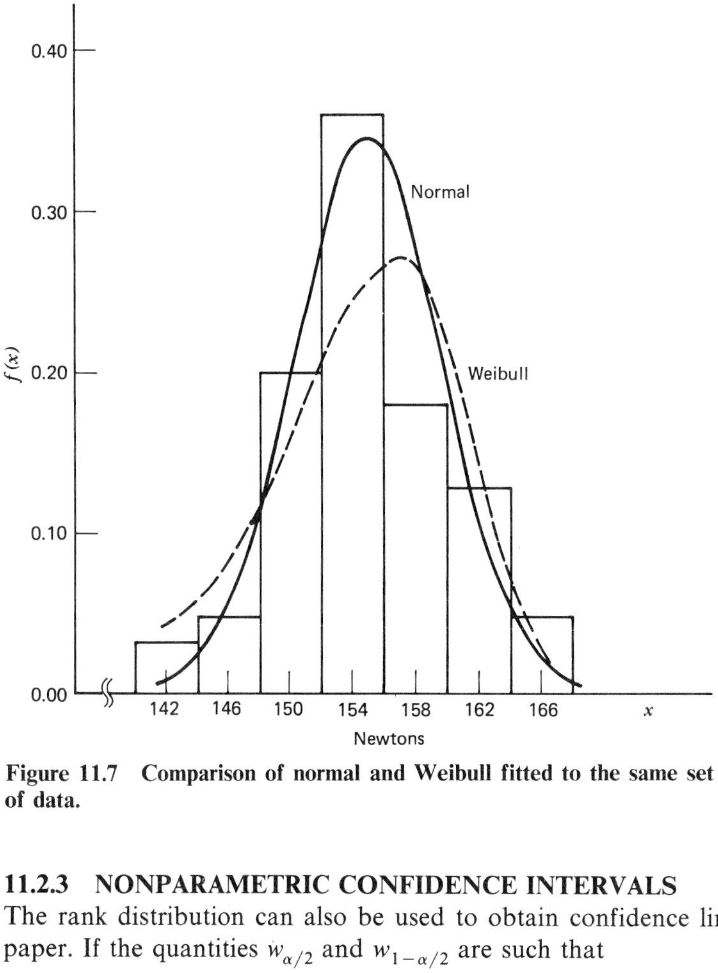

Figure 11.7 Comparison of normal and Weibull fitted to the same set of data.

11.2.3 NONPARAMETRIC CONFIDENCE INTERVALS

The rank distribution can also be used to obtain confidence limits on Weibull paper. If the quantities $w_{\alpha/2}$ and $w_{1-\alpha/2}$ are such that

$$\int_0^{w_{1-\alpha/2}} g({}_n p_j)\, d\,{}_n p_j = \frac{\alpha}{2} \tag{11.24}$$

and

$$\int_{w_{\alpha/2}}^{1} g({}_n p_j)\, d\,{}_n p_j = \frac{\alpha}{2} \tag{11.25}$$

then

$$\int_{w_{1-\alpha/2}}^{w_{\alpha/2}} g({}_n p_j)\, d\,{}_n p_j = 1-\alpha \tag{11.26}$$

The quantities $w_{1-\alpha/2}$, $w_{\alpha/2}$ are termed $100(1-\alpha)\%$ nonparametric confidence limits. These limits are nonparametric because although the underlying distribution is Weibull, that knowledge will not be used to construct them.

Calculation of the confidence limits from the rank distribution creates the same integration problem as was encountered in finding the median value. However, tables have been provided in Appendix VIII. Also it should be pointed out that since the rank distribution is a beta distribution, a well-known transformation can be applied to obtain an **F** distributed random variable (see Section 13.2 for the development and application of this transformation in a similar situation). For $\alpha \geqslant 0.50$, the confidence limit is found by

$$w_\alpha = \frac{(j/(n-j+1))}{F_{1-\alpha,2(n-j+1),2j} + j/(n-j+1)}$$

and for $\alpha < 0.50$

$$w_\alpha = \frac{(j/(n-j+1))F_{\alpha,2j,2(n-j+1)}}{1+(j/(n-j+1))F_{\alpha,2j,2(n-j+1)}}$$

where F_{α,n_1,n_2} is a value from an F distribution with n_1 and n_2 degrees of freedom and is such that $P[\mathbf{F}_{n_1,n_2} \geqslant F_{\alpha,n_1,n_2}] = \alpha$. Here w_α is a value from the rank distribution with parameters n and j such that $P[\mathbf{p} \geqslant w_\alpha] = \alpha$.

Table 11.9 gives the 5% and 95% ranks for a sample of size 5. Now in plotting these rank values if one assumes that the plotted straight line represents the true Weibull distribution, then any deviation from this straight line can be regarded as sampling error. So any points not falling on the straight line can be projected horizontally to the line. The 5% and 95% ranks are then projected about the straight line.The plotting procedure for the confidence limits is illustrated in Figure 11.8. Point A denotes the plot obtained from the first failure time in a sample of size 5. Recall that the first median rank value for a sample of size 5 is 12.9%, hence point A was plotted at 12.9% on the ordinate. Here, based on all of the sample points, the plot of the population line misses point A. Thus to plot confidence limits about the population line, point A must be projected

Table 11.9 Example of values for confidence limits

NUMBER	5% RANK	95% RANK
1	1.02%	45.07
2	7.64	65.74
3	18.92	81.08
4	34.26	92.36
5	54.93	98.98

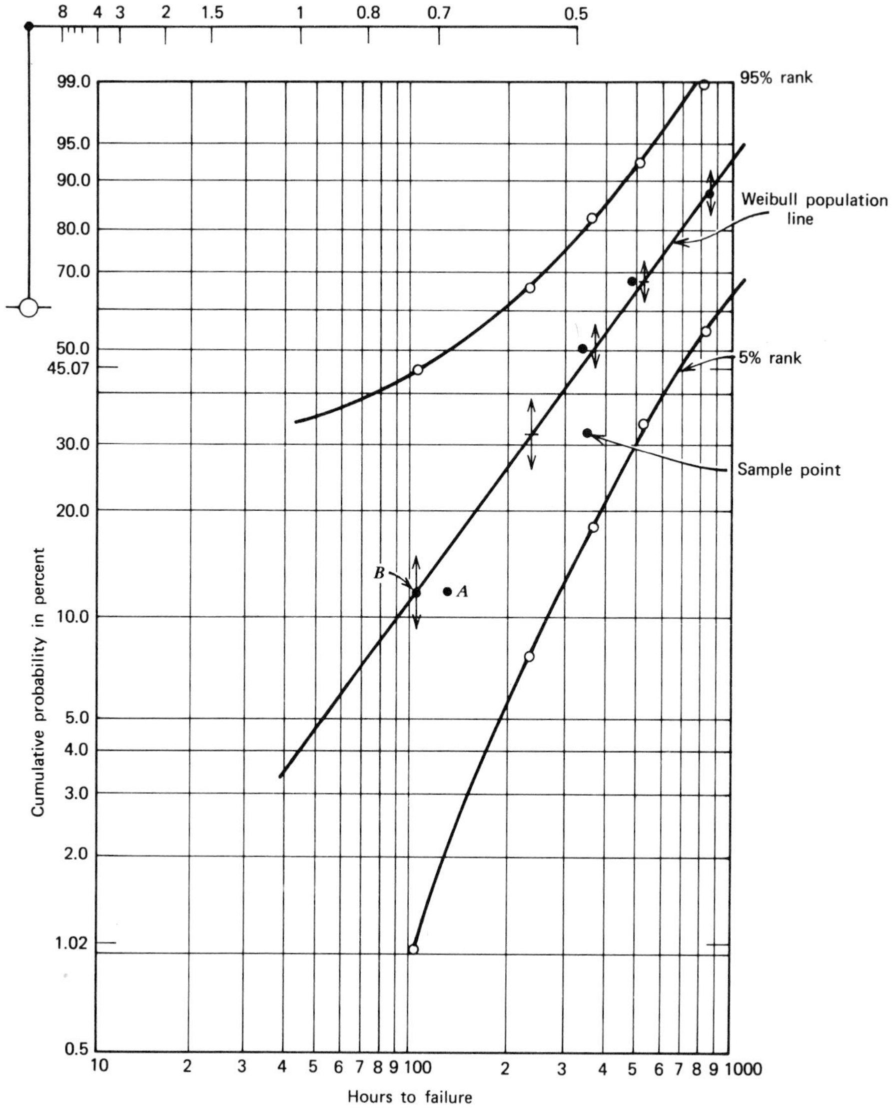

Figure 11.8 Confidence limit construction on Weibull paper.

horizontally to the population line. This projection defines point B on the population line.

Now from Table 11.9, the 5% and 95% ranks for the first failure in five are found to be 1.02% and 45.07%, respectively. These values are projected vertically about the population line at point B.

If the confidence limits are projected about the original sample points, the confidence limits will take an irregular rather than a bell shape.

The confidence limits as plotted in Figure 11.8 define a 90% confidence band about the population line. For example, consider 200 hours on the abscissa; here we can say that we are 90% confident that from 6 to 61% of the population will fail by 200 hours. Of course, the extreme width of this interval can be attributed to the small sample size.

Looking at this another way, consider the 50% failure point on the ordinate. Here we can say that we are 90% confident that 50% of the population will fail in the interval from 140 to 770 hours.

If we use only one confidence band in Figure 11.8 then the confidence level is 95%. For example, consider the 90% failure point on the ordinate, which is equivalent to a reliability level of 10%. We can say that we are 95% confident that 10% of the population will survive beyond 480 hours. This is a 95% one-sided lower confidence limit on the 10% reliability.

As a second example using one limit, consider 300 hours on the abscissa. Here we are 95% confident that less than 75% of the population will fail by 300 hours.

11.2.4 THE NONZERO MINIMUM LIFE CASE

There arise situations when the minimum life is not zero, which can be readily recognized on Weibull paper. For a minimum life $\delta > 0$ the cumulative function was previously given by Equation 11.1. The necessary transformation to a straight-line equation would be

$$\ln\left(\ln\frac{1}{1-F(x)}\right) = \beta\ln(x-\delta) - \beta\ln(\theta-\delta) \qquad (11.27)$$

Thus if we subtract the minimum life from each observation (x) the result will be a straight line. However, when we plot on Weibull paper, we may not know à priori that the minimum life will produce a characteristic line that will curve downward at the lower end.

Consider the data in Table 11.10, which represent the life of precision grinder wheels measured in number of pieces produced. The data is plotted in Figure 11.9, using median ranks. The downward curve is evident. Now we know

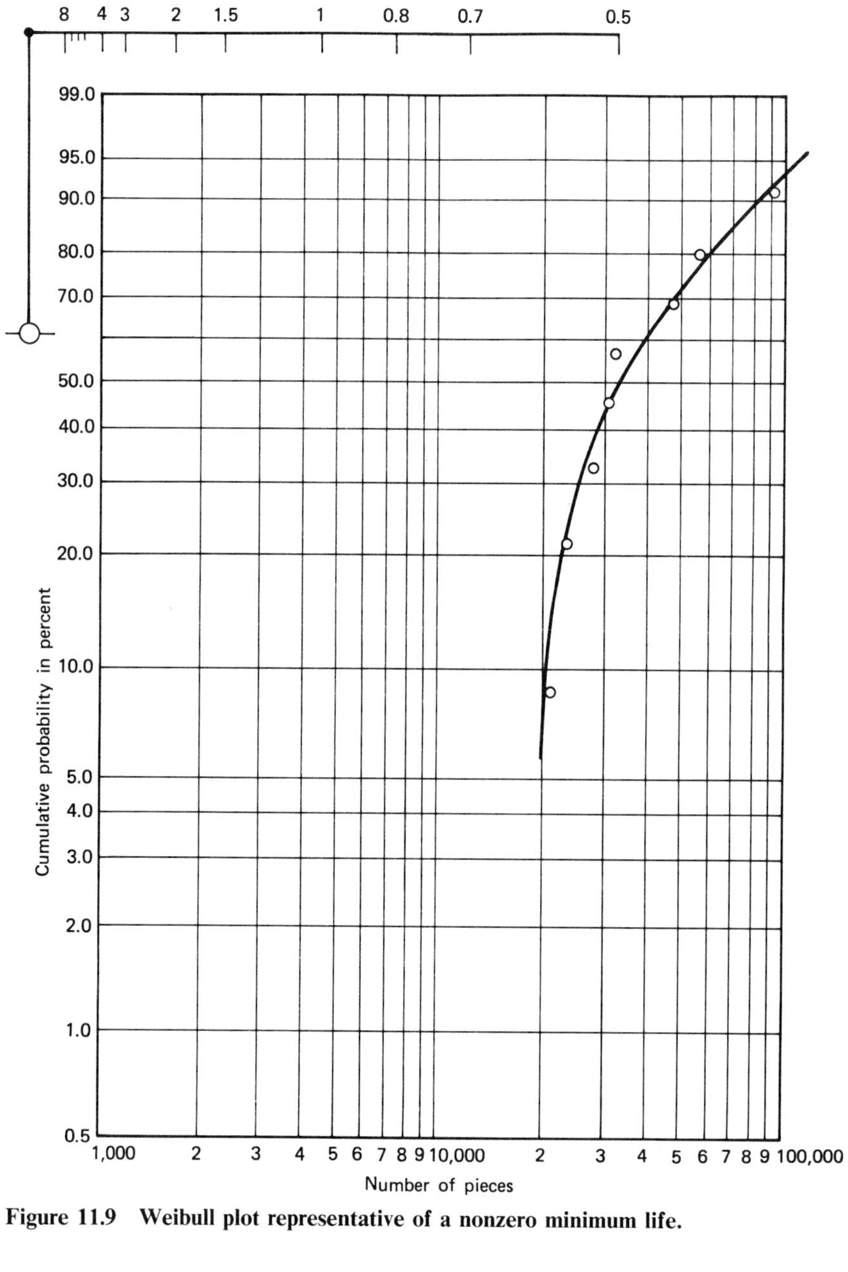

Figure 11.9 Weibull plot representative of a nonzero minimum life.

Table 11.10 Grinding wheel life

WHEEL NUMBER	PIECES PER WHEEL
1	22,000
2	25,000
3	30,000
4	33,000
5	35,000
6	52,000
7	63,000
8	104,000

that the minimum life (δ) is located somewhere between zero and the lowest reading, which in this case, is 22,000 pieces.

We could select the lowest value in the sample as an estimator for the minimum life, but this estimator is biased upward (see Appendix 11.A). It is more convenient to arbitrarily select a value slightly less than the minimum sample value (say $0.90x_1$). For example, in this case we arbitrarily select

$$\hat{\delta} = 19{,}600$$

We use this as a first estimate and subtract $\hat{\delta}$ from each value of the original data. Then we replot this data. If the first estimate $\hat{\delta}$ is too large, the plot will now curve upward; if it is too small the line will still curve downward. Some trial and error adjustment is usually necessary.

Figure 11.10 shows the adjusted data to which a straight line has been fitted reasonably well. The parameter estimates are

$$\hat{\delta} = 19{,}600$$

$$\hat{\theta} = 44{,}100$$

and

$$\hat{\beta} = 0.84$$

Hence the estimated reliability function is

$$R(x) = \exp\left[-\left(\frac{x - 19{,}600}{24{,}500}\right)^{0.84}\right], \quad x \geqslant 19{,}600 \tag{11.28}$$

If we wanted to graphically estimate the point at which 10% of the grinder wheels will wear out, we would find 2,000 pieces on Figure 11.10 and add the minimum life to obtain an estimate of 21,600 pieces. The confidence limits can be placed about the straight line as before.

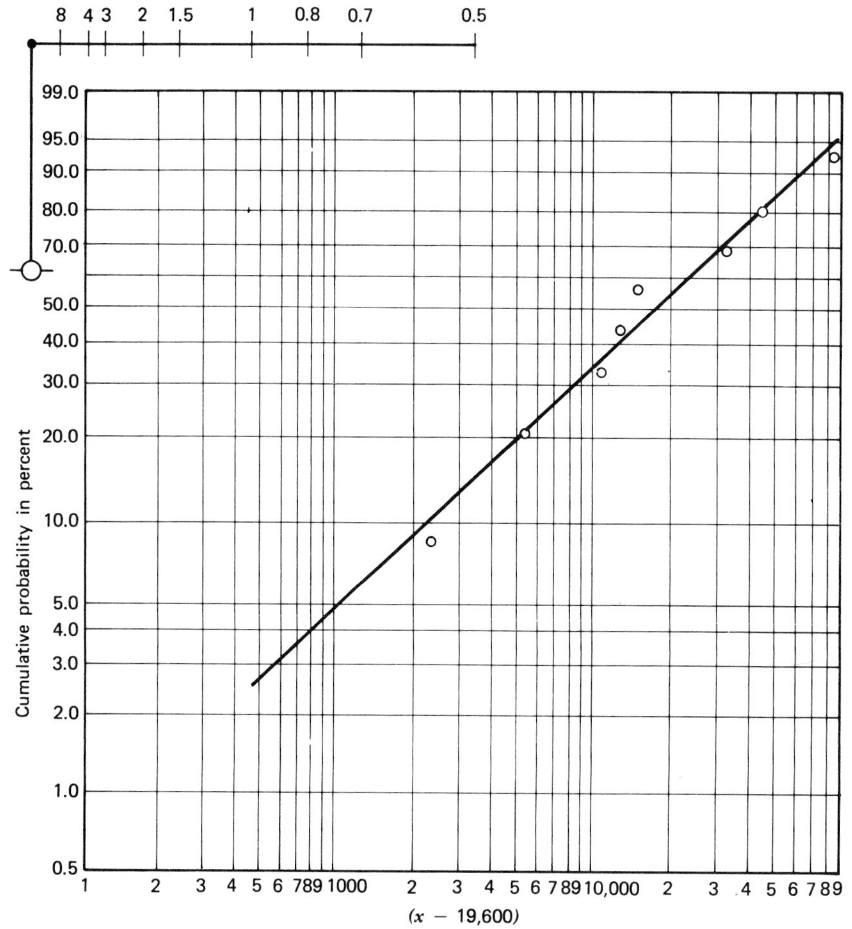

Figure 11.10 Data adjusted for nonzero minimum life.

11.2.5 GRAPHICAL ANALYSIS WITH SUSPENDED ITEMS

Occasionally items are taken off test for reasons other than failure. For instance, a test may be stopped if a test stand breaks down or a test vehicle becomes involved in an accident. Or, we may want to purposely place more items on test than we intend to fail in order to decrease the testing time.

Suspended data is handled by assigning an average order number to each failure time. If we tested five items to failure, the ordered failure times would have integer order numbers of one through five. Earlier we plotted in this fashion and either calculated the median ranks using Equation 11.22 or found them in Appendix VIII.

Table 11.11 Data containing a suspended item

FAILURE OR SUSPENSION	HOURS ON TEST
Failure (F_1)	84
Suspension (S_1)	91
Failure (F_2)	122
Failure (F_3)	274

Now suppose that four items are placed on test with results as given in Table 11.11. The table shows that the first failure was at 84 hours. Then at 91 hours an item was taken off of test for reasons other than failure (that is, it was suspended). Two more failures occurred at 122 and 274 hours.

If the suspended item had continued to failure, one of the following three outcomes would have resulted:

I	II	III
F_1	F_1	F_1
$S_1 \rightarrow F$	F_2	F_2
F_2	$S_1 \rightarrow F$	F_3
F_3	F_3	$S_1 \rightarrow F$

In other words, the suspended item could have failed in any of the above indicated positions, which would produce a particular ordering of the failure times. The average position or order number will be assigned to each failure time for plotting

In this example, the first observed failure time will always be in the first position and thus has the order number $j = 1$. However, for the second failure time there are two ways that it could have had an order number of $j = 2$, and one way that it would have had an order number of $j = 3$; thus the average order number is

$$j = \frac{3 + 2(2)}{3} = 2.33$$

We use these average position values to calculate the median rank. For example, the median rank for the second failure time (F_2) would be

$$\frac{2.33 - 0.3}{4 + 0.4} = 0.461$$

Table 11.12, summarizes the data for plotting on Weibull paper.

Table 11.12 Suspended item data

HOURS ON TEST	POSITION	MEDIAN RANK
84	1	0.159
122	2.33	0.461
274	3.67	0.766

Obviously, finding all of the sequences for a mixture of several suspensions and failures and then calculating the average order numbers is a time-consuming process. Fortunately a simplified formula is available for calculating the order numbers [17]. The formula produces what is termed a new increment. The new increment I is given by

$$I = \frac{(n+1) - (\text{previous order number})}{1 + (\text{number of items following suspended set})} \tag{11.29}$$

We will illustrate the use of Equation 11.29 for the data given in Table 11.13. Here, although 10 items were originally placed on test, only six failures resulted.

The first two failure times would have order numbers 1 and 2 respectively. For the third failure a new increment must be calculated. Applying Equation 11.29 yields

$$I = \frac{(10+1) - 2}{1 + (6)} = 1.29$$

Adding I to the previous order number, 2, gives the order number of 3.29 to the third failure.

We continue with the same increment until another suspended item is encountered. Thus the order number for the fourth failure is $3.29 + 1.29 = 4.58$.

Table 11.13 Suspended test data

HOURS ON TEST	SEQUENCE	STATUS
544	F_1	Failure
663	F_2	Failure
802	S_1	Suspension
827	S_2	Suspension
897	F_3	Failure
914	F_4	Failure
939	S_3	Suspension
1,084	F_5	Failure
1,099	F_6	Failure
1,202	S_4	Suspension

Applying the formula for the fifth failure gives

$$I = \frac{(10+1) - 4.58}{1 + (3)} = 1.60$$

The position for the fifth failure is 4.58 + 1.60 = 6.18, and the position for the sixth failure is 6.18 + 1.60 = 7.78. The final data for plotting is given in Table 11.14. Either the mean or the median rank can be used to plot the failure times on Weibull paper.

Table 11.14 Ranks for suspended test data

HOURS	POSITION	MEDIAN RANK	MEAN RANK
544	1.00	0.067	0.091
663	2.00	0.163	0.182
897	3.29	0.288	0.299
914	4.58	0.411	0.416
1,084	6.18	0.565	0.562
1,099	7.78	0.719	0.707

Confidence limits can also be obtained by interpolation. By using the positions shown in Table 11.14 and a sample size of 10 and by referring to Appendix VIII one can determine the confidence limits. Table 11.15 gives the interpolated confidence limits for the data in Table 11.14.

Table 11.15 Confidence limits for suspended test data

HOURS	POSITION	5% RANK	95% RANK
544	1.00	0.51	25.89
663	2.00	3.68	39.42
897	3.29	10.55	53.48
914	4.58	19.20	65.87
1,084	6.18	31.97	79.06
1,099	7.78	47.12	89.89

EXAMPLE 11.2

A fleet of cars is being monitored to determine failure rates of water pumps. At the time of the analysis four failures had occurred, and several water pumps had accumulated kilometers without failing. Management would like to know whether the pump is adequate for a 24,000 kilometer warranty period. The fleet data on the water pump are given in the Table 11.16.

Table 11.16 Water pump failure data

KILOMETERS	KILOMETERS
4,010	22,112
4,731	23,110
4,812	24,020, pump failure
8,657	25,004
12,550, pump failure	25,112
14,992	26,002
16,121	26,179, pump failure
16,437	26,842
20,740	30,529, pump failure
21,021	

In this situation we can use the suspended data method for plotting. The calculations for the order numbers follow. For the failure time of 12,550 kilometers

$$I = \frac{(19+1)-0}{1+15} = 1.250$$

and the order number is 1.250. For the failure time of 24,020 kilometers

$$I = \frac{(19+1)-1.250}{1+7} = 2.344$$

and the order number is 1.250+2.344=3.594. For the failure time of 26,179 kilometers

$$I = \frac{(19+1)-3.594}{1+3} = 4.102$$

and the order number is 3.594+4.102=7.696. For the last failure at 30,529 kilometers

$$I = \frac{20-7.696}{1+1} = 6.152$$

and the order number is 7.696+6.152=13.848.

The approximation given in Equation 11.22 is then used to calculate the median ranks from the order numbers. The results are shown in Table 11.17 and plotted in Figure 11.11.

Confidence limits can also be placed on the graph by interpolation from the confidence limits for a sample size of 19. Here we will place a one-sided 95% limit on our estimates. These values are also plotted on the graph.

Table 11.17 Water pump data with median ranks

KILOMETERS	ORDER NUMBER	MEDIAN RANK
12,050	1.250	4.66%
24,020	3.594	16.15
26,179	7.696	38.12
30,529	13.848	69.84

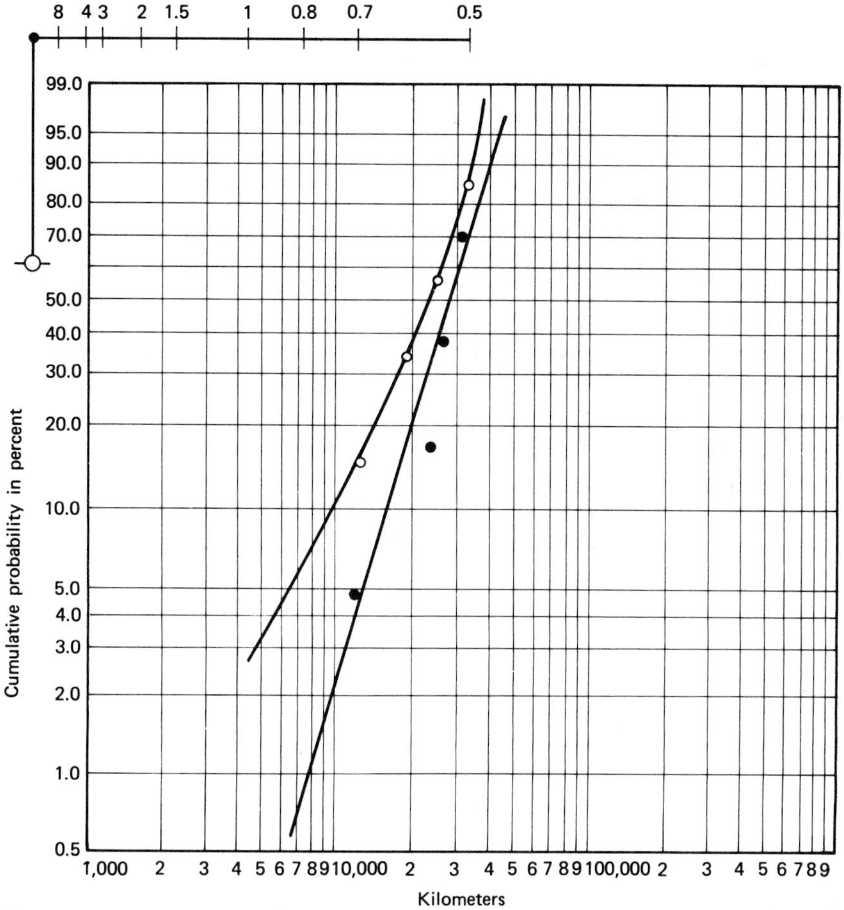

Figure 11.11 Water pump failure data.

ORDER NUMBER	95% RANK	INTERPOLATED VALUE
1	14.75%	
1.250		16.80%
2	22.96	
3	29.95	
3.594		33.67
4	36.21	
7	52.89	
7.696		56.47
8	58.04	
13	81.13	
13.848		84.62
14	85.25	

Suppose we want to quote the kilometer at which 10% of the population will have failed. The point estimate is 17,000 kilometers. We could also say that 90% of the population will survive 10,000 kilometers with 95% confidence.

Since in warranty work on most commercial products a failure rate of 1% is considered high, we might take a look at this point on the graph. At about 7,800 kilometers we can expect 1% of the population to fail. Here we would have trouble guessing at the lower confidence limit. By the end of the 24,000 kilometers warranty period we can say that we are 95% confident that at most 46% of the population will have failed.

However one looks at it, this water pump appears to be inadequate and should definitely be improved. Here we were able to reach this decision on relatively little data as is frequently provided by prototype testing.

EXAMPLE 11.3

A fleet of 100 city police patrol cars is being periodically monitored for the failure of a certain engine component. The accumulated information on the failure of this noncritical component is summarized in Table 11.18. At this time, we wish to estimate the Weibull failure distribution for the component.

Table 11.18 Suspended kilometers data

KILOMETERS	NUMBER NOT FAILED	NUMBER FAILED
2,000	18	3
4,000	26	4
6,000	17	3
8,000	16	2
10,000	7	1
Above 10,000	3	—

The above data give component failures thus far in intervals of 2,000 kilometers. For example, components in three vehicles failed in the first 2,000 kilometer interval, while 18 vehicles with less than 2,000 kilometers are still operating without a failure.

In the analysis of this situation we will assume that all failures in any kilometer interval occur at the end of that interval. We will employ the suspended data method.

We calculate the first new increment for which we have 18 suspensions:

$$I_1 = \frac{(100+1)-0}{1+82} = 1.217$$

Hence the order number for the last of the three failures is $0+3(1.217)=3.651$. This order number is used to calculate the median rank for plotting at 2,000 kilometers.

The next new increment is

$$I_2 = \frac{101-3.651}{1+53} = 1.803$$

The order number is $3.651+4(1.803)=10.862$. Continuing in this fashion we get

$$I_3 = \frac{101-10.862}{1+32} = 2.731$$

The order number is 10.862+(3)2.731=19.056. For the next interval

$$I_4 = \frac{101 - 19.056}{1 + 13} = 5.856$$

The order number is 19.056+2(5.853)=30.762. And finally,

$$I_5 = \frac{101 - 30.762}{1 + 4} = 14.048$$

and the order number is 30.762+(1)14.048=44.810.

The data for plotting is summarized in Table 11.19, and the resulting plot is shown in Figure 11.12.

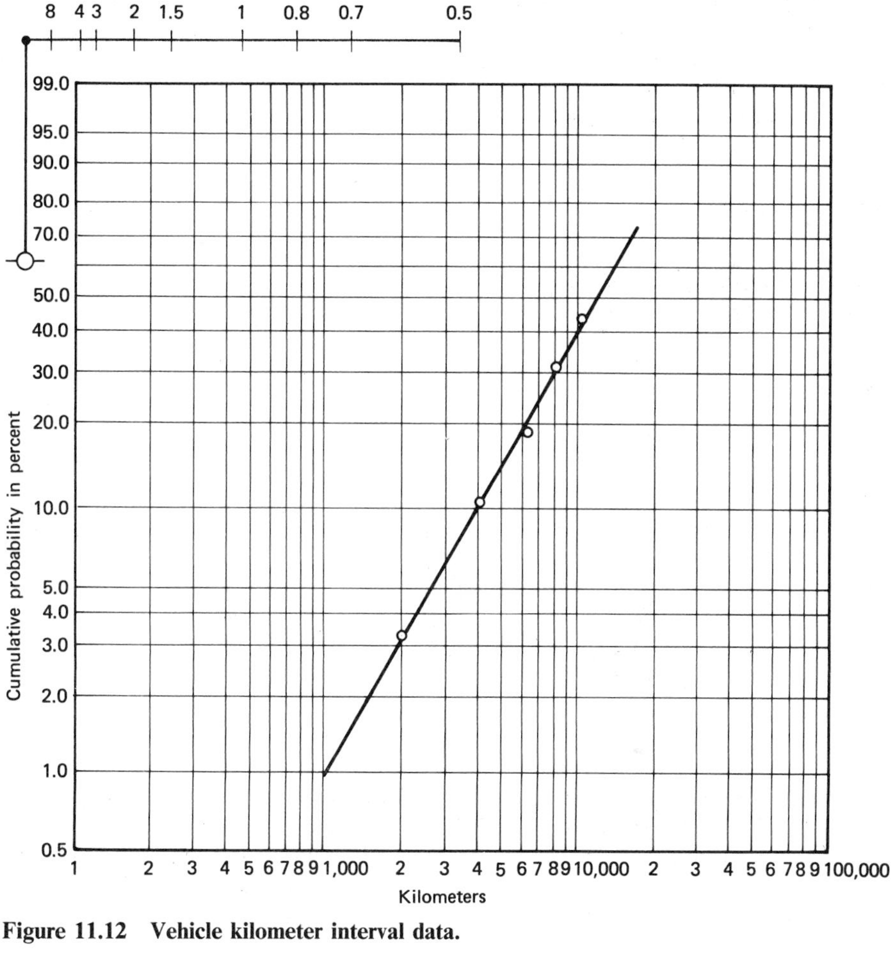

Figure 11.12 Vehicle kilometer interval data.

Table 11.19 Suspended kilometer data

KILOMETERS	ORDER NUMBER	MEDIAN RANK
2,000	3.651	3.34%
4,000	10.862	10.52
6,000	19.056	18.68
8,000	30.762	30.34
10,000	44.810	44.33

The following example illustrates the problem in comparing samples from two Weibull distributions.

EXAMPLE 11.4

Table 11.20 contains coded data representing the shear strength of brass and steel brake rivets. The data are plotted in Figure 11.13. As this figure shows, the Weibull plots diverge at the higher shear strength levels. In reliability studies this can mean that one component will be better than another for low life reliability but that the choice of components will change for high life reliability.

Table 11.20 Shear strength test data for brass and steel brake rivets (coded data)

BRASS RIVETS		STEEL RIVETS	
SHEAR STRENGTH	MEDIAN RANK	SHEAR STRENGTH	MEDIAN RANK
4.2	4.8%	5.7	5.6%
4.5	11.7	7.4	13.6
4.7	18.6	7.4	21.7
4.8	25.6	7.8	29.8
4.9	32.6	8.3	37.9
5.0	39.5	9.2	45.9
5.1	46.5	9.6	54.0
5.2	53.5	11.0	62.1
5.4	60.5	11.2	70.2
5.7	67.4	11.3	78.3
5.7	74.4	12.5	86.4
5.9	81.4	12.9	94.4
6.4	88.3		
6.8	95.2		

There is no precise statistic test for a significant difference between two samples if the populations are assumed to be Weibull. To overcome this problem, it has been suggested that confidence limits be placed about both lines, and then conclude that a significant difference exists where the confidence limits do not overlap. However, this procedure is incorrect, and the problem is still open for further research.

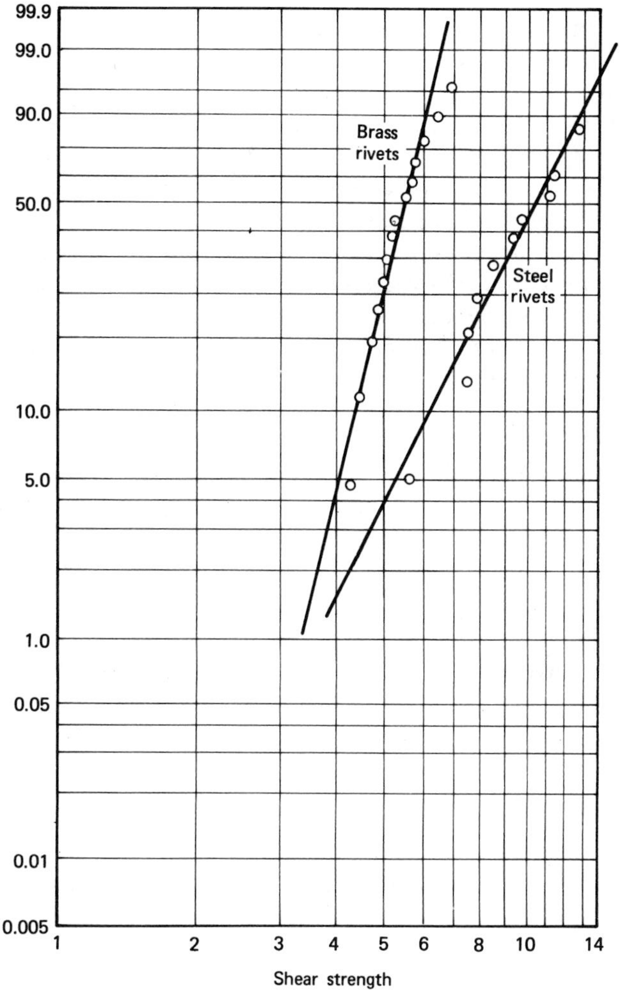

Figure 11.13 Comparison of two Weibull populations.

11.3 STATISTICAL ESTIMATION PROCEDURES

The application of statistical inference to the Weibull distribution followed the development of the graphical techniques. The traditional statistical inference procedures were applied to the Weibull distribution (or, for that matter, to any other distribution) because they are more objective than the inherently subjective graphical methods. In statistical estimation procedures the properties of an estimator can be determined and the accuracy of estimation assessed.

However, it is not easy to apply the traditional statistical inference procedures to the Weibull distribution. For example, applying the well-known maximum likelihood method of parameter estimation results in equations that cannot be solved explicitly, but must be solved using an iterative procedure. Further problems result with the maximum likelihood method; for example, with censored samples the resulting estimators may not have the property of minimum variance. Some of the same problems are also encountered in applying the method of matching moments for parameter estimation.

The method used for parameter estimation in the Weibull distribution involves essentially a weighting of the observations. The weights developed are such that the resulting estimators have the desirable property of minimum variance. The development of this procedure is primarily the work of Mann [22–31]. A discussion of the parameter estimation procedure follows.

11.3.1 PARAMETER ESTIMATION

Let $\mathbf{t}$ be a random variable with a two-parameter Weibull distribution as given by Equation 11.2. If we make the transformation $\mathbf{x} = \ln \mathbf{t}$, the cumulative distribution for $\mathbf{x}$ will be

$$F(x) = 1 - \exp\left[-\exp\left(\frac{x-u}{b}\right)\right], \qquad -\infty < x < \infty \tag{11.30}$$

where $u = \ln\theta$ and $b = 1/\beta$. This, of course, is the cumulative distribution of smallest extreme values. Now if we are interested in estimating $\mathbf{x}$ for a given reliability R, then Equation 11.30 can be rearranged as

$$x = u + b\left[\ln(\ln 1/R)\right] \tag{11.31}$$

which is the familiar linear equation $y = A + Bx$ with parameters u and b.

Let $\mathbf{t}_1, \mathbf{t}_2, \ldots, \mathbf{t}_r$ represent the ordered failure times from the two-parameter Weibull. We define $\mathbf{x}_i = \ln \mathbf{t}_i$. The parameters u and b in Equation 11.31 are then estimated by

$$\tilde{u} = \sum_{i=1}^{r} a_i x_i \tag{11.32}$$

and

$$\tilde{b} = \sum_{i=1}^{r} c_i x_i \tag{11.33}$$

where the a_i's and c_i's are linear weighting factors (given in Appendix IX) for the appropriate values of n and r. This procedure simply applies the concept of linear estimation to the parameters u and b.

The estimators for the original Weibull parameters are

$$\tilde{\theta} = e^{\tilde{u}} \tag{11.34}$$

and

$$\tilde{\beta} = \frac{1}{\tilde{b}} \tag{11.35}$$

Also

$$\tilde{R}(t) = \exp\left[-\left(\frac{t}{\tilde{\theta}}\right)^{\tilde{\beta}}\right] \tag{11.36}$$

The estimators are termed best linear invariant (BLI) estimators. They are "best" in the sense that they give uniformly smaller error* than any other linear estimator, and they are invariant under location and scalar transformations.

The resulting estimators are not unbiased. Unbiased estimators for u and b can, however, be obtained by the following simple linear transformations

$$b^* = \tilde{b}/[1 - E(LB)] \tag{11.37}$$

$$u^* = \tilde{u} + b^* E(CP) \tag{11.38}$$

where $E(LB)$ and $E(CP)$ are given in Appendix IX.

EXAMPLE 11.5

The data in Table 11.21 represent cycles to failure for automotive hood torsion bars. In this instance 10 torsion bars were placed on test, and the test was terminated at the time of the seventh failure. The weighting factors for $n = 10$ and $r = 7$ are obtained from Appendix IX and given in Table 11.21.

We compute

$$\tilde{u} = \sum_{i=1}^{7} a_i x_i = 9.493392$$

and

$$\tilde{b} = \sum_{i=1}^{7} c_i x_i = 0.479225$$

Hence the estimates for the Weibull parameters are

$$\tilde{\theta} = e^{\tilde{u}} = 13{,}272 \text{ cycles}$$

*Here error is taken as (expected mean square) $\times \beta^2$.

Table 11.21 Hood torsion bar test data

CYCLES-TO-FAILURE (t)	$x=\ln t$	a_i	c_i
2,670	7.889834	−0.022198	−0.124170
5,810	8.667336	−0.006909	−0.126894
7,220	8.884610	0.013224	−0.118392
7,410	8.910586	0.037994	−0.100924
9,600	9.169518	0.068153	−0.073988
12,240	9.412465	0.105164	−0.035501
13,680	9.523690	0.804572	0.579868

and

$$\tilde{\beta}=\frac{1}{\tilde{b}}=2.07$$

The reliability function is estimated by

$$\tilde{R}(t)=\exp\left[-\left(\frac{t}{13{,}272}\right)^{2.07}\right], \qquad t\geqslant 0 \tag{11.39}$$

Now if the useful life of a hood torsion bar is assumed to be 200 cycles, then the reliability is estimated by Equation 11.39 as $\tilde{R}(200)=0.9998$, for a single torsion bar. However, two bars are used for each hood, and if the reliability is defined as the successful operation of both torsion bars, then the system reliability would be $R_s=(0.9998)^2=0.9996$. This means that there will be about four failures per 10,000 vehicles.

Suppose we wanted to estimate the number of cycles at which 1% of the population is expected to fail. Then Equation 11.31 leads to

$$\tilde{x}_{0.01}=9.4933392+0.479225\ln[\ln(1/0.99)]=7.288885$$

or

$$t_{0.01}=e^{7.288885}=1{,}464 \text{ cycles}$$

It is apparent now that the table of weighting factors makes this method very easy to use.

11.3.2 CONFIDENCE LIMITS

Confidence limits for the Weibull parameters were developed by simulation of the appropriate statistic. Let us first consider the parameter b. The statistic $W=\tilde{b}/b$ is tabulated in Appendix X. We can choose specific values of W, say $W_{1-\alpha/2}$ and $W_{\alpha/2}$ such that

$$P\left[W_{1-\alpha/2}\leqslant\frac{\tilde{b}}{b}\leqslant W_{\alpha/2}\right]=1-\alpha \tag{11.40}$$

The inequality can be rearranged as

$$\frac{\tilde{b}}{W_{\alpha/2}} \leqslant b \leqslant \frac{\tilde{b}}{W_{1-\alpha/2}} \tag{11.41}$$

which is a $100(1-\alpha)\%$ two-sided confidence limit on b. From Equations 11.35 and 11.41 we can write the $100(1-\alpha)\%$ two-sided confidence limit on the Weibull slope β as

$$\frac{W_{1-\alpha/2}}{\tilde{b}} \leqslant \beta \leqslant \frac{W_{\alpha/2}}{\tilde{b}} \tag{11.42}$$

EXAMPLE 11.6

Reconsider Example 11.5 concerning the hood torsion bars. We wish to set a 90% two-sided confidence interval on the Weibull slope β.

From Appendix X, for $n=10$ and $r=7$ we find

$$W_{0.05}=1.46$$

and

$$W_{0.95}=0.42$$

Thus substituting into Equation 11.42, we obtain

$$\frac{0.42}{0.479225} \leqslant \beta \leqslant \frac{1.46}{0.479225}$$

or

$$0.88 \leqslant \beta \leqslant 3.05$$

which is the desired confidence interval.

Consider finding the value x_R for a specified reliability R. The statistic

$$V_R = \frac{(\tilde{u} - x_R)}{\tilde{b}} \tag{11.43}$$

is tabulated in Appendix XII. We can choose the lower and the upper limits such that

$$P\left[V_{R,1-\alpha/2} \leqslant \frac{(\tilde{u} - x_R)}{\tilde{b}} \leqslant V_{R,\alpha/2}\right] = 1-\alpha \tag{11.44}$$

which can be rearranged to put limits around x_R as

$$\tilde{u} - \tilde{b}\, V_{R,\alpha/2} \leqslant x_R \leqslant \tilde{u} - \tilde{b}\, V_{R,1-\alpha/2} \tag{11.45}$$

The tables in Appendix XII contain the values of V_R for $R=0.90$, 0.95, and 0.99, for various values of n and r.

The $100(1-\alpha)\%$ two-sided confidence interval on the Weibull life **t** would then be

$$\exp\left[\tilde{u}-\tilde{b}\,V_{R,\alpha/2}\right] \leqslant t_R \leqslant \exp\left[\tilde{u}-\tilde{b}\,V_{R,\alpha/2}\right] \tag{11.46}$$

EXAMPLE 11.7

Reconsider the torsion bar example (Example 11.5) for a 90% one-sided lower confidence limit on the cycles to failure for a reliability of 99%. From Appendix XII for $R=0.99$, $\alpha=0.10$, $n=10$, and $r=7$ we find

$$V_{0.99,0.10}=8.99$$

Hence

$$9.493392-(0.479225)8.99 \leqslant x_{0.99}$$

or

$$x_{0.99} \geqslant 5.18$$

and for the Weibull distribution of hood torsion bars, $t \geqslant 179$ cycles.

Confidence limits can be set on the characteristic life parameter by using the statistic $V=(\tilde{u}-u)/\tilde{b}$, which is tabulated in Appendix XI. The probability statement would be

$$P\left[V_{1-\alpha/2} \leqslant \frac{(\tilde{u}-u)}{\tilde{b}} \leqslant V_{\alpha/2}\right]=1-\alpha \tag{11.47}$$

which, after rearrangement, yields

$$\tilde{u}-\tilde{b}\,V_{\alpha/2} \leqslant u \leqslant \tilde{u}-\tilde{b}\,V_{1-\alpha/2} \tag{11.48}$$

Since $\theta=e^u$, these limits lead to the following bounds on θ

$$\exp\left(\tilde{u}-\tilde{b}\,V_{\alpha/2}\right) \leqslant \theta \leqslant \exp\left(\tilde{u}-\tilde{b}\,V_{1-\alpha/2}\right) \tag{11.49}$$

EXAMPLE 11.8

Again considering the hood torsion bar example, for $n=10$, $r=7$, and $\alpha=0.10$ we have from Appendix XI

$$V_{0.05}=0.70$$

and

$$V_{0.95}=1.08$$

which, when substituted into Equation 11.49, yields

$$\exp[9.493392-(0.479225)(0.70)] \leqslant \theta \leqslant \exp[9.493392-(0.479225)(-1.08)]$$

That is, the desired 90% confidence interval on the characteristic life θ is

$$9{,}489 \text{ cycles} \leqslant \theta \leqslant 22{,}269 \text{ cycles}$$

The tables for the statistic V_R can also be used to set a lower confidence limit on the reliability $R(t^*)$ for a particular failure time t^*. This is achieved by making the transformation $x_R^* = \ln t^*$ and calculating a value of the statistic $V_R^* = (\tilde{u} - x_R^*)/\tilde{b}$. Then, for a specified level of significance α, we find R^* corresponding to $V_{R,1-\alpha}^*$. The value R^* is the desired $100(1-\alpha)\%$ lower confidence limit on the reliability R at time t^*.

EXAMPLE 11.9

Reconsider the hood torsion bar example. We now wish to set a 95% lower confidence limit on the reliability at 200 cycles.

We have

$$t^* = 200 \text{ cycles}$$

and hence

$$x_R^* = \ln 200 = 5.30$$

and by Equation 11.43

$$V_R^* = \frac{(9.493392 - 5.30)}{0.479225} = 8.75$$

Consulting Appendix XII, for $n=10$, $r=7$, and $\alpha=0.05$, we find

$$V_{0.95,0.95} = 6.83$$

and

$$V_{0.99,0.95} = 10.66$$

Interpolating for $V_{R,0.95}^* = 8.75$ yields

$$R(200 \text{ cycles}) \geqslant 0.97$$

11.4 A GOODNESS-OF-FIT TEST FOR THE TWO-PARAMETER WEIBULL DISTRIBUTION

A goodness-of-fit test specifically for the Weibull distribution has been developed by Mann et al. [31]. Since the test is applicable to a specific distribution, it is expected to be more powerful than any of the general goodness-of-fit tests.

The null hypothesis in this case is, of course, that the population of interest is two-parameter Weibull distributed. If the null hypothesis is indeed rejected, then other distributions should be considered, including the three-parameter Weibull.

The application of the test is rather straightforward and proceeds as follows: Let $\mathbf{t}_1, \mathbf{t}_2, \ldots, \mathbf{t}_r$ represent the first r ordered failure times resulting from placing n items on test and truncating the test at the time of the rth failure ($r \leqslant n$). Define $\mathbf{x}_i$ as $\mathbf{x}_i = \ln \mathbf{t}_i$ for $i = 1, 2, \ldots, r$. Then the test statistic is

$$S = \frac{\displaystyle\sum_{i=[r/2]+1}^{r-1} \left[\frac{(x_{i+1} - x_i)}{M_i} \right]}{\displaystyle\sum_{i=1}^{r-1} \left[\frac{(x_{i+1} - x_i)}{M_i} \right]} \tag{11.50}$$

where $[r/2]$ denotes the greatest integer $\leqslant r/2$; for example, if $r = 7$, then $[r/2] = 3$. The values for the M_i's are found in Appendix XIII, along with the critical values for S.

EXAMPLE 11.10

Reconsider the data on the grinder wheel life given in Table 11.10. This example was previously used to illustrate a Weibull distribution with a nonzero minimum life. Let us now apply the goodness-of-fit test to this data. The calculations are shown in Table 11.22.

We have $[\frac{r}{2}] + 1 = 5$ and

$$S = \frac{2.848957}{3.674627} = 0.78$$

From Appendix XIII, for $n = r = 8$ and $\alpha = 0.05$, the critical value of S is 0.71. Thus we must reject the hypothesis that the data is from a two-parameter Weibull distribution. This is consistent with our previous conclusion, which we reached using subjective judgment in the graphical analysis.

Table 11.22 Grinding wheel life data

t_i	$x_i = \ln t_i$	M_i	$x_{i+1} - x_i$	$(x_{i+1} - x_i)/M_i$
22,000	9.998798	1.068252	0.127833	0.119666
25,000	10.126631	0.577339	0.182323	0.315796
30,000	10.308953	0.422889	0.095310	0.225379
33,000	10.404263	0.356967	0.058841	0.164835
35,000	10.463103	0.334089	0.395896	1.185001
52,000	10.859000	0.349907	0.191891	0.548406
63,000	11.050890	0.449338	0.501256	1.115544
104,000	11.552146			

Table 11.23 Grinding wheel data adjusted for nonzero minimum life

t_i	$x_i = \ln t_i$	$x_{i+1} - x_i$	$(x_{i+1} - x_i)/M_i$
2,400	7.783224	0.810930	0.7592
5,400	8.594154	0.655407	1.1352
10,400	9.249561	0.253449	0.5993
13,400	9.503010	0.139113	0.3897
15,400	9.642123	0.743791	2.2263
32,400	10.385914	0.392301	1.1212
43,400	10.678215	0.665108	1.4802
84,400	11.343323		

In our graphical analysis, we subtracted 19,600 from each observation and concluded that the resulting data followed the two-parameter Weibull distribution. Table 11.23 shows the adjusted data and the calculations for performing the goodness-of-fit test on this adjusted data.

In this case

$$S = \frac{4.8277}{7.7111} = 0.63$$

By checking the critical values of S in Appendix XIII, we see that the calculated statistic would not be significant at the 95%, or even the 90% level.

11.5 SAVINGS IN TEST TIME DUE TO CENSORING

Finding the reduction in testing time resulting from Type II censoring is considerably more difficult for the Weibull than for the exponential. If ${}_n\mathbf{x}_j$ is the jth ordered failure time where n items are on test, and if the test is to be suspended at the time of the jth failure, then $E({}_n\mathbf{x}_j)$ is the expected time to complete testing. Appendix 11.A is related to the development of $E({}_n\mathbf{x}_j)$ but using it directly involves difficult computations. However, there is an easy way to calculate the ratio $E({}_n\mathbf{x}_r)/E({}_r\mathbf{x}_r)$ for the Weibull distribution. Recall that this ratio would represent the savings in test time achieved by increasing the number of items on test to n and suspending the test at the time of the rth failure.

We define

$$\lambda_{r,n} = E(\mathbf{t}_{r,n})/E(\mathbf{t}_{r,r}) \tag{11.51}$$

where $\mathbf{t}$ is exponentially distributed. Values for $\lambda_{r,n}$ are found in Table 10.14. The corresponding ratio for a Weibull distribution with a slope of β is approximated by $(\lambda)^{1/\beta}$. For example, if $n=8$ and $r=6$, from Table 10.14 we find $\lambda=0.50$, and for a Weibull distribution with $\beta=3$ the savings in test time is $(0.50)^{1/3}=0.79$ or 79%.

11.6 SUMMARY

The Weibull distribution has a certain amount of versatility because it can assume a variety of distributional forms, but it also has definite limitations. Thus contrary to popular belief many situations exist in which the distribution does not fit "well." The Weibull skewness and kurtosis (which are indicative of the shape of a distribution) are directly related, and one implies the other. There are certain values of skewness and kurtosis that the Weibull cannot attain which the log normal or gamma distributions can attain. Fortunately, in most engineering applications the Weibull is usually close enough to provide reasonable guidance.

The rapid and highly visual graphical estimation procedure that can be used with the Weibull gives this distribution a great advantage in applications. However, for small sample sizes the graphical procedures can produce bias, particularly in estimating low percentiles of the distribution. Unfortunately, low percentile estimates from small samples are the rule rather than the exception in engineering applications. In this situation, a better procedure would be to estimate the Weibull parameters using the statistical estimation procedures of Section 11.3. Then a visual representation of the distribution can be plotted on Weibull paper using the estimated parameters and percentile estimates made from the resulting graph.

EXERCISES

1 A system has a Weibull time to failure distribution with parameters $\beta=3$ and $\theta=200$ kilometers. (a) Find the 150 kilometer reliability. (b) Find the hazard function. (c) Find the mean time to failure. (d) What is the probability that the system survives beyond its mean time to failure?

2 Find the kurtosis for the Weibull distribution.

3 Find the coefficient of variation for the Weibull distribution and plot this as a function of the shape parameter.

4 Sixteen vehicle water pumps were tested to failure. The results, in kilometers, follow:

30,254	56,636
35,110	62,130
39,606	63,144
43,112	63,552
49,038	64,065
49,245	66,868
50,539	72,132
53,418	75,068

Using the graphical procedure: (a) Estimate the parameters for the Weibull reliability function; (b) find the 95% lower confidence limit on the kilometers at which 20% of the population is expected to fail.

5 Eight leaf springs were tested to failure on an accelerated life test. The results, in cycles, follow:

8,712	79,000
21,915	110,200
39,400	151,208
54,613	204,312

Using the graphical estimation procedures find: (a) The Weibull shape parameter and characteristic life; (b) the 95% lower confidence limit for the 20,000 cycle reliability. (c) With 95% confidence find the number of cycles that will be exceeded by 50% of the population.

6 Reconsider the data in Exercises 4 and 5 of Chapter 10 and (a) run the Weibull goodness-of-fit test; (b) use graphical procedures to answer the questions posed in Problem 4. (c) Repeat part b using statistical estimation procedures.

7 Using the data in Table 2.1, graphically estimate the parameters for the Weibull hazard function and compare the values to those given in the Table 2.2.

8 The data below are from the testing of ball bearings. The data are ordered according to the observed endurance life.*

BEARING NUMBER	ENDURANCE IN 10^6 REVOLUTIONS	BEARING NUMBER	ENDURANCE IN 10^6 REVOLUTIONS
16	17.88	4	68.64
10	28.92	6	68.64
5	33.00	25	68.88 (Suspended)
19	41.52	22	84.12
9	42.12	17	93.12
11	45.60	7	98.64
15	48.48	23	105.12
12	51.84	24	105.84
20	51.96	8	125.04
18	54.12	21	127.92
13	55.56	14	173.40 (Suspended)
1	67.80		

*Data taken with permission from "Statistical Investigation of the Fatigue Life of Deep-Grooved Ball Bearings," by J. Lieblein and M. Zelen, *Journal of Research of the National Bureau of Standards*, Vol. 57, No. 5. November 1956.

Using graphical techniques: (a) Estimate the Weibull slope and characteristic life; (b) estimate the median life; (c) estimate the B-10 life (point T at which $P(t \leqslant T) = 0.10$); (d) estimate the mean life.

9 Consider the following data where S and F indicate suspension and failure, respectively. The data represent hours to failure. The test was terminated at the time of the last recorded failure (56.2 hr) with 19 units still on test.

1.0 *S*	17.7 *S*	31.9 *S*	44.2 *S*
2.2 *F*	18.4 *S*	36.4 *F*	44.9 *S*
3.6 *F*	19.8 *S*	36.8 *F*	45.3 *F*
3.8 *F*	22.7 *F*	37.1 *S*	47.4 *S*
4.7 *S*	23.5 *S*	37.4 *F*	48.5 *S*
11.2 *F*	25.0 *S*	38.0 *S*	49.6 *S*
11.3 *F*	26.1 *S*	39.2 *F*	50.4 *S*
13.1 *F*	27.7 *F*	40.6 *S*	50.9 *F*
13.3 *S*	28.9 *F*	43.7 *F*	53.2 *S*
15.0 *S*	29.5 *S*	44.0 *S*	55.5 *S*
15.8 *S*	30.2 *S*	44.1 *S*	56.2 *F*

(a) Using graphical procedures estimate the reliability function. (b) Estimate the number of hours at which 10% of the population will fail.

10 The following data are from a fleet of semitractors and relate to a particular bearing failure in the blower used on the diesel engine. The bearing can be changed at the company garage for \$80. A bearing failure while on the road costs approximately \$300 to repair, since spare parts and trained mechanics are not always available. Using this data suggest a maintenance policy that will minimize repair costs.

23,600 km	122,000
24,200	128,000
26,000	130,400
27,600	135,040
28,200	168,900
30,900	177,440
36,400	178,200
51,600	185,000
60,000	200,200
60,242	238,000
71,400	264,200
86,000	344,000
99,800	362,000
116,000	

11 Using logarithmic graph paper, develop graphical estimation procedures for the exponential distribution similar to those for the Weibull. Apply the procedure to answer the questions posed in Problem 8 of Chapter 10.

12 Twenty-four vehicles have completed a 74,000 kilometer scheduled run at an automobile company's proving grounds. Proving ground kilometers are assumed to be equivalent to customer kilometers.

During the test eight radiator hoses failed at the kilometers given below.

KILOMETERS TO FAILURE
2,760
3,700
7,100
17,220
29,500
48,400
52,600
65,000

Estimate the number of replacements during the 24,000 kilometer warranty period assuming two million vehicles will be sold, and write a brief engineering report of this situation for the chief engineer in charge of this car 'ine.

13 A designated test fleet of vehicles is in operation in a high corrosion producing area in northeastern United States. Specifically, 100 vehicles were monitored for evidence of corrosion failures of certain underbody components. After two years the following corrosion failure data had been collected:

KILOMETERS	NUMBER OF FAILURES
$M \leqslant 10{,}000$	11
$10{,}000 < M \leqslant 20{,}000$	26
$20{,}000 < M \leqslant 30{,}000$	28
$30{,}000 < M \leqslant 40{,}000$	20
$40{,}000 < M \leqslant 50{,}000$	14
$50{,}000 < M \leqslant 60{,}000$	1

Estimate the 24,000 kilometer reliability using the Weibull distribution as a model.

14 Passenger car pitman arms from the steering system were tested in the laboratory. The arms were cycled to failure by applying a completely reversing constant load stress to the lower end of the arm with the upper end held fixed.

Thus the arms failed in fatigue. The results from ten arms follow:

CYCLES TO FAILURE	CYCLES TO FAILURE
2,200	19,500
5,200	21,000
8,600	27,800
8,900	35,200
9,700	59,900

Use the Weibull paper to estimate: (a) the mean life; (b) the reliability function; (c) the cycles at which 10% of the population is expected to fail.

BIBLIOGRAPHY

1 Bain, L. J., "Inference Based on Censored Sampling from the Weibull or Extreme-Value Distribution," *Technometrics*, Vol. 14, No. 3, August 1972.

2 Benard, A. and Bos-Levenbach, E. C., "Het Uitzetten Van Waarnemingen op Waarschijnlykheidspaper," *Statistica*, Vol. 7, 1953.

3 Billmann, B. R., Antle, C. E. and Bain, L. J., "Statistical Inference from Censored Weibull Samples," *Technometrics*, Vol. 14, No. 4, November 1972.

4 Cohen, A. C., "Maximum Likelihood Estimation in the Weibull Distribution Based on Complete and on Censored Samples," *Technometrics*, Vol. 7, No. 4, November 1965.

5 D'Agostino, R. B., "Linear Estimation of the Weibull Parameters," *Technometrics*, Vol. 13, No. 1, February 1971.

6 Danziger, L., "Planning Censored Life Tests for Estimation of the Hazard Rate of a Weibull Distribution with Prescribed Precision," *Technometrics*, Vol. 12, No. 12, May 1970.

7 Dubey, S. D., "Asymptotic Properties of Several Estimators of Weibull Parameters," *Technometrics*, Vol. 7, No. 3, August 1965.

8 Engelhardt, M. and Bain, L. J., "Some Results on Point Estimation for the Two-Parameter Weibull or Extreme-Value Distribution," *Technometrics*, Vol. 16, No. 1, February 1974.

9 Falls, L. W., "Estimation of Parameters in Compound Weibull Distributions," *Technometrics*, Vol. 12, No. 2, May 1970.

10 Gumbel, E. J., *Statistics of Extremes*, Columbia University Press, New York, 1958.

11 Hager, H. W., Bain, L. J., and Antle, C. E., "Reliability Estimation for the Generalized Gamma Distribution and Robustness of the Weibull Model," *Technometrics*, Vol. 13, No. 3, August 1971.

12 Harter, H. L. and Moore, A. H., "Point and Interval Estimators, Based on *m* Order Statistics, for the Scale Parameter of a Weibull Population with Known Shape Parameter," *Technometrics*, Vol. 7, No. 3, August 1965.

13 Harter, H. L. and Moore, A. H., "Maximum-Likelihood Estimation of the Parameters of Gamma and Weibull Populations from Complete and from Censored Samples," *Technometrics*, Vol. 7, No. 4, November 1965.

14 Hogg, R. V. and Craig, A. T., *Introduction to Mathematical Statistics*, Second Edition, Macmillan, New York, 1965.

15 Jaech, J. L., "Estimation of Weibull Distribution Shape Parameters When No More Than Two Failures Occur per Lot," *Technometrics*, Vol. 6, No. 4, November 1964.

16 Johnson, L. G., *Some Statistical Techniques in Analyzing Fatigue Data*, Detroit, Research Laboratories Division, General Motors Corporation, Report No. ME1-68, January 20, 1950.

17 Johnson, L. G., *Theory and Technique of Variation Research*, Elsevier Publishing Co., New York, N. Y., 1964.

18 Johns, M. V., Jr. and Lieberman, G. J., "An Exact Asymptotically Efficient Confidence Bound for Reliability in the Case of the Weibull Distribution," *Technometrics*, Vol. 8, No. 1, February 1966.

19 Kao, J. H. K., "A Summary of Some New Techniques on Failure Analysis," Proceedings of the Sixth National Symposium on Reliability and Quality Control; Washington, D. C. January 11–13, 1960.

20 Kao, J. H. K., "A Graphical Estimation of Mixed Weibull Parameters in Life-Testing of Electron Tubes," *Technometrics*, Vol. 1, No. 4, November 1959.

21 Kimball, B. F., "On the Choice of Plotting Positions on Probability Paper," *Journal of the American Statistical Association*, Vol. 55, September 1960.

22 Mann, N. R., *Results on Location and Scale Parameter Estimation with Application to the Extreme-Value Distribution*, Dayton, Ohio, Wright Patterson Air Force Base, Aerospace Research Laboratories, ARL 67-0023, February 1967.

23 Mann, N. R., "Tables for Obtaining the Best Linear Invariant Estimates of Parameters of the Weibull Distribution," *Technometrics*, Vol. 9, No. 4, November 1967.

24 Mann, N. R., *Results on Statistical Estimation and Hypothesis Testing with Application to the Weibull and Extreme-Value Distributions*, Dayton, Ohio, Wright Patterson Air Force Base, Aerospace Research Laboratories, ARL 68-0068, April 1968.

25 Mann, N. R., "Point and Interval Estimation Procedures for the Two-Parameter Weibull and Extreme-Value Distribution," *Technometrics*, Vol. 10, No. 2, May 1968.

26 Mann, N. R., *Estimation of Location and Scale Parameters under Various Models of Censoring and Truncation*, Dayton, Ohio, Wright Patterson Air Force Base, Aerospace Research Laboratories, ARL 70-0026, February 1970.

27 Mann, N. R., "Cramer-Rao Efficiencies of Best Linear Invariant Estimators of Parameters of the Extreme-Value Distribution Under Type II Censoring from Above," *SIAM Journal of Applied Mathematics*, Vol. 17, No. 6, November 1969.

28 Mann, N. R., "Optimum Estimators for Linear Functions of Location and Scale Parameters," *Annals of Mathematical Statistics*, Vol. 40, No. 6, 1969.

29 Mann, N. R., "Estimators and Exact Confidence Bounds for Weibull Parameters Based on a Few Ordered Observations," *Technometrics*, Vol. 12, No. 2, May 1970.

30 Mann, N. R., "Best Linear Invariant Estimation for Weibull Parameters under Progressive Censoring," *Technometrics*, Vol. 13, No. 3, August 1971.

31 Mann, N. R., Fertig, K. W., and Scheuer, E. M., *Tolerance Bounds and a New Goodness-of-Fit for Two-Parameter Weibull or Extreme-Value Distribution* (*with Tables for Censored Samples of Size* 3[1]25), Dayton, Ohio, Wright Patterson Air Force Base, Aerospace Research Laboratories, ARL 71-0077, May 1971.

32 Menon, M. V., "Estimation of the Shape and Scale Parameters of the Weibull Distribution," *Technometrics, Vol.* 5, *No.* 2, *May* 1963.

33 Mood, A. M. and Graybill, F. A., *Introduction to the Theory of Statistics, Second Edition*, McGraw-Hill, New York, 1963.

34 Qureishi, A. S., Nabavian, K. J., and Ahanen, J. D., "Sampling Inspection Plans for Discriminating Between Two Weibull Processes," *Technometrics*, Vol. 7, No. 4, November 1965.

35 Thoman, D. R., Bain, L. J., and Antle, C. E., "Inferences on the Parameters of the Weibull Distribution," *Technometrics*, Vol. 11, No. 3, August 1969.

36 Thoman, D. R., Bain, L. J., and Antle, C. E., "Maximum Likelihood Estimation, Exact Confidence Intervals for Reliability, and Tolerance Limits in the Weibull Distribution," *Technometrics*, Vol. 12, No. 2, May 1970.

37 White, J. S., "The Moments of Log-Weibull Order Statistics," *Technometrics*, Vol. 11, No. 2, May 1969.

APPENDIX
RELATED MATHEMATICAL DERIVATIONS

APPENDIX

11.A
EXPECTED WAITING TIME FOR THE *j*th FAILURE

Let $\mathbf{x}_1 \leqslant \mathbf{x}_2 \leqslant \cdots \leqslant \mathbf{x}_n$ represent an ordered sample from a two-parameter Weibull distribution. The p.d.f. for $\mathbf{x}_j$ is then given by

$$f_j(x) = \frac{n!}{(j-1)!(n-j)!} F(x)^{j-1}(1-F(x))^{n-j} f(x), \qquad x \geqslant 0 \quad (11.A.1)$$

where $F(x)$ is the cumulative Weibull distribution and $f(x) = F'(x)$. If we use the identity

$$F(x)^{j-1} = \left[1-(1-F(x))\right]^{j-1}$$

and the binomial expansion, we may write Equation 11.A.1 as

$$f_j(x) = \sum_{i=0}^{j-1} \frac{(-1)^i n!}{(j-1)!(n-j)!} \binom{j-1}{i} (1-F(x))^{n-j+i} f(x) \quad (11.A.2)$$

The expected value of $\mathbf{x}_j$ is then given by

$$E(\mathbf{x}_j) = \sum_{i=0}^{j-1} \frac{(-1)^i n!}{(j-1)!(n-j)!} \binom{j-1}{i} \int_0^\infty x\left[1-F(x)\right]^{n-j+i} f(x)\,dx \quad (11.A.3)$$

Substituting the standard Weibull function for $F(x)$ and $f(x)$ the integral in Equation 11.A.3 becomes

$$\int_0^\infty \frac{\beta}{\theta^\beta} x^\beta e^{-(n+i-j+1)\left(\frac{x}{\theta}\right)^\beta} dx \quad (11.A.4)$$

If we let

$$z = (n+i-j+1)\left(\frac{x}{\theta}\right)^\beta$$

then Equation 11.A.4 converts to

$$\theta\left(\frac{1}{n+i-j+1}\right)^{1/\beta+1} \int_0^\infty z^{1/\beta} e^{-z}\,dz \quad (11.A.5)$$

which can be readily recognized as a gamma function. Hence

$$E(\mathbf{x}_j) = \theta\Gamma\left(\frac{1}{\beta}+1\right)\sum_{i=0}^{j-1}\frac{(-1)^i n!}{(j-1)!(n-j)!}\binom{j-1}{i}\left(\frac{1}{n-j+i+1}\right)^{1/\beta+1} \tag{11.A.6}$$

Now for example, the expected value of $\mathbf{x}_1$ would be obtained by letting $j=1$, in Equation 11.A.6, which on simplification yields

$$E(\mathbf{x}_1) = \left(\frac{1}{n}\right)^{1/\beta}\theta\Gamma\left(\frac{1}{\beta}+1\right) \tag{11.A.7}$$

If one is interested in a three-parameter Weibull variable $\mathbf{Y}$ with a characteristic life θ^*, a shape parameter β, and a minimum life δ, one can make the transformation $\mathbf{x}=\mathbf{Y}-\delta$. The new variable $\mathbf{x}$ will have a two-parameter Weibull distribution with a shape parameter $\theta=\theta^*-\delta$. Clearly,

$$E(\mathbf{Y}) = E(\mathbf{x}) + \delta$$

Where $E(\mathbf{x})$ is given by Equation 11.A.6 with θ replaced by $\theta^*-\delta$. For example, the expected value of $\mathbf{Y}_1$, the smallest of the ordered observations would be

$$E(\mathbf{Y}_1) = \delta + (\theta^* - \delta)\left(\frac{1}{n}\right)^{1/\beta}\Gamma\left(\frac{1}{\beta}+1\right) \tag{11.A.8}$$

Chapter 12 sequential life testing

Sequential life testing is a hypothesis testing situation in which the course of action is reassessed as observations become available. As soon as enough information is obtained on which to base a decision, the test is discontinued. Thus the sample size is not fixed in advance but depends on the observations as they become available. The principal purpose of hypothesis testing is to determine whether the product meets a reliability goal rather than to estimate the MTBF value. Sequential testing is appealing in that fewer observations should be required than for fixed sample size hypothesis testing.

The sequential sampling procedure will provide rules for making one of three possible decisions as each observation becomes available. The decisions are (1) to accept the null hypothesis, (2) to reject the null hypothesis, and (3) to obtain additional information by taking another observation.

A brief and straightforward presentation of the theoretical developments for the sequential testing procedure is contained in Section 12.1. The development of O.C. curves for sequential tests is included. This section is useful if a sequential test is needed for distributions other than those considered in the applications of Section 12.2.

Section 12.2 is concerned with the application of sequential testing. The distributions considered include the binomial, exponential, and Poisson. Various testing schemes are illustrated, as are the O.C. curves. Section 12.3 briefly considers the problem of terminating a sequential test before a decision is reached.

The next section is concerned with the basic theory of sequential testing. Those interested primarily in applications can proceed directly to Section 12.2.

12.1 THE THEORY OF SEQUENTIAL TESTING

This section provides a brief introduction to the theoretical basis of sequential testing. The coverage is sufficient to provide a basic understanding, and those desiring further depth should consult the reference list.

Let us consider the simple null hypothesis

$$H_0 \colon \theta = \theta_0$$

against the alternative hypothesis

$$H_1 \colon \theta = \theta_1$$

From the definitions for Type I and Type II errors in hypothesis testing we define

$$P(H_1|H_0) = \alpha \qquad \text{and} \qquad P(H_0|H_1) = \beta \tag{12.1}$$

where $P(H_i|H_j)$ is the probability of accepting H_i when H_j is true.

The development of a sequential test utilizes the likelihood ratio $L_{1,n}/L_{0,n}$, where $L_{k,n} = L(x_1, x_2, \ldots, x_n|\theta_k)$. The function $L_{k,n}$ is called the likelihood function with θ_k as the true value of the parameter. Before a sample is taken, $L_{k,n}$ is the p.d.f. for the random variables $(\mathbf{x}_1, \mathbf{x}_2, \ldots, \mathbf{x}_n)$ if θ_k is the true value of the parameter. Since the $\mathbf{x}_i$'s are independent, the joint p.d.f. for the sample would be the product of the marginal p.d.f.'s. Thus

$$L_{k,n} = \prod_{i=1}^{n} f(x_i|\theta_k) \tag{12.2}$$

where $f(x_i|\theta_k)$ is the p.d.f. for the random variables $\mathbf{x}_i$ and θ_k is the parameter of interest. Here $L_{k,n}$ is the p.d.f. for the sample.

Once values for the $\mathbf{x}_i$'s are obtained by taking a random sample, then $L_{k,n}$ is no longer a p.d.f., but is considered a function of the unknown parameter θ_k. $L_{k,n}$ is now called a likelihood function. When selecting a value for θ_k we reason that if θ_k is equal to the true value of the parameter, then $L_{k,n}$ should attain its maximum value. If we select θ_k such that $L_{k,n}$ is a maximum then this value of θ_k is termed the maximum likelihood estimator for θ_k.

Let us now return to the likelihood ratio $L_{1,n}/L_{0,n}$. In sequential testing this ratio is called the sequential probability ratio since n is not fixed in advance. If θ_1 is closer to the true value of the parameter than θ_0, then $L_{1,n} > L_{0,n}$ and the ratio becomes large. Thus, it would seem reasonable that a bound B could be determined such that if $L_{1,n}/L_{0,n} \geqslant B$ then we would reject H_0. Also, a bound A could be determined such that if $L_{1,n}/L_{0,n} \leqslant A$ then we would accept H_0. So this procedure is defined by the following rules:

1. If $L_{1,n}/L_{0,n} \leqslant A$, accept H_0
2. If $L_{1,n}/L_{0,n} \geqslant B$, reject H_0
3. If $A < L_{1,n}/L_{0,n} < B$, take one more observation

To determine the bounds let us define W_0 and W_1 as two regions of the parameter space such that if $(\mathbf{x}_1, \mathbf{x}_2, \ldots, \mathbf{x}_n) \in W_1$ we accept H_1, and if

$(\mathbf{x}_1,\mathbf{x}_2,\ldots,\mathbf{x}_n) \in W_0$ we accept H_0. Now, recall that $L_{0,n}$ is the p.d.f. for $(\mathbf{x}_1,\mathbf{x}_2,\ldots,\mathbf{x}_n)$ when $\theta=\theta_0$, then

$$P(H_1|H_0)=\int\int_{W_1}\cdots\int L_{0,n}dx_1dx_2\cdots dx_n \tag{12.3}$$

or in general we can say

$$P(H_i|H_j)=\int\int_{W_i}\cdots\int L_{j,n}dx_1dx_2\cdots dx_n, \qquad i=0,1;\, j=0,1 \tag{12.4}$$

which gives us four equations including Equation 12.3.

From Equation 12.4 we have

$$P(H_0|H_1)=\int\int_{W_0}\cdots\int L_{1,n}dx_1dx_2\cdots dx_n \tag{12.5}$$

and in region W_0 we accept H_0 when H_1 is true if $L_{1,n} \leqslant AL_{0,n}$ as per procedural rule (1). Substituting the fact that $L_{1,n} \leqslant AL_{0,n}$ into Equation 12.5 we have

$$P(H_0|H_1) \leqslant A\int\int_{W_0}\cdots\int L_{0,n}dx_1dx_2\cdots dx_n \tag{12.6}$$

But by referring back to Equation 12.4 and substituting for the multiple integral on the right-hand side of Equation 12.6 we have

$$P(H_0|H_1) \leqslant AP(H_0|H_0) \tag{12.7}$$

By similar reasoning for region W_1 we can conclude that

$$P(H_1|H_1) \geqslant BP(H_1|H_0) \tag{12.8}$$

If we assume that the probability is one that the procedure eventually terminates, which will result in either accepting or rejecting H_0, then

$$P(H_0|H_i)+P(H_1|H_i)=1, \qquad i=0,1 \tag{12.9}$$

or

$$P(H_0|H_0)=1-P(H_1|H_0) \tag{12.10}$$

Substituting Equation 12.10 into Equation 12.7 gives

$$P(H_0|H_1) \leqslant A\left[1-P(H_1|H_0)\right] \tag{12.11}$$

From the definition of errors of the first and second kind as defined in Equation

12.1 we have

$$\beta \leqslant A(1-\alpha) \tag{12.12}$$

Also from Equation 12.9,

$$P(H_1|H_1) = 1 - P(H_0|H_1) \tag{12.13}$$

and, combining this result with Equation 12.8, we have

$$[1 - P(H_0|H_1)] \geqslant BP(H_1|H_0) \tag{12.14}$$

or

$$(1-\beta) \geqslant B(\alpha) \tag{12.15}$$

If in Equations 12.12 and 12.15 the inequalities are replaced by equalities, then values for the bounds A and B can be determined. To do this we must assume that when the test is discontinued, $L_{1,n}/L_{0,n}$ is exactly equal to one of the boundary values. Since n is actually discrete, this will not be precisely true; however, analytical investigations have shown that, in practice, the error is insignificant because α and β are both small. Thus, the values for A and B are approximated by

$$A = \frac{\beta}{1-\alpha} \tag{12.16}$$

and

$$B = \frac{1-\beta}{\alpha} \tag{12.17}$$

The error of this approximation can be investigated by rewriting Equations 12.12 and 12.15 as

$$A \geqslant \frac{\beta}{1-\alpha} \tag{12.18}$$

and

$$B \leqslant \frac{1-\beta}{\alpha} \tag{12.19}$$

By using equalities in Equations 12.18 and 12.19 we have used a value for A that is smaller than the exact value and a value for B that is larger than the exact value. Let us define α' and β' as the error probabilities of the first and second kinds, respectively, resulting from using $A = \beta/(1-\alpha)$ and $B = (1-\beta)/\alpha$. Then substituting into Equations 12.18 and 12.19 we have

$$\frac{\beta}{1-\alpha} \geqslant \frac{\beta'}{1-\alpha'} \tag{12.20}$$

and

$$\frac{1-\beta}{\alpha} \leqslant \frac{1-\beta'}{\alpha'} \tag{12.21}$$

Rewriting these inequalities

$$\frac{\beta'}{1-\alpha'} \leqslant \frac{\beta}{1-\alpha} \tag{12.22}$$

and

$$\frac{\alpha'}{1-\beta'} \leqslant \frac{\alpha}{1-\beta} \tag{12.23}$$

Now since $(1-\alpha')<1$, $(1-\beta')<1$ and both are positive, we can say that

$$\beta' \leqslant \frac{\beta}{1-\alpha} \tag{12.24}$$

and

$$\alpha' \leqslant \frac{\alpha}{1-\beta} \tag{12.25}$$

We can now see the consequences of our choice of α and β on the true errors (α',β'). If, for example, $\alpha=0.05$ and $\beta=0.10$, then the true errors are bounded by $\alpha' \leqslant 0.056$ and $\beta' \leqslant 0.105$. Theoretically, if we multiply Equation 12.22 by $(1-\alpha)(1-\alpha')$ and Equation 12.23 by $(1-\beta)(1-\beta')$ and add the resulting inequalities we find that

$$\alpha'+\beta' \leqslant \alpha+\beta \tag{12.26}$$

This means that in addition to the bounds given by Equations 12.24 and 12.25 we also know that at least one of the inequalities $\alpha' \leqslant \alpha$ or $\beta' \leqslant \beta$ must hold. So using the approximate bounds will at most cause one of the error probabilities to increase.

The assumption that the test eventually terminates can easily be shown to be valid. If we define the variable

$$z_i = \ln \frac{f(x_i|\theta_1)}{f(x_i|\theta_0)}$$

then a sequence of observations $\mathbf{x}_1, \mathbf{x}_2, \ldots$ generates a sequence $\mathbf{z}_1, \mathbf{z}_2, \ldots$, where the $\mathbf{z}_i$'s are independent just as are the $\mathbf{x}_i$'s. By procedural rule (3) the sequential test will continue as long as $\ln A < \mathbf{z}_i < \ln B$. Note that by our definitions for A and B, $\ln B$ is positive and $\ln A$ is negative, and if we let $C = (\ln B - \ln A)$, then $C > 0$.

If any one $\mathbf{z}_i$ falls outside of the interval $(-C, C)$, the test will be discontinued. Let us define $P = P(-C < \mathbf{z}_i < C)$, then for n observations the probability that all n of the $\mathbf{z}_i$'s fall in the interval is P^n. Since $P < 1$, as n increases the probability that no $\mathbf{z}_i$ falls outside of the continue region approaches zero. Thus, the test must eventually terminate. Here we have, of course, ignored the possibility that the sum of the $\mathbf{z}_i$'s could fall outside of the continue region before any one of the $\mathbf{z}_i$ values did.

12.1.1 THEORETICAL CONSIDERATIONS FOR OBTAINING THE OPERATING CHARACTERISTIC CURVE

Consider two simple hypotheses designated H_i and H_j. Define a constant h such that if

$$(L_{1,n}/L_{0,n})^h = L_{i,n}/L_{j,n} \tag{12.27}$$

then continue region for the sequential test for H_i and H_j can be written as

$$A^h < \frac{L_{i,n}}{L_{j,n}} < B^h, \qquad \text{if } h > 0 \tag{12.28}$$

or

$$B^h < L_{i,n}/L_{j,n} < A^h, \qquad \text{if } h < 0 \tag{12.29}$$

where $A = \beta/(1-\alpha)$ and $B = (1-\beta)/\alpha$ as previously obtained [See 5, 6].

Let us first consider the case where $h > 0$ and the inequality given in Equation 12.28 applies. The sequential probability ratio $L_{i,n}/L_{j,n}$ could be used to develop a sequential test for H_i and H_j. Let the errors of the first and second kind for this test be denoted by α' and β', respectively. Then from Equation 12.28 we have

$$\frac{\beta'}{1-\alpha'} = A^h \tag{12.30}$$

and

$$\frac{1-\beta'}{\alpha'} = B^h \tag{12.31}$$

Solving Equations 12.30 and 12.31 for α', we find

$$\alpha' = \frac{1 - A^h}{B^h - A^h} \tag{12.32}$$

This is the (approximate) probability of accepting H_j when H_i is true. Since Equation 12.32 is true for any i and j then this is equivalent to

$$P[\text{accept } H_1 | H_0] = \frac{1 - A^h}{B^h - A^h} \tag{12.33}$$

An identical result will be obtained where $h < 0$ and Equation 12.29 is used.

The O.C. curve is then the complement of Equation 12.33 and is

$$Pa = \frac{B^h - 1}{B^h - A^h} \tag{12.34}$$

However, determining h for a particular H_1 and H_0 is still a problem. Fortunately this problem has been resolved, and the method will be described. Equation 12.27 suggests that

$$g(x|\theta) = \left[\frac{f(x|\theta_1)}{f(x|\theta_0)}\right]^h f(x|\theta) \tag{12.35}$$

and $g(x|\theta)$ is a p.d.f. Depending on whether **x** is discrete or continuous, the left hand side of Equation 12.35 should sum or integrate to unity.

Let us assume that **x** is a continuous random variable and define

$$\phi(h) = \int_{-\infty}^{\infty} \left[\frac{f(x|\theta_1)}{f(x|\theta_0)}\right]^h f(x|\theta) dx \tag{12.36}$$

The general scheme for finding the O.C. curve is to find $\phi(h)$ from Equation 12.36. The proof of the existence of h is based on $\phi(h) = 1$, [5, 6]. So then set $\phi(h) = 1$ and solve for θ. Now an arbitrarily chosen h gives a value θ from Equation 12.36 that allows $P(\theta) = Pa$ to be found from Equation 12.34

EXAMPLE 12.1

Let us consider the exponential distribution

$$f(x|\lambda) = \lambda e^{-\lambda x}, \qquad x \geqslant 0$$

From Equation 12.36

$$\phi(h) = \int_0^{\infty} \left[\frac{\lambda_1 e^{-\lambda_1 x}}{\lambda_0 e^{-\lambda_0 x}}\right]^h \lambda e^{-\lambda x} dx$$

or

$$\phi(h)=\frac{\lambda(\lambda_1/\lambda_0)^h}{h(\lambda_1-\lambda_0+\lambda)}$$

Setting $\phi(h)$ equal to unity and solving for λ gives

$$\lambda=\frac{h(\lambda_0-\lambda_1)}{1-(\lambda_1/\lambda_0)^h}$$

By combining this result with Equation 12.34, we can calculate the O.C. curve for a sequential test using the exponential distribution.

12.2 THE SEQUENTIAL TESTING PROCEDURE

From a computational standpoint the sequential testing procedure is relatively easy. In this section, we present the method of testing and provide various specific applications.

The theory of sequential testing is developed using two simple hypotheses (that is, hypotheses of the type $H_0:\theta=\theta_0$ and $H_1:\theta=\theta_1$, where $\theta_1\neq\theta_0$). Unfortunately, these simple hypotheses are rarely encountered in a practical application. We can, however, view this problem as being no different than the one encountered in ordinary hypothesis testing. Suppose we are interested in considering $H_0:\theta=\theta_0$ and $H_1:\theta>\theta_0$, and we want our Type I error to be α. The probability of a Type II error is no longer a fixed value but is a function of θ, the true value of the parameter. If $f(x|\theta)$ is a continuous function of x and θ, then the Type II error will approach $1-\alpha$ as θ approaches θ_0 and will decrease as θ becomes larger than θ_0. Again it is emphasized that this is no different than the situation in ordinary hypothesis testing where we must decide on a value θ_1 such that if $\theta=\theta_1$ we want a Type II error probability no greater than β. Thus the sequential test can be developed using θ_0 and θ_1, and this will give a Type I error of α when $\theta=\theta_0$ and a Type II error less than or equal to β whenever $\theta\geqslant\theta_1$. Of course, the same logic is used for the hypotheses $H_0:\theta\leqslant\theta_0$ and $H_1:\theta>\theta_0$; here the probability of Type I error is α when $\theta=\theta_0$.

To perform a sequential test, the quantities θ_0, θ_1, α, and β must be specified. Then with $f(x|\theta)$ as the underlying p.d.f. with parameter θ, the sequential test is developed as follows:

1. Compute A, where $A=\beta/(1-\alpha)$
2. Compute B, where $B=(1-\beta)/\alpha$
3. Take a random observation x_i, from $f(x|\theta)$ and compute

$$\lambda_i=f(x_i|\theta_1)/f(x_i|\theta_0)$$

4. If $\lambda\leqslant A$, accept H_0
5. If $\lambda\geqslant B$, reject H_0

6. If $A < \lambda < B$, then take another observation x_j, and compute

$$\lambda_j = \prod_{i=1}^{j} f(x_i|\theta_1)/f(x_i|\theta_0) \tag{12.37}$$

7. Repeat steps 4, 5, and 6 until the process terminates.

These computations are often made easier if the continue region in step 6 is taken as $\ln A < \ln\lambda < \ln B$, and similarly in steps 4 and 5. The inequalities can then usually be rearranged to a more convenient form.

12.2.1 APPLICATION TO THE BINOMIAL DISTRIBUTION

In the binomial distribution the outcome of the test is classified into one of two possible categories. For example, consider a laboratory test where the product is cycled a fixed number of times, and the product is said to either fail or survive the test. For a particular product placed on test the point binomial describes the probability of an occurrence. The point binomial is given by

$$P(x) = p^x(1-p)^{1-x}, \qquad x = \begin{cases} 0 \text{ if success} \\ 1 \text{ if failure} \end{cases} \tag{12.38}$$

where p = the probability of failure for a product.

Let the null hypothesis be

$$H_0: p \leqslant p_0$$

and the alternate be

$$H_1: p > p_0$$

Here p_0 is a value of p such that if $p = p_0$ then the probability of accepting H_0 is $(1-\alpha)$. Also let p_1 be a value of p such that $p_1 > p_0$ and for $p = p_1$ the probability of accepting H_0 is β. These quantities (α, p_0, p_1, and β) define the sequential test. The test can now be developed by setting up the sequential probability ratio.

For n observations the sequential probability ratio as given by Equation 12.37 becomes

$$\left(\frac{p_1}{p_0}\right)^y \left(\frac{1-p_1}{1-p_0}\right)^{n-y} \tag{12.39}$$

where $y = \sum_{i=1}^{n} x_i$, which is the total number of failures in n trials.

Then the continue region becomes

$$A < \left(\frac{p_1}{p_0}\right)^y \left(\frac{1-p_1}{1-p_0}\right)^{n-y} < B \tag{12.40}$$

where $A=\beta/(1-\alpha)$ and $B=(1-\beta)/\alpha$. By taking the natural logarithm of each term in this inequality and solving for the relationship between n and y we have

$$\frac{n}{D}\ln\left(\frac{1-p_0}{1-p_1}\right)-\frac{1}{D}\ln\left(\frac{1-\alpha}{\beta}\right)\leqslant y\leqslant\frac{n}{D}\ln\left(\frac{1-p_0}{1-p_1}\right)+\frac{1}{D}\ln\left(\frac{1-\beta}{\alpha}\right) \quad (12.41)$$

where

$$D=\ln\left[\left(\frac{p_1}{p_0}\right)\left(\frac{1-p_0}{1-p_1}\right)\right] \quad (12.42)$$

and recall that

$$p_0<p_1$$

Let the inequality in Equation 12.41 be represented as

$$A_n<y<B_n \quad (12.43)$$

Now by our procedural rules we accept H_0 if $y\leqslant A_n$, reject H_0 if $y \geqslant B_n$, and take an additional observation if $A_n<y<B_n$.

The boundaries of Equation 12.43 graph as parallel straight lines, and a sequential testing graph can be constructed as shown in Figure 12.1. The test results can then be plotted on this graph to provide a visual representation of the test progress.

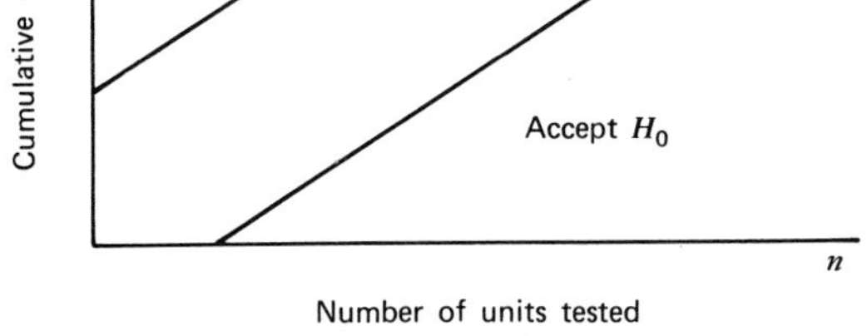

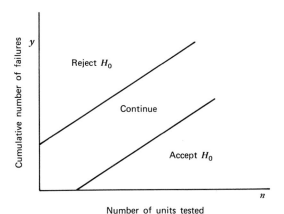

Figure 12.1 A sequential testing graph.

The entire O.C. curve for this test can also be determined. The probability of accepting H_0 when p is the true fraction failing is

$$Pa(p) = \frac{B^h - 1}{B^h - A^h} \tag{12.44}$$

where

$$p = \frac{1 - \left(\dfrac{1-p_1}{1-p_0}\right)^h}{\left(\dfrac{p_1}{p_0}\right)^h - \left(\dfrac{1-p_1}{1-p_0}\right)^h} \tag{12.45}$$

To obtain the O.C. curve for this test, Equation 12.45 is first used with an arbitrarily selected value of h to compute p. Then Equation 12.44 is used to calculate the probability of acceptance for the value of p. Of course, for $p = p_0$, $Pa(p_0) = 1 - \alpha$ and for $p = p_1$, $Pa(p_1) = \beta$.

The expected number of observations to reach a decision is given by

$$E(p,n) = \frac{Pa(p)\ln A + (1 - Pa(p))\ln B}{p\ln\left(\dfrac{p_1}{p_0}\right) + (1-p)\ln\left(\dfrac{1-p_1}{1-p_0}\right)}. \tag{12.46}$$

where $Pa(p)$ is given by Equation 12.44. The plot of $E(p,n)$ as a function of p will yield the curve shown in Figure 12.2, where the maximum lies between p_0 and p_1.

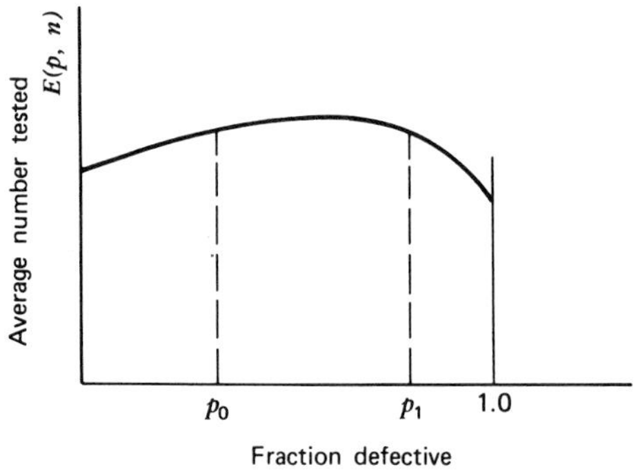

Figure 12.2 Average sample number curve.

EXAMPLE 12.2

Suppose we have a hypothesis testing situation given by

$$H_0: p \leqslant 0.001$$

$$H_1: p > 0.001$$

where the binomial distribution applies. In analyzing this situation we agree upon $\alpha = 0.05$, $\beta = 0.10$ for $p_0 = 0.001$ and $p_1 = 0.020$. We will now specify a sequential test for this situation.

Substituting into Equation 12.41, the continue region becomes

$$0.0064n - 3.151 < y < 0.0064n + 0.9587$$

This inequality is graphed and shown in Figure 12.3. Here it can be seen that the fastest acceptance route requires 493 successes in a row while the fastest rejection route requires only that the first unit tested result in a failure.

The boundaries are a function of the closeness of p_1 to p_0 and of the relative magnitudes of p_1 and p_0. To illustrate this, Figure 12.4 shows a sequential test for $p_0 = 0.001$ and $p_1 = 0.100$. Here the difference between p_0 and p_1 is more than five times the difference of the previous test. This test requires only 92 successes in a row for the fastest route to acceptance and four failures for the fastest route to rejection. This can now be compared to the 493 successes and one failure for the previous test.

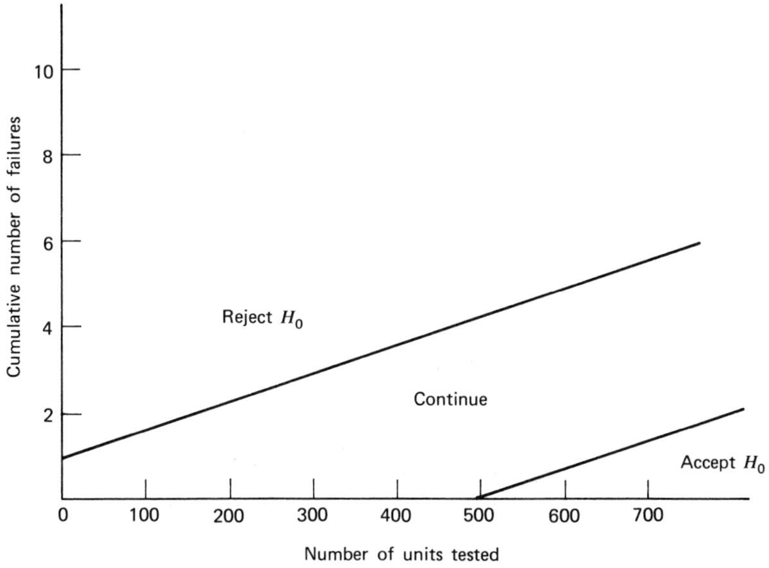

Figure 12.3 Sequential test for H_0: $p \leqslant 0.001$ and H_1: $p > 0.001$ with $p_1 = 0.020$.

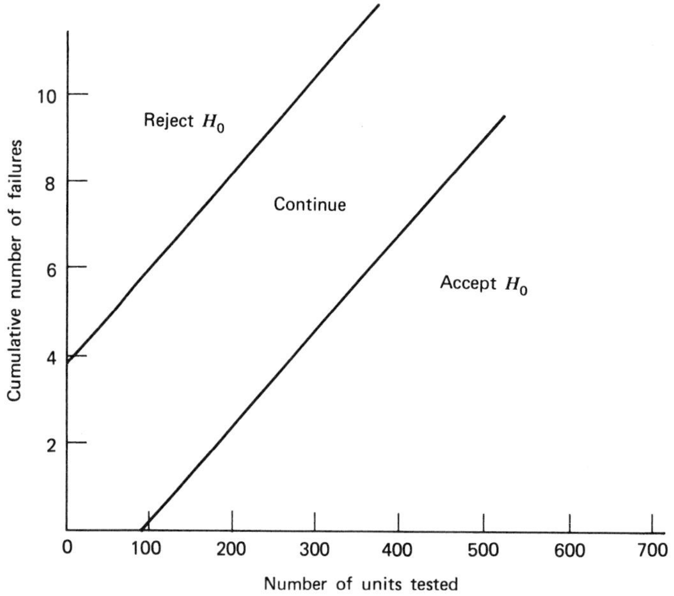

Figure 12.4 Sequential test for H_0: $p \leqslant 0.001$ and H_1: $p > 0.001$ with $p_1 = 0.100$.

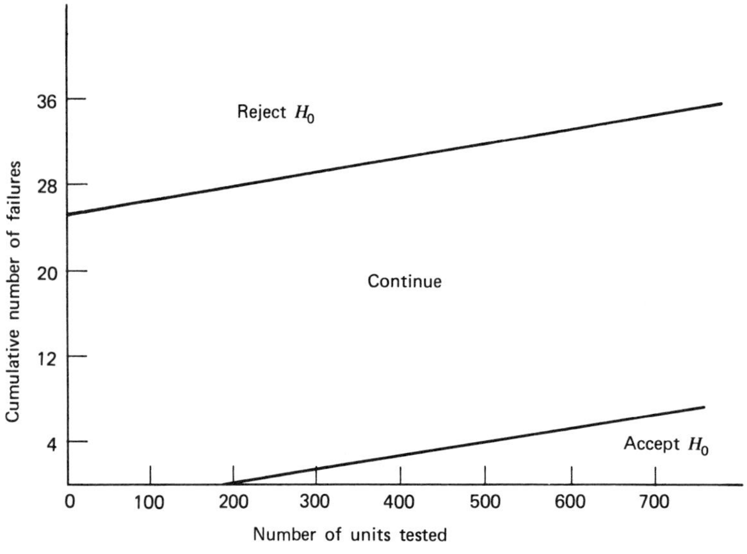

Figure 12.5 Sequential test for H_0: $p \leqslant 0.050$ and H_1: $p > 0.050$ with $p_1 = 0.100$.

Now consider still another test where $p_0 = 0.050$ and $p_1 = 0.100$. This sequential test is shown in Figure 12.5. Note in this graph that the ordinate scale is changed from the previous graphs. In this situation, the fastest routes to acceptance and rejection of H_0 are 175 successes or 25 failures, respectively. Here apparently the increase in both successes and failures over the previous test is due to the closeness of p_0 to p_1; however, the number of successes needed for acceptance has decreased because p_0 has increased.

The above illustrates a rather complex problem, and the three tests described are summarized in Table 12.1. Generally as p_0 and p_1 become close together the parallel accept and reject lines will move farther apart because it becomes more difficult to distinguish between p_0 and p_1. This can be seen by comparing the difference between the fastest routes to success and failure from test I to test II and from test II to test III. Also as p_0 becomes larger relative to p_1 the slope of the lines will increase as evident by comparing Figures 12.3 and 12.4.

Table 12.1 Comparison of Three Sequential Tests for the Binomial

TEST	p_0	p_1	p_1/p_0	$p_1 - p_0$	FASTEST ROUTE SUCCESS	FAILURE
I	0.001	0.020	20	0.019	493	1
II	0.001	0.100	100	0.099	92	4
III	0.050	0.100	2	0.050	175	25

Returning to Example 12.2, Table 12.2 shows values of p and $Pa(p)$ as calculated from Equations 12.44 and 12.45, respectively, for arbitrarily chosen values of h. Figure 12.6 is an O.C. curve for this test obtained by plotting the tabled values.

Table 12.2 Calculated Values for the O.C. Curve for Example 12.2

h	p	$Pa(p)$	R
−2.00	0.035	0.011	.965
−1.00	0.020	0.100	.980
−0.75	0.016	0.167	.984
−0.50	0.012	0.268	.988
−0.30	0.010	0.375	.990
−0.20	0.008	0.435	.992
−0.02	0.007	0.549	.993
0.07	0.006	0.606	.994
0.20	0.005	0.680	.995
0.60	0.002	0.863	.996
1.00	0.001	0.950	.999

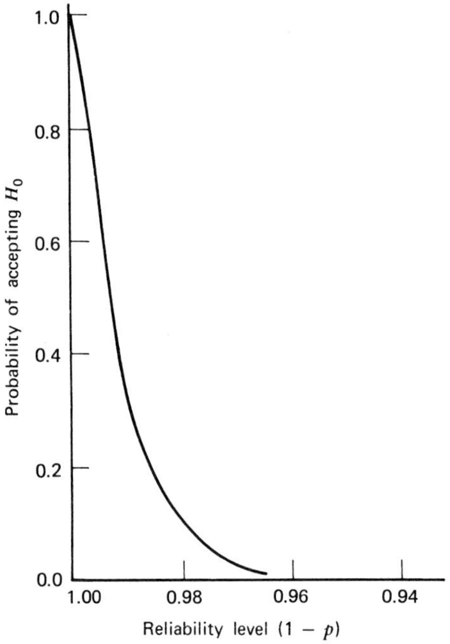

Figure 12.6 O. C. curve for the sequential test H_0: $p \leqslant 0.001$ and H_1: $p > 0.001$ with $p_1 = 0.020$.

12.2.2 APPLICATION TO THE EXPONENTIAL DISTRIBUTION

The testing situation can be varied somewhat, and the first situation to be considered is where single items are placed on test in sequence. Or this might also be thought of as obtaining kilometers between failures when testing a vehicle. Thus the result of a test will be ($\mathbf{x}_i$) the time to failure for the ith unit tested.

The exponential distribution is given by

$$f(x) = \frac{1}{\theta} e^{-x/\theta}, \qquad x \geqslant 0$$

where $\theta =$ the mean time to failure. The hypothesis testing situation will be

$$H_0\colon \theta \geqslant \theta_0$$
$$H_1\colon \theta < \theta_0$$

If $\theta = \theta_0$ the probability of accepting H_0 is set at $(1-\alpha)$ and if $\theta = \theta_1$ where $\theta_1 < \theta_0$ the probability of accepting H_0 is set at a low level β. Once the decision on these quantities is made then the test is defined.

The sequential probability ratio becomes

$$\prod_{i=1}^{n} \frac{(1/\theta_1)\exp(-x_i/\theta_1)}{(1/\theta_0)\exp(-x_i/\theta_0)} = \left(\frac{\theta_0}{\theta_1}\right)^n \exp\left[-\left(\frac{1}{\theta_1}-\frac{1}{\theta_0}\right)\right]\sum_{i=1}^{n} x_i$$

Substituting into the inequality for the continue region

$$\frac{\beta}{1-\alpha} < \left(\frac{\theta_0}{\theta_1}\right)^n \exp\left[-\left(\frac{1}{\theta_1}-\frac{1}{\theta_0}\right)\right]\sum_{i=1}^{n} x_i < \frac{1-\beta}{\alpha}$$

Taking the natural logarithm and rearranging

$$\frac{n\ln(\theta_0/\theta_1)-\ln((1-\beta)/\alpha)}{(1/\theta_1-1/\theta_0)} < \sum x_i < \frac{n\ln(\theta_0/\theta_1)+\ln((1-\alpha)/\beta)}{(1/\theta_1-1/\theta_0)} \tag{12.47}$$

In the rearrangement note that $(1/\theta_1-1/\theta_0)>0$.

The O.C. curve for the test, that is, the probability of accepting H_0 when θ is the true parameter value denoted as $P(\theta)$, is calculated from the pair of equations

$$P(\theta)=\frac{B^h-1}{B^h-A^h} \tag{12.48}$$

and

$$\theta=\frac{(\theta_0/\theta_1)^h-1}{h(1/\theta_1-1/\theta_0)} \tag{12.49}$$

where $A=\beta/(1-\alpha)$ and $B=(1-\beta)/\alpha$. First Equation 12.49 is used to determine θ for an arbitrarily selected value of h, then $P(\theta)$ can be found from Equation 12.48. The value h can be any real number, and meaningful selections are made by trial and error.

EXAMPLE 12.3

Suppose that we are interested in life testing a new product to see if it meets a standard of 1,000 hours. We select $\theta_0=1{,}000$ hours with $\alpha=0.05$ and decide on $\theta_1=500$ hours with $\beta=0.10$.

Substituting into Equation 12.47 gives a continue region defined by

$$\frac{n\ln 2-\ln(0.90/0.05)}{(1/500-1/1000)} < \sum x_i < \frac{n\ln 2+\ln(0.95/0.10)}{(1/500-1/1000)}$$

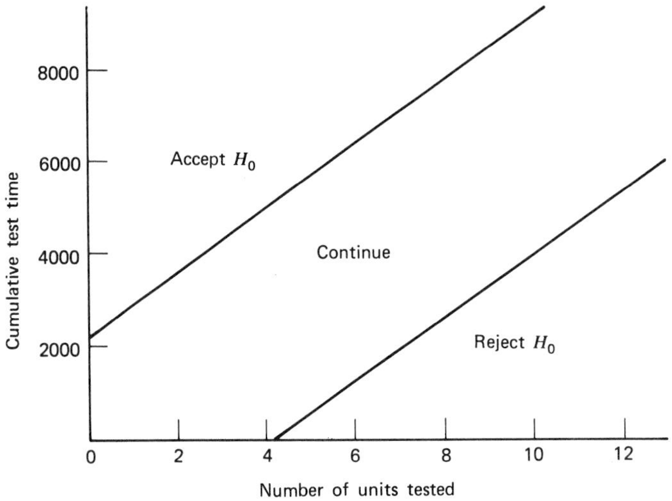

Figure 12.7 Sequential life testing graph for the exponential distribution.

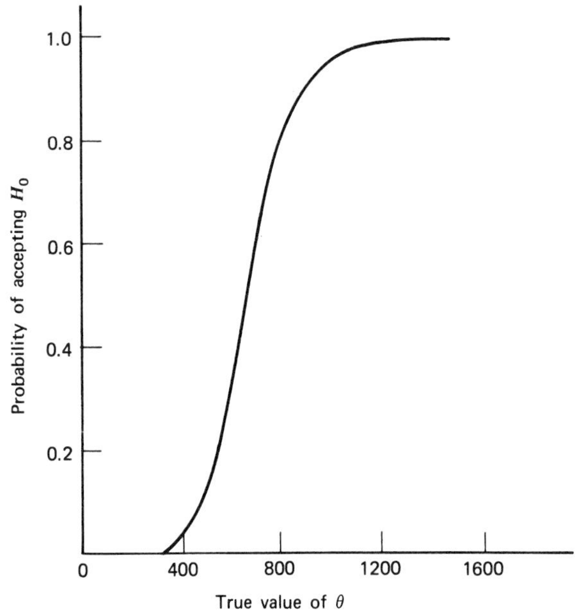

Figure 12.8 Operating characteristic curve for a sequential test.

which reduces to

$$693n - 2890 < \sum x_i < 693n + 2251$$

A sequential test graph can then be developed from this inequality as shown in Figure 12.7 with the O.C. curve for this test given in Figure 12.8 as obtained from the values in Table 12.3.

The O.C. curve for a test should be studied carefully because it assesses the risks involved in the decision process. Note in the previous example that the conclusion that the design goal had been met was made on acceptance of the null hypothesis. This type of decision process is frowned upon because if the decision is made on rejection of the null hypothesis then one is sure that there is statistical significance. It is important to be aware of the errors involved in the decision process; decisions based on either the null or alternate hypothesis that do not consider the magnitude of the errors is a dangerous and unreasonable procedure. When analyzing the decision-making procedure both the Type I and Type II errors should be considered.

12.2.3 APPLICATION TO THE EXPONENTIAL DISTRIBUTION: RECORDING TOTAL TEST TIME VERSUS TOTAL NUMBER OF FAILURES

In the previous sequential test, devices were individually tested to failure. We now consider a second approach. Recall that the total accumulated test time and total number of observed failures can be described probabilistically by using the relationship between the exponential and Poisson distributions (Sections 10.1 and 10.A). This permits design of a sequential test that is applicable to a test situation where failed units are replaced or where kilometers are accumulated on several test vehicles.

Specifically, if the time-to-failure distribution is exponential with failure rate λ, then for an interval of time of length T, the number of failures **r** is a

Table 12.3 Values for the O.C. Curve for Example 12.3

h	$p(\theta)$	θ
2.000	0.9969	1500
1.000	0.9500	1000
0.500	0.8275	828
0.250	0.7110	757
0.100	0.6242	718
0.001	0.5625	693
−0.500	0.2681	586
−1.000	0.1000	500
−2.000	0.0110	375

Poisson distributed random variable with parameter λT. The p.d.f. for **r** is

$$P(r|\lambda)=\frac{(\lambda T)^r e^{-\lambda T}}{r!}, \qquad r=0,1,2,\ldots$$

and the sequential probability ratio would be $P(r|\lambda_1)/P(r|\lambda_0)$.

Let us now consider a specific set of hypotheses given by

$$H_0{:}\lambda \geqslant \lambda_0 \qquad \text{and} \qquad H_1{:}\lambda < \lambda_0$$

In this situation, the rejection of H_0 implies that the failure rate is significantly less than λ_0. Here perhaps λ_0 is the standard based on an old design, and production of the new design is not to be considered unless it represents significant improvement in reliability.

To define the sequential test the values α, β, and λ_1 must also be specified. Here λ_1 is a value such that $\lambda_1 < \lambda_0$. Then the sequential probability ratio is

$$(\lambda_1/\lambda_0)^r \exp[-T(\lambda_1-\lambda_0)]$$

and the continue region becomes

$$\frac{T(\lambda_0-\lambda_1)-\ln\left(\dfrac{1-\beta}{\alpha}\right)}{\ln(\lambda_0/\lambda_1)} < r < \frac{T(\lambda_0-\lambda_1)+\ln\left(\dfrac{1-\alpha}{\beta}\right)}{\ln(\lambda_0/\lambda_1)} \tag{12.50}$$

The O.C. curve for the test is calculated from

$$P(\lambda)=\frac{B^h-1}{B^h-A^h} \tag{12.51}$$

with $A=\beta/(1-\alpha)$ and $B=(1-\beta)/\alpha$, as previously defined, and

$$\lambda=\frac{h(\lambda_0-\lambda_1)}{1-(\lambda_1/\lambda_0)^h} \tag{12.52}$$

Here $P(\lambda)$ is the probability of accepting H_0 when λ is the true value of the failure rate. Again these equations are solved for arbitrarily chosen values of h.

EXAMPLE 12.4

Let us assume that we are interested in setting up a sequential test for the hypotheses H_0: $\lambda \geqslant 0.002$ and H_1: $\lambda < 0.002$. We specify $\alpha=0.05$, $\beta=0.10$, and $\lambda_1=0.001$. Using Equation 12.50 we have

$$14.4\times 10^{-4}T-4.1699 < r < 14.4\times 10^{-4}T+3.2479$$

Here the independent variable (T) would be plotted on the abscissa of the sequential graph.

Continuing with this example, the O.C. curves are calculated from

$$\lambda = \frac{0.001h}{1-(0.5)^h}$$

and

$$P(\lambda) = \frac{18^h - 1}{18^h - 0.105^h}$$

The calculated values for the O.C. curve as obtained from the above equations are given in Table 12.4. The O.C. curve is graphed in Figure 12.9.

12.2.4 BATCH TESTING USING THE EXPONENTIAL DISTRIBUTION

It should now be recognized that many different testing schemes can be developed. In this testing situation the time to failure will be taken as exponentially distributed just as in the previous situation. However, a batch of n items will be simultaneously placed on test, and the test will be truncated after a length of time T resulting in r failures ($r \geqslant 0$). At this time a decision will be made either to accept, reject, or place another batch of n items on test.

For the exponential distribution with failure rate λ, the probability of an item failing prior to the end of a time interval T is given by

$$F(T) = 1 - e^{-\lambda T}, \qquad T \geqslant 0$$

Then the probability of observing r failures among the n items on test is binomially distributed and is given by

$$P(r|\lambda) = \binom{n}{r}\left[1 - e^{-\lambda T}\right]^r \left[e^{-\lambda T}\right]^{n-r}, \qquad r = 0, 1, \ldots, n$$

Suppose we are interested in the hypotheses

$$H_0: \lambda \geqslant \lambda_0 \qquad \text{and} \qquad H_1: \lambda < \lambda_0$$

and that we have specified α, β, and λ_1, ($\lambda_1 < \lambda_0$). Then if k is the number of

Table 12.4 Calculations for the O.C. curve of Example 12.4

h	$P(\lambda)$	λ
2.000	0.9969	0.0027
1.000	0.9500	0.0020
0.500	0.8275	0.0017
0.250	0.7110	0.0016
0.100	0.6242	0.0015
0.001	0.5625	0.0014
−0.500	0.2681	0.0012
−1.000	0.1000	0.0010
−2.000	0.0110	0.0007

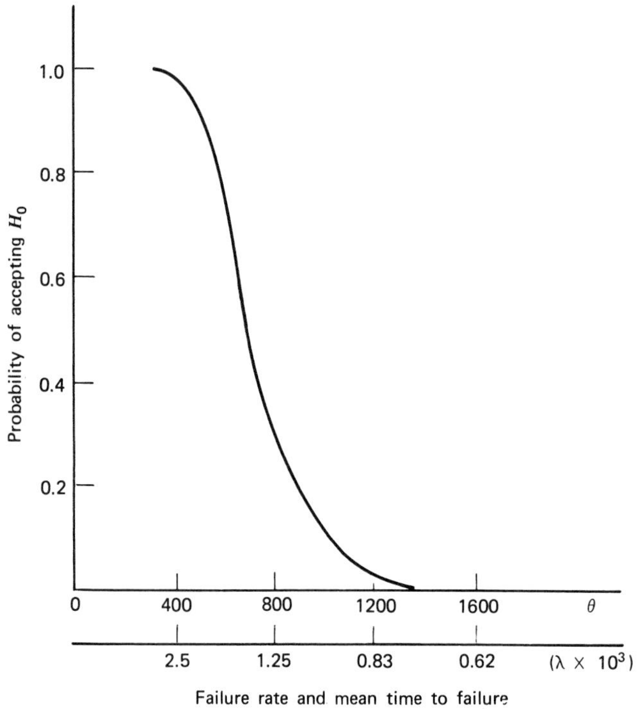

Figure 12.9 O. C. curve for a sequential testing situation using the exponential distribution.

batches of size n placed on test, the continue region becomes

$$\frac{(\lambda_0-\lambda_1)Tnk-\ln\left(\frac{1-\beta}{\alpha}\right)}{D}<\sum_{i=1}^{k} r_i<\frac{(\lambda_0-\lambda_1)Tnk+\ln\left(\frac{1-\alpha}{\beta}\right)}{D} \tag{12.53}$$

where

$$D=T(\lambda_0-\lambda_1)+\ln(1-e^{-\lambda_0 T})-\ln(1-e^{-\lambda_1 T}) \tag{12.54}$$

In this case if the upper limit is exceeded then H_0 is rejected.

12.3 A PROCEDURE FOR EARLY TRUNCATION

Test situations are occasionally encountered where the test must be truncated after a fixed time or number of observations. A line partitioning the sequential graph can be drawn as shown in Figure 12.10. This line is drawn through the

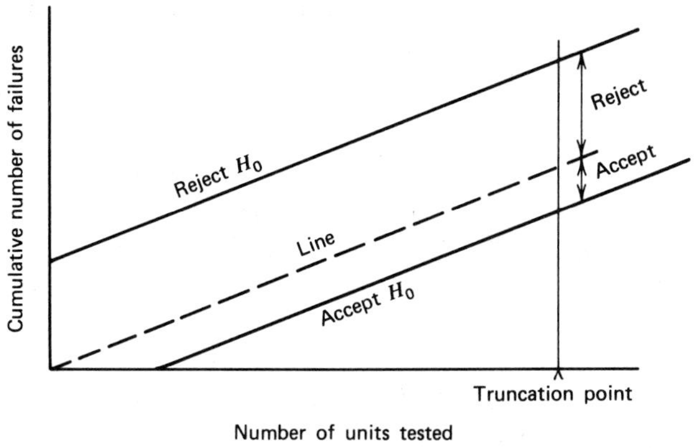

Figure 12.10 A truncation procedure for sequential testing.

origin of the graph parallel to the accept and reject lines. Then when the truncation point is reached the decision to accept or reject H_0 simply depends on which side of the line the final outcome lies. Obviously this procedure changes the α and β levels of the original test; however, the change is slight if the truncation point is large.

Actually the expected number of observations or expected testing time as appropriate for a particular situation should be calculated for both $\theta = \theta_0$ and $\theta = \theta_1$. Then the truncation point should be beyond 3 or 4 times this average to produce only a slight change in α or β. We have not given the formulas to calculate the expected number of observations; however, these are available in the References 4, 5, and 6.

12.4 SUMMARY

The obvious advantage of sequential testing is the savings in test time with a resulting savings in cost. This is an advantage of paramount importance; however, in practice the sequential procedure is frequently regarded with suspicion. In sequential testing, one must face up to all of the errors present in hypothesis testing while in nonsequential hypothesis testing one can live in happy ignorance. Unfortunately, too frequently nonsequential testing consists of taking a small sample, calculating a test statistic, not rejecting the null hypothesis, and then reaching a conclusion of major importance. Fortunately, one cannot do this with sequential testing. Also please note that one should not do this in any hypothesis testing situation, and the consequences of Type II error should always be considered, since it is always present.

The mechanics of the sequential method as presented in Section 12.2 are fairly easy to adapt to testing situations other than those specifically considered in this section. However, in order to develop the O.C. curve for the test, Equations 12.34 and 12.36 of Section 12.2 must be used.

EXERCISES

1 In automotive vehicle testing the exponential distribution is to be used to model the kilometers between failures. Given the hypothesis testing situation H_0: $\theta \leqslant \theta_0$ and H_1: $\theta > \theta_0$ with $\theta_0 = 16{,}000$ km, $\theta_1 = 24{,}000$ km, $\alpha = 0.05$ and $\beta = 0.10$; (a) develop the sequential graph; (b) develop the O.C. curve for the test.

2 For the situation in Problem 1 suppose that you were given the hypothesis testing situation H_0: $\theta \geqslant \theta_0$ and H_1: $\theta < \theta_0$ with $\theta_0 = 24{,}000$ km, $\theta_1 = 16{,}000$ km, $\alpha = 0.05$ and $\beta = 0.10$, develop the O.C. curve for the test and compare to Problem 1.

3 A new passenger car engine is being developed with a MTBF design goal of 45,000 kilometers. You are asked to design a sequential test that can be used during the road testing of these engines to determine if the design goal has been met. Design the test and indicate the decisions that you had to make and the rationale that you used for your decisions.

4 An accelerated laboratory test rotates and stresses ball joints in a deleterious environment for 300 hours. This test is said to be representative of a specific amount of customer use. The test fixtures available can simultaneously accommodate 20 ball joints. Each ball joint tested will be classified as either failing or surviving the test. The hypotheses of interest are $H_0{:}p \geqslant p_0$ and $H_1{:}p < p_0$. (a) Develop the sequential decision rules for this testing situation with $p_0 = 0.001$, $p_1 = 0.0001$, $\alpha = 0.05$, and $\beta = 0.10$ and draw the sequential graph; label the accept and reject regions. (b) Using the concepts from Section 12.2.1, develop the O.C. curve for the test.

5 Suppose that we are interested in developing a sequential test for the Weibull distribution with p.d.f.

$$f(x) = \frac{\beta}{\theta}\left(\frac{x}{\theta}\right)^{\beta-1} \exp\left[-\left(\frac{x}{\theta}\right)^{\beta}\right], \qquad x \geqslant 0$$

Assume that β is known and use the fact that $\mathbf{y} = \mathbf{x}^{\beta}$ is exponentially distributed to develop a sequential test for the characteristic life parameter θ.

BIBLIOGRAPHY

1 Bazovsky, I., *Reliability Theory and Practice*, Prentice-Hall, Englewood Cliffs, N.J., 1961.

2 Epstein, B. and M. Sobel, "Sequential Life Tests in the Exponential Case," *Annals of Mathematical Statistics*, Vol. XXVI, 1955.

3 Hald, A., *Statistical Theory With Engineering Applications*, John Wiley & Sons, New York, 1952.

4 Johnson, N. L. and F. C. Leone, *Statistics and Experimental Design in Engineering and the Physical Sciences—Volume II*, John Wiley & Sons, New York, 1964.

5 Mood, A. M. and F. A. Graybill, *Introduction to the Theory of Statistics*, Second Edition, McGraw-Hill, New York, 1963.

6 Wald, A., "Sequential Tests of Statistical Hypotheses," *Annals of Mathematical Statistics*, Vol. XVI, No. 2, 1945.

7 Wald, A., *Sequential Analysis*, John Wiley & Sons, New York, 1947.

Chapter 13 bayesian reliability in design and testing

In the design of a product, reliability goals are established, and the designer, drawing on a wealth of past experience, strives to meet or exceed these goals. However, the traditional approach to statistical inference does not account for this past experience. Thus a large amount of hard data obtained by testing is necessary to demonstrate a reliability level with a high degree of confidence.

Frequently, the designer will express a high degree of confidence in his design after relatively little testing. Furthermore, the designer may also be skeptical of the results produced by traditional statistical inference, since these results may not conform to his past experiences (here the non-Bayesian approach is termed traditional). For example, suppose that a designer redesigns an air conditioning system. This application is identical to the previous year's model, and the specific objective of the redesign is to improve the reliability. Here if we estimate with a given level of confidence a failure rate that is higher than the previous year's failure rate, the designer may simply not believe our results. And unless considerable hard data is available, disagreement can easily occur. Also, in such situations, management intuitively feels that relatively little testing is needed. So the reliability engineer is placed in a difficult position—he cannot rely on the traditional statistical inference approach to produce believable results.

One answer to this dilemma is provided by Bayesian statistics, which combines subjective judgment or experience with hard data to provide estimates similar to those obtained from the traditional statistical inference approach. However, the Bayesian approach is controversial because no experimental or analytical methods exist for the quantification of the belief in the performance of a new system, and in order to use the Bayesian approach this subjective information must be quantified.

The basic concepts of the Bayesian analytic procedure are presented in Section 13.1. This section first develops Bayes' theorem, then considers p.d.f.'s and Bayes' theorem in Section 13.1.1, Bayesian point estimation in Section

13.1.2, and Bayesian confidence intervals in Section 13.1.3. For those already versed in Bayesian analysis, Section 13.1 can be skipped.

The Bayesian approach for the binomial testing situation is considered in Section 13.2. In this situation component testing results in either success or failure. Point estimators and confidence limits for the fraction failing or for the reliability are developed. A short discussion of selection of the prior distribution is also included.

Section 13.3 presents the same concepts as Section 13.2, but this section uses the exponential distribution as the model for time to failure.

The Bayesian approach can be applied to sequential testing, and Bayesian sequential testing is presented in Section 13.4. Also included is a procedure for developing a truncated sequential test.

The material of Sections 13.5 and 13.6 contains thoughts on the application of Bayesian analysis at the design stage. Section 13.5 presents various decision situations that might be encountered at the design stage and illustrates the use of the Bayesian approach for solution. Section 13.6 considers a component subjected to a stress environment and illustrates how knowledge about the environment can be used to estimate reliability.

13.1 THE BAYESIAN APPROACH TO STATISTICAL INFERENCE

The Bayesian approach to statistical inference is based on a theorem first presented by the Reverend Thomas Bayes, an English minister who lived in the eighteenth century. Bayes' basic theorem was later modified by Laplace, and this modified version is used today and is commonly referred to as Bayes' theorem.

In order to demonstrate the development of this theorem we will first illustrate the concept of conditional probability. Consider the Venn diagram of Figure 13.1. Here A and B are two events of interest defined on a sample space, Ω. Let $P(A)$, $P(B)$ and $P(A \cap B)$ be the probabilities assigned to events A, B, and $A \cap B$, respectively.

It can be recognized that given the information that the event B has occurred, the probability of occurrence of any event outside of event B is zero.

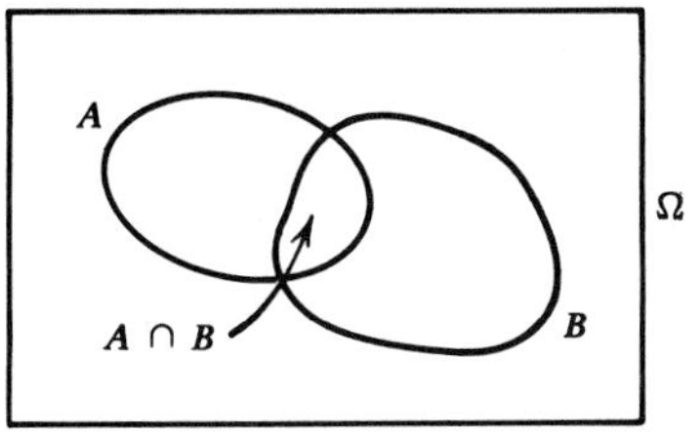

Figure 13.1 Venn diagram.

The resulting total sample space is now restricted to event B. Thus, the probability that event A has occurred is given by

$$P(A|B)=\frac{P(A\cap B)}{P(B)} \tag{13.1}$$

where, of course, $P(B)>0$. $P(A|B)$ is referred to as the conditional probability of A given B.

From Equation 13.1 we can also say

$$P(A\cap B)=P(B)P(A|B) \tag{13.2}$$

or by considering $P(B|A)$ we have

$$P(A\cap B)=P(A)P(B|A) \tag{13.3}$$

Combining Equations 13.1 and 13.3 we can write

$$P(A|B)=P(A)\frac{P(B|A)}{P(B)} \tag{13.4}$$

Here $P(A)$ is the prior probability of the event A before the information about B becomes available, and $P(A|B)$ is the posterior probability of A based on the information. This is a version of Bayes' theorem.

The conditional probability theorem can be extended to obtain another version of Bayes' theorem. Consider the Venn diagram in Figure 13.2 where the events $H_1, H_2, \ldots, H_k$ are defined on a sample space Ω. Let the events $H_1, H_2, \ldots, H_k$ form a partitioning of Ω, which means that $\cup_{i=1}^{k} H_i=\Omega$ and H_i, H_j are mutually exclusive for all i,j where $i \neq j$.

Let B be an event of interest also defined on Ω. At this point we will consider the probability of B occurring in terms of the H_i's. We see that

$$\Omega=H_1 \cup H_2 \cup \ldots \cup H_k$$

and, of course

$$B\cap\Omega=B$$

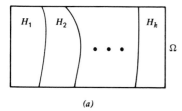

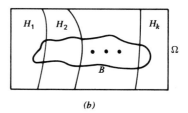

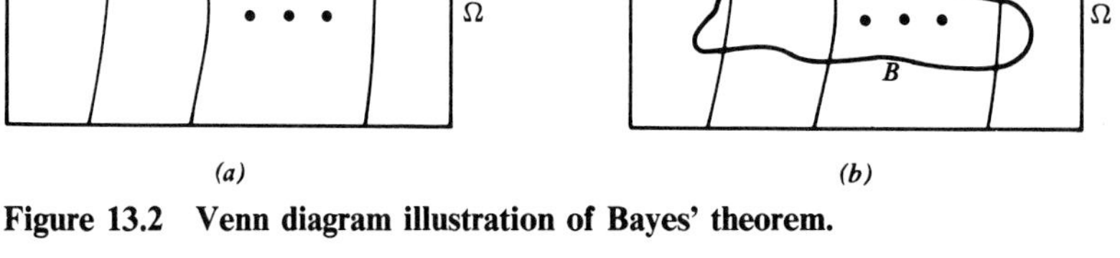

Figure 13.2 Venn diagram illustration of Bayes' theorem.

thus

$$B = B \cap (H_1 \cup H_2 \cup \ldots \cup H_k)$$

using the distributive law

$$B = (B \cap H_1) \cup (B \cap H_2) \cup \ldots \cup (B \cap H_k)$$

Since the H_i's are mutually exclusive, then $(B \cap H_i)$ and $(B \cap H_j)$ are mutually exclusive for all i,j where $i \neq j$.

Thus the probability of B can be written

$$P(B) = P(B \cap H_1) + P(B \cap H_2) + \ldots + P(B \cap H_k) \tag{13.5}$$

Then applying Equation 13.2 to Equation 13.5 gives

$$P(B) = \sum_{i=1}^{k} P(H_i)P(B|H_i) \tag{13.6}$$

EXAMPLE 13.1

As an example of an application of Equation 13.6, suppose we are presented with two indistinguishable urns. We are informed that one urn contains 30 red balls and 70 green balls while the second urn contains 50 red balls and 50 green balls. We mentally label these urns I and II, respectively. We are told to select an urn at random and to randomly draw one ball from this urn. Suppose we want to know the probability that the ball drawn is red.

Let B = the event that the ball drawn is red
H_i = the event that the ith urn is selected $(i = 1, 2)$

Then applying Equation 13.6 we write

$$P(B) = P(H_1)P(B|H_1) + P(H_2)P(B|H_2)$$

Here we see that H_1, H_2 form a partitioning of S as required for application of this equation. Substituting in the numerical values we obtain

$$P(B) = \frac{1}{2}\left(\frac{30}{100}\right) + \frac{1}{2}\left(\frac{50}{100}\right) = 0.40$$

Now reconsider Figure 13.2 and let us concern ourselves with one of the events H_m. Specifically, we are interested in $P(H_m|B)$. Applying Equation 13.1 for conditional probability, we obtain

$$P(H_m|B) = \frac{P(H_m \cap B)}{P(B)}, \qquad P(B) > 0 \tag{13.7}$$

By substituting Equation 13.6 for $P(B)$ and by also rewriting the numerator slightly, we obtain

$$P(H_m|B)=\frac{P(H_m)P(B|H_m)}{\sum_{i=1}^{k} P(H_i)P(B|H_i)} \tag{13.8}$$

This is Bayes' theorem and is also sometimes referred to as the Bayes-Laplace rule. Equation 13.8 represents an inverse sort of reasoning from effect to cause.

EXAMPLE 13.2

Let us return to our previous example concerning the two urns. Suppose an independent observer randomly draws a ball from one of the urns and the ball turns out to be red. We now ask, "What is the probability that urn I was selected, given the information that a red ball was drawn?"

Using the events as previously defined in Example 13.1 and applying Equation 13.8 we have

$$P(H_1|B)=\frac{P(H_1)P(B|H_1)}{P(H_1)P(B|H_1)+P(H_2)P(B|H_2)}=\frac{\frac{1}{2}\left(\frac{30}{100}\right)}{0.40}=0.375$$

Here we see that $P(H_1)$, the prior probability for event H_1, is 0.50 and $P(H_1|B)$, the posterior probability based on the information, has decreased to 0.375.

EXAMPLE 13.3

Now let us pose another very hypothetical example to further illustrate application of the concepts. Suppose that we are concerned with the reliability level of a new and yet untested system. We feel that the system might have one of two possible reliability levels, denoted by R_1 and R_2. Based on past experience we believe that the system may have a reliability level $R_1=0.95$, but if the system designer has miscalculated on a certain factor in question, then the reliability may be at a lower value, $R_2=0.75$.

We will express our confidence in the system designer by assigning an 80% chance that level R_1 has been attained, thus leaving a 20% chance that level R_2 has been attained.

Now suppose that we test one system and find it to operate successfully. We would like to know the probability that reliability level R_1 has been attained.

Let us define

R_1 = the event that reliability level R_1 has been attained

S_i = the event that the ith system tested results in a success

Then we want $P(R_1|S_1)$, which by Equation 13.8 is

$$P(R_1|S_1)=\frac{P(R_1)P(S_1|R_1)}{P(R_1)P(S_1|R_1)+P(R_2)P(S_1|R_2)}$$

and by substituting the numerical values we find that

$$P(R_1|S_1)=\frac{(0.80)(0.95)}{(0.80)(0.95)+(0.20)(0.75)}=0.835$$

Let us assume that a second system is tested, and it is also successful. We now want

$$P(R_1|S_1\cap S_2)=\frac{P(R_1)P(S_1\cap S_2|R_1)}{P(R_1)P(S_1\cap S_2|R_1)+P(R_2)P(S_1\cap S_2|R_2)}$$

and this gives

$$P(R_1|S_1\cap S_2)=\frac{(0.80)(0.95\times 0.95)}{(0.80)(0.95\times 0.95)+(0.20)(0.75\times 0.75)}=0.865$$

We can see how the probability of the event R_1 is updated by application of Bayes' theorem as new information becomes available.

13.1.1 CONDITIONAL DISTRIBUTIONS AND BAYES' THEOREM

The above results can be extended to joint probability distributions. Consider the joint p.d.f. $f(x_1,x_2)$ for the random variables $\mathbf{x}_1$ and $\mathbf{x}_2$; then the conditional p.d.f. for $\mathbf{x}_1$ given x_2 is

$$k(x_1|x_2)=\frac{f(x_1,x_2)}{f_2(x_2)} \tag{13.9}$$

where $f_2(x_2)$ is the marginal p.d.f. for $\mathbf{x}_2$.

Now by reapplying Equation 13.9 we can also write $f(x_1,x_2)$ as

$$f(x_1,x_2)=h(x_1)\,g(x_2|x_1) \tag{13.10}$$

where $h(x_1)$ is the marginal p.d.f. for $\mathbf{x}_1$. Substituting this into Equation 13.9 gives

$$k(x_1|x_2)=\frac{h(x_1)\,g(x_2|x_1)}{f_2(x_2)}, \qquad f_2(x_2)>0 \tag{13.11}$$

This is the equivalent of Bayes' theorem. Here we can find $f_2(x_2)$ from

$$f_2(x_2)=\int_{x_1} f(x_1,x_2)\,dx_1=\int_{x_1} h(x_1)\,g(x_2|x_1)\,dx_1 \tag{13.12}$$

if $\mathbf{x}_1$ is a continuous random variable. If $\mathbf{x}_1$ is discrete then the integral is replaced by a summation.

For continuous random variables, finding the marginal p.d.f. $f_2(x_2)$ as given in Equation 13.12 may be difficult because the right-hand side of Equation 13.12

may not be integrable by other than numerical means. This is, in fact, true unless $h(x_1)$, the prior p.d.f., and $g(x_2|x_1)$, the conditional p.d.f., are compatibly selected. There are p.d.f.'s that go together; that is, Equation 13.12 is integrable and such that the posterior p.d.f. is of the same form as the prior. These p.d.f.'s are termed natural conjugates.

The above conditional probability argument can be used as another approach to statistical inference, and this approach has been termed Bayesian estimation. To demonstrate the approach, let **x** have a p.d.f. $f(x)$, which is dependent on θ. In the traditional statistical inference approach θ is an unknown parameter and hence is a constant. We now describe our prior belief in the value of θ by a p.d.f. $h(\theta)$. This amounts to quantitatively assessing subjective judgment and really should not be confused with the so-called objective probability assessment derived from the long-term frequency approach. Thus, essentially θ will now be treated as a random variable $\boldsymbol{\theta}$ with p.d.f. $h(\theta)$.

Consider a random sample $\mathbf{x}_1, \mathbf{x}_2, \ldots, \mathbf{x}_n$ from $f(x)$ and define a statistic **y** as a function of this random sample. Then there exists a conditional p.d.f. $g(y|\theta)$ for **y** given θ. The joint p.d.f. for **y** and $\boldsymbol{\theta}$ is

$$f(\theta, y) = h(\theta)\, g(y|\theta) \tag{13.13}$$

and if $\boldsymbol{\theta}$ is continuous, then

$$f_2(y) = \int_{\theta} h(\theta)\, g(y|\theta)\, d\theta \tag{13.14}$$

is the marginal p.d.f. for the statistic **y**. The conditional p.d.f. for $\boldsymbol{\theta}$ given the information **y** is

$$k(\theta|y) = \frac{h(\theta)\, g(y|\theta)}{f_2(y)}, \qquad f_2(y) > 0 \tag{13.15}$$

Again, this is simply a form of Bayes' theorem. Here $h(\theta)$ is the prior p.d.f. that expresses our belief in the value of θ before the hard data (y) became available. Then $k(\theta|y)$ is the posterior p.d.f. of $\boldsymbol{\theta}$ given the hard data (y).

In Equation 13.15, if $h(\theta)$, the prior p.d.f. of $\boldsymbol{\theta}$ which expresses the belief in the location of the parameter θ, could be objectively determined, then the Bayesian approach would not be controversial. However, since we are concerned with an untested system and with beliefs held by individuals, $h(\theta)$ must be a subjectively determined distribution, and this subjectivity naturally leads to controversy.

Returning to Equation 13.15, the change in the shape of the prior p.d.f. $h(\theta)$ to the posterior p.d.f. $k(\theta|y)$ due to the information is a result of the product of $g(y|\theta)$ and $h(\theta)$ because $f_2(y)$ is simply a normalization constant for a fixed y.

Or we can say that

$$k(\theta|y) \propto g(y|\theta)h(\theta)$$

The quantity $g(y|\theta)$ is termed the likelihood and uses the information supplied by the sample (y) to change the prior $h(\theta)$.

EXAMPLE 13.4

We will now demonstrate the application of Equation 13.15 by considering a more realistic approach to our previous example. Let us say that we have designed a system that we feel has a fair reliability level. Our belief in the reliability level is expressed by the p.d.f.

$$h(r) = 4r^3, \qquad 0 \leqslant r \leqslant 1 \tag{13.16}$$

where **r** is the reliability level. This p.d.f. is shown in Figure 13.3.

Now let us assume that we test one system and it succeeds. We must find $k(r|S_1)$, the posterior p.d.f. for the reliability level **r** given the event of one success, S_1.

From Equation 13.13 we have

$$f(r, S_1) = h(r)\,g(S_1|r)$$

For this particular situation, the probability of a success given a reliability level is

$$g(S_1|r) = r$$

and then combining $h(r)$ and $g(S_1|r)$ gives

$$f(r, S_1) = 4r^4$$

We now find the marginal of S_1 by

$$f_2(S_1) = \int_0^1 4r^4\,dr = \frac{4}{5}$$

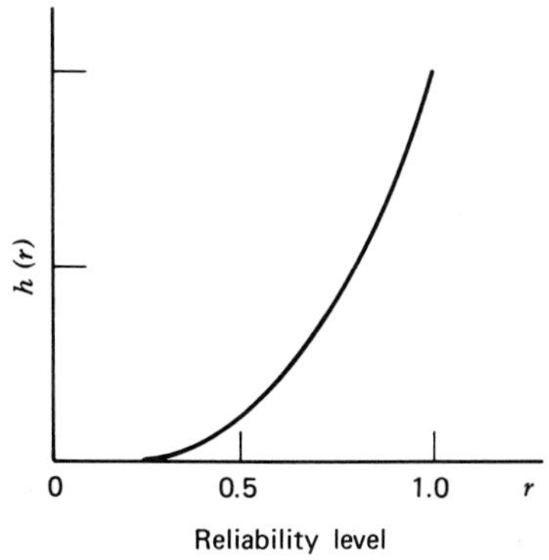

Figure 13.3 Prior probability density function.

Substituting into Equation 13.15 we have

$$k(r|S_1)=\frac{4r^4}{\left(\frac{4}{5}\right)}=5r^4, \qquad 0\leqslant r\leqslant 1$$

Thus this is the posterior distribution of **r** given the hard evidence S_1.

13.1.2 BAYESIAN POINT ESTIMATION

The Bayesian approach yields a posterior p.d.f. $k(\theta|y)$ which expresses our belief in the value of the parameter of interest based on the hard evidence y. In order to obtain a point estimate of $\boldsymbol{\theta}$, we must select one representative value from this posterior p.d.f. Possible suggestions are either the mean or median of $k(\theta|y)$ and both are justifiable from a decision theory standpoint. We will use the mean as the Bayesian point estimator for $\boldsymbol{\theta}$. That is, the Bayesian point estimator for a parameter $\boldsymbol{\theta}$ is

$$\hat{\theta}_b=E(\boldsymbol{\theta}|y)=\int_{\theta}\theta k(\theta|y)\,d\theta \tag{13.17}$$

13.1.3 BAYESIAN CONFIDENCE LIMITS

Since in reliability work the concern is usually on a one-sided confidence limit, that approach will be taken here. From $k(\theta|y)$ the posterior p.d.f. for $\boldsymbol{\theta}$, we can find either $u(y)$ or $v(y)$ such that

$$\int_{-\infty}^{u(y)}k(\theta|y)\,d\theta=\alpha \tag{13.18}$$

or

$$\int_{v(y)}^{\infty}k(\theta|y)\,d\theta=\alpha \tag{13.19}$$

Then $u(y)$ is a Bayesian $(1-\alpha)$ lower confidence limit for θ while $v(y)$ is a $(1-\alpha)$ upper confidence limit for θ. It should be noted in this instance that since $k(\theta|y)$ is a p.d.f., then $P[\boldsymbol{\theta}>u(y)]=P[\boldsymbol{\theta}<v(y)]=1-\alpha$.

EXAMPLE 13.5

Let us reconsider the previous example and first find the Bayesian point estimate of **r**, the reliability, as well as a lower confidence limit for it.

We previously found that the posterior p.d.f. was

$$k(r|S_1)=5r^4, \qquad 0\leqslant r\leqslant 1$$

Then the point estimate of the reliability is

$$\hat{R}_b=\int_0^1(r)5r^4\,dr=\frac{5}{6}$$

Now let us consider a Bayesian 97.5% lower confidence limit on the reliability. We have

$$0.025 = \int_0^{R_L} 5r^4\,dr$$

which yields a lower confidence limit of $R_L = 0.478$. Thus we can conclude with 97.5% confidence (or probability) that $\mathbf{r} \geqslant 0.478$.

We next consider some specific applications of the Bayesian approach to several common life testing situations.

13.2 BINOMIAL DISTRIBUTION TESTING SITUATION

The binomial distribution is applicable to a success-failure testing situation. For example, devices might be placed on test for a fixed number of cycles, and if a device survives the test, this is termed a success.

If n devices are placed on test and $\mathbf{y}$ is the random variable representing the number of failures, the p.d.f. of $\mathbf{y}$ is binomial and is given by

$$g(y|p) = \binom{n}{y} p^y (1-p)^{n-y}, \qquad y = 0, 1, 2, \ldots, n \tag{13.20}$$

where p is the probability of failure for a device.

Let us describe the prior belief in $\mathbf{p}$, the fraction failing by a beta distribution which also happens to be the natural conjugate, and this means that the posterior will be beta. Then

$$h(p) = \frac{\Gamma(\delta+\rho)}{\Gamma(\delta)\Gamma(\rho)} p^{\delta-1}(1-p)^{\rho-1}, \qquad 0 \leqslant p \leqslant 1 \tag{13.21}$$

where δ and ρ are constants that determine the shape of the distribution.

We want

$$f(y,p) = h(p)\,g(y|p)$$

which in this case is

$$f(y,p) = \binom{n}{y} \frac{\Gamma(\delta+\rho)}{\Gamma(\delta)\Gamma(\rho)} p^{\delta+y-1}(1-p)^{\rho+n-y-1} \tag{13.22}$$

with $0 \leqslant p \leqslant 1$ and $y = 0, 1, 2, \ldots, n$. The marginal is found from Equation 13.22 by

$$f_1(y) = \binom{n}{y} \frac{\Gamma(\delta+\rho)}{\Gamma(\delta)\Gamma(\rho)} \int_0^1 p^{\delta+y-1}(1-p)^{\rho+n-y-1}\,dp$$

This integral is recognizable as a beta function, and we obtain

$$f_1(y)=\binom{n}{y}\frac{\Gamma(\delta+\rho)\Gamma(\delta+y)\Gamma(\rho+n-y)}{\Gamma(\delta)\Gamma(\rho)\,\Gamma(\delta+\rho+n)}, \qquad y=0,1,2,\ldots,n \tag{13.23}$$

which happens to be called a hyperbinomial distribution.

Then the posterior p.d.f. for **p** is found by

$$k(p|y)=\frac{h(p)\,g(y|p)}{f_1(y)}, \qquad f_1(y)>0$$

which in this case is

$$k(p|y)=\frac{\Gamma(\delta+\rho+n)}{\Gamma(\delta+y)\Gamma(\rho+n-y)}\,p^{\delta+y-1}(1-p)^{\rho+n-y-1}, \qquad 0\leqslant p\leqslant 1 \tag{13.24}$$

This is a beta distribution with parameters $(\delta+y)$ and $(\rho+n-y)$.

The Bayesian point estimator for **p** is $E(\mathbf{p}|y)$ as obtained from the conditional density function defined in Equation 13.24. We write this as

$$\hat{p}_b=\frac{\delta+y}{\delta+\rho+n} \tag{13.25}$$

where n is the number of devices on test, and y is the observed number of failures.

A Bayesian confidence limit for **p** can be most easily obtained by applying the well-known relationship between the beta and F distributions. This relationship will now be applied.

If we let $r_1=\rho+n-y$ and $r_2=\delta+y$, Equation 13.24 can be written as

$$k(p|y)\,dp=\frac{\Gamma(r_1+r_2)}{\Gamma(r_1)\Gamma(r_2)}\,p^{r_2-1}(1-p)^{r_1-1}\,dp, \qquad 0\leqslant p\leqslant 1 \tag{13.26}$$

Consider a random variable **F** as defined by the transformation

$$p=\frac{1}{1+(r_1/r_2)F}$$

then

$$|dp|=\frac{(r_1/r_2)\,dF}{\left[1+(r_1/r_2)F\right]^2}$$

These values may be substituted into Equation 13.26 and the result is

$$f(F)\,dF=\frac{\Gamma(r_1+r_2)}{\Gamma(r_1)\Gamma(r_2)}\left(\frac{r_1}{r_2}\right)^{r_1}\left[\frac{1}{1+(r_1/r_2)F}\right]^{r_1+r_2}(F)^{r_1-1}\,dF$$

This is an F distribution with degrees of freedom $\nu_1=2r_1$ and $\nu_2=2r_2$.

Now define P_α as a value of the random variable $\mathbf{p}$ such that

$$P[\mathbf{p}\leqslant P_\alpha]=1-\alpha$$

If $G(p)$ is the cumulative distribution for $\mathbf{p}$, we have

$$G(P_\alpha)=P\left[\mathbf{F}_{\nu_1,\nu_2}>\frac{r_2}{r_1}\left(\frac{1}{P_\alpha}-1\right)\right]=1-\alpha$$

or

$$F_{1-\alpha,\nu_1,\nu_2}=\frac{r_2}{r_1}\left(\frac{1}{P_\alpha}-1\right) \tag{13.27}$$

Since $F_{1-\alpha,\nu_1,\nu_2}$ is not given in standard tables, we use the reciprocal relationship for F distributed random variables and solve Equation 13.27 for P_α obtaining

$$P_\alpha=\frac{F_{\alpha,2(\delta+y),2(\rho+n-y)}}{F_{\alpha,2(\delta+y),2(\rho+n-y)}+[(\rho+n-y)/(\delta+y)]} \tag{13.28}$$

which is the upper limit on $\mathbf{p}$.

Proceeding in a similar manner the lower limit is found to be

$$P_{1-\alpha}=\frac{1}{1+\left(\dfrac{\rho+n-y}{\delta+y}\right)F_{\alpha,2(\rho+n-y),2(\delta+y)}} \tag{13.29}$$

Then, for example, using Equation 13.29, the parameter $\mathbf{p}$ is contained in the interval given by

$$P_{1-\alpha}\leqslant\mathbf{p}\leqslant 1$$

with a $(1-\alpha)$ probability. Or this is termed a one-sided lower Bayesian confidence limit for the fraction failing p, with a confidence coefficient of $(1-\alpha)$. Equation 13.28 could be used in a similar fashion to obtain an upper confidence limit.

EXAMPLE 13.6

Let us consider a situation where p is the fraction failing and where we want to estimate p by placing devices on test for a fixed period of time. Let the beta prior be defined as

$$h(p)=20(1-p)^{19}, \qquad 0 \leqslant p \leqslant 1$$

here $\rho=20$ and $\delta=1$. This distribution is shown in Figure 13.4. From this figure it appears that there is a belief that p is small.

Now let us assume that 20 components are tested and one failure results. A point estimate and a 95% Bayesian upper confidence limit for the fraction failing is desired.

From Equation 13.25 the point estimate is

$$\hat{p}_b=\frac{1+1}{1+20+20}=0.0488$$

This can be compared to the answer using traditional statistical inference which is

$$\hat{p}=\frac{y}{n}=\frac{1}{20}=0.0500$$

Here the point estimates are reasonably close.

For the 95% upper confidence limit, the tabled F value is

$$F_{\alpha, 2(\delta+y), 2(\rho+n-y)}=F_{0.05,4,78}=2.49$$

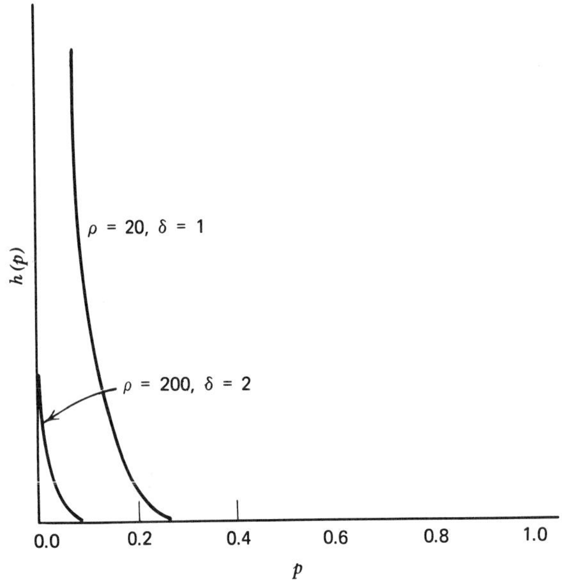

Figure 13.4 Examples of beta prior distributions.

Substituting into Equation 13.28 gives an upper limit on p as

$$p_{0.05} = \frac{2.49}{2.49 + (39/2)} = 0.113$$

So we can say that

$$p \leqslant 0.113$$

with 95% probability. Or, equivalently, if we are interested in reliability, then

$$0.887 < R$$

To compare the Bayesian confidence limit with a non-Bayesian limit for the binomial distribution, the traditional confidence limits are now given.

13.2.1 NON-BAYESIAN AND EXACT CONFIDENCE LIMITS FOR THE BINOMIAL

Confidence limits for the binomial testing situation are given by

$$p_{\alpha/2} = \frac{(y+1)F_{\alpha/2, 2(y+1), 2(n-y)}}{(n-y) + (y+1)F_{\alpha/2, 2(y+1), 2(n-y)}} \tag{13.30}$$

which is the upper limit and

$$p_{1-\alpha/2} = \frac{y}{y + (n-y+1)F_{\alpha/2, 2(n-y+1), 2y}} \tag{13.31}$$

which is the lower limit. These are exact limits for the binomial parameter p and are developed in a manner similar to the Bayesian limits of Equations 13.28 and 13.29. That is, it is well known that the binomial sum and beta integral can be shown to be equivalent. Then the beta integral can be equated to the F distribution just as was done for the Bayesian limits.

EXAMPLE 13.7

The conventional statistical confidence limit for the previous problem where $n = 20$ units were tested with $y = 1$ failure is found from Equation 13.30. The tabled F value as needed for this application is

$$F_{0.05, 4, 38} = 2.62$$

which gives

$$p_{0.05} = \frac{2(2.62)}{19 + 2(2.62)} = 0.216$$

So the 95% confidence limit for p is

$$p \leqslant 0.216$$

or

$$0.784 < R$$

As can be seen, the Bayesian limit produces a slightly more advantageous lower bound on reliability from the producer's standpoint.

EXAMPLE 13.8

Let us consider another example where the beta prior distribution has parameters $\beta = 200$ and $\alpha = 2$. This beta distribution is also shown in Figure 13.4.

Now let us assume that 20 components are tested with no failures, and we wish to assess the reliability.

Using the Bayesian approach the point estimate of the fraction failing as obtained by Equation 13.25 is

$$\hat{p} = \frac{2+0}{2+200+20} = 0.0090$$

or

$$R = 0.9910$$

Here the conventional approach could not be applied to obtain a point estimate because there were no failures.

Now consider a 95% lower confidence limit for the reliability. Using Equation 13.28 we obtain the upper limit on the fraction failing as

$$p_{0.05} = \frac{F_{0.05,4,440}}{F_{0.05,4,440} + (440/4)} = 0.0211$$

or

$$p \leqslant 0.0211$$

and hence

$$R > 0.9799$$

Comparing this to the conventional approach we use Equation 13.30 to obtain

$$p_{0.05} = \frac{F_{0.05,2,40}}{20 + F_{0.05,2,40}} = \frac{3.23}{23.23} = 0.1390$$

or

$$R > 0.8610$$

Here the Bayesian prior distribution has provided information that yielded a much higher lower bound on reliability.

13.2.2 PROBLEMS IN SELECTION OF THE BETA PRIOR

The selection of the beta prior distribution is not an easy task, and this selection has a direct effect on the subsequent predictions. We now point out some consequences of parameter selection for the prior in hopes that this might provide some guidance to the practitioner.

The beta distribution as given by Equation 13.21 has a mean of

$$\mu_p = \frac{\delta}{(\delta+\rho)} \tag{13.32}$$

After testing n items and obtaining y failures, the Bayesian estimator for the fraction failing is

$$\hat{p}_b = \frac{\delta+y}{\delta+\rho+n}$$

Then Equation 13.32 might be viewed as the subjective estimate of the fraction failing prior to testing. However, let us suppose that the initial subjective estimate is $p=0.10$. It can be seen from Equation 13.32 that many combinations of δ and ρ exist that would provide a mean of 0.10. The first two rows of Table 13.1 give various combinations of δ and ρ such that $\delta/(\delta+\rho)=0.10$.

Let us now consider a situation where 10 items are tested with no failures. This test result would seem to reinforce our initial subjective estimate of $p=0.10$. For this situation, $\hat{p}_1$ in Table 13.1 is the resulting Bayesian point estimate. It can be seen that as δ and ρ are made larger, the test results have less effect on the resulting estimate.

Now let us consider the same prior estimate of $p=0.10$, but when the 10 items are placed on test, 10 failures occur. Here, intuition would tell us that we are in serious trouble and that our initial estimate was completely in error. Let us see what the Bayesian estimator tells us. In Table 13.1, $\hat{p}_2$ gives the estimates for this situation. Here it can be seen that the large values of δ and ρ give a prior distribution that is so strong that the completely contradictory test results do not, to any significant extent, change the prior estimate. The strong prior can be attributed to the fact that the variance of a beta distribution is given by $(\delta\rho)/(\delta+\rho)^2(\delta+\rho+1)$ and as both δ and ρ increase, the variance decreases. A small variance in the prior implies a strong belief in the subjectively estimated fraction failing before testing.

Clearly, large values of δ and ρ have a strong influence on the prior distribution even if test results are running contrary to the prior estimate. This suggests that if one wants the test results to have an impact, then small values

Table 13.1 Illustration of prior parameter selection on the Bayesian estimate

δ	$\frac{1}{2}$	1	3	10	20
$\delta+\rho$	5	10	30	100	200
$\hat{p}_1$	0.03	0.05	0.075	0.091	0.095
$\hat{p}_2$	0.70	0.55	0.325	0.182	0.143

for δ and ρ are in order. It also suggests that blindly applying the Bayesian procedure, even when test results are running contrary to reason, can produce erroneous estimates.

13.3 EXPONENTIAL DISTRIBUTION TESTING SITUATION

In this situation the exponential distribution is used to represent time to failure. If the failure rate is λ, then for a total cumulative test time T the number of failures **r** is Poisson distributed. That is

$$P(r|\lambda)=\frac{(T\lambda)^r e^{-T\lambda}}{r!}, \qquad r=0,1,2,\ldots \tag{13.33}$$

The parameter of interest is λ and our prior belief in this parameter will be described by a gamma distribution, which is the natural conjugate for the Poisson. The gamma prior p.d.f. is

$$h(\lambda)=\frac{\rho^{\delta}\lambda^{\delta-1}e^{-\rho\lambda}}{\Gamma(\delta)}, \qquad \lambda\geqslant 0, \quad \delta\geqslant 0, \quad \rho\geqslant 0 \tag{13.34}$$

The posterior p.d.f. for $\boldsymbol{\lambda}$ given r failures over a time interval T will now be found using Bayes' theorem. That is, we must find

$$k(\lambda|r)=\frac{h(\lambda)P(r|\lambda)}{f_2(r)} \tag{13.35}$$

Recognize that the joint p.d.f. for **r** and $\boldsymbol{\lambda}$ is found from

$$f(r,\lambda)=h(\lambda)P(r|\lambda)$$

which in this instance becomes

$$f(r,\lambda)=\frac{\rho^{\delta}T^r}{\Gamma(\delta)\Gamma(r+1)}\lambda^{\delta+r-1}e^{-\lambda(\rho+T)}, \qquad \lambda\geqslant 0 \tag{13.36}$$

Then the marginal p.d.f. for **r** is

$$f_2(r)=\frac{\rho^{\delta}T^r}{\Gamma(\delta)\Gamma(r+1)}\int_0^{\infty}\lambda^{\delta+r-1}e^{-\lambda(\rho+T)}\,d\lambda \tag{13.37}$$

If we let $u=\lambda(\rho+T)$ then Equation 13.37 becomes

$$f_2(r)=\frac{\rho^{\delta}T^r}{\Gamma(\delta)\Gamma(r+1)(\rho+T)^{\delta+r}}\int_0^{\infty}u^{\delta+r-1}e^{-u}\,du \tag{13.38}$$

The integral is now easily recognizable as a gamma function. Thus

$$f_2(r)=\frac{\rho^{\delta}T^{r}\Gamma(\delta+r)}{\Gamma(\delta)\Gamma(r+1)(\delta+T)^{\delta+r}},\qquad r=0,1,2,\ldots \tag{13.39}$$

We can substitute Equations 13.36 and 13.39 into Equation 13.35 to obtain the posterior distribution for $\boldsymbol{\lambda}$. This gives

$$k(\lambda|r)=\frac{(\rho+T)^{\delta+r}}{\Gamma(\delta+r)}\lambda^{\delta+r-1}e^{-\lambda(\rho+T)},\qquad \lambda\geqslant 0 \tag{13.40}$$

This should be recognized as a gamma p.d.f. with parameters $(\rho+T)$ and $(\delta+r)$.

The Bayesian point estimator for $\boldsymbol{\lambda}$ is the mean of the gamma posterior p.d.f. as given by Equation 13.40 and is

$$\hat{\lambda}_b=\frac{\delta+r}{\rho+T} \tag{13.41}$$

13.3.1 CONFIDENCE LIMITS IN THE EXPONENTIAL CASE

In order to obtain the upper or lower confidence limit for λ, the posterior distribution as given by Equation 13.40 must be used to find either λ_u or λ_L such that

$$\int_{\lambda_u}^{\infty}k(\lambda|r)\,d\lambda=\alpha$$

or

$$\int_{0}^{\lambda_L}k(\lambda|r)\,d\lambda=\alpha$$

Then λ_u and λ_L, respectively, define a $100(1-\alpha)\%$ one-sided upper or lower Bayesian confidence limit for λ.

Since the posterior p.d.f. is gamma, a simple transformation defined by

$$z=2\lambda(\rho+T) \tag{13.42}$$

and applied to the p.d.f. in Equation 13.40 will produce a random variable $\mathbf{z}$ which is chi-square distributed with $2(\delta+r)$ degrees of freedom. Thus, tabled chi-square values can be used to obtain confidence limits. Considering the transformation as given in Equation 13.42 we can write

$$P\left[\chi^2_{1-\alpha,2(\delta+r)}\leqslant 2\boldsymbol{\lambda}(\rho+T)\right]=1-\alpha \tag{13.43}$$

and rearranging the inequality

$$\frac{\chi^2_{1-\alpha,2(\delta+r)}}{2(\rho+T)} \leqslant \lambda \tag{13.44}$$

This is a $100(1-\alpha)\%$ lower Bayesian confidence limit for the failure rate λ. Similarly, the upper limit is

$$\lambda \leqslant \frac{\chi^2_{\alpha,2(\delta+r)}}{2(\rho+T)}$$

13.3.2 THOUGHTS ON SELECTING THE PARAMETERS FOR THE GAMMA PRIOR DISTRIBUTION

In order to apply Bayesian analysis, all objective and subjective information must be integrated to produce a distribution that describes the belief as to the value of the failure rate for some new and yet untested system. Although data pertaining to past systems may be available for study, it still remains that one must project one's belief about a future system, and this is the crux of Bayesian analysis. For, however one arrives at the prior distribution, it is still subject to controversy, and there is virtually no way that the prior can be verified because subjective belief about future situations cannot be sampled.

The parameters for the gamma prior p.d.f. can be viewed in the same fashion as the parameters for the beta prior in the binomial testing situation. For example, consider a situation where the prior estimate for the failure rate is $\lambda = 0.0001$. Recall that the Bayesian estimator is

$$\hat{\lambda}_b = \frac{\delta + r}{\rho + T}$$

which allows using δ/ρ as the prior estimate of the failure rate.

In Table 13.2 various combinations of δ and ρ are taken such that $\delta/\rho = 0.0001$ to illustrate the effect of this selection on the resulting Bayesian estimate.

Let us first consider a test that results in $T = 20{,}000$ units of time and $r = 0$ failures. This outcome seems to reinforce the prior belief that $\lambda = 0.0001$. For this situation, $\hat{\lambda}_1$ is the resulting Bayesian estimate. As can be seen, for high values of δ and ρ the hard data have little effect on the prior estimate.

Table 13.2 Effect of prior parameter selection in the case of the gamma

δ	1	5	100	1000
ρ	10,000	50,000	10^6	10^7
$\hat{\lambda}_1$	0.033×10^{-3}	0.07×10^{-3}	0.098×10^{-3}	0.0998×10^{-3}
$\hat{\lambda}_2$	0.7×10^{-3}	0.358×10^{-3}	0.118×10^{-3}	0.102×10^{-3}

Now let us consider a second situation where the outcome from testing is $T=20{,}000$ time units with $r=20$ failures. Intuitively, the test result seems to be contrary to the prior estimate. The resulting Bayesian estimate is designated as $\hat{\lambda}_2$ in Table 13.2. Here again we see that high values for δ and ρ produce an estimator that is controlled by the prior belief. This is because the variance of the gamma prior is given by δ/ρ^2 and large values for ρ produce a small variance.

It is possible to establish the parameters for the gamma prior by subjectively estimating two performance levels for the system. These estimated performance levels might be expressed as two percentiles of the gamma distribution because two percentiles uniquely determine the parameters of the gamma. For example, one might estimate a median failure rate $\lambda_{0.50}$ such that there is believed to be a 50-50 chance that the system's failure rate will be more or less than this value. Then a second failure rate must be estimated; say, for example, the 95th percentile, $\lambda_{0.95}$, which is a value of the failure rate such that there is believed to be only five chances out of 100 that the system's failure rate will exceed this value. Then for the gamma prior as given by Equation 13.34 using successive integration by parts we can say that

$$H(\lambda)=1-\sum_{i=0}^{\delta-1}\frac{(\rho\lambda)^i\exp(-\rho\lambda)}{i!} \tag{13.45}$$

where δ is restricted to integer values and, of course, $H(\lambda)$ is the cumulative distribution function. Substituting in the estimated failure rates $\lambda_{0.50}$ and $\lambda_{0.95}$ would give two equations

$$0.95=H(\lambda_{0.95}) \tag{13.46}$$

and

$$0.50=H(\lambda_{0.50}) \tag{13.47}$$

These two equations can then be solved for δ and ρ, the required parameters. However, the solution process is cumbersome since one must resort to an iterative procedure.

EXAMPLE 13.9

Let us assume that the air conditioning system for an automobile is being redesigned. For the new system the design engineer estimates

$$\hat{\lambda}_{0.50}=2.1\times10^{-5}\text{ failures/km}$$

and

$$\hat{\lambda}_{0.95}=3.3\times10^{-5}\text{ failures/km}$$

From 36,000 kilometers of testing, two failures result. A point and interval estimate of the system's failure rate is desired.

To solve this problem the gamma prior must first be found. Using an iterative technique for solving Equations 13.46 and 13.47, we find that a shape parameter of 12 and scale parameter of 554,896 best fits the subjective estimates. Here it should be immediately recognized that these high values for the parameters of the gamma prior will essentially control the estimation process.

The Bayesian point estimate of the reliability is found by substituting into Equation 13.41 and is

$$\hat{\lambda}_b = \frac{12+2}{554{,}896+36{,}000} = 2.37 \times 10^{-5} \text{ failures/km}$$

The Bayesian 95% upper confidence limit is found from Equation 13.44. The necessary tabled chi-square value is

$$\chi^2_{0.05,2(14)} = 41.337$$

and the Bayesian confidence interval is

$$\lambda \leqslant \frac{41.337}{2(554{,}896+36{,}000)}$$

which gives

$$\lambda \leqslant 3.50 \times 10^{-5} \text{ failures/km}$$

In terms of kilometers between failures, this implies that

$$28{,}571 \text{ km} \leqslant \theta$$

Now, for example, the 95% lower confidence limit for the 12,000 km reliability would be

$$R(12{,}000) \geqslant \exp(-12{,}000/28{,}571)$$

or

$$R(12{,}000) \geqslant 0.657$$

It is left to the reader to compare these results to those that would be obtained using the traditional approach.

13.4 THE BAYESIAN APPROACH TO SEQUENTIAL TESTING

The Bayesian approach to sequential testing uses a posterior probability statement that is continually updated as new test data become available. To illustrate the Bayesian sequential approach, we use the test situation considered in Section 13.3. Recall that in Section 13.3 the time between failures was exponential, and the prior distribution for the failure rate was gamma.

Here we will consider the null hypothesis

$$H_0 : \lambda \leqslant \lambda_0$$

with the alternate hypothesis

$$H_1 : \lambda > \lambda_0$$

The test criterion will be based on the posterior probability $P(\lambda > \lambda_0)$. Intuitively, if this probability becomes large then we would suspect that H_1 was true. This reasoning leads to the following decision rules:

1. Accept H_0 if $P(\lambda > \lambda_0) \leqslant \beta$
2. Reject H_0 if $P(\lambda > \lambda_0) \geqslant 1 - \alpha$
3. Continue if $\beta < P(\lambda > \lambda_0) < 1 - \alpha$

Now for this test situation, the posterior p.d.f. $k(\lambda|r)$ is given by Equation 13.40 and the required probability is

$$P(\lambda > \lambda_0) = \int_{\lambda_0}^{\infty} k(\lambda|r)\, d\lambda$$

Invoking the transformation given in Equation 13.42 allows use of the chi-square distribution, and the test criterion becomes

1. Accept H_0 if $T \geqslant \dfrac{\chi^2_{\beta, 2(\delta + r)}}{2\lambda_0} - \rho$

2. Reject H_0 if $T \leqslant \dfrac{\chi^2_{1-\alpha, 2(\delta + r)}}{2\lambda_0} - \rho$

3. Continue testing as long as the following inequality holds

$$\frac{\chi^2_{1-\alpha, 2(\delta + r)}}{2\lambda_0} - \rho < T < \frac{\chi^2_{\beta, 2(\delta + r)}}{2\lambda_0} - \rho \qquad (13.48)$$

EXAMPLE 13.10

Consider a situation where a design goal of 12,000 km has been set for an automotive subsystem. The system designer's prior estimate of the system's adequacy to meet the goal is expressed by a gamma distribution with $\delta = \frac{1}{2}$ and $\rho = \lambda_0$ where $\lambda_0 = 1/12{,}000$ km. Then the gamma prior is

$$h(\lambda) = \frac{(\lambda_0/\lambda)^{0.5}}{\sqrt{\pi}} e^{-(\lambda)\lambda_0}, \qquad \lambda \geqslant 0$$

We will design a sequential test for this situation using $\alpha = 0.05$ and $\beta = 0.10$. In order to develop the test boundaries, we must calculate

$$\frac{\chi^2_{1-\alpha, 2(\delta + r)}}{2\lambda_0} - \rho$$

Table 13.3 Calculations for a Bayesian sequential test

r	$2(\delta + r)$	$\chi^2_{0.95, 2(\delta+r)}$	REJECT BOUNDARY	$\chi^2_{0.10, 2(\delta+r)}$	ACCEPT BOUNDARY
0	1	0.004	24	2.706	16,236
1	3	0.352	2,112	6.251	37,506
2	5	1.145	6,870	9.236	55,416
3	7	2.167	13,002	12.017	72,102
4	9	3.325	19,950	14.684	88,104
5	11	4.575	27,450	17.275	103,650
6	13	5.892	35,352	19.812	118,872
7	15	7.261	43,566	22.307	133,842
8	17	8.672	52,032	24.769	—
9	19	10.117	60,702	27.204	—
10	21	11.591	69,546	29.625	—

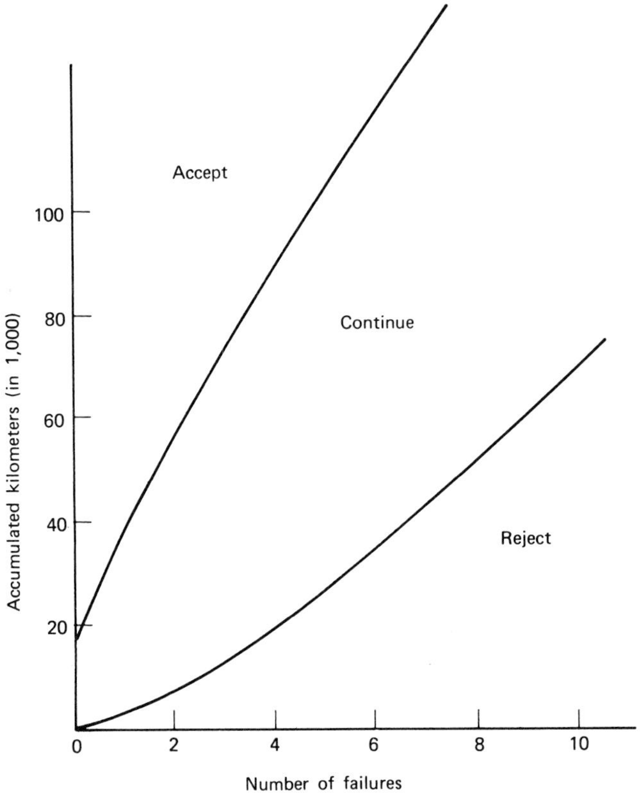

Figure 13.5 A Bayesian sequential test graph.

and

$$\frac{\chi^2_{\beta,2(\delta+r)}}{2\lambda_0}-\rho$$

for various values of r. The results of these calculations are given in Table 13.3 where, since the quantity $\rho=1/12{,}000$ is small, it was neglected.

The calculated boundary values can then be used to make a sequential graph on which test results can be plotted as they become available. Such a graph is shown in Figure 13.5.

13.4.1 TRUNCATING THE BAYESIAN SEQUENTIAL TEST

The Bayesian sequential test can have a rather large continue region allowing the test to require considerable time to terminate. However, the test can be truncated in a rather natural manner that is consistent with non-Bayesian sequential testing.

Truncation of the Bayesian sequential test requires selecting two values, λ_1 and λ_0 such that $\lambda_1<\lambda_0$. These values are analogous to those considered in non-Bayesian sequential testing. Here if $P(\boldsymbol{\lambda}>\lambda_0)<\beta$, a small quantity, then most likely $\boldsymbol{\lambda}\leqslant\lambda_0$, an acceptable value, and we wish to accept the product. On the other hand, if $P(\boldsymbol{\lambda}<\lambda_1)<\alpha$, a small quantity, then most likely $\boldsymbol{\lambda}\geqslant\lambda_1$, and we wish to reject the product.

The region $\lambda_1<\boldsymbol{\lambda}<\lambda_0$ is a compromise region that must be agreed upon by consumer and producer. For example, suppose a tracked vehicle is to have a MTBF of 700 km, or equivalently a failure rate of $\lambda=1/700$. Now demonstrating the concept in terms of the MTBF ($\theta=1/\lambda$), the producer and consumer must agree on two acceptable values, θ_1 and θ_0, in order to provide for a shorter truncated test. The consumer might agree to accept the product if it is shown that $P(\boldsymbol{\theta}<610)<0.10$. That is, there is less than a 10% chance that $\theta<610$ km. Also, the producer might agree to a compromise such that his product would be rejected if it is shown that $P(\boldsymbol{\theta}>720)<0.05$, which means that probably $\theta\leqslant 720$ km. Here we now have $\lambda_1=1/720$ and $\lambda_0=1/610$ with $\alpha=0.05$ and $\beta=0.10$.

The continue region for the test is then

$$P(\boldsymbol{\lambda}>\lambda_0)>\beta \qquad \text{and} \qquad P(\boldsymbol{\lambda}<\lambda_1)>\alpha \tag{13.49}$$

When Equation 13.49 is violated, the test is terminated with an accept decision if $P(\boldsymbol{\lambda}>\lambda_0)<\beta$ or with a reject decision if $P(\boldsymbol{\lambda}<\lambda_1)<\alpha$.

Now consider $P(\boldsymbol{\lambda}>\lambda_0)=\beta$ which will give one boundary of the continue region. Here $P(\boldsymbol{\lambda}>\lambda_0)$ is the posterior probability that would be obtained from Equation 13.40. However, using the chi-square relationship as previously defined in Equation 13.42 would give

$$2\lambda_0(\rho+T)=\chi^2_{\beta,2(\delta+r)} \tag{13.50}$$

and using analogous reasoning for $P(\lambda<\lambda_1)=\alpha$ we have

$$2\lambda_1(\rho+T)=\chi^2_{1-\alpha,2(\delta+r)} \tag{13.51}$$

Then the continue region is

$$\frac{\chi^2_{1-\alpha,2(\delta+r)}}{2\lambda_1}<T+\rho<\frac{\chi^2_{\beta,2(\delta+r)}}{2\lambda_0} \tag{13.52}$$

Now the lower limit of this region can be rearranged slightly, giving

$$\left(\frac{\lambda_0}{\lambda_1}\right)\frac{\chi^2_{1-\alpha,2(\delta+r)}}{2\lambda_0}<T+\rho<\frac{\chi^2_{\beta,2(\delta+r)}}{2\lambda_0} \tag{13.53}$$

The relationship between these limits and the previous limits (Equation 13.48) is now obvious. That is, the lower limit of the previous sequential test is multiplied by the constant λ_0/λ_1 which is greater than unity, and this will cause the boundaries to meet.

The maximum number of failures to test termination can be found by equating the point at which the boundaries meet. This point is

$$\chi^2_{\beta,2(\delta+r^*)}=\left(\frac{\lambda_0}{\lambda_1}\right)\chi^2_{1-\alpha,2(\delta+r^*)} \tag{13.54}$$

where r^* is the maximum number of failures. Now recall in Example 10.16 where a normal approximation to a chi-square variable was given as

$$\chi^2_{p,\nu}\approx\left(\frac{1}{2}\right)\left(z_p+\sqrt{2\nu-1}\,\right)^2 \tag{13.55}$$

where z_p is the $(1-p)$th percentile of the standard normal distribution. Applying this approximation to Equation 13.54 and solving for r^* gives

$$r^*=\left(\frac{1}{4}\right)\left[\frac{z_\beta+(\lambda_0/\lambda_1)^{1/2}z_\alpha}{(\lambda_0/\lambda_1)^{1/2}-1}\right]^2+1-\delta \tag{13.56}$$

EXAMPLE 13.11

To demonstrate a truncated Bayesian sequential test, we use $\lambda_0=0.005$, $\lambda_1=0.002$, $\alpha=0.05$, and $\beta=0.10$. For the gamma prior we will assume that $\delta=2$ and $\rho=350$.

The calculations for the truncated sequential test are given in Table 13.4. Using these calculations the sequential graph of Figure 13.6 was developed. Here it can be seen that the test must terminate by nine failures or at a maximum of about 3,080 accumulated kilometers.

Table 13.4 Calculations for a truncated Bayesian sequential test

r	$2(\delta+r)$	$\chi^2_{0.10,2(\delta+r)}$	$\dfrac{\chi^2_{0.10,2(\delta+r)}}{2\lambda_0}$	$\chi^2_{0.95,2(\delta+r)}$	$\dfrac{\chi^2_{0.95,2(\delta+r)}}{2\lambda_1}$
0	4	7.779	778	0.711	178
1	6	10.645	1,064	1.635	409
2	8	13.362	1,336	2.733	683
3	10	15.987	1,599	3.940	985
4	12	18.549	1,855	5.226	1,306
5	14	21.064	2,106	6.571	1,643
6	16	23.542	2,354	7.962	1,990
7	18	25.989	2,599	9.390	2,348
8	20	28.412	2,841	10.851	2,713
9	22	30.813	3,081	12.338	3,084
10	24	33.196	3,320	13.848	3,462

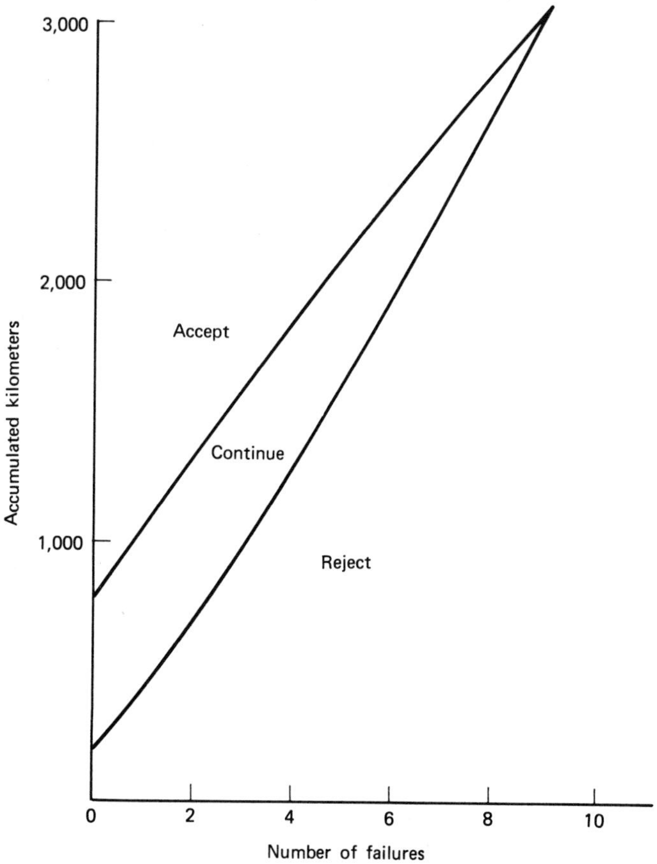

Figure 13.6 A truncated Bayesian sequential test graph.

13.5 APPLICATIONS OF BAYES' THEOREM IN DESIGN RELIABILITY

Designing for reliability requires a quantitative assessment of all engineering uncertainty. In some design situations, past data are available. The major contribution of Bayesian statistical decision theory is that it provides a procedure that includes professional information and information from past data in decision making for the present design. This idea is illustrated by the following simple examples, which are based on a paper by Cornell [3].

EXAMPLE 13.12

Suppose a design engineer has developed a new mechanical system that has never been built or tested before. The engineer believes, based on his previous experience and intuition, that if the system has been designed properly to meet the performance criteria, the time to failure is normally distributed with a mean μ of 50,000 kilometers. If the system is improperly designed, the mean life may be 30,000 kilometers. Based on his experience, the engineer has good confidence in his design. A priori, he says that the probability that the design has a mean life of 50,000 kilometers is 0.80, and hence 0.20 is the probability that the design has a mean life of 30,000 kilometers.

A single prototype is built and tested in a simulated environment that duplicates as nearly as possible the actual environment. The system is tested, but economic considerations dictate that the testing be stopped at 40,000 kilometers. The engineer also says that, based on past experience, it is known that the standard deviation for the life of the system is 10% of the mean life.

The objective is to predict the reliability of the system and find the confidence for the prediction at this design stage.

Let A be the event that the system has been tested and operated successfully for 40,000 kilometers. Let B_1 be the hypothesis that the mean life is 50,000 kilometers and B_2 be the hypothesis that the mean life is 30,000 kilometers.

From the design engineer's a priori estimates, we have

$$P(\mu = 50{,}000) = 0.8$$

$$P(\mu = 30{,}000) = 0.2$$

Then, to compute $P(A|B_1)$, if $\mathbf{t}$ is the life of the system, we have

$$P(A|B_1) = P(\mathbf{t} > 40{,}000 | B_1) \tag{13.57}$$

where $\mathbf{t}$ is a normally distributed random variable with a mean of 50,000 kilometers and a standard deviation of 5,000 kilometers. Thus, the value of the standard normal variable is

$$z = \frac{40{,}000 - 50{,}000}{5{,}000} = -2.0$$

and hence $P(A|B_1) = 0.9772$ from the standard normal tables. Also, to compute $P(A|B_2)$, we have

$$z = \frac{40{,}000 - 30{,}000}{3{,}000} = +3.33$$

and hence $P(A|B_2)=0.00045$ from the standard normal tables. Thus, using Bayes' theorem

$$P(B_1|A)=\frac{P(B_1)P(A|B_1)}{P(B_1)P(A|B_1)+P(B_2)P(A|B_2)}$$

$$=\frac{0.8\times 0.9772}{0.8\times 0.9972+0.2\times 0.00045}=0.9999 \qquad (13.58)$$

and

$$P(B_2|A)=\frac{P(B_2)P(A|B_2)}{P(B_1)P(A|B_1)+P(B_2)P(A|B_2)}=1-P(B_1|A) \qquad (13.59)$$

Hence,

$$P(B_2|A)=0.0001$$

Suppose, later on, after production, a randomly selected system is desired for an application where it is to perform for 35,000 kilometers. Then, the reliability of the system will be 0.9987 as computed by

$$z=\frac{35{,}000-50{,}000}{5{,}000}=-3.0$$

with a confidence of 0.9999, since this is $P(B_1|A)$ as obtained from above. However, we see that before the testing, the reliability for the system to perform for 35,000 kilometers would have been 0.9987 with a confidence of 0.8, as obtained from the engineer's original beliefs. After the testing, using Bayes' theorem, the confidence for the predicted reliability has been increased to 0.9999.

EXAMPLE 13.13

We are interested in the design reliability verification of a new, expensive system that has never been designed, built, or tested before. If the design team did a successful job meeting the performance criteria, the time to failure of the system has an exponential distribution with a mean of 100,000 kilometers (i.e., $1/\lambda_1=100{,}000$ km). If they neglected or overlooked some important aspect of the design, the mean time to failure is significantly less and is 10,000 kilometers (i.e., $1/\lambda_2=10{,}000$). The reliability engineer is not in a state of total ignorance as to the true mean. He knows the character of this design with respect to the state of the design art, and he knows the past record of this particular design team. He expresses this information by stating that his prior probability that the design is satisfactory is 0.9. Hence, the engineer hypothesizes that

$$P[\lambda=\lambda_1]=0.9$$

and

$$P[\lambda=\lambda_2]=0.1$$

A single specimen is tested in a simulated environment and the test continues for 30,000 kilometers, at which time it is terminated for economic reasons. The engineer can combine these two sources of information into a single statement by using Bayes' theorem.

Let A be the event of observing a lifetime in excess of 30,000 kilometers and the likelihood of this is $e^{-30{,}000\lambda}$. Hence, the posterior probabilities of the two hypotheses by using Bayes' theorem are

$$P[\lambda=\lambda_1|A]=\frac{P(\lambda=\lambda_1)P(A|\lambda=\lambda_1)}{\sum_{i=1}^{2} P(\lambda=\lambda_i)P(A|\lambda=\lambda_i)}$$

$$=\frac{0.9\exp\left[\dfrac{-30{,}000}{100{,}000}\right]}{0.9\exp\left[\dfrac{-30{,}000}{100{,}000}\right]+0.1\exp\left[\dfrac{-30{,}000}{10{,}000}\right]}$$

$$=\frac{0.6667}{0.6667+0.0049}=0.9927 \tag{13.60}$$

and similarly

$$P[\lambda=\lambda_2|A]=\frac{0.0049}{0.6667+0.0049}=0.0073$$

Thus, even the inconclusive test result has significantly altered the engineer's degree of belief in the second hypothesis.

If the system is asked to perform for 10,000 kilometers, its reliability R is either $e^{-10{,}000/100{,}000}=0.905$, or $e^{-10{,}000/10{,}000}=0.368$, depending on the value of λ. Hence, the engineer can state, based on his judgment and limited data, that the system reliability is 0.905 with a probability of 0.9927.

A more practical question, the question whose answer would decide whether the design is satisfactory, is "What is the probability that the system will perform satisfactorily for 10,000 km?" The engineer's answer must include all possible reliabilities with their associated confidence probabilities. Using this concept, we can compute the Bayesian reliability $\tilde{R}$ as follows:

$$\tilde{R}=P[\text{successful performance}]$$

$$=\sum_i [R|\lambda=\lambda_i]P[\lambda=\lambda_i]$$

$$=0.905\times 0.9927+0.368\times 0.0073=0.9011$$

Before testing, the number was $(0.905\times 0.9)+(0.368\times 0.1)=0.851$. Hence, the numbers 0.851 and 0.9011 might be called the prior and posterior reliabilities. We should recognize that these prior and posterior reliabilities state the system reliability in a way that reflects not only the inherent uncertainty but also the statistical uncertainty associated with an uncertain parameter value. So, both the professional and observed data influence the degree of uncertainty in system reliability.

13.6 DESIGN FOR RELIABILITY OF A COMPONENT SUBJECTED TO RANDOM STRESSES

Consider a component that is subjected to repeated stresses. The interest is in the reliability of the component subjected to this stress environment. Prior knowledge will be assumed for the magnitude and intensity of the stresses.

Let the intensity of the stresses from cycle to cycle be random-independent (see Chapter 8) with an exponential distribution with parameter λ. Let the strength be fixed and known to be y_0. Further, let the number of stress occurrences be expressed by a random variable that is Poisson distributed. Hence, we have the following information:

1. Stress **x**: $f(x)=\lambda e^{-\lambda x}, \quad \lambda>0, x\geqslant 0$ (13.61)
2. Strength **y**: $y=y_0$, a constant, $\quad y_0>0$ (13.62)
3. Stress occurrences $\mathbf{N_T}$: $P[\mathbf{N_T}=n]=\dfrac{(\alpha T)^n e^{-\alpha T}}{n!}, \quad n=0,1,2,3,\ldots$ (13.63)

where

$$\mathbf{N_T}=\text{number of stress occurrences in time } T$$

$$\alpha=\text{average number of stress occurrences per unit time}$$

$$1/\lambda=\text{average stress intensity}$$

Suppose we model the above problem physically and conduct an experiment to obtain test data. From the test we find that in a time period of 6 units ($T=6$), 60 stress occurrences, $x_1, x_2, \ldots, x_{60}$ ($N_6=60$) were observed. Let the total of these 60 stress occurrences be 78 MPa. The value of y_0 is given as 9.1 MPa.

The non-Bayesian approach for this problem is to use the classical statistical methods and obtain point estimates of the reliability of the component at time T based on the observed data. From Equation 8.51, we have

$$R(T)=e^{-\alpha T(1-R)} \tag{13.64}$$

where

$$R=\int_0^{y_0} f(x)\,dx=1-e^{-\lambda y_0} \tag{13.65}$$

Thus

$$R(T)=\exp(-\alpha T e^{-\lambda y_0}), \qquad T\geqslant 0 \tag{13.66}$$

which means that the reliability has the form of an extreme value distribution.

The point estimates from the test data are

$$\hat{\alpha}=60/6=10 \text{ occurrences/unit time} \tag{13.67}$$

and

$$\hat{\lambda}=\frac{1}{\sum_{i=1}^{60}\frac{x_i}{60}}=\frac{60}{78}=0.77 \tag{13.68}$$

Substituting the values of $\hat{\alpha}$ and $\hat{\lambda}$ into Equation 13.66, we have

$$R(T)=\exp-(10Te^{-0.77\times 9.1})=e^{-0.009127T} \tag{13.69}$$

Thus, for example

$$R(1)=0.990922 \qquad \text{and} \qquad R(10)=0.912835$$

Now suppose we have some prior knowledge (before the experiment was conducted) that we wish to include to develop a "better" estimate of $R(T)$ than the one given by Equation 13.69. This can be done by using the Bayesian approach. Based on our past experience we feel that the average rate of stress occurrences is 8 per unit time with the tolerance being 30% of the average (here tolerance is taken as three standard deviations) and the average value of $\boldsymbol{\lambda}$ is $1/1.5$ with a tolerance of 40% of the average value. Thus

$$\mu_\alpha=8, \qquad \sigma_\alpha=\frac{0.30\times 8}{3}=0.8$$

$$\mu_\lambda=0.67, \qquad \sigma_\lambda=\frac{0.40\times 0.67}{3}=0.088$$

Using the Bayesian approach, the parameters α and λ are not known and are assumed to be random variables. The mean and standard deviation of these random variables have been estimated above based on past experience. Also, our prior belief in these parameters is expressed by stating that $\boldsymbol{\alpha}$ and $\boldsymbol{\lambda}$ follow a gamma distribution. If a random variable $\mathbf{x}$ follows the gamma distribution, then the density function $h(x)$ of $\mathbf{x}$ is

$$h(x)=\frac{\rho^x x^{\delta-1}e^{-\rho x}}{\Gamma(\delta)}, \qquad \rho>0,\ \delta>0,\ x\geqslant 0 \tag{13.70}$$

and

$$\mu_{\mathbf{x}}=\frac{\delta}{\rho} \qquad \text{and} \qquad \sigma_{\mathbf{x}}^2=\frac{\delta}{\rho^2}$$

Thus, for $\boldsymbol{\alpha}$ we have

$$\mu_\alpha = 8 = \frac{\delta_\alpha}{\rho_\alpha} \qquad \text{and} \qquad \sigma_\alpha = 0.8 = \sqrt{\delta_\alpha/(\rho_\alpha)^2}$$

which yields

$$\delta_\alpha = 100 \qquad \text{and} \qquad \rho_\alpha = 12.5$$

Similarly for $\boldsymbol{\lambda}$, we have

$$\mu_\lambda = 0.67 = \frac{\delta_\lambda}{\rho_\lambda} \qquad \text{and} \qquad \alpha_\lambda = 0.088 = \sqrt{\delta_\lambda/(\rho_\lambda)^2}$$

which yields

$$\delta_\lambda = 56.25 \qquad \text{and} \qquad \rho_\lambda = 84.375$$

It was proven in Section 13.3 that the posterior distribution of $\boldsymbol{\lambda}$ is also gamma with modified parameters. Indicating Bayesian quantities from the posterior distribution by a tilda ($\sim$) over the symbol, we have

$$\tilde{\mu}_\alpha = \frac{\delta_\alpha + n}{\rho_\alpha + T} = \frac{100+60}{12.5+6} = 8.65$$

and

$$\tilde{\sigma}_\alpha^2 = \frac{\delta_\alpha + n}{(\rho_\alpha + T)^2} = \frac{100+60}{(12.5+6)^2} = 0.4675$$

or

$$\tilde{\sigma}_\alpha = 0.684$$

We observed a sequence of n stresses denoted by $\mathbf{x}_1, \mathbf{x}_2, \ldots, \mathbf{x}_n$. Then the sample average is

$$\bar{x} = \frac{1}{n} \sum_{i=1}^{n} x_i$$

and the prior p.d.f. for $\boldsymbol{\lambda}$ is

$$h(\lambda) = \frac{\rho^\delta \lambda^{\delta-1} e^{-\rho\lambda}}{\Gamma(\delta)}, \qquad \lambda \geqslant 0$$

So the density function of the random sample $\mathbf{x}_1,\ldots,\mathbf{x}_n$ given λ is

$$k(x_1,\ldots,x_n|\lambda)=\lambda e^{-\lambda x_1}\lambda e^{-\lambda x_2}\cdots\lambda e^{-\lambda x_n}=\lambda^n e^{-n\lambda\bar{x}} \tag{13.71}$$

The joint density function for the sample and $\boldsymbol{\lambda}$ is

$$\begin{aligned} f(x_1,\ldots,x_n,\lambda)&=h(\lambda)k(x_1,\ldots,x_n|\lambda)\\ &=\frac{\rho^{\delta}}{\Gamma(\delta)}\lambda^{\delta-1}e^{-\rho\lambda}\lambda^n e^{-n\lambda\bar{x}}\\ &=\frac{\rho^{\delta}}{\Gamma(\delta)}\lambda^{n+\delta-1}e^{-\lambda(\rho+n\bar{x})} \end{aligned} \tag{13.72}$$

The marginal density function for the sample is

$$\begin{aligned} f(x_1,\ldots,x_n)&=\frac{\rho^{\delta}}{\Gamma(\delta)}\int_0^{\infty}\lambda^{n+\delta-1}e^{-\lambda(\rho+n\bar{x})}\,d\lambda\\ &=\frac{\rho^{\delta}}{\Gamma(\delta)}\cdot\frac{\Gamma(n+\delta)}{(\rho+n\bar{x})^{n+\delta}} \end{aligned} \tag{13.73}$$

Thus, the posterior density function of $\boldsymbol{\lambda}$ given the sample is

$$\begin{aligned} \tilde{h}(\lambda|x_1,\ldots,x_n)&=\frac{f(x_1,\ldots,x_n,\lambda)}{f(x_1,\ldots,x_n)}\\ &=\frac{(\rho+n\bar{x})^{n+\delta}\lambda^{n+\delta-1}e^{-\lambda(\rho+n\bar{x})}}{\Gamma(n+\delta)} \end{aligned} \tag{13.74}$$

which is again the gamma density function with parameters $(n+\delta)$ and $(\rho+n\bar{x})$. Thus

$$\tilde{\mu}_{\lambda}=\frac{\delta_{\lambda}+n}{\rho_{\lambda}+\sum_{i=1}^{n}x_i}=\frac{56.25+60}{84.375+78}=0.716$$

and

$$\tilde{\sigma}_{\lambda}^2=\frac{\delta_{\lambda}+n}{\left(\rho_{\lambda}+\sum_{i=1}^{n}x_i\right)^2}=\frac{116.25}{(162.375)^2}=0.00441$$

Using the mean values from the posterior distributions as the Bayesian point estimators for $\boldsymbol{\alpha}$ and $\boldsymbol{\lambda}$, we have $\tilde{\alpha}=8.65$ and $\tilde{\lambda}=0.716$. The Bayesian point estimate of $R(T)$ using Equation 13.66 is

$$\tilde{R}(T)=\exp(-8.65\,Te^{-0.716(9.1)})=e^{-0.0128T} \tag{13.75}$$

Thus, based on the prior information, we now have

$$\tilde{R}(1)=0.9872 \qquad \text{and} \qquad \tilde{R}(10)=0.8798$$

A $100(1-\alpha)\%$ two-sided confidence interval can be established for both $\boldsymbol{\alpha}$ and $\boldsymbol{\lambda}$. The procedure is identical to the development in Section 13.3.1. Using the transformation of the gamma random variable to take advantage of χ^2 tables, we have

$$P\left[\chi^2_{1-\alpha/2,2(\delta_\lambda+n)} \leqslant 2\boldsymbol{\lambda}(\rho_\lambda+n\bar{x}) \leqslant \chi^2_{\alpha/2,2(\delta_\lambda+n)}\right]=1-\alpha$$

and rearranging the inequality

$$\frac{\chi^2_{1-\alpha/2,2(\delta_\lambda+n)}}{2(\rho_\lambda+n\bar{x})} \leqslant \boldsymbol{\lambda} \leqslant \frac{\chi^2_{\alpha/2,2(\delta_\lambda+n)}}{2(\rho_\lambda+n\bar{x})}$$

This is a $100(1-\alpha)\%$ two-sided Bayesian confidence interval for λ. Thus,

$$\frac{\chi^2_{1-\alpha/2,2(\delta_\lambda+n)}}{2\left(\rho_\lambda+\sum_{i=1}^{n} x_i\right)} \leqslant \boldsymbol{\lambda} \leqslant \frac{\chi^2_{\alpha/2,2(\delta_\lambda+n)}}{2\left(\rho_\lambda+\sum_{i=1}^{n} x_i\right)} \tag{13.76}$$

Or, for a 95% confidence interval

$$\frac{\chi^2_{0.975,232.5}}{324.75} \leqslant \boldsymbol{\lambda} \leqslant \frac{\chi^2_{0.025,232.5}}{324.75}$$

Note that in this example the transformation used is not completely valid since the original gamma distribution is not necessarily restricted to having integer values for δ. Thus, in this example, the chi-square values of Equation 13.76 do not have integer degrees of freedom. As long as the values for the degrees of freedom are large, rounding to an integer will produce negligible error.

Recall from Example 10.16 that the normal approximation for a chi-square variable is

$$\chi^2_{\alpha,\nu} \approx \frac{\left[z_\alpha + \sqrt{2\nu - 1}\,\right]^2}{2}$$

where z is the standard normal deviate. Applying this to obtain the limits for λ, we have

$$0.59 \leqslant \lambda \leqslant 0.85$$

Or the average stress intensity has limits of

$$1.18 \leqslant \frac{1}{\lambda} \leqslant 1.69$$

The $100(1-\beta)\%$ confidence interval for α, the average number of stress occurrences, can be obtained in a similar fashion by

$$\frac{\chi^2_{1-\beta/2,2(\delta_\alpha+n)}}{2(\rho_\alpha+T)} \leqslant \alpha \leqslant \frac{\chi^2_{\beta/2,2(\delta_\alpha+n)}}{2(\rho_\alpha+T)}$$

Upon substituting the values one will find that

$$\frac{\chi^2_{0.975,320}}{37} \leqslant \alpha \leqslant \frac{\chi^2_{0.025,320}}{37}$$

or

$$7.35 \leqslant \alpha \leqslant 10.03$$

A confidence interval for $R(T)$ can be obtained by combining the intervals for α and λ. Let us denote the $100(1-\alpha')\%$ confidence limits for λ by $\lambda_1 \leqslant \lambda \leqslant \lambda_2$ and the $100(1-\beta)\%$ confidence limits for α by $\alpha_1 \leqslant \alpha \leqslant \alpha_2$. Then the $100(\alpha'+\beta - \alpha'\beta)\%$ confidence interval for $R(T)$ is given by

$$\exp\left[-\alpha_2 T e^{-\lambda_1 y_0}\right] \leqslant R(T) \leqslant \exp\left[-\alpha_1 T e^{-\lambda_2 y_0}\right]$$

For example, using the previously computed limits for λ and α, the confidence limits for $R(10)$ are

$$\exp\left[-10.03(10)e^{-0.59(9.1)}\right] \leqslant R(10) \leqslant \exp\left[-7.35(10)e^{-0.85(9.1)}\right]$$

or

$$0.627 \leqslant R(10) \leqslant 0.968$$

with the confidence coefficient being $1-0.05-0.05+0.05^2=0.9025$.

13.7 SUMMARY

In this chapter the Bayesian approach utilizing prior information has been applied to the binomial and exponential distributions. Procedures for point estimates, confidence intervals, and sequential testing have been developed. Several different decision situations concerning reliability at the design stage were also considered. In all of these analyses, it was assumed that subjective information could be quantified somehow and represented in the form of a prior distribution. This process of quantifying subjective belief is the crux of Bayesian decision making. Today, there exists no feasible way to consistently go about extracting this information from the minds of individuals.

The concept of a strong prior distribution also presents a problem that should be carefully considered when applying the Bayesian estimation procedure. If a strong prior is used, then the estimation procedure should be employed with caution. However, somewhat of a dilemma exists in that it may actually take a strong prior to best describe the prior belief. In this case, there is some question as to why test at all, since the test results will have little impact on the estimation. Possibly, when a strong prior is used and if test results seem inconsistent with the prior belief, then it would seem appropriate to forget about estimation in the Bayesian sense and to proceed to correct the problem that is causing the failures. This happens to be precisely the procedure followed by many design engineers.

What this underlines is that although the Bayesian estimation procedure is intuitively attractive, it does have difficulties and it should be applied with caution.

EXERCISES

1 Two identical urns each contain a mixture of 10,000 red and blue poker chips. One urn, which we mentally designate as Urn I, contains 4,000 red chips while the second urn (Urn II) contains 6,000 red chips. One urn is selected at random and 10 chips are randomly sampled from the urn. (a) What is the probability that the sample will contain four red chips? (b) Given the information that the sample does contain four red chips, what is the probability that Urn I was selected?

2 A chest has three drawers. The first drawer contains two pennies, the second contains a dime and a penny, and the third contains two dimes. A drawer is chosen at random and a coin is randomly drawn from it. What is the probability that the coin remaining in the drawer is a penny given that the coin drawn is a dime?

3 Reconsider Example 13.3 and suppose that we simultaneously place 10 systems on test resulting in two failures. Now find the probability that reliability level R_1 has been attained.

4 Suppose that we are interested in assessing the reliability of a pyrotechnic device. The device is destructively tested, resulting in either a pass or failure in the test. Ten devices are placed on test resulting in two failures. Use the Bayesian approach in answering the following: (a) Assume that the device either possesses reliability level $R_1=0.99$ or $R_2=0.80$ with prior probabilities of 0.90 and 0.10, respectively. Find the posterior probability for each reliability level. What would you estimate the reliability level to be based on the given information? (b) Assume that the prior distribution for the reliability level (**r**) is uniform given by $f(r)=1,\ 0 \leqslant r \leqslant 1$. Find the posterior distribution and estimate the reliability. (c) Assume that the prior distribution for the reliability is beta with $\delta=4.0$ and $\rho=4.1$, and estimate the reliability. (d) For part (c), find the 90% lower confidence limit on the reliability.

5 The p.d.f. for the life of a device is given by

$$f(t)=\frac{1}{4}e^{-t}+\frac{3}{2}e^{-2t}, \qquad t \geqslant 0$$

(a) Find the probability that a device will survive at least another hour given that it has been in operation for exactly 1 hour and has not failed. (b) Find the probability that a device will last at least 2 hours. (c) Explain the reason for the difference in the two answers.

6 Consider a binomial testing situation where the prior distribution is uniform given by

$$h(p)=1, \qquad 0 \leqslant p \leqslant 1$$

where **p** is the fraction failing. (a) Find the posterior p.d.f. (b) Find the point estimator for **p**. (c) Find the upper confidence limit for p.

7 Seven automobiles are each run over a 36,000 kilometer test schedule. The testing produced a total of 19 failures. Assuming an exponential failure distribution and a gamma prior with parameters $\rho=30{,}000$ and $\delta=3$, answer the following: (a) What is the Bayesian point estimate for the MTBF? (b) What is the 90% lower confidence limit on the 10,000 kilometer reliability?

8 A tracked vehicle is to be tested. A Bayesian sequential test will be used with $\theta_0=700$ km, $\theta_1=900$ km, $\alpha=0.05$, and $\beta=0.10$. The gamma prior for the failure rate has parameters $\delta=2$ and $\rho=10$. (a) Develop a truncated sequential test and draw the sequential graph. (b) Estimate the maximum number of failures to test termination.

9 Reconsider the test situation as given in Section 13.3 where testing is binomial with a beta prior distribution. Develop a sequential test for this situation.

10 A component must have an expected life of 50,000 kilometers. The time to failure of the component is exponentially distributed. This component has never been built before and hence no field data are available to estimate the expected or mean life of the component. The engineer gives the following confidence levels for four values of the mean life:

30,000	0.1
40,000	0.2
50,000	0.5
60,000	0.2
	1.0

A single prototype is built and tested. The test is terminated at 45,000 kilometers for economic reasons. Develop the Bayesian reliability statements for the above component.

11 Consider the Bayesian design reliability problem given in Section 13.6. Suppose the expected value of α is 0.40 and the standard deviation of $\alpha = 0.06$. In the last 1,000 time units, 55 occurrences are observed when the stresses exceed a specified limit. (a) Find the posterior density function of the parameter α. (b) Now suppose that the parameter λ is also gamma distributed with a mean of 5.0 and standard deviation of 0.85. A sample of size 50 is taken for stress magnitudes and the sample average is found to be 7.5. Find the posterior density function of the parameter λ. Prove all the relationships which you use for finding your answer. (c) Develop the 95% two-sided Bayesian confidence intervals for the parameters α and λ.

BIBLIOGRAPHY

1 Barnett, V. D., "A Bayesian Sequential Life Test," *Technometrics*, Vol. 15, No. 2, May 1972.

2 Bayes, T., "An Essay Towards Solving a Problem in the Doctrine of Chances," *Biometrika*, Volume 45, 1958.

3 Cornell, C. A., "Bayesian Statistical Decision Theory and Reliability-Based Design," International Conference on Structural Safety and Reliability of Engineering Structures, Washington, D.C., April 1969.

4 Deely, J. J. and W. J. Zimmer, "Some Comparisons of Bayesian and Classical Confidence Intervals in the Exponential Case," *Proceedings of the Sixth National Symposium on Reliability and Quality Control*, January 11–13, 1960, Washington, D.C.

5 DeHardt, J. H. and H. D. McLaughlin, "Using Bayesian Methods to Select a Design with Known Reliability Without a Confidence Coefficient," *Annals of Reliability and Maintainability*, 1966.

6 Grohowski, G., W. C. Hausman, and L. R. Lamberson, "An Application of the Bayesian Statistical Inference Approach to Automotive Reliability Estimation." *Journal of Quality Technology*, Vol. 8, No. 4, October 1976.

7 Hamburg, M., "Bayesian Decision Theory and Statistical Quality Control," *Industrial Quality Control*, Vol. 19, No. 6, December 1962.

8 Hartley, H. O., "In Dr. Bayes' Consulting Room," *The American Statistician*, Vol. 17, No. 1, February 1963.

9 Hogg, R. V. and A. T. Craig, *Introduction to Mathematical Statistics*, Second Edition, Macmillan, New York, 1965.

10 Lindley, D. V., *Bayesian Statistics, A Review*, Society for Industrial and Applied Mathematics, Philadelphia, Pa., Copyright, 1972.

11 MacFarland, W. J., "Use of Bayes Theorem in Its Discrete Formulation for Reliability Estimation Purposes," *Proceedings of the Sixth National Symposium on Reliability and Quality Control*, January 11–13, 1960, Washington, D.C.

12 Mood, A. M. and F. A. Graybill, *Introduction to the Theory of Statistics*, McGraw-Hill, New York, 1963.

13 Olsson, J. E., "Implementation of a Bayesian Reliability Measurement Program," *Proceedings of the Sixth National Symposium on Reliability and Quality Control*, January 11–13, 1960, Washington, D.C.

14 Pratt, J. W., H. Raiffa, and R. Schlaifer, *Introduction to Statistical Decision Theory*; McGraw-Hill, New York, 1965.

15 Raiffa, H. and R. Schlaifer, *Applied Statistical Decision Theory*, School of Business, Harvard University, Boston, Mass., 1961.

16 Weir, W. T., "Bayesian Reliability Evaluation?" *Proceedings of the Sixth National Symposium on Reliability and Quality Control*, January 11–13, 1960, Washington, D.C.

Chapter 14 reliability optimization

In this chapter we discuss reliability allocation models and reliability optimization in design. The objective of this allocation is to use the reliability model to assign reliability to the subsystems so as to achieve a specified reliability goal for the system. Reliability allocation models are developed and techniques to solve the problem are given in Section 14.1. The dynamic programming technique is used to solve the complex reliability apportionment problem and is given in Section 14.2. A simple introduction to the dynamic programming method is also provided. The solution to engineering design and reliability problems generally requires a compromise among several objectives. The characteristics of the design depend on the fixed as well as the adjustable design parameters. Optimization problems for probabilistic design methodology are developed and different solution methods are discussed in Section 14.3. The optimization problem may be to maximize the reliability (subject to certain constraints on the amounts of resources available for the control of the parameters) or to minimize the cost of resources to be spent to control the parameters (subject to the constraint that the reliability of the system must meet a specified reliability goal). Some prior knowledge of the optimization methods is assumed for the understanding of the material. Some basic optimization concepts are explained whenever the context demands. The details on the optimization methods discussed here can be found in various books on optimization theory, which are given in the bibliography [3, 5, 6, 9, 13].

14.1 RELIABILITY ALLOCATION

Reliability and design engineers must translate overall system performance, including reliability, into component performance, including reliability. The process of assigning reliability requirements to individual components to attain the specified system reliability is called *reliability allocation*. This allocation problem is complex for several reasons, among which are: the role a component plays for the functioning of the system, the method of accomplishing this function, the complexity of the component, and the reliability of the component changing with the type of function to be performed. The problem is further complicated by the lack of detailed information on many of these factors early in the system design phase.

Some of the advantages of the reliability allocation program are:

1. The reliability allocation program forces the designer to understand and develop the relationships between component, subsystem, and system reliabilities. This leads to an understanding of the basic reliability problems inherent in the design.
2. The designer is obliged to consider reliability equally with other system parameters such as weight, cost and performance characteristics.
3. The contractor is required to meet the reliability goals in military contracts. In this situation, the reliability allocation program results in improved design, manufacturing methods, and testing procedures.

The allocation of specified system reliability R^* to the component reliability requires solving the following inequality

$$f(R_1, R_2, \ldots, R_n) \geqslant R^* \tag{14.1}$$

where R_i = the reliability allocated to the ith unit
f = the functional relationship between the components and the system

For series and parallel systems, the functional relationship f is well known. This relationship is complex for other system configurations and may not even be known mathematically for some complex systems. If we are interested in reliability as a function of time, Equation 14.1 may be generalized by considering R^* and R_i, $i=1,\ldots,n$ as functions of time t.

Most of the basic reliability allocation models are based on the assumption that component failures are independent, the failure of any component results in system failure (i.e., the system is composed of components in series), and that the failure rates of the components are constant. These assumptions lead to the following equation as a special case of Equation 14.1:

$$R_1(t)R_2(t)\ldots R_n(t) \geqslant R^*(t) \tag{14.2}$$

Let λ_i = failure rate of the ith component and λ^* = failure rate of the system. Then Equation 14.2 becomes

$$e^{-\lambda_1 t}e^{-\lambda_2 t}\ldots e^{-\lambda_n t} \geqslant e^{-\lambda^* t}$$

or

$$\lambda_1 + \lambda_2 + \cdots + \lambda_n \leqslant \lambda^*$$

Thus the reliability allocation for an overall requirement in terms of $R^*(t)$ or λ^* can be performed by similar approaches. Discussion of some of the reliability allocation models follows.

14.1.1 EQUAL APPORTIONMENT TECHNIQUE

The equal apportionment technique assigns equal reliabilities to all the subsystems in order to acheive a specified level of reliability for the total system. The system is assumed to consist of n subsystems in series. The main drawback of this method is that the subsystem reliability goals are not assigned in accordance with the degree of difficulty associated with meeting them.

Let R^* be the required system reliability and R_i be the reliability for the subsystem i. Then

$$R^* = \prod_{i=1}^{n} R_i$$

or

$$R_i = (R^*)^{1/n}, \qquad i = 1, 2, \ldots, n \tag{14.3}$$

EXAMPLE 14.1

Consider a communication system consisting of three subsystems (the transmitter, the receiver, and the coder), each of which must function if the communication system is to function. Assume that each subsystem is equally expensive to develop. What reliability requirement should be assigned to each subsystem in order to meet a system requirement of 0.8573?

Using Equation 14.3 we have

$$R_i = (R^*)^{1/n} = (0.8573)^{1/3} = 0.95$$

Thus a reliability requirement of 0.95 should be assigned to each subsystem of the communication system.

14.1.2 THE ARINC APPORTIONMENT TECHNIQUE

In the ARINC method we assume that the subsystems are in series with constant failure rates, that any subsystem failure causes a system failure, and that the subsystem mission times equal the system mission time. This apportionment technique requires the expression of the required reliability in terms of the failure rates. The objective is to choose λ_i^*'s such that

$$\sum_{i=1}^{n} \lambda_i^* \leqslant \lambda^* \tag{14.4}$$

where λ_i^* is the failure rate allocated to subsystem i, $i=1,\ldots,n$, and λ^* is the required system failure rate.

The following steps summarize this technique:

1. Determine the subsystem failure rates (λ_i) from the past data, observed or estimated.

2. Assign a weighting factor (ω_i) to each subsystem according to the failure rates determined in step 1, where ω_i is given by

$$\omega_i = \frac{\lambda_i}{\sum_{i=1}^{n} \lambda_i}, \qquad i=1,\ldots,n \tag{14.5}$$

Thus ω_i represents the relative failure vulnerability of the ith component and

$$\sum_{i=1}^{n} \omega_i = 1$$

3. Compute the subsystem failure rate requirements using

$$\lambda_i^* = \omega_i \lambda^*, \qquad i=1,\ldots,n$$

assuming equality holds in Equation 14.4.

It is clear that this method allocates the new failure rates based on relative weighting factors that are functions of the past failure rates of the subsystems.

EXAMPLE 14.2

Consider a system composed of three subsystems with the estimated failure rates of $\lambda_1=0.005$, $\lambda_2=0.003$, and $\lambda_3=0.001$ failure per hour, respectively. The system has a mission time of 20 hours. A system reliability of 0.95 is required. Find the reliability requirements for the subsystems.

Using Equation 14.5, we compute the weighting factors:

$$\omega_1 = \frac{0.005}{0.005+0.003+0.001} = 0.555$$

$$\omega_2 = \frac{0.003}{0.005+0.003+0.001} = 0.333$$

$$\omega_3 = \frac{0.001}{0.005+0.003+0.001} = 0.111$$

We know that

$$R^*(20) = \exp[-\lambda^*(20)] = 0.95$$

or

$$\lambda^* = 0.00256 \text{ failure per hour}$$

Hence the failure rates for the subsystems are

$$\lambda_1^* = \omega_1 \lambda^* = 0.555 \times 0.00256 = 0.00142$$

and similarly

$$\lambda_2^* = 0.333 \times 0.00256 = 0.000852$$

$$\lambda_3^* = 0.111 \times 0.00256 = 0.000284$$

The corresponding apportioned reliabilities for the subsystem are

$$R_1^*(20) = \exp[-0.00142(20)] = 0.97$$

$$R_2^*(20) = \exp[-0.000852(20)] = 0.98$$

$$R_3^*(20) = \exp[-0.000284(20)] = 0.99$$

14.1.3 THE AGREE ALLOCATION METHOD

The AGREE allocation method [1] is more sophisticated than the previous methods. This method is based on component or subsystem complexity and explicitly considers the relationship between component and system failure.The AGREE formula is used to determine the minimum MTBF for each component required to meet the system reliability. The components are supposed to have constant failure rates that are independent of each other and they operate in series with respect to their effect on system success.

Component complexity is defined in terms of modules and their associated circuitry. Examples of a module are an electron tube, a transistor, or a magnetic tape; a diode is considered a half-module. It is recommended that for digital computers (where the module count is high), the count should be reduced because failure rates for digital parts are generally far lower than for radio-radar types. The importance factor of a unit or subsystem is defined in terms of the probability of system failure if the particular subsystem fails. The importance factor of 1 means that the subsystem must operate for the system to operate successfully, and the importance factor of 0 means that the failure of the subsystem has no effect on system operation.

The allocation assumes that each module makes an equal contribution to system success. An equivalent requirement is that each module has the same failure rate. Making the observation that $e^{-x} \approx 1 - x$, when x is very small, the allocated failure rate to the ith unit is given by

$$\lambda_i = \frac{N_i[-\ln R^*(t)]}{N\omega_i t_i}, \quad i = 1, 2, \ldots, n \tag{14.6}$$

where t = mission time, or the required system operation time
t_i = time units for which the ith subsystem will be required to operate during t units of system operation ($0 < t_i \leq t$)
N_i = number of modules in ith subsystem
N = total number of modules in the system $= \Sigma N_i$
ω_i = importance factor for the ith subsystem
$= P$[system failure | subsystem i fails]
$R^*(t)$ = required system reliability for operation time t

The allocated reliability for the ith subsystem for t_i operating time units is given by

$$R_i(t_i) = 1 - \frac{1 - \left[R^*(t)\right]^{N_i/N}}{\omega_i} \tag{14.7}$$

The AGREE formula will lead to distorted allocation if the importance factor for a certain unit is very low. It is a good approximation if ω_i is close to one for each subsystem.

EXAMPLE 14.3

A system consisting of four subsystems is required to demonstrate a reliability level of 0.95 for 10 hours of continuous operation. Subsystems 1 and 3 are essential for the successful operation of the system. Subsystem 2 has to function for only 9 hours for the operation of the system, and its importance factor is 0.95. Subsystem 4 has an importance factor of 0.90 and must function for 8 hours for the system to function. Solve the reliability allocation problem by AGREE method using the data in Table 14.1.

We have

$$N = \sum_{i=1}^{4} N_i = 210$$

The minimum acceptable failure rates for the subsystems are given by Equation 14.6

Table 14.1

SUBSYSTEM NUMBER (i)	NUMBER OF MODULES (N_i)	IMPORTANCE FACTOR (ω_i)	OPERATING TIME (t_i)
1	15	1.00	10
2	25	0.95	9
3	100	1.00	10
4	70	0.90	8

and these are

$$\lambda_1 = \frac{15[-\ln 0.95]}{(210)(1.0)(10)} = 0.000366$$

$$\lambda_2 = \frac{25[-\ln 0.95]}{(210)(0.95)(9)} = 0.000714$$

$$\lambda_3 = \frac{100[-\ln 0.95]}{(210)(1.0)(10)} = 0.002442$$

$$\lambda_4 = \frac{70[-\ln 0.95]}{(210)(0.90)(8)} = 0.002377$$

Thus, the allocated subsystem reliabilities are, using Equation 14:7

$$R_1(10) = 1 - \frac{1-(0.95)^{15/210}}{1} = 0.99635$$

$$R_2(9) = 1 - \frac{1-(0.95)^{25/210}}{0.95} = 0.99274$$

$$R_3(10) = 1 - \frac{1-(0.95)^{100/210}}{1} = 0.97587$$

$$R_4(8) = 1 - \frac{1-(0.95)^{70/210}}{0.90} = 0.98116$$

As a check, we have the system reliability as

$$\begin{aligned} R^* &= (0.99635)(0.99274)(0.97587)(0.98116) \\ &= 0.94723 \end{aligned}$$

which is slightly less than the specified reliability. This is a result of the approximate nature of the formula and to the importance factors being less than 1.0 for subsystems 2 and 4.

14.1.4 EFFORT MINIMIZATION ALGORITHM

Let $R_1, \ldots, R_n$ denote the subsystem reliabilities. Assuming the subsystems to be in series, the system reliability R is given by

$$R = \prod_{i=1}^{n} R_i \tag{14.8}$$

Suppose the required reliability R^* for the system is greater than the present reliability R. From Equation 14.8 it is clear that to achieve $R^* > R$, the reliability of at least one subsystem must be increased. To do this, a certain amount of expenditure of "effort" is needed. Some typical examples of this "effort" are: further engineering development, extra manpower, extensive testing, new technology.

Let the effort function, denoted by $G(R_i, R_i^*), i=1,\ldots,n$, be defined to be the amount of effort needed to increase the reliability of the ith subsystem from a level R_i to a new level R_i^*. The effort function $G(x,y)$, $y > x \geqslant 0$, is assumed to satisfy the following conditions:

1. $G(x,y) \geqslant 0$
2. $G(x,y)$ is nondecreasing in y for a fixed value of x and nonincreasing in x for a fixed value of y; that is

$$G(x,y) \leqslant G(x,y+\Delta y), \qquad \Delta y > 0$$

$$G(x,y) \geqslant G(x+\Delta x,y), \qquad \Delta x > 0$$

3. $G(x,y)$ is additive, that is

$$G(x,y)+G(y,z)=G(x,z), \qquad x<y<z$$

4. $G(0,y)$ has a derivative $h(y)$ such that $yh(y)$ is strictly increasing in y, $0<y<1$.

Now we can readily define the following optimization problem:

$$\min \sum_{i=1}^{n} G(R_i, R_i^*)$$

subject to

$$\prod_{i=1}^{n} R_i^* \geqslant R^* \tag{14.9}$$

It can be shown that the above optimization problem has the following unique solution:

$$R_i^* = \begin{cases} R_0^* & \text{if} \quad i \leqslant k_0 \\ R_i & \text{if} \quad i > k_0 \end{cases} \tag{14.10}$$

where the subsystems are assumed to have been reordered with nondecreasing reliabilities; that is, it is true that

$$R_1 \leqslant R_2 \leqslant \cdots \leqslant R_n$$

The number k_0 is the maximum value of j such that

$$R_j < \left[\frac{R^*}{\prod_{i=j+1}^{n+1} R_i} \right]^{1/j} = r_j, \qquad j = 1, 2, \ldots, n \tag{14.11}$$

where $R_{n+1} = 1$ by definition. The number R_0^* is given by

$$R_0^* = \left[\frac{R^*}{\prod_{j=k_0+1}^{n+1} R_j} \right]^{1/k_0} \tag{14.12}$$

Equation 14.12 can be rearranged so that the system reliability R^* is given by

$$R^* = (R_0^*)^{k_0} \prod_{j=k_0+1}^{n+1} R_j \tag{14.13}$$

EXAMPLE 14.4

A system consists of three subsystems A, B, and C, all of which must function without failure in order to achieve a system success. The system reliability requirement has been set at 0.75. The predicted subsystem reliabilities are $R_A = 0.90$, $R_B = 0.80$, $R_C = 0.85$. How can we apportion the reliabilities to the subsystems such that the total effort spent on the system improvement is minimal, but, at the same time, the system reliability goal is met?

We assume identical effort functions for the three subsystems. Reordering the subsystems in nondecreasing order of their reliabilities yields

$$R_1 = R_B = 0.8$$

$$R_2 = R_C = 0.85$$

$$R_3 = R_A = 0.90$$

Now we find k_0 which is the maximum value of j such that

$$R_j < \left[\frac{R^*}{\prod_{i=j+1}^{n+1} R_i} \right]^{1/j}$$

For $j = 1$

$$R_1 = 0.8 < \left[\frac{0.75}{0.85 \times 0.9 \times 1} \right]^1 = 0.980$$

For $j=2$

$$R_2=0.85<\left[\frac{0.75}{0.9\times 1}\right]^{1/2}=0.912$$

For $j=3$

$$R_3=0.9<\left[\frac{0.75}{1}\right]^{1/3}=0.908$$

Hence

$$k_0=3$$

which yields

$$R_0^*=\left[\frac{R^*}{\prod_{j=k_0+1}^{n+1} R_j}\right]^{1/k_0}$$

$$=\left[\frac{0.75}{1}\right]^{1/3}=0.908$$

Therefore the reliability of each of the three subsystems would be increased to 0.908.

14.2 A DYNAMIC PROGRAMMING APPROACH TO RELIABILITY APPORTIONMENT

The effort minimization algorithm requires that all subsystems be subject to the same effort function. If this requirement cannot be met, a dynamic programming approach may prove efficient in determining the reliability apportionment with minimum expenditure of effort. Thus the dynamic programming technique can be used when the subsystems are subject to different but measurable effort functions.

Consider a proposed system consisting of n subsystems, each of which is to be developed independently. These subsystems are to function independently and are assumed to be in series. We wish to quantify the reliability goal to be set for each subsystem so that the system goal will be attained with a minimum expenditure of the development effort. We define

$n=$ number of subsystems
$\bar{y}=$ system reliability goal, $0<\bar{y}<1$
$x_i=$ reliability level of subsystem i at the present state of development, $0\leqslant x_i\leqslant 1$
$y_i=$ reliability goal set for subsystem i, $x_i\leqslant y_i\leqslant 1$
$G_i(x_i,y_i)=$ units of development effort required to raise the reliability level of subsystem i from x_i to y_i, $x_i\leqslant y_i$
$y_i^*=$ optimal reliability goal apportioned to subsystem i that minimizes the total development effort

The optimization problem can be formulated as follows:

$$\min \sum_{i=1}^{n} G_i(x_i, y_i) \tag{14.14}$$

subject to

$$\prod_{i=1}^{n} y_i \geqslant \bar{y}$$

and

$$0 \leqslant x_i \leqslant y_i \leqslant 1, \qquad i = 1, 2, \ldots, n$$

This problem can be converted to a dynamic programming problem in the following fashion. The problem is shown schematically in Figure 14.1 to aid the reader who is not familiar with dynamic programming and details are given in [9].

Stages
Identify each of the n subsystems as a stage. Thus a reliability apportionment decision can be made sequentially at each stage.

State Variable
Define the set S_k of all possible states s_k at stage k. Then we have the following relationship among the state variables:

$$1 = s_n \geqslant s_{n-1} \geqslant \cdots \geqslant s_1 \geqslant s_0 = \bar{y}$$

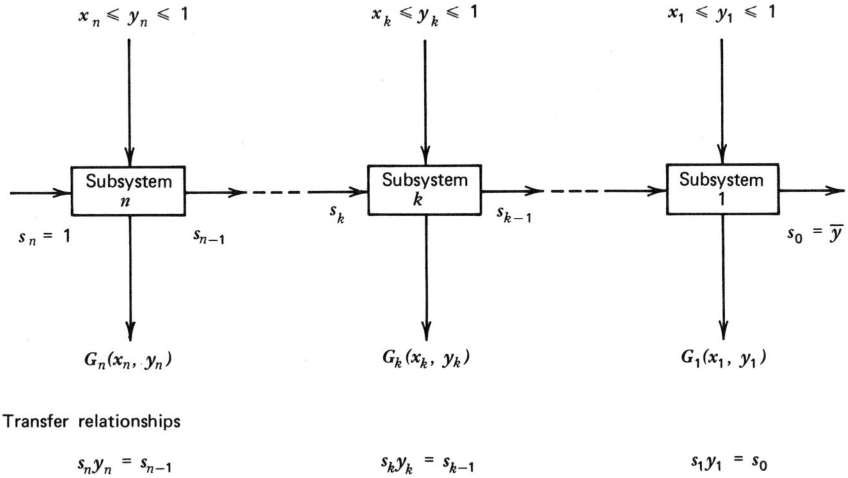

Figure 14.1 Sequential block diagram for dynamic programming.

The state variables indicate how much reliability may be allocated for the stage in order to meet the system reliability goal.

Decision Variable

Define the set D_k of all possible decision alternatives $d_k = y_k$ at stage k such that

$$x_k \leqslant y_k \leqslant 1, \qquad k = 1,2,\ldots,n$$

Transformation Function

$$T_k(s_k, d_k) = s_k y_k = s_{k-1}, \qquad k = 1,2,\ldots,n$$

Return Function

$$R_k(s_k, d_k) = G_k(s_k, y_k), \qquad k = 1,2,\ldots,n$$

Let $f_k(s_k)$ be the optimal return function for stage k. Then the general recursion equation for the dynamic programming problem is (for details see Reference 9)

$$f_k(s_k) = \min_{d_k} \{G_k(s_k, d_k) + f_{k-1}(s_{k-1})\}, \qquad k = 1,2,\ldots,n \qquad (14.15)$$

and $f_0(s_0) = 0$. The above equation is used recursively backward to solve the problem.

EXAMPLE 14.5

A system has three independent subsystems. The system can function successfully if and only if each of the three subsystems functions properly. The system reliability requirement is 0.95. Based on engineering analysis and historical data for similar systems, the estimated reliability levels of the subsystems are 0.93, 0.95, and 0.96. What reliability goal should be set for each subsystem in order to minimize the total expenditure of effort if the estimated effort functions are as shown in Table 14.2?

Table 14.2 Effort functions

y_1	$G_1(0.93, y_1)$	y_2	$G_2(0.95, y_2)$	y_3	$G_3(0.96, y_3)$
0.93	0		—		—
0.94	0.5		—		—
0.95	1.0	0.95	0		—
0.96	1.5	0.96	2	0.96	0
0.97	2.5	0.97	6	0.97	3
0.98	4.5	0.98	12	0.98	9
0.99	20.0	0.99	22	0.99	32
0.995	45.0	0.995	40	0.995	65

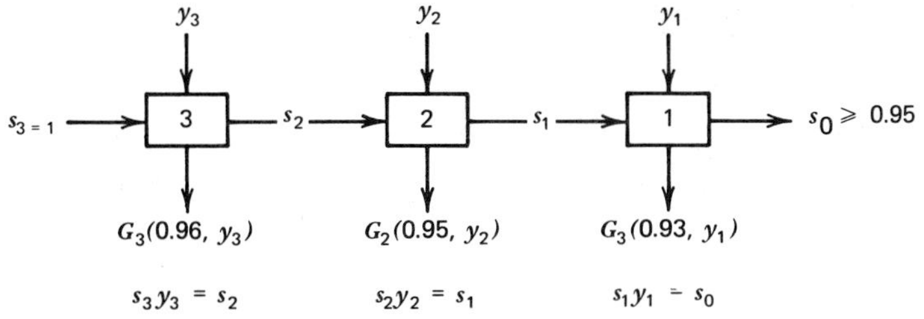

Figure 14.2 Block diagram.

Figure 14.2 shows the sequential block diagram for this example.
The recursion equations are:

$$f_1(s_1) = \min_{\{y_1 \geqslant 0.95/s_1\}} \left[G_1(0.93, y_1) \right]$$

$$f_2(s_2) = \min_{\{y_2 \geqslant 0.95/s_2 y_1\}} \left[G_2(0.95, y_2) + f_1(s_1) \right]$$

$$f_3(s_3) = \min_{\{y_3 \geqslant 0.95/s_3 y_2\}} \left[G_3(0.96, y_3) + f_2(s_2) \right] \tag{14.16}$$

First we develop the state transformation tables that give us the set of all possible input states for various stages. Since the system reliability goal is 0.95, no stage can have a reliability less than 0.95.

Table 14.3 Transformation at Stage 3

$T_3(s_3, y_3) = s_3 y_3 = y_3 = s_2$

y_3	0.96	0.97	0.98	0.99	0.995
s_2	0.96	0.97	0.98	0.99	0.995

Table 14.4 Transformation at Stage 2

s_2 \ y_2	$s_1 = s_2 y_2$					
	0.95	0.96	0.97	0.98	0.99	0.995
0.96	0.9120	0.9216	0.9302	0.9408	0.9504	0.9552
0.97	0.9215	0.9312	0.9409	0.9506	0.9603	0.9651
0.98	0.9310	0.9408	0.9506	0.9604	0.9702	0.9751
0.99	0.9405	0.9504	0.9603	0.9702	0.9801	0.9850
0.995	0.9452	0.9552	0.96515	0.9751	0.9850	0.9900

Table 14.5 Transformation at Stage 1

s_1 \ y_1	$s_1 y_1 = s_0$						
	0.94	0.95	0.96	0.97	0.98	0.99	0.995
0.9552							0.9504
0.9603						0.9506	0.9554
0.9604						0.9507	0.9555
0.9651						0.9554	0.9602
0.9702					0.9507	0.9604	0.9653
0.9751					0.9555	0.9653	0.9702
0.9801				0.9506	0.9604	0.9702	0.9751
0.9850				0.9550	0.9653	0.9702	0.9751
0.9900			0.9504	0.9603	0.9702	0.9801	0.9850

Values of s_1 less than 0.95 are not feasible and hence need not be considered. The infeasible values of the state variables are shown in Table 14.5 as blank entries.

Only the values of s_0 greater than 0.95 are shown in Table 14.5.

We will now develop the optimal return functions. Using Equation 14.15, we have

$$f_1(s_1) = \min_{y_1} \{ G_1(s_1, y_1) + f_0(s_0) \}$$

$$= \min_{y_1} G_1(s_1, y_1)$$

since $f_0(s_0)$ is zero. Next we construct Table 14.6 to compute $f_1(s_1)$.

Proceeding to the next stage, we write

$$f_2(s_2) = \min_{y_2} \{ G_2(s_2, y_2) + f_1(s_1) \}$$

which leads to Table 14.7.

Table 14.6

s_1 \ y_1	$G_1(s_1, y_1)$							
	0.94	0.95	0.96	0.97	0.98	0.99	0.995	$f_1(s_1)$
0.9552							45	45
0.9603						20	45	20
0.9604						20	45	20
0.9651						20	45	20
0.9702					4.5	20	45	4.5
0.9751					4.5	20	45	4.5
0.9801				2.5	4.5	20	45	2.5
0.9850				2.5	4.5	20	45	2.5
0.9900			1.5	2.5	4.5	20	45	1.5

Table 14.7

s_2 \ y_2	$Q_2(s_2,y_2)=G_2(s_2,y_2)+f_1(s_1)$						
	0.95	0.96	0.97	0.98	0.99	0.995	$f_2(s_2)$
0.97						40+20	60.0
0.98					22+4.5	40+4.5	26.5
0.99				12+4.5	22+2.5	40+2.5	16.5
0.995			6+20	12+4.5	22+2.5	40+1.5	16.5

Table 14.8

s_3 \ y_3	$Q_3(s_3,y_3)=G_3(s_3,y_3)+f_2(s_2)$					
	0.96	0.97	0.98	0.99	0.995	$f_3(s_3)$
1		3+60	9+26.5	32+16.5	65+16.5	35.5

Finally, we have

$$f_3(s_3)=\min_{y_3}\{G_3(s_3,y_3)+f_2(s_2)\}$$

which is tabulated in Table 14.8.
Table 14.8 shows that the minimum value of the total effort is 35.5. Also, this is achieved when $y_3^*=0.98$. Tracing backward, from Table 14.7, we find that $y_2^*=0.99$ when $f_2(s_2)=26.5$. This mean that $f_1(s_1)=4.5$ and hence from Table 14.6, $y_1^*=0.98$.

EXAMPLE 14.6

The reliability analysis of a system shows that the four subsystems for this system have a parallel structure, as shown in Figure 14.3. The present estimates of the reliability levels of the four subsystems are 0.900, 0.980, 0.940, and 0.960. The requirements for the system reliability are 0.9999990. Compute what reliability goals each subsystem should have in order to minimize the total expenditure of effort if the estimated effort functions are as given in Table 14.9.

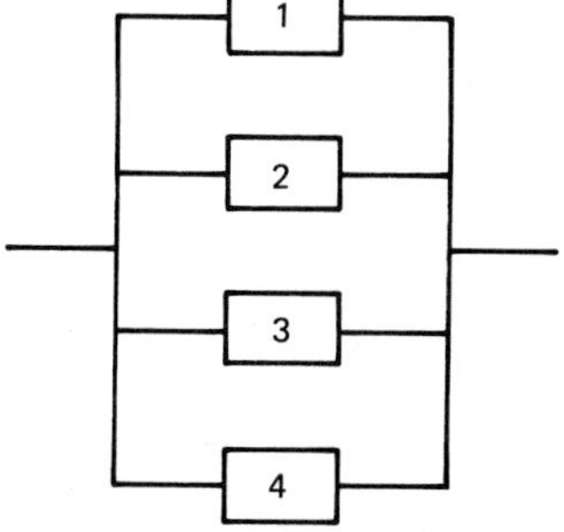

Figure 14.3 Parallel structure.

Table 14.9 Effort Functions

y_1	$G_1(0.90,y_1)$	y_2	$G_2(0.98,y_2)$	y_3	$G_3(0.94,y_3)$	y_4	$G_4(0.96,y_4)$
0.90	0.00						
0.91	0.80						
0.92	1.30						
0.93	2.50						
0.94	4.80			0.94	0.0		
0.95	7.40			0.95	3.5		
0.96	10.50			0.96	7.0	0.96	0.0
0.97	15.00			0.97	11.5	0.97	2.5
0.98	24.00	0.98	0	0.98	19.6	0.98	7.8
0.99	35.00	0.99	17.5	0.99	28.6	0.99	11.5
0.995	48.00	0.995	25.0	0.995	42.5	0.995	28.0
0.999	82.00	0.999	41.0	0.999	61.4	0.999	41.5

Example 14.6 can be transformed to a problem with its structure similar to Example 14.5. This is done by considering the unreliability levels of the subsystems. Let y_i be the reliability level for the ith subsystem. Then we want

$$1-(1-y_1)(1-y_2)(1-y_3)(1-y_4) \geqslant 0.9999990$$

Let

$$z_i = 1 - y_i$$

then

$$\prod_{i=1}^{4} z_i \leqslant 0.000001$$

Hence, the optimization problem is

$$\min \sum_{i=1}^{4} G_i(x_i, (1-z_i))$$

subject to

$$\prod_{i=1}^{4} z_i = \bar{z} \leqslant 0.000001 \tag{14.17}$$

This problem presented by Equation 14.17 is similar to the problem presented in Equation 14.14 and hence can be solved by dynamic programming. The optimal solution is $z_i^* = 0.08$, $z_2^* = 0.02$, $z_3^* = 0.06$, and $z_4^* = 0.01$ and the value of the minimum effort is 12.8 units. The above values result in a system reliability of 0.99999904.

EXAMPLE 14.7

The time to failure **t** of a component has exponential distribution with a mean time to failure equal to T. Thus the reliability of the component at time t is

$$R(t) = \exp(-t/T) \tag{14.18}$$

The cost to produce the component is a function of the mean time to failure T. This is expressed by a power function of the form

$$C(T)=cT^{a} \tag{14.19}$$

where c and a are given constants. Suppose we are interested in the reliability of the component at a specified time t_0 or t_0 is the mission time. Then

$$R=R(t_0)=\exp(-t_0/T)$$

or

$$T=-t_0/\ln R \tag{14.20}$$

Let us consider two designs with mean times to failure given by T_1 and T_2, respectively, and where $T_1<T_2$. Then, using Equation 14.20, we have

$$C(T_1)=cT_1^a=ct_0^a(-\ln R_1)^{-a} \tag{14.21}$$

and

$$C(T_2)=cT_2^a=ct_0^a(-\ln R_2)^{-a} \tag{14.22}$$

where $R_1=\exp(-t_0/T_1)$ and $R_2=\exp(-t_0/T_2)$. Thus, the effort function $G(R_1,R_2)$, which gives the cost or effort of improving the component reliability from R_1 to R_2, $R_2>R_1$, is given by

$$G(R_1,R_2)=ct_0^a\left[(-\ln R_2)^{-a}-(-\ln R_1)^{-a}\right] \tag{14.23}$$

Let us consider a system consisting of n components in series. Then the effort function G_i for the ith component is given by

$$G_i(R_1,R_2)=c_it_0^{a_i}\left[(-\ln R_2)^{-a_i}-(-\ln R_1)^{-a_i}\right] \tag{14.24}$$

The cost functions for the four components in terms of T are estimated and given below:

$$C_1(T)=0.7573\,T^{0.4307}, \qquad 10{,}000\leqslant T\leqslant 50{,}000$$

$$C_2(T)=0.0001\,T^{1.2549}, \qquad 20{,}000\leqslant T\leqslant 48{,}000$$

$$C_3(T)=0.00000275\,T^{1.6131}, \qquad 15{,}000\leqslant T\leqslant 45{,}000$$

$$C_4(T)=0.000037\,T^{1.3588}, \qquad 25{,}000\leqslant T\leqslant 60{,}000$$

These cost functions result in the following boundary points for the functions:

$$C_1(10{,}000)=40.0, \qquad C_1(50{,}000)=80.0$$

$$C_2(20{,}000)=25.0, \qquad C_2(48{,}000)=75.0$$

$$C_3(15{,}000)=15.0, \qquad C_3(45{,}000)=90.0$$

$$C_4(25{,}000)=35.0, \qquad C_4(60{,}000)=115.0$$

We are interested in the analysis of the system reliability at 30,000 time units. Given the range for mean time to failure for the components, the upper and lower bounds for reliabiliy of the components at 30,000 can be computed using Equation 14.18.

Component 1

$$\text{Lower bound for } R_1 = \exp\left(-\frac{30{,}000}{10{,}000}\right) = 0.0498$$

$$\text{Upper bound for } R_1 = \exp\left(-\frac{30{,}000}{50{,}000}\right) = 0.5488$$

Component 2

$$\text{Lower bound for } R_2 = \exp\left(-\frac{30{,}000}{20{,}000}\right) = 0.2231$$

$$\text{Upper bound for } R_2 = \exp\left(-\frac{30{,}000}{48{,}000}\right) = 0.5353$$

Component 3

$$\text{Lower bound for } R_3 = \exp\left(-\frac{30{,}000}{15{,}000}\right) = 0.1353$$

$$\text{Upper bound for } R_3 = \exp\left(-\frac{30{,}000}{45{,}000}\right) = 0.5134$$

Component 4

$$\text{Lower bound for } R_4 = \exp\left(-\frac{30{,}000}{25{,}000}\right) = 0.3012$$

$$\text{Upper bound for } R_4 = \exp\left(-\frac{30{,}000}{60{,}000}\right) = 0.6065$$

The present design results in reliabilities 0.20, 0.25, 0.20, and 0.35 of components 1, 2, 3, and 4, respectively. We thus have the necessary information to develop the reliability improvement cost functions using Equation 14.24 for all the components. After developing the cost functions, we can solve the problem of finding optimal reliability goals for the four components that will minimize the total cost by the technique discussed before. Suppose we are given a reliability goal for the system as 0.9000. The four components are in parallel and the present reliability of the system is given by

$$1-(1-0.20)(1-0.25)(1-0.20)(1-0.35)=0.688$$

The above problem was solved by the method discussed in Examples 14.5 and 14.6 and the optimal solution is as below:

Minimum total cost $=84.95$

$$R_1^* = 0.55,\ R_2^* = 0.50,\ R_3^* = 0.20, \text{ and } R_4^* = 0.45$$

14.3 OPTIMIZATION IN PROBABILISTIC DESIGN

The solution to engineering design and reliability problems requires a compromise among several objectives. The characteristics of the design depend on the fixed as well as the adjustable design parameters. It is evident from the reliability computations in Chapter 6 that reliability is a function of strength and stress distributions, and these distributions have certain parameters associated with them. It follows, then, that reliability is a function of these parameters. It may be possible to control the values of some of these parameters, but not without the expenditure of certain resources. We may wish to maximize the reliability subject to certain constraints on amounts of resources available for control of the parameters. Alternatively, we may wish to minimize the cost of resources subject to the constraint that the reliability of the system must meet a specified level.

14.3.1 NORMALLY DISTRIBUTED STRENGTH AND STRESS

First we consider the case in which strength and stress are independent and normally distributed. From Equation 6.18, we know that the reliability depends on the value of the lower limit of the integral. If we want to maximize reliability, the lower limit of the integral should have as low a value as possible.

Let $c_1(\mu_{\mathbb{S}})$ denote the cost function for the mean strength. A higher mean value for strength may require that we use better materials, employ different heat treatment processes, or exercise better control on the manufacturing processes for the materials, which would naturally raise the cost. Thus $c_1(\mu_{\mathbb{S}})$ is a monotonically increasing function of $\mu_{\mathbb{S}}$. From the reliability point of view, lower values of $\sigma_{\mathbb{S}}$ for symmetric distributions are desirable. To reduce $\sigma_{\mathbb{S}}$ we need to control the factors that introduce variability in the strength such as an unsmooth surface finish, notch effects, or a heterogeneous internal structure. Thus the cost function $c_4(\sigma_{\mathbb{S}})$ is a monotonically decreasing function of $\sigma_{\mathbb{S}}$.

Let $c_3(\mu_s)$ and $c_4(\sigma_s)$ denote the cost functions associated with the mean value and the standard deviation of the stress, respectively. Lower values of μ_s and σ_s will obviously lead to higher reliability. In order to reduce these values, it may be necessary to increase the dimensions of the component with less dimensional variability, to exercise better control on the operational forces acting on the component. Thus $c_3(\mu_s)$ and $c_4(\sigma_s)$ are both monotonically decreasing functions of μ_s and σ_s, respectively.

Now let us consider the optimization problem where we wish to minimize the total cost, subject to the constraint that the component must have a certain desired level of reliability. It is

$$\min TC = c_1(\mu_{\mathbb{S}}) + c_2(\sigma_{\mathbb{S}}) + c_3(\mu_s) + c_4(\sigma_s)$$

subject to (14.25)

$$\frac{\mu_{\mathbb{S}} - \mu_s}{\sqrt{\sigma_{\mathbb{S}}^2 + \sigma_s^2}} \geqslant z$$

where z is determined by the coupling equation for a specified level of reliability. Note that z in Equation 14.25 is the negative of z in the coupling equation. The Lagrangian function associated with the problem presented by Equation 14.25 is

$$L(\mu_{\mathfrak{s}}, \sigma_{\mathfrak{s}}, \mu_s, \sigma_s, \lambda) = c_1(\mu_{\mathfrak{s}}) + c_2(\sigma_{\mathfrak{s}}) + c_3(\mu_s) + c_4(\sigma_s) + \lambda\left[\mu_{\mathfrak{s}} - \mu_s - z\left(\sigma_{\mathfrak{s}}^2 + \sigma_s^2\right)^{1/2}\right] \tag{14.26}$$

To find the locally optimal solutions, we differentiate the Lagrangian with respect to the variables and equate to zero, that is

$$\frac{\partial L}{\partial \mu_{\mathfrak{s}}} = \frac{\partial c_1(\mu_{\mathfrak{s}})}{\partial \mu_{\mathfrak{s}}} + \lambda = 0 \tag{14.27a}$$

$$\frac{\partial L}{\partial \mu_s} = \frac{\partial c_3(\mu_s)}{\partial \mu_s} - \lambda = 0 \tag{14.27b}$$

$$\frac{\partial L}{\partial \sigma_{\mathfrak{s}}} = \frac{\partial c_2(\sigma_{\mathfrak{s}})}{\partial \sigma_{\mathfrak{s}}} - \lambda z \sigma_{\mathfrak{s}}(\sigma_{\mathfrak{s}}^2 + \sigma_s^2)^{-1/2} = 0 \tag{14.27c}$$

$$\frac{\partial L}{\partial \sigma_s} = \frac{\partial c_4(\sigma_s)}{\partial \sigma_s} - \lambda z \sigma_s(\sigma_{\mathfrak{s}}^2 + \sigma_s^2)^{-1/2} = 0 \tag{14.27d}$$

$$\frac{\partial L}{\partial \lambda} = \mu_{\mathfrak{s}} - \mu_s - z\sqrt{\sigma_{\mathfrak{s}}^2 + \sigma_s^2} = 0 \tag{14.27e}$$

The simultaneous solution of the above system of five equations in five unknowns yields all the local optima. We may then evaluate the objective function for all these local solutions and choose the globally optimal solution. In some design problems, the various cost functions may be strictly monotonic, in which case, for the optimal solution the inequality in Equation 14.25 would be satisfied as an equality. This is because higher values of the left-hand side of the inequality cost more, and hence the optimization process will reduce it to the lowest specified z value.

In order to find the globally optimal solution, we next study the structure of the above problem. Let us first note the constraint function

$$g(\sigma_{\mathfrak{s}}, \sigma_s, \mu_{\mathfrak{s}}, \mu_s) = z\sqrt{\sigma_{\mathfrak{s}}^2 + \sigma_s^2} - \mu_{\mathfrak{s}} + \mu_s$$

The Hessian matrix for this function is

$$\begin{bmatrix} z\sigma_s^2\left(\sigma_S^2+\sigma_s^2\right)^{-3/2} & -z\sigma_S\sigma_s\left(\sigma_S^2+\sigma_s^2\right)^{-3/2} & 0 & 0 \\ -z\sigma_S\sigma_s\left(\sigma_S^2+\sigma_s^2\right)^{-3/2} & z\sigma_S^2\left(\sigma_S^2+\sigma_s^2\right)^{-3/2} & 0 & 0 \\ 0 & 0 & 0 & 0 \\ 0 & 0 & 0 & 0 \end{bmatrix}$$

We can easily verify that the Hessian matrix is a positive semidefinite matrix for $z > 0$ and hence the constraint function is convex. If all the cost functions are convex, then Equation 14.25 becomes a convex programming probelm, which means that any local optimum is a global optimum [3]. In order to solve the optimization problem, we have to solve the system of Equations 14.27a through 14.27e.

As illustrated by the example in the next section, the optimization problem can be reduced to a search problem in one variable. If we fix the value of μ_S, then using Equations 14.27a and 14.27b we can uniquely determine μ_s by solving

$$\lambda = -\frac{\partial c_1(\mu_S)}{\partial \mu_S} = \frac{\partial c_3(\mu_s)}{\partial \mu_s} \qquad (14.28)$$

Then from Equation 14.27e the value of $\sigma_S^2 + \sigma_s^2$ can be computed since μ_S and μ_s are now known. The standard deviations σ_S and σ_s can then be determined from Equations 14.27c and 14.27d, respectively. Thus we can compute the total cost for these values of the variables. It can be easily proved that the total cost function is a convex function of μ_S. We can evaluate the total cost function for different values of μ_S and search for the global optimum by any one of the various well-known search methods.

EXAMPLE 14.8

The data on the functions for a system are as given below. The units for strength and stress are MPa and the units for cost are \$100.

μ_S	5	10	15	20	25	30	35	40
$c_1(\mu_S)$	2.23	6.32	11.62	17.89	25.00	32.86	41.41	50.60
σ_S	1	2	3	4	5	6	7	8
$c_2(\sigma_S)$	100	43.53	26.76	18.95	14.50	11.65	9.68	8.25
μ_s	5	10	15	20	25	30	35	40
$c_3(\mu_s)$	38.07	25.12	19.69	16.57	14.50	12.99	11.85	10.93
σ_s	1	2	3	4	5	6	7	8
$c_4(\sigma_s)$	50	30.78	23.17	18.95	16.21	14.26	12.81	11.66

It is often possible to develop analytical expressions for various cost functions that fit the discrete data. The following cost functions represent the above data.

$$c_1(\mu_{\mathbf{S}}) = 0.2\mu_{\mathbf{S}}^{1.5}$$

$$c_2(\sigma_{\mathbf{S}}) = 100\sigma_{\mathbf{S}}^{-1.2}$$

$$c_3(\mu_s) = 100\mu_s^{-0.6}$$

$$c_4(\sigma_s) = 50\sigma_s^{-0.7}$$

These four cost functions satisfy the convexity requirement because they all have positive second partial derivatives.

The specified reliability for which the system is to be designed is 0.990. From normal tables, using the coupling equation, we find $z = 2.33$. Thus the optimization problem is

$$\min TC = 0.2\mu_{\mathbf{S}}^{1.5} + 100\sigma_{\mathbf{S}}^{-1.2} + 100\mu_s^{-0.6} + 50\sigma_s^{-0.7}$$

subject to (14.29)

$$\mu_{\mathbf{S}} - \mu_s - 2.33\left(\sigma_{\mathbf{S}}^2 + \sigma_s^2\right)^{1/2} \geqslant 0$$

The system of equations corresponding to Equations 14.27a through 14.27e is

$$0.3\mu_{\mathbf{S}}^{0.5} + \lambda = 0 \tag{14.30a}$$

$$-60\mu_s^{-1.6} - \lambda = 0 \tag{14.30b}$$

$$-120\sigma_{\mathbf{S}}^{-2.2} - 2.33\lambda(\sigma_{\mathbf{S}}^2 + \sigma_s^2)^{-1/2}\sigma_{\mathbf{S}} = 0 \tag{14.30c}$$

$$-35\sigma_s^{-1.7} - 2.33\lambda(\sigma_{\mathbf{S}}^2 + \sigma_s^2)^{-1/2}\sigma_s = 0 \tag{14.30d}$$

$$\mu_{\mathbf{S}} - \mu_s - 2.33(\sigma_{\mathbf{S}}^2 + \sigma_s^2)^{1/2} = 0 \tag{14.30e}$$

From Equations 14.30a and 14.30b we have

$$\mu_s = 27.424\mu_{\mathbf{S}}^{-0.3125} \tag{14.31}$$

and from Equations 14.30c and 14.30d we have

$$\sigma_s = 0.6336\sigma_{\mathbf{S}}^{1.185} \tag{14.32}$$

Combining Equations 14.32 and 14.30e yields

$$\sigma_{\mathbf{S}}^2 + 0.4014\sigma_{\mathbf{S}}^{2.3704} = \left(\frac{\mu_{\mathbf{S}} - \mu_s}{2.33}\right)^2 \tag{14.33}$$

Because of the monotonicity properties of the polynomial given by Equation 14.33, the unique value of $\sigma_{\mathbf{s}}$ that satisfies Equation 14.33 can be found.

Based on these relationships, Table 14.10 has been developed. From this table, it is clear that the optimal value of $\mu_{\mathbf{s}}$ lies between 26 and 28 ksi. The Fibonacci search method [5] was used to find the optimal values by searching between 26 and 28. The optimal value of $\mu_{\mathbf{s}}$ may also be found by a trial-and-error method or by a simple heuristic search method under the assumption that the function is convex. It is left up to the reader to consider other search methods. The optimal value of $\mu_{\mathbf{s}}$ was found to lie between 27.00 and 27.25. Some of the evaluations are shown in Table 14.11. The plot of TC against $\mu_{\mathbf{s}}$ is shown in Figure 14.4.

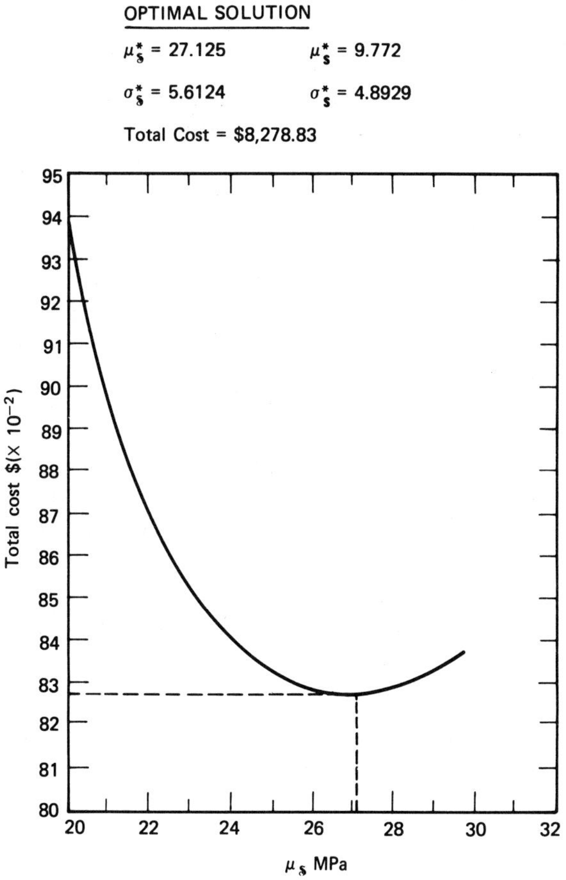

Figure 14.4 Total cost curve as a function of mean strength ($\mu_{\mathbf{S}}$).

Table 14.10

μ_S	μ_s	$\mu_S-\mu_s$	$\sigma_S^2+\sigma_s^2$	σ_S	σ_s	$c_1(\mu_S)$	$c_3(\mu_s)$	$c_2(\sigma_S)$	$c_4(\sigma_s)$	TC
15	11.7656	3.2344	1.9270	1.1625	0.7574	11.6190	22.7841	83.4696	60.7355	178.6082
20	10.7540	9.2460	15.7469	3.1252	2.4454	17.8885	24.0468	25.4769	26.7378	94.1500
25	10.0297	14.9703	41.2809	4.895	4.1621	25.0000	25.0742	14.8695	18.4269	83.3706
26	9.9075	16.0925	47.7018	5.2345	4.5064	26.5149	25.2593	13.7199	17.4297	82.9237
27	9.7913	17.2087	54.5484	5.5695	4.8502	28.0592	25.4388	12.7356	16.5553	82.7889
28	9.6807	18.3193	61.8167	5.9019	5.1951	29.6324	25.6127	11.8798	15.7781	82.9031
30	9.4742	20.5258	77.6047	6.5549	5.8831	32.8634	25.9462	10.4742	14.4626	83.7464

Table 14.11

μ_S	μ_s	$\mu_S-\mu_s$	$\sigma_S^2+\sigma_s^2$	σ_S	σ_s	$c_1(\mu_S)$	$c_3(\mu_s)$	$c_2(\sigma_S)$	$c_4(\sigma_s)$	TC
27.000	9.7913	17.2087	54.5484	5.5695	4.8502	28.0592	25.4388	12.7356	16.5553	82.7889
27.125	9.7772	17.3478	55.4341	5.6124	4.8929	28.2543	25.4608	12.6189	16.4543	82.7881
27.250	9.7632	17.4868	56.3261	5.6539	4.9327	28.4498	25.4827	12.5078	16.3610	82.8013
27.375	9.7492	17.6258	57.2248	5.6950	4.9782	28.6458	25.5046	12.3996	16.2562	82.8062
27.500	9.7354	17.7646	58.1301	5.7355	5.0220	28.8422	25.5263	12.2946	16.1568	82.8199
27.625	9.7216	17.9034	59.0419	5.7790	5.0654	29.0391	25.5480	12.1836	16.0598	82.8305

Let us consider a practical example. A bar in tension is to be designed to withstand random loadings. The bar can be made of different materials with different ultimate tensile strength. The mean strength of the material and the strength variability can be changed by heat treatment or some other control of the manufacturing process which costs money. Based on the information on cost of materials, heat treatment, and manufacturing processes, cost functions $c_1(\mu_S)$ and $c_2(\sigma_S)$ can be established. The dimensions and tolerances of the bar will determine the cost functions $c_3(\mu_s)$ and $c_4(\sigma_s)$. Higher tolerances will mean more variablility in the stress in the bar. Reducing the tolerances will need a better control of the manufacturing processes and hence will incur higher cost. In an actual design situation for a tension element the data on all the factors mentioned above are necessary to construct the cost functions such as those given in the numerical Example 14.8.

EXAMPLE 14.9

Let us consider the problem which is in some sense dual to Equation 14.29. We wish to maximize the reliability for the normal case peresented in Chapter 6, subject to certain resource constraints; that is

$$\max z = (\mu_S - \mu_s)(\sigma_S^2 + \sigma_s^2)^{-1/2}$$

subject to (14.34)

$$c_1(\mu_S) + c_2(\sigma_S) + c_3(\mu_s) + c_4(\sigma_s) \leqslant r$$

where r denotes the amount of resources available.

This optimization problem can be solved by a method similar to the one used in Example 14.8. We write the Lagrangian function as

$$L(\mu_{\mathbf{S}},\mu_s,\sigma_{\mathbf{S}},\sigma_s;\lambda)=(\mu_{\mathbf{S}}-\mu_s)(\sigma_{\mathbf{S}}^2+\sigma_s^2)^{-1/2}+\lambda[c_1(\mu_{\mathbf{S}})+c_2(\sigma_{\mathbf{S}})+c_3(\mu_s)+c_4(\sigma_s)-r]$$

To find the locally optimal solutions, we need to solve the following system of equations:

$$\frac{\partial L}{\partial \mu_{\mathbf{S}}}=(\sigma_{\mathbf{S}}^2+\sigma_s^2)^{-1/2}+\lambda\frac{\partial c_1(\mu_{\mathbf{S}})}{\partial \mu_{\mathbf{S}}}=0 \tag{14.35a}$$

$$\frac{\partial L}{\partial \mu_s}=-(\sigma_{\mathbf{S}}^2+\sigma_s^2)^{-1/2}+\lambda\frac{\partial c_3(\mu_s)}{\partial \mu_s}=0 \tag{14.35b}$$

$$\frac{\partial L}{\partial \sigma_{\mathbf{S}}}=-\sigma_{\mathbf{S}}(\mu_{\mathbf{S}}-\mu_s)(\sigma_{\mathbf{S}}^2+\sigma_s^2)^{-3/2}+\lambda\frac{\partial c_2(\sigma_{\mathbf{S}})}{\partial \sigma_{\mathbf{S}}}=0 \tag{14.35c}$$

$$\frac{\partial L}{\partial \sigma_s}=-\sigma_s(\mu_{\mathbf{S}}-\mu_s)(\sigma_{\mathbf{S}}^2+\sigma_s^2)^{-3/2}+\lambda\frac{\partial c_4(\sigma_s)}{\partial \sigma_s}=0 \tag{14.35d}$$

$$\frac{\partial L}{\partial \lambda}=c_1(\mu_{\mathbf{S}})+c_2(\sigma_{\mathbf{S}})+c_3(\mu_s)+c_4(\sigma_s)-r=0 \tag{14.35e}$$

From Equations 14.35a and 14.35b we have

$$\frac{\partial c_1(\mu_{\mathbf{S}})}{\partial(\mu_{\mathbf{S}})}=-\frac{\partial c_3(\mu_s)}{\partial(\mu_s)} \tag{14.36}$$

and from Equations 14.35c and 14.35d we have

$$\frac{1}{\sigma_{\mathbf{S}}}\frac{\partial c_2(\sigma_{\mathbf{S}})}{\partial \sigma_{\mathbf{S}}}=\frac{1}{\sigma_s}\frac{\partial c_4(\sigma_s)}{\partial \sigma_s} \tag{14.37}$$

The above relations are used as before to reduce the problem to a search problem in one variable. To insure that any local optimal solution is a global optimal solution, we have to prove that the objective function in Equation 14.34 is a concave function. This can be proved by studying the Hessian matrix of the objective function, which is

$$\begin{bmatrix}
\dfrac{(\mu_{\mathbf{S}}-\mu_s)(2\sigma_{\mathbf{S}}^2-\sigma_s^2)}{(\sigma_{\mathbf{S}}^2+\sigma_s^2)^{5/2}} & \dfrac{3\sigma_{\mathbf{S}}\sigma_s(\mu_{\mathbf{S}}-\mu_s)}{(\sigma_{\mathbf{S}}^2+\sigma_s^2)^{5/2}} & \dfrac{-\sigma_{\mathbf{S}}}{(\sigma_{\mathbf{S}}^2+\sigma_s^2)^{3/2}} & \dfrac{\sigma_{\mathbf{S}}}{(\sigma_{\mathbf{S}}^2+\sigma_s^2)^{3/2}} \\[2ex]
\dfrac{3\sigma_{\mathbf{S}}\sigma_s(\mu_{\mathbf{S}}-\mu_s)}{(\sigma_{\mathbf{S}}^2+\sigma_s^2)^{5/2}} & \dfrac{(\mu_{\mathbf{S}}-\mu_s)(2\sigma_s^2-\sigma_{\mathbf{S}}^2)}{(\sigma_{\mathbf{S}}^2+\sigma_s^2)^{5/2}} & \dfrac{-\sigma_s}{(\sigma_{\mathbf{S}}^2+\sigma_s^2)} & \dfrac{\sigma_s}{(\sigma_{\mathbf{S}}^2+\sigma_s^2)^{3/2}} \\[2ex]
\dfrac{-\sigma_{\mathbf{S}}}{(\sigma_{\mathbf{S}}^2+\sigma_s^2)^{3/2}} & \dfrac{-\sigma_s}{(\sigma_{\mathbf{S}}^2+\sigma_s^2)^{3/2}} & 0 & 0 \\[2ex]
\dfrac{\sigma_{\mathbf{S}}}{(\sigma_{\mathbf{S}}^2+\sigma_s^2)^{3/2}} & \dfrac{\sigma_s}{(\sigma_{\mathbf{S}}^2+\sigma_s^2)^{3/2}} & 0 & 0
\end{bmatrix}$$

Thus this matrix is negative semidefinite if $(2\sigma_{\mathrm{S}}^2 - \sigma_s^2) \geqslant 0$. Hence the objective function in Equation 14.34 is concave if we have $\sigma_{\mathrm{S}}/\sigma_s \geqslant 1/\sqrt{2}$, in which case a local optimum will be the global optimum [3, 5].

Consider the following numerical example with the same cost functions as in Example 14.8.

$$\max z = (\mu_{\mathrm{S}} - \mu_s)(\sigma_{\mathrm{S}} + \sigma_s)^{-1/2}$$

subject to (14.38)

$$0.2\mu_{\mathrm{S}}^{1.5} + 100\sigma_{\mathrm{S}}^{-1.2} + 100\mu_s^{-0.6} + 50\sigma_s^{-0.7} \leqslant 75$$

Using Equations 14.36 and 14.37, we have

$$\mu_s = 27.4248\mu_{\mathrm{S}}^{-0.3125}$$

and

$$\sigma_s = 0.6336\sigma_{\mathrm{S}}^{1.185}$$

and from Equation 14.35e we have

$$100\sigma_{\mathrm{S}}^{-1.2} + 68.8183\sigma_{\mathrm{S}}^{-0.8296} = 75 - \left[0.2\mu_{\mathrm{S}}^{1.5} + 100\mu_s^{-0.6}\right]$$

We use these equations to construct Table 14.12. It is clear that once we select a value for μ_{S}, we can uniquely determine the values of μ_s, σ_{S}, and σ_s. The dichotomous search technique [3, 5] was used to develop the table using values of μ_{S} between 20 and 30 MPa.

The reader may search for the optimal solution by trial and error under the assumption that the function is convex and any local optimal solution is a global optimal solution.

The optimal solution based on Table 14.12 is: $\mu_{\mathrm{S}}^* = 24.9380$, $\mu_s^* = 10.0375$, $\sigma_{\mathrm{S}}^* = 6.52508$ and $\sigma_s^* = 5.8513$; $z^* = 1.7001$ and $R^* = 0.95543$.

Table 14.12

EVALUATION NUMBER	μ_{S}	μ_s	σ_{S}	σ_s	z
1	24.5000	10.0932	6.3440	5.6595	1.6946
2	25.5000	10.0406	6.7620	6.1041	1.6970
3	22.5000	10.3654	5.6120	4.8941	1.6296
4	23.5000	10.2255	5.9520	5.2474	1.6729
5	23.3750	10.2426	5.9070	5.2004	1.6687
6	24.3750	10.1094	6.2890	5.6014	1.6838
7	23.9380	10.1667	6.1150	5.4182	1.6858
8	24.9380	10.0375	6.5250	5.8513	1.7001
9	24.2190	10.1297	6.2270	5.5359	1.6909
10	25.2190	10.0024	6.6590	5.9940	1.6984

It is clear from Table 14.12 that it is true that $\sigma_{\mathbf{S}}/\sigma_{\mathbf{s}} \geqslant 1/\sqrt{2}$. Hence the above solution is a globally optimal solution.

14.3.2 NORMALLY DISTRIBUTED STRENGTH AND EXPONENTIALLY DISTRIBUTED STRESS

We now discuss the optimization technique when the strength is normal with parameters $(\mu_{\mathbf{S}}, \sigma_{\mathbf{S}})$ and the stress is exponential with parameter $\lambda_{\mathbf{s}}$. As discussed before, the cost function $c_1(\mu_{\mathbf{S}})$ is a monotonically increasing function of $\mu_{\mathbf{S}}$, and $c_2(\sigma_{\mathbf{S}})$ is a monotonically increasing function of $\sigma_{\mathbf{s}}$. We assume that $c_3(\lambda_{\mathbf{s}})$ is a monotonically increasing function of $\lambda_{\mathbf{s}}$. In addition, these cost functions are assumed to be generally convex. Let us now consider the following reliability maximization problem (ignoring $\Phi(\cdot)$ terms in Equation 6.30).

$$\max R = 1 - \exp\left[-\mu_{\mathbf{S}}\lambda_{\mathbf{s}} + \frac{1}{2}\lambda_{\mathbf{s}}^2\sigma_{\mathbf{S}}^2 \right]$$

subject to (14.39)

$$c_1(\mu_{\mathbf{S}}) + c_2(\sigma_{\mathbf{S}}) + c_3(\lambda_{\mathbf{s}}) \leqslant r$$

where r represents the total resources available for allocation. The Lagrangian approach was used in an attempt to solve the problem presented by Equation 14.39 but in this case it was found impossible to reduce it to a search problem in one variable. Thus we were forced to solve a system of four nonlinear equations in four unknowns, simultaneously. The technique of geometric programming [10, 13] was found to be useful in reducing it to a search problem in one variable (instead of four in the Lagrangian method).

EXAMPLE 14.10

$$\max R = 1 - \exp\left[-\left(\mu_{\mathbf{S}}\lambda_{\mathbf{s}} - \frac{1}{2}\lambda_{\mathbf{s}}^2\sigma_{\mathbf{S}}^2 \right) \right]$$

subject to (14.40)

$$0.1\mu_{\mathbf{S}}^{2.1} + 100\sigma_{\mathbf{S}}^{-1.1} + 50\lambda_{\mathbf{s}}^{0.6} \leqslant 75$$

To solve by geometric programming, we let $\mu_{\mathbf{S}} = x_1$, $\sigma_{\mathbf{S}} = x_2$, and $\lambda_{\mathbf{s}} = x_3$. Then Equation 14.40 is equivalent to the following optimization problem:

$$\min y = -x_1x_3 + \frac{1}{2}x_2^2x_3^2$$

subject to (14.41)

$$(1.3333 \times 10^{-3})x_1^{2.1} + 1.3333x_2^{-1.1} + 0.6667x_3^{0.6} \leqslant 1$$

The dual variables (see Appendix 14.A) must satisfy the following system of equations:

$$\begin{aligned} -\omega_{01}+\omega_{02} &= \alpha_0 \\ -\omega_{01} \quad +2.1\omega_{11} &= 0 \\ 2\omega_{02} \quad -1.1\omega_{12} &= 0 \\ -\omega_{01}+2\omega_{02} \quad +0.6\omega_{13} &= 0 \end{aligned} \tag{14.42}$$

The degree of difficulty for the problem presented by Equation 14.41 is $5-(3+1)=1$. Thus the present problem can be reduced to a search problem in one variable. It was found that $\alpha_0=+1$ was not possible and hence $\alpha_0=-1$ was used in the following computations. The system of Equations 14.42 was solved for different values of ω_{01}. Once ω_{01} is fixed, the values of the remaining dual variables are determined uniquely from Equation 14.42. The dual function as given in the appendix is evaluated for these dual variables in Table 14.13. Our objective is to obtain the maximum value of this dual function.

Table 14.13

ω_{01}	DUAL FUNCTION $= d(\omega, \alpha_0)$
1.6	−5.1917
1.5	−4.5254
1.4	−4.2961
1.3	−4.4775

From Table 14.13, it is clear that the optimal value lies between 1.3 and 1.5. The Fibonacci search was used to find the optimal value of ω_{01} between the interval $[1.3, 1.5]$. The optimal value of ω_{01} was found to be $\omega_{01}^*=1.3952$. The values of the remaining variables are

$$\omega_{10}^*=\alpha_1(\omega_{11}^*+\omega_{13}^*)=2.4568$$

$$d^*=y^*=-4.2953$$

The values of the primal variables are found by using the equations in the appendix. These values are:

$$x_1^*=12.7133,\ x_2^*=3.8747,\ \text{and}\ x_3^*=0.4659$$

Thus the optimal solution to Equation 14.39 is $\mu_{\bar{s}}^*=12.7133$, $\sigma_{\bar{s}}^*=3.8747$, $\lambda_s^*=0.4659$, or $\mu_s^*=2.1276$, and $R^*=0.9864$.

14.4 SUMMARY

In this chapter reliability allocation optimization models were presented. In order to solve these models, dynamic programming, nonlinear programming using the Lagrangian method, and geometric programming techniques were used. Further details on these techniques can be found in References 3, 5, 6, 9, 10, 13.

EXERCISES

1 Show how you would derive Equation 14.4. State clearly any assumptions you make.

2 A system consists of five subsystems in series. The system reliability goal is 0.990 for 10 hours of operation. The necessary information for the subsystem is given below:

SUBSYSTEM NUMBER i	NUMBER OF MODULES (N_i)	IMPORTANCE FACTOR (ω_i)	OPERATING TIME (t_i)
1	25	1.00	10
2	80	0.97	9
3	45	1.00	10
4	60	0.93	7
5	70	1.00	10

Compute the reliability goal for the subsystems using the AGREE method.

3 Prove that the optimization problem given by Equation 14.9 has the solution of the form given by Equation 14.10. State clearly any assumptions you may make in the derivation.

4 A system consists of four subsystems that must function if the system has to function properly. The system reliability goal is 0.950. All the four subsystems have identical reliability improvement effort functions. The estimated subsystem reliabilities at the present time are 0.75, 0.85, 0.90, 0.95. What reliability goal should be apportioned to the subsystems so as to minimize the total effort spent on the system improvement?

5 Assume that the system in Example 14.5 consists of three subsystems that have a parallel structure. The system reliability goal is 0.9999950. Using the effort functions given in Example 14.5, determine what reliability goal should be set for each subsystem in order to minimize the total expenditure of effort.

6 Solve Example 14.6 and verify the optimal solution given in the example.

7 Solve Example 14.7 and verify the optimal solution given in the example.

8 Assume that the four components in Example 14.7 are in series and the system reliability requirements at 30,000 is 0.070. Solve the effort minimization problem using the data given in Example 14.7.

9 A system consists of N subsystems. Let R_j denote the reliability for the subsystem j, $j=1,2,\ldots,N$. Then the system reliability R_S is given by

$$R_S = \prod_{j=1}^{N} R_j = \prod_{j=1}^{N} \left(1 - \bar{R}_j\right)$$

The reliability of each subsystem is improved by adding redundant components. Let n_j be the number of redundant components for subsystem j. Thus, the subsystem unreliability $\bar{R}_j$ is related to individual component unreliability $\bar{r}_j$ for subsystem j and is given by

$$\bar{R}_j = \bar{r}_j^{(n_j+1)}, \qquad n_j \geqslant 0$$

Let c_j be the cost of a component for subsystem j and w_j be the weight of a component for subsystem j. Let C be the total resources available for cost and W be the maximum additional weight allowed for the system. Then, we have the following optimization problem:

$$\max R_S = \prod_{j=1}^{N} \left(1 - \bar{r}_j^{(n_j+1)}\right)$$

subject to

$$\sum_{j=1}^{N} c_j n_j \leqslant C$$

$$\sum_{j=1}^{N} w_j n_j \leqslant W$$

$$n_j \geqslant 0$$

$$n_j \text{ integer}$$

Solve the above optimization problem with the following data:

SUBSYSTEM j	COST c_j	WEIGHT w_j	RELIABILITY r_j	RESOURCES AVAILABLE
1	1.1	4.0	0.85	
2	2.5	3.8	0.75	$C=40$
3	3.8	7.5	0.90	$W=60$
4	4.1	9.5	0.80	

10 A production process contains a machine that, under heavy usage, deteriorates so rapidly in both quality and output so that it is inspected at the end of every day. The state of the machine is classified into the following four states after inspection:

STATE	CONDITION
1	Good—as new
2	Operable—minor deterioration
3	Operable—major deterioration
4	Inoperable—output of unacceptable quality

If the machine is inoperable, it takes one day to replace it at a cost of \$10,000, and the cost of lost production for that day is \$3,000. Hence, the total cost incurred is \$13,000. Costs for other states are

STATE	EXPECTED COST OF PRODUCING DEFECTIVE ITEMS
1	\$ 0
2	3,000
3	7,000

The stochastic process has the following transition matrix under the above maintenance policy:

	FOR POLICY P_a			
STATE	1	2	3	4
1	0	$\frac{3}{4}$	$\frac{3}{16}$	$\frac{1}{16}$
2	0	$\frac{5}{8}$	$\frac{1}{4}$	$\frac{1}{8}$
3	0	0	$\frac{1}{2}$	$\frac{1}{2}$
4	1	0	0	0

There are other maintenance policies. Let us call the above maintenance policy as P_a and the others P_b, P_c, and P_d.

POLICY	ACTION
P_a	Replace in state 4
P_b	Replace in state 4 overhaul in state 3
P_c	Replace in states 3 and 4
P_d	Replace in states 2, 3, and 4.

The associated transition matrices are:

FOR POLICY P_b			
0	$\frac{3}{4}$	$\frac{3}{16}$	$\frac{1}{16}$
0	$\frac{5}{8}$	$\frac{1}{4}$	$\frac{1}{8}$
0	1	0	0
1	0	0	0

FOR POLICY P_c			
0	$\frac{3}{4}$	$\frac{3}{16}$	$\frac{1}{16}$
0	$\frac{5}{8}$	$\frac{1}{4}$	$\frac{1}{8}$
1	0	0	0
1	0	0	0

FOR POLICY P_d			
0	$\frac{3}{4}$	$\frac{3}{16}$	$\frac{1}{16}$
1	0	0	0
1	0	0	0
1	0	0	0

Cost data

DECISION	STATE	EXPECTED COST OF DEFECTIVE ITEMS	COST OF MAINTENANCE	COST FROM LOST PRODUCTION	TOTAL COST PER DAY
1. Leave	1	0	0	0	0
machine	2	3,000	0	0	3,000
as is	3	7,000	0	0	7,000
	4	∞	0	0	∞
2. Overhaul	3	0	5,000	3,000	8,000
3. Replace	2,3, OR 4	0	10,000	3,000	13,000

Find an optimal policy for this problem when the planning horizon is infinite. Also find an optimal policy when we use a discounting factor $\beta = 0.95$.

11 Consider the design reliability problem when both stress (s) and strength ($\mathfrak{S}$) are normally distributed. The reliability goal for the component is 0.990. The cost functions for the four parameters are:

$$c_1(\mu_{\mathfrak{S}}) = 0.0002\mu_{\mathfrak{S}}^{1.135}, \qquad 30{,}000 \leqslant \mu_{\mathfrak{S}} \leqslant 75{,}000 \text{ MPa}$$

$$c_2(\sigma_{\mathfrak{S}}) = 800\sigma_{\mathfrak{S}}^{-0.475}, \qquad 1{,}000 \leqslant \sigma_{\mathfrak{S}} \leqslant 10{,}000 \text{ MPa}$$

$$c_3(\mu_s) = 8997\mu_s^{-0.513}, \qquad 10{,}000 \leqslant \mu_s \leqslant 68{,}000 \text{MPa}$$

$$c_4(\sigma_s) = 366\sigma_s^{-0.358}, \qquad 500 \leqslant \sigma_s \leqslant 7{,}500 \text{ MPa}$$

Find the values of the four parameters $\mu_{\mathfrak{S}}$, μ_s, $\sigma_{\mathfrak{S}}$, and σ_s that will minimize the total cost subject to the constraints that the component reliability goal is achieved.

12 Consider the cost functions given in Problem 11. Determine values for the four parameters that will give maximum value for the reliability of the component, subject to the constraint that the total cost must be less than or equal to 100 units.

13 Time to failure of a component is exponentially distributed with MTBF of θ. Thus

$$R(t)=\exp(-t/\theta), \qquad t \geqslant 0,\ \theta > 0$$

The cost to produce the component is a function of θ and follows the following relationship

$$C(\theta)=k\theta^{a}$$

where k and a are given constants. Suppose we are designing a system that consists of three components in series. The constants k and a for the three components are to be determined from the following boundary relationships for the component cost:

$$C_1(10)=20.0 \qquad C_2(40)=45.0$$
$$C_2(20)=15.0 \qquad C_2(50)=30.0$$
$$C_3(15)=30.0 \qquad C_3(35)=50.0$$

The units of θ are thousands of kilometers. The system is to be designed with an MTBF of 10. Solve the problem of minimizing the cost of the system and simultaneously achieving the reliability goal in terms of MTBF.

BIBLIOGRAPHY

1 AGREE Report, "Reliability of Military Electronic Equipment," Office of the Assistant Secretary of Defence, Washington, D. C., GPO, 1957.

2 American Society for Metals (ASM), Vol. I, *Properties and Selection of Materials;* Vol. II, *Heat Treatment, Cleaning and Finishing;* Vol. III., *Machining.* All volumes published by *American Society for Metals,* Metals Park, Ohio, 1969.

3 Beveridge, G. S. C. and R. S. Schecter, *Optimization: Theory and Practice,* McGraw-Hill, New York, 1970.

4 Haugen, E., *Probabilistic Approaches to Design,* John Wiley & Sons, New York, 1968.

5 Himmelblau, D. M., *Applied Nonlinear Programming,* McGraw-Hill, New York, 1973.

6 Kapur, K. C., "Optimization in Design by Reliability," *AIIE Transactions*, Vol 7, No 2, June 1975.

7 Kececioglu, D. and D. Cormier, "Designing a Specified Reliability Directly into a Component," Proceedings of the Third Annual Aerospace Reliability and Maintainability Conference, Statler-Hilton Hotel, Washington, D. C., June 29–July 1, 1967, pp. 546–565.

8 Lipson, C., N. J. Sheth, and R. L. Disney, "Reliability Prediction-Mechanical Stress/Strength Interference," Rome Air Development Center, Technical Report No. RADC-TR-77-810, March 1967.

9 Nemhauser, G. L., *Introduction to Dynamic Programming,* John Wiley & Sons New York 1966.

10 Passy, V. and D. J. Wilde, "Generalized Polynomial Optimization," *SIAM J. Appl. Math.,* Vol. 15, September 1967, pp. 1344–1356.

11 Quality Assurance, Reliability Handbook, AMCP 702-3, U.S. Materiel Command, October 1968.

12 Shooman, M. L., *Probabilistic Reliability: An Engineering Approach,* McGraw-Hill, New York, 1968.

13 Taraman, S. I. and K. C. Kapur, "Optimization Considerations in Design Reliability, by Stress-Strength Interference Theory." IEEE Transactions on Reliability, Vol R-24, No. 2, June 1975.

14 Tribus, M., *Rational Descriptions, Decisions and Designs,* Pergamon Press, New York, 1970.

APPENDIX 14.A

GEOMETRIC PROGRAMMING WITH POSYNOMIALS [10]

The formulation of a general geometric programming problem is

$$\min g_0(x)$$

subject to

$$0 < \alpha_m g_m^{\alpha_m} \leqslant 1, \qquad m = 1, 2, \ldots, M$$

where

$$g_m = \sum_{t=1}^{T_m} \alpha_{mt} c_{mt} \prod_{n=1}^{N} x_n^{a_{mtn}}, \qquad m = 0, 1, \ldots, M$$

$\alpha_{mt} = \pm 1$, $c_{mt} > 0$, $x_n > 0$, and a_{mtn} are any real numbers.

This problem can be solved by working with a set of dual variables ω_{mt}, one set for each of the constraint functions $\{g_m(x)\}$, which satisfy

1. $0 \leqslant \omega_{mt} < \infty, \qquad m = 0, 1, \ldots, M$
2. Normality condition,

$$\sum_{t=1}^{T_0} \alpha_{0t} \omega_{0t} = \alpha_0$$

3. N linear orthogonality conditions,

$$\sum_{m=0}^{M} \sum_{t=1}^{T_m} \alpha_{mt} a_{mtn} \omega_{mt} = 0, \qquad n = 1, 2, \ldots, N$$

4. M linear inequality constraints,

$$\omega_{m0} = \alpha_m \sum_{t=1}^{T_m} \alpha_{mt} \omega_{mt} \geqslant 0, \qquad m = 1, 2, \ldots, M$$

In the foregoing conditions, α_0 is not specified in advance and hence must be chosen to satisfy the constraints. The degree of difficulty of the generalized geometric programming problem is

$$\sum_{m=0}^{M} T_m - (N + 1)$$

where

$$T=\left(\sum_{m=0}^{M} T_m\right)=\text{number of dual variables}$$

The dual function is given as

$$d(\omega,\alpha_0)=\alpha_0\left[\prod_{m=0}^{M}\prod_{t=1}^{T_m}\left(\frac{c_{mt}\omega_{m0}}{\omega_{mt}}\right)^{\alpha_{mt}\omega_{mt}}\right]^{\alpha_0}$$

where $\omega_{00}=1$.

For every solution x^* where $g_0(x^*)$ is a local minimum, there exists a set of dual variables α_0^*, ω^*, satisfying conditions 1 through 4, such that

$$d(\omega^*,\alpha_0^*)=g_0(x^*)$$

Once the optimal dual variables are found, the corresponding values of the primal variables x are calculated from

$$c_{0t}\prod_{n=1}^{N} x_n^{a_{0tn}}=\omega_{0t}\alpha_0 g_0^*, \qquad t=1,2,\ldots,T_0$$

and

$$c_{mt}\prod_{n=1}^{N} x_n^{a_{mtn}}=\frac{\omega_{mt}}{\omega_{m0}}, \qquad t=1,2,\ldots,T_m,\ m=1,\ldots,M$$

appendixes

appendix I ordinate height values for the normal distribution

$$\phi(z) = \frac{1}{\sqrt{2\pi}} e^{-z^2/2} \text{ for } 0.00 \leqslant z \leqslant 4.99.$$

z	.00	.01	.02	.03	.04	.05	.06	.07	.08	.09
.0	.3989	.3989	.3989	.3988	.3986	.3984	.3982	.3980	.3977	.3973
.1	.3970	.3965	.3961	.3956	.3951	.3945	.3939	.3932	.3925	.3918
.2	.3910	.3902	.3894	.3885	.3876	.3867	.3857	.3847	.3836	.3825
.3	.3814	.3802	.3790	.3778	.3765	.3752	.3739	.3725	.3712	.3697
.4	.3683	.3668	.3653	.3637	.3621	.3605	.3589	.3572	.3555	.3538
.5	.3521	.3503	.3485	.3467	.3448	.3429	.3410	.3391	.3372	.3352
.6	.3332	.3312	.3292	.3271	.3251	.3230	.3209	.3187	.3166	.3144
.7	.3123	.3101	.3079	.3056	.3034	.3011	.2989	.2966	.2943	.2920
.8	.2897	.2874	.2850	.2827	.2803	.2780	.2756	.2732	.2709	.2685
.9	.2661	.2637	.2613	.2589	.2565	.2541	.2516	.2492	.2468	.2444
1.0	.2420	.2396	.2371	.2347	.2323	.2299	.2275	.2251	.2227	.2203
1.1	.2179	.2155	.2131	.2107	.2083	.2059	.2036	.2012	.1989	.1965
1.2	.1942	.1919	.1895	.1872	.1849	.1826	.1804	.1781	.1758	.1736
1.3	.1714	.1691	.1669	.1647	.1626	.1604	.1582	.1561	.1539	.1518
1.4	.1497	.1476	.1456	.1435	.1415	.1394	.1374	.1354	.1334	.1315
1.5	.1295	.1276	.1257	.1238	.1219	.1200	.1182	.1163	.1145	.1127
1.6	.1109	.1092	.1074	.1057	.1040	.1023	.1006	.09893	.09728	.09566
1.7	.09405	.09246	.09089	.08933	.08780	.08628	.08478	.08329	.08183	.08038
1.8	.07895	.07754	.07614	.07477	.07341	.07206	.07074	.06943	.06814	.06687
1.9	.06562	.06438	.06316	.06195	.06077	.05959	.05844	.05730	.05618	.05508
2.0	.05399	.05292	.05186	.05082	.04980	.04879	.04780	.04682	.04586	.04491
2.1	.04398	.04307	.04217	.04128	.04041	.03955	.03871	.03788	.03706	.03626
2.2	.03547	.03470	.03394	.03319	.03246	.03174	.03103	.03034	.02965	.02898
2.3	.02833	.02768	.02705	.02643	.02582	.02522	.02763	.02406	.02349	.02294
2.4	.02239	.02186	.02134	.02083	.02033	.01984	.01936	.01888	.01842	.01797

$$\phi(z)=\frac{1}{\sqrt{2\pi}}\,e^{-z^2/2} \text{ for } 0.00 \leqslant z \leqslant 4.99.$$

z	.00	.01	.02	.03	.04	.05	.06	.07	.08	.09
2.5	.01753	.01709	.01667	.01625	.01585	.01545	.01506	.01468	.01431	.01394
2.6	.01358	.01323	.01289	.01256	.01223	.01191	.01160	.01130	.01100	.01071
2.7	.01042	.01014	$.0^{2}9871$	$.0^{2}9606$	$.0^{2}9347$	$.0^{2}9094$	$.0^{2}8846$	$.0^{2}8605$	$.0^{2}8370$	$.0^{2}8140$
2.8	$.0^{2}7915$	$.0^{2}7697$	$.0^{2}7483$	$.0^{2}7274$	$.0^{2}7071$	$.0^{2}6873$	$.0^{2}6679$	$.0^{2}6491$	$.0^{2}6307$	$.0^{2}6127$
2.9	$.0^{2}5953$	$.0^{2}5782$	$.0^{2}5616$	$.0^{2}5454$	$.0^{2}5296$	$.0^{2}5143$	$.0^{2}4993$	$.0^{2}4847$	$.0^{2}4705$	$.0^{2}4567$
3.0	$.0^{2}4432$	$.0^{2}4301$	$.0^{2}4173$	$.0^{2}4049$	$.0^{2}3928$	$.0^{2}3810$	$.0^{2}3695$	$.0^{2}3584$	$.0^{2}3475$	$.0^{2}3370$
3.1	$.0^{2}3267$	$.0^{2}3167$	$.0^{2}3070$	$.0^{2}2975$	$.0^{2}2884$	$.0^{2}2794$	$.0^{2}2707$	$.0^{2}2623$	$.0^{2}2541$	$.0^{2}2461$
3.2	$.0^{2}2384$	$.0^{2}2309$	$.0^{2}2236$	$.0^{2}2165$	$.0^{2}2096$	$.0^{2}2029$	$.0^{2}1964$	$.0^{2}1901$	$.0^{2}1840$	$.0^{2}1780$
3.3	$.0^{2}1723$	$.0^{2}1667$	$.0^{2}1612$	$.0^{2}1560$	$.0^{2}1508$	$.0^{2}1459$	$.0^{2}1411$	$.0^{2}1364$	$.0^{2}1319$	$.0^{2}1275$
3.4	$.0^{2}1232$	$.0^{2}1191$	$.0^{2}1151$	$.0^{2}1112$	$.0^{2}1075$	$.0^{2}1038$	$.0^{2}1003$	$.0^{3}9689$	$.0^{3}9358$	$.0^{3}9037$
3.5	$.0^{3}8727$	$.0^{3}8426$	$.0^{3}8135$	$.0^{3}7853$	$.0^{3}7581$	$.0^{3}7317$	$.0^{3}7061$	$.0^{3}6814$	$.0^{3}6575$	$.0^{3}6343$
3.6	$.0^{3}6119$	$.0^{3}5902$	$.0^{3}5693$	$.0^{3}5490$	$.0^{3}5294$	$.0^{3}5105$	$.0^{3}4921$	$.0^{3}4744$	$.0^{3}4573$	$.0^{3}4408$
3.7	$.0^{3}4248$	$.0^{3}4093$	$.0^{3}3944$	$.0^{3}3800$	$.0^{3}3661$	$.0^{3}3526$	$.0^{3}3396$	$.0^{3}3271$	$.0^{3}3149$	$.0^{3}3032$
3.8	$.0^{3}2919$	$.0^{3}2810$	$.0^{3}2705$	$.0^{3}2604$	$.0^{3}2506$	$.0^{3}2411$	$.0^{3}2320$	$.0^{3}2232$	$.0^{3}2147$	$.0^{3}2065$
3.9	$.0^{3}1987$	$.0^{3}1910$	$.0^{3}1837$	$.0^{3}1766$	$.0^{3}1698$	$.0^{3}1633$	$.0^{3}1569$	$.0^{3}1508$	$.0^{3}1449$	$.0^{3}1393$
4.0	$.0^{3}1338$	$.0^{3}1286$	$.0^{3}1235$	$.0^{3}1186$	$.0^{3}1140$	$.0^{3}1094$	$.0^{3}1051$	$.0^{3}1009$	$.0^{4}9687$	$.0^{4}9299$
4.1	$.0^{4}8926$	$.0^{4}8567$	$.0^{4}8222$	$.0^{4}7890$	$.0^{4}7570$	$.0^{4}7263$	$.0^{4}6967$	$.0^{4}6683$	$.0^{4}6410$	$.0^{4}6147$
4.2	$.0^{4}5894$	$.0^{4}5652$	$.0^{4}5418$	$.0^{4}5194$	$.0^{4}4979$	$.0^{4}4772$	$.0^{4}4573$	$.0^{4}4382$	$.0^{4}4199$	$.0^{4}4023$
4.3	$.0^{4}3854$	$.0^{4}3691$	$.0^{4}3535$	$.0^{4}3386$	$.0^{4}3242$	$.0^{4}3104$	$.0^{4}2972$	$.0^{4}2845$	$.0^{4}2723$	$.0^{4}2606$
4.4	$.0^{4}2494$	$.0^{4}2387$	$.0^{4}2284$	$.0^{4}2185$	$.0^{4}2090$	$.0^{4}1999$	$.0^{4}1912$	$.0^{4}1829$	$.0^{4}1749$	$.0^{4}1672$
4.5	$.0^{4}1598$	$.0^{4}1528$	$.0^{4}1461$	$.0^{4}1396$	$.0^{4}1334$	$.0^{4}1275$	$.0^{4}1218$	$.0^{4}1164$	$.0^{4}1112$	$.0^{4}1062$
4.6	$.0^{4}1014$	$.0^{5}9684$	$.0^{5}9248$	$.0^{5}8830$	$.0^{5}8430$	$.0^{5}8047$	$.0^{5}7681$	$.0^{5}7331$	$.0^{5}6996$	$.0^{5}6676$
4.7	$.0^{5}6370$	$.0^{5}6077$	$.0^{5}5797$	$.0^{5}5530$	$.0^{5}5274$	$.0^{5}5030$	$.0^{5}4796$	$.0^{5}4573$	$.0^{5}4360$	$.0^{5}4156$
4.8	$.0^{5}3961$	$.0^{5}3775$	$.0^{5}3598$	$.0^{5}3428$	$.0^{5}3267$	$.0^{5}3112$	$.0^{5}2965$	$.0^{5}2824$	$.0^{5}2690$	$.0^{5}2561$
4.9	$.0^{5}2439$	$.0^{5}2322$	$.0^{5}2211$	$.0^{5}2105$	$.0^{5}2003$	$.0^{5}1907$	$.0^{5}1814$	$.0^{5}1727$	$.0^{5}1643$	$.0^{5}1563$

appendix II the cumulative normal distribution function

$$\Phi(z) = \frac{1}{\sqrt{2\pi}} \int_{-\infty}^{z} e^{-x^2/2}\,dx \text{ for } 0.00 \leqslant z \leqslant 4.99.$$

z	.00	.01	.02	.03	.04	.05	.06	.07	.08	.09
.0	.5000	.5040	.5080	.5120	.5160	.5199	.5239	.5279	.5319	.5359
.1	.5398	.5438	.5478	.5517	.5557	.5596	.5636	.5675	.5714	.5753
.2	.5793	.5832	.5871	.5910	.5948	.5987	.6026	.6064	.6103	.6141
.3	.6179	.6217	.6255	.6293	.6331	.6368	.6406	.6443	.6480	.6517
.4	.6554	.6591	.6628	.6664	.6700	.6736	.6772	.6808	.6844	.6879
.5	.6915	.6950	.6985	.7019	.7054	.7088	.7123	.7157	.7190	.7224
.6	.7257	.7291	.7324	.7357	.7389	.7422	.7454	.7486	.7517	.7549
.7	.7580	.7611	.7642	.7673	.7703	.7734	.7764	.7794	.7823	.7852
.8	.7881	.7910	.7939	.7967	.7995	.8023	.8051	.8078	.8106	.8133
.9	.8159	.8186	.8212	.8238	.8264	.8289	.8315	.8340	.8365	.8389
1.0	.8413	.8438	.8461	.8485	.8508	.8531	.8554	.8577	.8599	.8621
1.1	.8643	.8665	.8686	.8708	.8729	.8749	.8770	.8790	.8810	.8830
1.2	.8849	.8869	.8888	.8907	.8925	.8944	.8962	.8980	.8997	.90147
1.3	.90320	.90490	.90658	.90824	.90988	.91149	.91309	.91466	.91621	.91774
1.4	.91924	.92073	.92220	.92364	.92507	.92647	.92785	.92922	.93056	.93189
1.5	.93319	.93448	.93574	.93699	.93822	.93943	.94062	.94179	.94295	.94408
1.6	.94520	.94630	.94738	.94845	.94950	.95053	.95154	.95254	.95352	.95449
1.7	.95543	.95637	.95728	.95818	.95907	.95994	.96080	.96164	.96246	.96327
1.8	.96407	.96485	.96562	.96638	.96712	.96784	.96856	.96926	.96995	.97062
1.9	.97128	.97193	.97257	.97320	.97381	.97441	.97500	.97558	.97615	.97670
2.0	.97725	.97778	.97831	.97882	.97932	.97982	.98030	.98077	.98124	.98169
2.1	.98214	.98257	.98300	.98341	.98382	.98422	.98461	.98500	.98537	.98574
2.2	.98610	.98645	.98679	.98713	.98745	.98778	.98809	.98840	.98870	.98899
2.3	.98928	.98956	.98983	$.9^{2}0097$	$.9^{2}0358$	$.9^{2}0613$	$.9^{2}0863$	$.9^{2}1106$	$.9^{2}1344$	$.9^{2}1576$
2.4	$.9^{2}1802$	$.9^{2}2024$	$.9^{2}2240$	$.9^{2}2451$	$.9^{2}2656$	$.9^{2}2857$	$.9^{2}3053$	$.9^{2}3244$	$.9^{2}3431$	$.9^{2}3613$

$$\Phi(z) = \frac{1}{\sqrt{2\pi}} \int_{-\infty}^{z} e^{-x^2/2}\,dx \text{ for } 0.00 \leqslant z \leqslant 4.99.$$

z	.00	.01	.02	.03	.04	.05	.06	.07	.08	.09
2.5	$.9^{2}3790$	$.9^{2}3963$	$.9^{2}4132$	$.9^{2}4297$	$.9^{2}4457$	$.9^{2}4614$	$.9^{2}4766$	$.9^{2}4915$	$.9^{2}5060$	$.9^{2}5201$
2.6	$.9^{2}5339$	$.9^{2}5473$	$.9^{2}5604$	$.9^{2}5731$	$.9^{2}5855$	$.9^{2}5975$	$.9^{2}6093$	$.9^{2}6207$	$.9^{2}6319$	$.9^{2}6427$
2.7	$.9^{2}6533$	$.9^{2}6636$	$.9^{2}6736$	$.9^{2}6833$	$.9^{2}6928$	$.9^{2}7020$	$.9^{2}7110$	$.9^{2}7197$	$.9^{2}7282$	$.9^{2}7365$
2.8	$.9^{2}7445$	$.9^{2}7523$	$.9^{2}7599$	$.9^{2}7673$	$.9^{2}7744$	$.9^{2}7814$	$.9^{2}7882$	$.9^{2}7948$	$.9^{2}8012$	$.9^{2}8074$
2.9	$.9^{2}8134$	$.9^{2}8193$	$.9^{2}8250$	$.9^{2}8305$	$.9^{2}8359$	$.9^{2}8411$	$.9^{2}8462$	$.9^{2}8511$	$.9^{2}8559$	$.9^{2}8605$
3.0	$.9^{2}8650$	$.9^{2}8694$	$.9^{2}8736$	$.9^{2}8777$	$.9^{2}8817$	$.9^{2}8856$	$.9^{2}8893$	$.9^{2}8930$	$.9^{2}8965$	$.9^{2}8999$
3.1	$.9^{3}0324$	$.9^{3}0646$	$.9^{3}0957$	$.9^{3}1260$	$.9^{3}1553$	$.9^{3}1836$	$.9^{3}2112$	$.9^{3}2378$	$.9^{3}2636$	$.9^{3}2886$
3.2	$.9^{3}3129$	$.9^{3}3363$	$.9^{3}3590$	$.9^{3}3810$	$.9^{3}4024$	$.9^{3}4230$	$.9^{3}4429$	$.6^{3}4623$	$.9^{3}4810$	$.9^{3}4991$
3.3	$.9^{3}5166$	$.9^{3}5335$	$.9^{3}5499$	$.9^{3}5658$	$.9^{3}5811$	$.9^{3}5959$	$.9^{3}6103$	$.9^{3}6242$	$.9^{3}6376$	$.9^{3}6505$
3.4	$.9^{3}6631$	$.9^{3}6752$	$.9^{3}6869$	$.9^{3}6982$	$.9^{3}7091$	$.9^{3}7197$	$.9^{3}7299$	$.9^{3}7398$	$.9^{3}7493$	$.9^{3}7585$
3.5	$.9^{3}7674$	$.9^{3}7759$	$.9^{3}7842$	$.9^{3}7922$	$.9^{3}7999$	$.9^{3}8074$	$.9^{3}8146$	$.9^{3}8215$	$.9^{3}8282$	$.9^{3}8347$
3.6	$.9^{3}8409$	$.9^{3}8469$	$.9^{3}8527$	$.9^{3}8583$	$.9^{3}8637$	$.9^{3}8689$	$.9^{3}8739$	$.9^{3}8787$	$.9^{3}8834$	$.9^{3}8879$
3.7	$.9^{3}8922$	$.9^{3}8964$	$.9^{4}0039$	$.9^{4}0426$	$.9^{4}0799$	$.9^{4}1158$	$.9^{4}1504$	$.9^{4}1838$	$.9^{4}2159$	$.9^{4}2568$
3.8	$.9^{4}2765$	$.9^{4}3052$	$.9^{4}3327$	$.9^{4}3593$	$.9^{4}3848$	$.9^{4}4094$	$.9^{4}4331$	$.9^{4}4558$	$.9^{4}4777$	$.9^{4}4988$
3.9	$.9^{4}5190$	$.9^{4}5385$	$.9^{4}5573$	$.9^{4}5753$	$.9^{4}5926$	$.9^{4}6092$	$.9^{4}6253$	$.9^{4}6406$	$.9^{4}6554$	$.9^{4}6696$
4.0	$.9^{4}6833$	$.9^{4}6964$	$.9^{4}7090$	$.9^{4}7211$	$.9^{4}7327$	$.9^{4}7439$	$.9^{4}7546$	$.9^{4}7649$	$.9^{4}7748$	$.9^{4}7843$
4.1	$.9^{4}7934$	$.9^{4}8022$	$.9^{4}8106$	$.9^{4}8186$	$.9^{4}8263$	$.9^{4}8338$	$.9^{4}8409$	$.9^{4}8477$	$.9^{4}8542$	$.9^{4}8605$
4.2	$.9^{4}8665$	$.9^{4}8723$	$.9^{4}8778$	$.9^{4}8832$	$.9^{4}8882$	$.9^{4}8931$	$.9^{4}8978$	$.9^{5}0226$	$.9^{5}0655$	$.9^{5}1066$
4.3	$.9^{5}1460$	$.9^{5}1837$	$.9^{5}2199$	$.9^{5}2545$	$.9^{5}2876$	$.9^{5}3193$	$.9^{5}3497$	$.9^{5}3788$	$.9^{5}4066$	$.9^{5}4332$
4.4	$.9^{5}4587$	$.9^{5}4831$	$.9^{5}5065$	$.9^{5}5288$	$.9^{5}5502$	$.9^{5}5706$	$.9^{5}5902$	$.9^{5}6089$	$.9^{5}6268$	$.9^{5}6439$
4.5	$.9^{5}6602$	$.9^{5}6759$	$.9^{5}6908$	$.9^{5}7051$	$.9^{5}7187$	$.9^{5}7318$	$.9^{5}7442$	$.9^{5}7561$	$.9^{5}7675$	$.9^{5}7784$
4.6	$.9^{5}7888$	$.9^{5}7987$	$.9^{5}8081$	$.9^{5}8172$	$.9^{5}8258$	$.9^{5}8340$	$.9^{5}8419$	$.9^{5}8494$	$.9^{5}8566$	$.9^{5}8634$
4.7	$.9^{5}8699$	$.9^{5}8761$	$.9^{5}8821$	$.9^{5}8877$	$.9^{5}8931$	$.9^{5}8983$	$.9^{6}0320$	$.9^{6}0789$	$.9^{6}1235$	$.9^{6}1661$
4.8	$.9^{6}2067$	$.9^{6}2453$	$.9^{6}2822$	$.9^{6}3173$	$.9^{6}3508$	$.9^{6}3827$	$.9^{6}4131$	$.9^{6}4420$	$.9^{6}4696$	$.9^{6}4958$
4.9	$.9^{6}5208$	$.9^{6}5446$	$.9^{6}5673$	$.9^{6}5889$	$.9^{6}6094$	$.9^{6}6289$	$.9^{6}6475$	$.9^{6}6652$	$.9^{6}6821$	$.9^{6}6981$

Example: $\Phi(3.39) = 0.9996505$ $\Phi(0.98) = 0.8365$

appendix III reliability tables when stress has normal distribution and strength has weibull distribution

The values in the tables are failure probabilities
stress distribution—normal
strength distribution—weibull

$\beta = 1.00,\quad C = \dfrac{\theta - S_0}{\sigma_s},\quad A = \dfrac{S_0 - \mu_s}{\sigma_s}$

A \ C	10	15	20	25	30	35	40	45	50	55
.8	.0115	.0078	.0059	.0047	.0039	.0034	.0030	.0026	.0024	.0022
.6	.0160	.0109	.0082	.0066	.0055	.0047	.0042	.0037	.0033	.0030
.4	.0218	.0148	.0112	.0090	.0075	.0065	.0057	.0051	.0046	.0041
.2	.0290	.0197	.0149	.0120	.0100	.0086	.0076	.0067	.0061	.0055
.0	.0375	.0255	.0193	.0156	.0130	.0112	.0098	.0087	.0079	.0072
−.2	.0475	.0323	.0245	.0197	.0165	.0142	.0125	.0111	.0100	.0091
−.4	.0588	.0401	.0304	.0245	.0205	.0176	.0155	.0138	.0124	.0113
−.6	.0713	.0487	.0370	.0298	.0250	.0215	.0189	.0168	.0151	.0138
−.8	.0849	.0581	.0442	.0356	.0298	.0257	.0225	.0201	.0181	.0165
−1.0	.0994	.0681	.0518	.0418	.0351	.0302	.0265	.0236	.0213	.0194
−1.4	.1301	.0896	.0683	.0552	.0463	.0399	.0350	.0312	.0282	.0256
−1.8	.1620	.1121	.0857	.0693	.0582	.0502	.0441	.0393	.0355	.0323
−2.2	.1940	.1348	.1033	.0837	.0704	.0607	.0533	.0476	.0430	.0391
−2.6	.2252	.1574	.1209	.0981	.0826	.0713	.0627	.0559	.0505	.0460
−3.0	.2555	.1795	.1382	.1124	.0947	.0818	.0720	.0643	.0581	.0529
−3.4	.2847	.2010	.1553	.1265	.1067	.0922	.0812	.0725	.0656	.0598
−3.8	.3127	.2221	.1720	.1430	.1185	.1025	.0903	.0808	.0730	.0666
−4.2	.3397	.2425	.1884	.1540	.1302	.1127	.0994	.0889	.0804	.0734
−4.6	.3656	.2625	.2045	.1674	.1417	.1228	.1084	.0969	.0877	.0801
−5.0	.3904	.2819	.2202	.1806	.1530	.1328	.1172	.1049	.0950	.0867
−5.5	.4202	.3054	.2395	.1968	.1670	.1451	.1282	.1148	.1040	.0950
−6.0	.4484	.3282	.2583	.2127	.1808	.1572	.1390	.1246	.1129	.1032
−6.5	.4753	.3502	.2766	.2283	.1944	.1692	.1497	.1343	.1217	.1113
−7.0	.5009	.3715	.2944	.2436	.2077	.1809	.1603	.1438	.1305	.1194
−8.0	.5484	.4120	.3288	.2733	.2336	.2040	.1810	.1627	.1477	.1352
−9.0	.5914	.4500	.3616	.3018	.2588	.2264	.2012	.1811	.1646	.1508
−10.0	.6303	.4854	.3927	.3291	.2831	.2482	.2210	.1991	.1811	.1661

Appendix III (*Continued*)

$\beta = 1.00, \quad C = \dfrac{\theta - \mathbb{S}_0}{\sigma_s}, \quad A = \dfrac{\mathbb{S}_0 - \mu_s}{\sigma_s}$

$A \backslash C$	60	65	70	75	80	85	90	95	100
.8	.0020	.0018	.0017	.0016	.0015	.0014	.0013	.0013	.0012
.6	.0028	.0026	.0024	.0022	.0021	.0020	.0019	.0018	.0017
.4	.0038	.0035	.0033	.0031	.0029	.0027	.0025	.0024	.0023
.2	.0051	.0047	.0043	.0041	.0038	.0036	.0034	.0032	.0031
.0	.0066	.0061	.0056	.0053	.0049	.0047	.0044	.0042	.0040
−.2	.0084	.0077	.0072	.0067	.0063	.0059	.0056	.0053	.0050
−.4	.0104	.0096	.0089	.0083	.0078	.0074	.0069	.0066	.0063
−.6	.0126	.0117	.0109	.0101	.0095	.0090	.0085	.0080	.0076
−.8	.0151	.0140	.0130	.0121	.0114	.0107	.0101	.0096	.0091
−1.0	.0178	.0164	.0153	.0143	.0134	.0126	.0119	.0113	.0107
−1.4	.0235	.0218	.0202	.0189	.0177	.0167	.0158	.0150	.0142
−1.8	.0297	.0274	.0255	.0238	.0224	.0211	.0199	.0189	.0179
−2.2	.0360	.0332	.0309	.0289	.0271	.0255	.0241	.0229	.0218
−2.6	.0432	.0391	.0364	.0340	.0319	.0301	.0284	.0270	.0256
−3.0	.0486	.0450	.0419	.0391	.0367	.0346	.0327	.0610	.0295
−3.4	.0550	.0509	.0473	.0442	.0415	.0391	.0370	.0351	.0334
−3.8	.0612	.0567	.0527	.0493	.0463	.0437	.0413	.0392	.0372
−4.2	.0675	.0625	.0581	.0544	.0511	.0481	.0455	.0432	.0411
−4.6	.0737	.0682	.0635	.0594	.0558	.0526	.0498	.0472	.0449
−5.0	.0798	.0739	.0688	.0644	.0605	.0571	.0540	.0512	.0487
−5.5	.0875	.0810	.0755	.0706	.0664	.0626	.0592	.0562	.0535
−6.0	.0950	.0881	.0821	.0768	.0722	.0681	.3644	.0612	.0582
−6.5	.1025	.0951	.0886	.0829	.0780	.0736	.0696	.0661	.0629
−7.0	.1100	.1020	.0951	.0890	.0837	.0790	.0748	.0710	.0676
−8.0	.1247	.1157	.1079	.1011	.0951	.0898	.0850	.0807	.0768
−9.0	.1392	.1292	.1206	.1130	.1063	.1004	.0951	.0902	.0860
−10.0	.1534	.1425	.1330	.1247	.1174	.1109	.1051	.0999	.0951

Appendix III (*Continued*)

$\beta = 2.00, \quad C = \dfrac{\theta - \mathcal{S}_0}{\sigma_s}, \quad A = \dfrac{\mathcal{S}_0 - \mu_s}{\sigma_s}$

$A \backslash C$	10	15	20	25	30	35	40	45	50	55
.8	.0011	.0005	.0003	.0002	.0001	.0001	.0001	.0001	.0000	.0000
.6	.0017	.0008	.0004	.0003	.0002	.0001	.0001	.0001	.0001	.0001
.4	.0025	.0011	.0007	.0004	.0003	.0002	.0002	.0001	.0001	.0001
.2	.0035	.0016	.0009	.0007	.0004	.0003	.0002	.0002	.0001	.0001
.0	.0049	.0022	.0012	.0008	.0006	.0004	.0003	.0002	.0002	.0002
−.2	.0067	.0030	.0017	.0011	.0008	.0006	.0004	.0003	.0003	.0002
−.4	.0089	.0040	.0023	.0014	.0010	.0007	.0006	.0004	.0004	.0003
−.6	.0116	.0052	.0030	.0019	.0013	.0010	.0007	.0006	.0005	.0004
−.8	.0149	.0067	.0038	.0024	.0017	.0012	.0010	.0008	.0006	.0005
−1.0	.0188	.0085	.0048	.0031	.0021	.0016	.0012	.0009	.0008	.0006
−1.4	.0284	.0128	.0073	.0047	.0032	.0024	.0018	.0014	.0012	.0010
−1.8	.0407	.0185	.0105	.0067	.0047	.0034	.0026	.0021	.0017	.0014
−2.2	.0557	.0254	.0144	.0093	.0065	.0047	.0036	.0029	.0023	.0019
−2.6	.0733	.0336	.0191	.0123	.0086	.0063	.0048	.0038	.0031	.0026
−3.0	.0935	.0431	.0246	.0158	.0110	.0081	.0062	.0049	.0040	.0033
−3.4	.1159	.0538	.0308	.0198	.0138	.0102	.0078	.0062	.0050	.0041
−3.8	.1406	.0658	.0377	.0243	.0170	.0125	.0096	.0076	.0062	.0051
−4.2	.1671	.0789	.0453	.0293	.0205	.0151	.0116	.0092	.0074	.0061
−4.6	.1954	.0930	.0536	.0347	.0243	.0179	.0137	.0109	.0088	.0073
−5.0	.2251	.1082	.0626	.0406	.0284	.0210	.0161	.0127	.0103	.0086
−5.5	.2640	.1286	.0748	.0486	.0341	.0251	.0193	.0153	.0124	.0103
−6.0	.3043	.1504	.0879	.0573	.0402	.0297	.0228	.0181	.0147	.0121
−6.5	.3457	.1735	.1020	.0667	.0468	.0346	.0266	.0211	.0171	.0142
−7.0	.3876	.1977	.1170	.0767	.0539	.0399	.0307	.0244	.0198	.0164
−8.0	.4713	.2490	.1493	.0985	.0695	.0516	.0398	.0316	.0256	.0212
−9.0	.5525	.3032	.1845	.1226	.0869	.0646	.0499	.0396	.0322	.0267
−10.0	.6285	.3591	.2222	.1488	.1059	.0790	.0611	.0486	.0396	.0328

Appendix III (*Continued*)

$\beta = 2.00, \quad C = \dfrac{\theta - \mathbb{S}_0}{\sigma_s}, \quad A = \dfrac{\mathbb{S}_0 - \mu_s}{\sigma_s}$

$A \backslash C$	60	65	70	75	80	85	90	95	100
.8	.0000	.0000	.0000	.0000	.0000	.0000	.0000	.0000	.0000
.6	.0000	.0000	.0000	.0000	.0000	.0000	.0000	.0000	.0000
.4	.0001	.0001	.0001	.0000	.0000	.0000	.0000	.0000	.0000
.2	.0001	.0001	.0001	.0001	.0001	.0000	.0000	.0000	.0000
.0	.0001	.0001	.0001	.0001	.0001	.0001	.0001	.0001	.0001
−.2	.0002	.0002	.0001	.0001	.0001	.0001	.0001	.0001	.0001
−.4	.0003	.0002	.0002	.0002	.0001	.0001	.0001	.0001	.0001
−.6	.0003	.0003	.0002	.0002	.0002	.0002	.0001	.0001	.0001
−.8	.0004	.0004	.0003	.0003	.0002	.0002	.0002	.0002	.0002
−1.0	.0005	.0005	.0004	.0003	.0003	.0003	.0002	.0002	.0002
−1.4	.0008	.0007	.0006	.0005	.0005	.0004	.0004	.0003	.0003
−1.8	.0012	.0010	.0009	.0008	.0007	.0006	.0005	.0005	.0004
−2.2	.0016	.0014	.0012	.0010	.0009	.0008	.0007	.0006	.0006
−2.6	.0022	.0018	.0016	.0014	.0012	.0011	.0010	.0009	.0008
−3.0	.0028	.0024	.0020	.0018	.0016	.0014	.0012	.0011	.0010
−3.4	.0035	.0030	.0026	.0022	.0020	.0017	.0015	.0014	.0013
−3.8	.0043	.0036	.0031	.0027	.0024	.0021	.0019	.0017	.0015
−4.2	.0052	.0044	.0038	.0033	.0029	.0026	.0023	.0021	.0019
−4.6	.0061	.0052	.0045	.0039	.0035	.0031	.0027	.0025	.0022
−5.0	.0072	.0061	.0053	.0046	.0041	.0036	.0032	.0029	.0026
−5.5	.0086	.0074	.0064	.0055	.0049	.0043	.0038	.0035	.0031
−6.0	.0102	.0087	.0075	.0066	.0058	.0051	.0046	.0041	.0037
−6.5	.0119	.0102	.0088	.0077	.0067	.0060	.0053	.0048	.0043
−7.0	.0138	.0118	.0101	.0088	.0078	.0069	.0062	.0055	.0050
−8.0	.0179	.0153	.0132	.0115	.0101	.0090	.0080	.0072	.0065
−9.0	.0225	.0192	.0166	.0145	.0127	.0113	.0101	.0090	.0082
−10.0	.0277	.0236	.0204	.0178	.0157	.0139	.0124	.0111	.0100

Appendix III (*Continued*)

$\beta = 3.00, \quad C = \dfrac{\theta - \mathbb{S}_0}{\sigma_s}, \quad A = \dfrac{\mathbb{S}_0 - \mu_s}{\sigma_s}$

A \ C	10	15	20	25	30	35	40	45	50	55
.8	.0001	.0000	.0000	.0000	.0000	.0000	.0000	.0000	.0000	.0000
.6	.0003	.0001	.0000	.0000	.0000	.0000	.0000	.0000	.0000	.0000
.4	.0004	.0001	.0000	.0000	.0000	.0000	.0000	.0000	.0000	.0000
.2	.0005	.0002	.0001	.0000	.0000	.0000	.0000	.0000	.0000	.0000
.0	.0008	.0002	.0001	.0001	.0000	.0000	.0000	.0000	.0000	.0000
−.2	.0011	.0003	.0001	.0001	.0000	.0000	.0000	.0000	.0000	.0000
−.4	.0016	.0005	.0002	.0001	.0001	.0000	.0000	.0000	.0000	.0000
−.6	.0022	.0007	.0003	.0001	.0001	.0001	.0000	.0000	.0000	.0000
−.8	.0030	.0009	.0004	.0002	.0001	.0001	.0000	.0000	.0000	.0000
−1.0	.0041	.0012	.0005	.0003	.0002	.0001	.0001	.0000	.0000	.0000
−1.4	.0069	.0021	.0009	.0004	.0003	.0002	.0001	.0001	.0001	.0000
−1.8	.0111	.0033	.0014	.0007	.0004	.0003	.0002	.0001	.0001	.0001
−2.2	.0169	.0051	.0022	.0011	.0006	.0004	.0003	.0002	.0001	.0001
−2.6	.0247	.0075	.0032	.0016	.0009	.0006	.0004	.0003	.0002	.0002
−3.0	.0349	.0106	.0045	.0023	.0013	.0008	.0006	.0004	.0003	.0002
−3.4	.0475	.0145	.0062	.0032	.0018	.0012	.0008	.0005	.0004	.0003
−3.8	.0630	.0193	.0082	.0042	.0024	.0015	.0010	.0007	.0005	.0004
−4.2	.0815	.0252	.0108	.0055	.0032	.0020	.0014	.0010	.0007	.0005
−4.6	.1031	.0322	.0138	.0071	.0041	.0026	.0017	.0012	.0009	.0007
−5.0	.1279	.0404	.0173	.0089	.0052	.0033	.0022	.0015	.0011	.0008
−5.5	.1634	.0524	.0225	.0116	.0067	.0043	.0029	.0020	.0015	.0011
−6.0	.2037	.0665	.0287	.0148	.0086	.0054	.0036	.0025	.0019	.0014
−6.5	.2485	.0828	.0360	.0186	.0108	.0068	.0046	.0032	.0023	.0018
−7.0	.2973	.1013	.0443	.0230	.0134	.0084	.0057	.0040	.0029	.0022
−8.0	.4039	.1454	.0645	.0336	.0196	.0124	.0083	.0059	.0043	.0032
−9.0	.5165	.1985	.0897	.0471	.0276	.0175	.0117	.0083	.0060	.0045
−10.0	.6269	.2600	.1202	.0636	.0374	.0327	.0160	.0112	.0082	.0062

Appendix III (*Continued*)

$\beta = 3.00, \quad C = \frac{\theta - \mathbb{S}_0}{\sigma_s}, \quad A = \frac{\mathbb{S}_0 - \mu_s}{\sigma_s}$

$A \backslash C$	60	65	70	75	80	85	90	95	100
.8	.0000	.0000	.0000	.0000	.0000	.0000	.0000	.0000	.0000
.6	.0000	.0000	.0000	.0000	.0000	.0000	.0000	.0000	.0000
.4	.0000	.0000	.0000	.0000	.0000	.0000	.0000	.0000	.0000
.2	.0000	.0000	.0000	.0000	.0000	.0000	.0000	.0000	.0000
.0	.0000	.0000	.0000	.0000	.0000	.0000	.0000	.0000	.0000
−.2	.0000	.0000	.0000	.0000	.0000	.0000	.0000	.0000	.0000
−.4	.0000	.0000	.0000	.0000	.0000	.0000	.0000	.0000	.0000
−.6	.0000	.0000	.0000	.0000	.0000	.0000	.0000	.0000	.0000
−.8	.0000	.0000	.0000	.0000	.0000	.0000	.0000	.0000	.0000
−1.0	.0000	.0000	.0000	.0000	.0000	.0000	.0000	.0000	.0000
−1.4	.0000	.0000	.0000	.0000	.0000	.0000	.0000	.0000	.0000
−1.8	.0001	.0000	.0000	.0000	.0000	.0000	.0000	.0000	.0000
−2.2	.0001	.0001	.0001	.0000	.0000	.0000	.0000	.0000	.0000
−2.6	.0001	.0001	.0001	.0001	.0000	.0000	.0000	.0000	.0000
−3.0	.0002	.0001	.0001	.0001	.0001	.0001	.0000	.0000	.0000
−3.4	.0002	.0002	.0001	.0001	.0001	.0001	.0001	.0001	.0000
−3.8	.0003	.0002	.0002	.0002	.0001	.0001	.0001	.0001	.0001
−4.2	.0004	.0003	.0003	.0002	.0002	.0001	.0001	.0001	.0001
−4.6	.0005	.0004	.0003	.0003	.0002	.0002	.0002	.0001	.0001
−5.0	.0006	.0005	.0004	.0003	.0003	.0002	.0002	.0002	.0001
−5.5	.0008	.0007	.0005	.0004	.0004	.0003	.0003	.0002	.0002
−6.0	.0011	.0009	.0007	.0006	.0005	.0004	.0003	.0003	.0002
−6.5	.0014	.0011	.0009	.0007	.0006	.0005	.0004	.0003	.0003
−7.0	.0017	.0013	.0011	.0009	.0007	.0006	.0005	.0004	.0004
−8.0	.0025	.0019	.0016	.0013	.0010	.0009	.0007	.0006	.0005
−9.0	.0035	.0027	.0022	.0018	.0015	.0012	.0010	.0009	.0008
−10.0	.0048	.0037	.0030	.0024	.0020	.0017	.0014	.0012	.0010

Appendix III (*Continued*)

$\beta = 4.00, \quad C = \dfrac{\theta - \mathbb{S}_0}{\sigma_s}, \quad A = \dfrac{\mathbb{S}_0 - \mu_s}{\sigma_s}$

A \ C	10	15	20	25	30	35	40	45	50	55
.8	.0000	.0000	.0000	.0000	.0000	.0000	.0000	.0000	.0000	.0000
.6	.0000	.0000	.0000	.0000	.0000	.0000	.0000	.0000	.0000	.0000
.4	.0001	.0000	.0000	.0000	.0000	.0000	.0000	.0000	.0000	.0000
.2	.0001	.0000	.0000	.0000	.0000	.0000	.0000	.0000	.0000	.0000
.0	.0001	.0000	.0000	.0000	.0000	.0000	.0000	.0000	.0000	.0000
−.2	.0002	.0000	.0000	.0000	.0000	.0000	.0000	.0000	.0000	.0000
−.4	.0003	.0001	.0000	.0000	.0000	.0000	.0000	.0000	.0000	.0000
−.6	.0005	.0001	.0000	.0000	.0000	.0000	.0000	.0000	.0000	.0000
−.8	.0007	.0001	.0000	.0000	.0000	.0000	.0000	.0000	.0000	.0000
−1.0	.0010	.0002	.0001	.0000	.0000	.0000	.0000	.0000	.0000	.0000
−1.4	.0018	.0004	.0001	.0000	.0000	.0000	.0000	.0000	.0000	.0000
−1.8	.0033	.0006	.0002	.0001	.0000	.0000	.0000	.0000	.0000	.0000
−2.2	.0055	.0011	.0003	.0001	.0001	.0000	.0000	.0000	.0000	.0000
−2.6	.0088	.0018	.0006	.0002	.0001	.0001	.0000	.0000	.0000	.0000
−3.0	.0136	.0027	.0009	.0004	.0002	.0001	.0001	.0000	.0000	.0000
−3.4	.0201	.0041	.0013	.0005	.0003	.0001	.0001	.0001	.0000	.0000
−3.8	.0289	.0059	.0019	.0008	.0004	.0002	.0001	.0001	.0000	.0000
−4.2	.0404	.0082	.0026	.0011	.0005	.0003	.0002	.0001	.0001	.0000
−4.6	.0051	.0113	.0036	.0015	.0007	.0004	.0002	.0001	.0001	.0001
−5.0	.0732	.0152	.0048	.0020	.0010	.0005	.0003	.0002	.0001	.0001
−5.5	.1015	.0214	.0068	.0028	.0014	.0007	.0004	.0003	.0002	.0001
−6.0	.1366	.0293	.0094	.0039	.0019	.0010	.0006	.0004	.0002	.0002
−6.5	.1788	.0392	.0126	.0052	.0025	.0014	.0008	.0005	.0003	.0002
−7.0	.2282	.0515	.0167	.0069	.0033	.0018	.0011	.0007	.0004	.0003
−8.0	.3466	.0839	.0275	.0114	.0055	.0030	.0017	.0011	.0007	.0005
−9.0	.4838	.1284	.0429	.0179	.0087	.0047	.0027	.0017	.0011	.0008
−10.0	.6252	.1862	.0638	.0267	.0130	.0070	.0041	.0026	.0017	.0012

Appendix III (*Continued*)

$\beta = 4.00,\quad C = \dfrac{\theta - \mathbb{S}_0}{\sigma_s},\quad A = \dfrac{\mathbb{S}_0 - \mu_s}{\sigma_s}$

A \ C	60	65	70	75	80	85	90	95	100
.8	.0000	.0000	.0000	.0000	.0000	.0000	.0000	.0000	.0000
.6	.0000	.0000	.0000	.0000	.0000	.0000	.0000	.0000	.0000
.4	.0000	.0000	.0000	.0000	.0000	.0000	.0000	.0000	.0000
.2	.0000	.0000	.0000	.0000	.0000	.0000	.0000	.0000	.0000
.0	.0000	.0000	.0000	.0000	.0000	.0000	.0000	.0000	.0000
−.2	.0000	.0000	.0000	.0000	.0000	.0000	.0000	.0000	.0000
−.4	.0000	.0000	.0000	.0000	.0000	.0000	.0000	.0000	.0000
−.6	.0000	.0000	.0000	.0000	.0000	.0000	.0000	.0000	.0000
−.8	.0000	.0000	.0000	.0000	.0000	.0000	.0000	.0000	.0000
−1.0	.0000	.0000	.0000	.0000	.0000	.0000	.0000	.0000	.0000
−1.4	.0000	.0000	.0000	.0000	.0000	.0000	.0000	.0000	.0000
−1.8	.0000	.0000	.0000	.0000	.0000	.0000	.0000	.0000	.0000
−2.2	.0000	.0000	.0000	.0000	.0000	.0000	.0000	.0000	.0000
−2.6	.0000	.0000	.0000	.0000	.0000	.0000	.0000	.0000	.0000
−3.0	.0000	.0000	.0000	.0000	.0000	.0000	.0000	.0000	.0000
−3.4	.0000	.0000	.0000	.0000	.0000	.0000	.0000	.0000	.0000
−3.8	.0000	.0000	.0000	.0000	.0000	.0000	.0000	.0000	.0000
−4.2	.0000	.0000	.0000	.0000	.0000	.0000	.0000	.0000	.0000
−4.6	.0000	.0000	.0000	.0000	.0000	.0000	.0000	.0000	.0000
−5.0	.0001	.0000	.0000	.0000	.0000	.0000	.0000	.0000	.0000
−5.5	.0001	.0001	.0000	.0000	.0000	.0000	.0000	.0000	.0000
−6.0	.0001	.0001	.0001	.0000	.0000	.0000	.0000	.0000	.0000
−6.5	.0002	.0001	.0001	.0001	.0000	.0000	.0000	.0000	.0000
−7.0	.0002	.0002	.0001	.0001	.0001	.0001	.0000	.0000	.0000
−8.0	.0003	.0003	.0002	.0001	.0001	.0001	.0001	.0001	.0000
−9.0	.0005	.0004	.0003	.0002	.0002	.0001	.0001	.0001	.0001
−10.0	.0008	.0006	.0004	.0003	.0003	.0002	.0002	.0001	.0001

Appendix III (*Continued*)

$\beta = 5.00, \quad C = \dfrac{\theta - S_0}{\sigma_s}, \quad A = \dfrac{S_0 - \mu_s}{\sigma_s}$

$A \backslash C$	10	15	20	25	30	35	40	45	50	55
.8	.0000	.0000	.0000	.0000	.0000	.0000	.0000	.0000	.0000	.0000
.6	.0000	.0000	.0000	.0000	.0000	.0000	.0000	.0000	.0000	.0000
.4	.0000	.0000	.0000	.0000	.0000	.0000	.0000	.0000	.0000	.0000
.2	.0000	.0000	.0000	.0000	.0000	.0000	.0000	.0000	.0000	.0000
.0	.0000	.0000	.0000	.0000	.0000	.0000	.0000	.0000	.0000	.0000
−.2	.0001	.0000	.0000	.0000	.0000	.0000	.0000	.0000	.0000	.0000
−.4	.0001	.0000	.0000	.0000	.0000	.0000	.0000	.0000	.0000	.0000
−.6	.0001	.0000	.0000	.0000	.0000	.0000	.0000	.0000	.0000	.0000
−.8	.0002	.0000	.0000	.0000	.0000	.0000	.0000	.0000	.0000	.0000
−1.0	.0003	.0000	.0000	.0000	.0000	.0000	.0000	.0000	.0000	.0000
−1.4	.0005	.0001	.0000	.0000	.0000	.0000	.0000	.0000	.0000	.0000
−1.8	.0010	.0001	.0000	.0000	.0000	.0000	.0000	.0000	.0000	.0000
−2.2	.0019	.0003	.0001	.0000	.0000	.0000	.0000	.0000	.0000	.0000
−2.6	.0033	.0004	.0001	.0000	.0000	.0000	.0000	.0000	.0000	.0000
−3.0	.0055	.0007	.0002	.0001	.0000	.0000	.0000	.0000	.0000	.0000
−3.4	.0089	.0012	.0003	.0001	.0000	.0000	.0000	.0000	.0000	.0000
−3.8	.0137	.0018	.0004	.0001	.0001	.0000	.0000	.0000	.0000	.0000
−4.2	.0206	.0028	.0007	.0002	.0001	.0000	.0000	.0000	.0000	.0000
−4.6	.0300	.0041	.0010	.0003	.0001	.0001	.0000	.0000	.0000	.0000
−5.0	.0426	.0058	.0014	.0005	.0002	.0001	.0000	.0000	.0000	.0000
−5.5	.0639	.0089	.0021	.0007	.0003	.0001	.0001	.0000	.0000	.0000
−6.0	.0924	.0131	.0031	.0010	.0004	.0002	.0001	.0001	.0000	.0000
−6.5	.1295	.0187	.0045	.0015	.0006	.0003	.0001	.0001	.0000	.0000
−7.0	.1761	.0263	.0063	.0021	.0008	.0004	.0002	.0001	.0001	.0001
−8.0	.2988	.0484	.0118	.0039	.0016	.0007	.0004	.0002	.0001	.0001
−9.0	.4546	.0828	.0205	.0068	.0027	.0013	.0006	.0004	.0002	.0001
−10.0	.6235	.1328	.0337	.0112	.0045	.0021	.0011	.0006	.0004	.0002

Taken from: C. Lipson, N. J. Sheth, and R. Disney: *Reliability Prediction-Mechanical Stress / Strength Inference,* Final Tech. Rep. RADC-TR-66-710, Rome Air Development Center, Research and Technology Division, Air Force Systems Command, Griffiss Air Force Base, New York, March 1967.

appendix IV reliability tables when stress has largest extreme value distribution and strength has Weibull distribution

Appendix IVA. Reliability values for type II L.E.V. stress and Weibull strength

$A = \dfrac{\theta_{\mathbb{S}}}{\theta_s} = 1.0, \quad C = \dfrac{\mathbb{S}_0 - s_0}{\theta_s} = 0.5$

$\beta_{\mathbb{S}} \backslash \beta_s$	1.0	2.0	3.0	4.0	5.0	6.0	7.0	8.0	9.0	10.0
1.0	0.440427	0.482383	0.509071	0.527856	0.541200	0.550930	0.558254	0.563932	0.568446	0.572113
2.0	0.462273	0.536687	0.590609	0.629214	0.657077	0.677534	0.692872	0.704627	0.713826	0.721165
3.0	0.475469	0.566076	0.635951	0.688020	0.726510	0.755107	0.776600	0.792994	0.805699	0.815703
4.0	0.483549	0.583475	0.662781	0.723179	0.768426	0.802237	0.827635	0.846898	0.861684	0.873183
5.0	0.488889	0.594683	0.679834	0.745383	0.794777	0.831720	0.859376	0.880202	0.896025	0.908176
6.0	0.492648	0.602401	0.691366	0.760191	0.812139	0.850923	0.879809	0.901389	0.917615	0.929922
7.0	0.495425	0.607996	0.699573	0.770555	0.824104	0.863958	0.893473	0.915348	0.931632	0.943836
8.0	0.497556	0.612219	0.705660	0.778113	0.832686	0.873156	0.902960	0.924882	0.941048	0.953032
9.0	0.499241	0.615510	0.710328	0.783816	0.839059	0.879877	0.909777	0.931619	0.947591	0.959316
10.0	0.500605	0.618143	0.714009	0.788246	0.843935	0.884939	0.914832	0.936534	0.952285	0.963747

$A = \dfrac{\theta_{\mathbb{S}}}{\theta_s} = 1.00, \quad C = \dfrac{\mathbb{S}_0 - s_0}{\theta_s} = 2.0$

$\beta_{\mathbb{S}} \backslash \beta_s$	1.0	2.0	3.0	4.0	5.0	6.0	7.0	8.0	9.0	10.0
1.0	0.699606	0.872192	0.946399	0.977289	0.990224	0.995722	0.998097	0.999138	0.999601	0.999809
2.0	0.702140	0.879543	0.953341	0.981982	0.992966	0.997208	0.998870	0.999530	0.999796	0.999905
3.0	0.705176	0.883602	0.956490	0.983919	0.994028	0.997755	0.999140	0.999661	0.999858	0.999934
4.0	0.707293	0.886034	0.958221	0.984924	0.994555	0.998016	0.999265	0.999719	0.999885	0.999946
5.0	0.708781	0.887625	0.959297	0.985525	0.994859	0.998162	0.999332	0.999749	0.999899	0.999953
6.0	0.709869	0.888739	0.960025	0.985920	0.995055	0.998253	0.999374	0.999768	0.999907	0.999956
7.0	0.710695	0.889560	0.960547	0.986198	0.995189	0.998315	0.999402	0.999780	0.999912	0.999958
8.0	0.711341	0.890188	0.960939	0.986402	0.995287	0.998360	0.999421	0.999788	0.999915	0.999960
9.0	0.711860	0.890684	0.961244	0.986559	0.995361	0.998393	0.999436	0.999794	0.999918	0.999961
10.0	0.712285	0.891084	0.961487	0.986683	0.995419	0.998419	0.999447	0.999799	0.999920	0.999961

$A = \frac{\theta_{\mathbb{S}}}{\theta_s} = 1.50, \quad C = \frac{\mathbb{S}_0 - s_0}{\theta_s} = 1.00$

$\beta_{\mathbb{S}} \backslash \beta_s$	1.0	2.0	3.0	4.0	5.0	6.0	7.0	8.0	9.0	10.0
1.0	0.612033	0.748128	0.820728	0.862993	0.889869	0.908215	0.921446	0.931402	0.939149	0.945341
2.0	0.631698	0.794637	0.881379	0.927403	0.952851	0.967687	0.976808	0.982695	0.986665	0.989444
3.0	0.642296	0.814069	0.903691	0.948557	0.971283	0.983176	0.989658	0.993347	0.995538	0.996894
4.0	0.648521	0.824165	0.914247	0.957710	0.978557	0.988733	0.993840	0.996489	0.997915	0.998710
5.0	0.552546	0.830197	0.920130	0.962475	0.982085	0.991234	0.995581	0.997696	0.998754	0.999299
6.0	0.655340	0.834157	0.923795	0.965292	0.984062	0.992557	0.996447	0.998258	0.999119	0.999538
7.0	0.657387	0.836938	0.926265	0.967116	0.985289	0.993343	0.996937	0.998561	0.999306	0.999653
8.0	0.658947	0.838989	0.928028	0.968377	0.986110	0.993851	0.997243	0.998742	0.999413	0.999716
9.0	0.660174	0.840561	0.929345	0.969294	0.986691	0.994201	0.997448	0.998860	0.999480	0.999755
10.0	0.661163	0.841802	0.930362	0.969938	0.987122	0.994455	0.997593	0.998941	0.999525	0.999780

$A = \frac{\theta_{\mathbb{S}}}{\theta_s} = 2.00, \quad C = \frac{\mathbb{S}_0 - s_0}{\theta_s} = 1.00$

$\beta_{\mathbb{S}} \backslash \beta_s$	1.0	2.0	3.0	4.0	5.0	6.0	7.0	8.0	9.0	10.0
1.0	0.647614	0.786332	0.853737	0.890715	0.913363	0.928450	0.939150	0.947106	0.953243	0.958115
2.0	0.673881	0.839279	0.915781	0.951970	0.970297	0.980308	0.986179	0.989942	0.992252	0.993911
3.0	0.686517	0.859679	0.936458	0.969665	0.984520	0.991535	0.995055	0.996932	0.997993	0.998627
4.0	0.693591	0.869722	0.945578	0.976642	0.989506	0.995022	0.997495	0.998662	0.999242	0.999545
5.0	0.698036	0.875504	0.950416	0.980043	0.991727	0.996432	0.998388	0.999232	0.999612	0.999790
6.0	0.701066	0.879202	0.953323	0.931960	0.992896	0.997121	0.998790	0.999468	0.999752	0.999875
7.0	0.703256	0.881747	0.955232	0.983159	0.993590	0.997508	0.999002	0.999584	0.999816	0.999911
8.0	0.704908	0.883597	0.956568	0.983967	0.994039	0.997748	0.999128	0.999650	0.999850	0.999929
9.0	0.706198	0.884998	0.957550	0.934543	0.994350	0.997908	0.999209	0.999690	0.999870	0.999939
10.0	0.707232	0.886094	0.958299	0.984972	0.994575	0.998021	0.999264	0.999717	0.999883	0.999945

Appendix IV-A (*Continued*)

$A = \dfrac{\theta_{\mathbb{S}}}{\theta_s} = 2.00, \quad C = \dfrac{\mathbb{S}_0 - s_0}{\theta_s} = 2.00$

$\beta_{\mathbb{S}} \backslash \beta_s$	1.0	2.0	3.0	4.0	5.0	6.0	7.0	8.0	9.0	10.0
1.0	0.746624	0.905983	0.963942	0.985600	0.994047	0.997467	0.998894	0.999504	0.999769	0.999887
2.0	0.756684	0.920493	0.974231	0.991381	0.996998	0.998909	0.999583	0.999830	0.999923	0.999959
3.0	0.762509	0.926670	0.978079	0.993344	0.997915	0.999319	0.999763	0.999908	0.999957	0.999974
4.0	0.766018	0.929931	0.979944	0.994231	0.998303	0.999481	0.999830	0.999935	0.999968	0.999978
5.0	0.768318	0.931906	0.981008	0.994711	0.998502	0.999561	0.999861	0.999947	0.999972	0.999980
6.0	0.769931	0.933218	0.981684	0.995004	0.998619	0.999605	0.999877	0.999953	0.999975	0.999981
7.0	0.771119	0.934146	0.982147	0.995199	0.998695	0.999634	0.999887	0.999957	0.999976	0.999981
8.0	0.772029	0.934837	0.982482	0.995337	0.998747	0.999652	0.999894	0.999959	0.999977	0.999982
9.0	0.772749	0.935369	0.982735	0.995438	0.998785	0.999666	0.999899	0.999961	0.999977	0.999982
10.0	0.773330	0.935791	0.982932	0.995516	0.998814	0.999676	0.999902	0.999962	0.999978	0.999982

$A = \dfrac{\theta_{\mathbb{S}}}{\theta_s} = 2.50, \quad C = \dfrac{\mathbb{S}_0 - s_0}{\theta_s} = 2.00$

$\beta_{\mathbb{S}} \backslash \beta_s$	1.0	2.0	3.0	4.0	5.0	6.0	7.0	8.0	9.0	10.0
1.0	0.763316	0.916444	0.968886	0.987795	0.995012	0.997893	0.999084	0.999589	0.999808	0.999904
2.0	0.776397	0.932856	0.979641	0.993525	0.997827	0.999229	0.999708	0.999880	0.999943	0.999967
3.0	0.783213	0.939429	0.983413	0.995325	0.998622	0.999569	0.999852	0.999940	0.999969	0.999978
4.0	0.737162	0.942771	0.985163	0.996095	0.998936	0.999692	0.999900	0.999959	0.999976	0.999981
5.0	0.789696	0.944744	0.986130	0.996495	0.999089	0.999749	0.999920	0.999966	0.999978	0.999982
6.0	0.791447	0.946031	0.986731	0.996733	0.999177	0.999780	0.999931	0.999970	0.999980	0.999982
7.0	0.792726	0.946930	0.987136	0.996887	0.999231	0.999799	0.999937	0.999972	0.999980	0.999983
8.0	0.793698	0.947592	0.987425	0.996995	0.999268	0.999811	0.999941	0.999973	0.999981	0.999983
9.0	0.794461	0.948098	0.987641	0.997073	0.999295	0.999819	0.999944	0.999974	0.999981	0.999983
10.0	0.795076	0.948497	0.987808	0.997133	0.999315	0.999826	0.999946	0.999974	0.999981	0.999983

Appendix IV-B. Reliability values for type I L.E.V. stress and Weibull strength

For $\beta_{\mathbb{S}} = 1.00$

$$A = \frac{\theta_{\mathbb{S}}}{\theta_s} \quad \text{and} \quad C = \frac{\mathbb{S}_0 - \delta_s}{\theta_s}$$

$C \backslash A$	1.00	1.25	1.50	1.75	2.00	2.25	2.50	2.75	3.00	3.25
−1.50	0.220605	0.278581	0.330518	0.376619	0.417471	0.453734	0.486034	0.514923	0.540873	0.564284
−1.00	0.343602	0.400128	0.443717	0.490605	0.526922	0.558623	0.586484	0.611132	0.633073	0.652718
−0.50	0.489891	0.538326	0.578828	0.613058	0.642297	0.667524	0.689490	0.708776	0.725835	0.741027
0.00	0.632108	0.669253	0.699780	0.725257	0.746812	0.765271	0.781247	0.795204	0.807499	0.818409
0.50	0.749756	0.775991	0.797324	0.814991	0.829850	0.842515	0.853435	0.862945	0.871302	0.878701
1.00	0.836662	0.854182	0.868336	0.880002	0.889778	0.898088	0.905236	0.911449	0.916899	0.921719
1.50	0.396263	0.907544	0.916622	0.924083	0.930322	0.935615	0.940162	0.944110	0.947569	0.950626
2.00	0.935255	0.942354	0.948053	0.952729	0.956633	0.959942	0.962783	0.965247	0.967405	0.969311
2.50	0.960027	0.964431	0.967961	0.970854	0.973269	0.975313	0.977067	0.978589	0.979920	0.981096
3.00	0.975483	0.978191	0.980360	0.982136	0.983618	0.984872	0.985948	0.986880	0.987697	0.988417
3.50	0.985020	0.986676	0.988002	0.989087	0.989992	0.990759	0.991415	0.991985	0.992483	0.992923
4.00	0.990865	0.991875	0.992683	0.993344	0.993896	0.994362	0.994762	0.995109	0.995412	0.995680
4.50	0.994433	0.995047	0.995539	0.995941	0.996276	0.996560	0.996803	0.997014	0.997199	0.997361
5.00	0.996606	0.996979	0.997277	0.997522	0.997726	0.997898	0.998046	0.998174	0.998286	0.998385
5.50	0.997926	0.998153	0.998334	0.998483	0.998606	0.998711	0.998801	0.998878	0.998946	0.999006
6.00	0.998728	0.998866	0.998976	0.999066	0.999141	0.999205	0.999259	0.999306	0.999348	0.999384
6.50	0.999215	0.999299	0.999366	0.999420	0.999466	0.999504	0.999537	0.999566	0.999591	0.999613
7.00	0.999511	0.999562	0.999602	0.999635	0.999663	0.999686	0.999706	0.999724	0.999739	0.999752
7.50	0.999690	0.999721	0.999746	0.999766	0.999782	0.999797	0.999809	0.999819	0.999829	0.999837
8.00	0.999799	0.999818	0.999833	0.999845	0.999855	0.999864	0.999871	0.999877	0.999883	0.999888

Appendix IV-B (*Continued*)

For $\beta_{\mathfrak{s}} = 2.00$

$A = \frac{\theta_{\mathfrak{s}}}{\theta_s}$ and $C = \frac{\mathfrak{S}_0 - \delta_s}{\theta_s}$

$C \backslash A$	1.00	1.25	1.50	1.75	2.00	2.25	2.50	2.75	3.00	3.25
−1.50	0.178603	0.246742	0.315060	0.380320	0.440771	0.495690	0.544978	0.588889	0.627850	0.662353
−1.00	0.328603	0.400397	0.466387	0.525570	0.577873	0.623713	0.663717	0.698572	0.728948	0.755456
−0.50	0.495244	0.558153	0.612944	0.660169	0.700669	0.735350	0.765068	0.790585	0.812561	0.831552
0.00	0.645980	0.694329	0.735081	0.769356	0.798207	0.822555	0.843179	0.860725	0.875722	0.888602
0.50	0.764035	0.798071	0.826195	0.849499	0.868891	0.885110	0.898752	0.910290	0.920106	0.928503
1.00	0.848077	0.870719	0.889205	0.904385	0.916929	0.927363	0.936100	0.943464	0.949711	0.955042
1.50	0.904361	0.918896	0.930678	0.940300	0.948217	0.954781	0.960262	0.964873	0.968776	0.972102
2.00	0.940039	0.949767	0.957133	0.963129	0.968050	0.972121	0.975516	0.978367	0.980778	0.982831
2.50	0.963476	0.969131	0.973682	0.977379	0.980409	0.982912	0.984997	0.986747	0.988227	0.989485
3.00	0.977644	0.981119	0.983911	0.936176	0.988030	0.989561	0.990836	0.991905	0.992808	0.993577
3.50	0.986357	0.983481	0.990186	0.991568	0.992699	0.993632	0.994409	0.995061	0.995611	0.996079
4.00	0.991686	0.992980	0.994019	0.994560	0.995548	0.996116	0.996589	0.996985	0.997320	0.997604
4.50	0.994935	0.995722	0.996353	0.996865	0.997283	0.997628	0.997915	0.998156	0.998359	0.998532
5.00	0.996911	0.997389	0.997773	0.998084	0.998338	0.998547	0.998722	0.998868	0.998991	0.999096
5.50	0.998112	0.998403	0.998635	0.998824	0.998978	0.999105	0.999211	0.999300	0.999375	0.999438
6.00	0.998841	0.999018	0.999159	0.999273	0.999367	0.999444	0.999508	0.999562	0.999608	0.999646
6.50	0.999234	0.999393	0.999477	0.999546	0.999603	0.999650	0.999689	0.999721	0.999749	0.999772
7.00	0.999553	0.999617	0.999669	0.999712	0.999746	0.999774	0.999798	0.999818	0.999835	0.999849
7.50	0.999716	0.999755	0.999786	0.999812	0.999833	0.999850	0.999864	0.999876	0.999887	0.999895
8.00	0.999814	0.999838	0.999857	0.999873	0.999886	0.999896	0.999905	0.999912	0.999918	0.999923

For $\beta_{\mathbf{S}} = 3.00$

$$A = \frac{\theta_{\mathbf{S}}}{\theta_s} \quad \text{and} \quad C = \frac{\mathbb{S}_0 - \delta_s}{\theta_s}$$

$C \backslash A$	1.00	1.25	1.50	1.75	2.00	2.25	2.50	2.75	3.00	3.25
−1.50	0.171489	0.242270	0.316359	0.389480	0.458723	0.522419	0.579835	0.630861	0.675774	0.715050
−1.00	0.330704	0.407848	0.480527	0.546804	0.605901	0.657792	0.702892	0.741829	0.775313	0.804046
−0.50	0.503539	0.571714	0.631893	0.684154	0.729030	0.767294	0.799787	0.827326	0.850654	0.870430
0.00	0.655843	0.708164	0.752611	0.790071	0.821500	0.847815	0.869840	0.888290	0.903773	0.916797
0.50	0.772559	0.809295	0.839773	0.864996	0.885860	0.903134	0.917464	0.929382	0.939324	0.947648
1.00	0.854411	0.878791	0.898730	0.915049	0.928432	0.939437	0.948516	0.956033	0.962283	0.967499
1.50	0.908691	0.924315	0.936982	0.947281	0.955682	0.962562	0.968218	0.972890	0.976764	0.979992
2.00	0.943461	0.953260	0.961164	0.967563	0.972767	0.977018	0.980506	0.983381	0.985763	0.987746
2.50	0.965263	0.971329	0.976206	0.980145	0.983342	0.985950	0.988086	0.989846	0.991303	0.992515
3.00	0.978757	0.982482	0.985471	0.987882	0.989836	0.991428	0.992733	0.993806	0.994694	0.995432
3.50	0.987042	0.989319	0.991144	0.992614	0.993805	0.994775	0.995568	0.996222	0.996762	0.997211
4.00	0.992106	0.993493	0.994604	0.995493	0.996223	0.996813	0.997295	0.997692	0.998020	0.998293
4.50	0.995191	0.996034	0.996710	0.997253	0.997694	0.998052	0.993345	0.998586	0.998785	0.998951
5.00	0.997067	0.997579	0.997990	0.998320	0.998587	0.998805	0.998983	0.999129	0.999250	0.999350
5.50	0.998207	0.998518	0.998767	0.998967	0.999130	0.999262	0.999370	0.999458	0.999532	0.999593
6.00	0.998899	0.999088	0.999239	0.999360	0.999459	0.999539	0.999604	0.999658	0.999703	0.999740
6.50	0.999319	0.999434	0.999525	0.999599	0.999659	0.999707	0.999747	0.999780	0.999807	0.999829
7.00	0.999574	0.999643	0.999699	0.999744	0.999780	0.999809	0.999833	0.999853	0.999870	0.999883
7.50	0.999728	0.999771	0.999804	0.999831	0.999853	0.999871	0.999886	0.999898	0.999908	0.999916
8.00	0.999822	0.999848	0.999868	0.999885	0.999898	0.999909	0.999918	0.999925	0.999931	0.999936

Appendix IV-B (*Continued*)

For $\beta_{\mathfrak{S}} = 4.00$

$A = \frac{\theta_{\mathfrak{S}}}{\theta_s}$ and $C = \frac{\mathfrak{S}_0 - \delta_s}{\theta_s}$

$C \backslash A$	1.00	1.25	1.50	1.75	2.00	2.25	2.50	2.75	3.00	3.25
−1.50	0.171120	0.243742	0.321091	0.398482	0.472451	0.540816	0.602470	0.657085	0.704841	0.746211
−1.00	0.334869	0.414980	0.491159	0.561053	0.623537	0.678355	0.725805	0.766489	0.801145	0.830537
−0.50	0.510289	0.581086	0.643911	0.698561	0.745432	0.785237	0.818815	0.847018	0.870646	0.890417
0.00	0.662574	0.716823	0.762983	0.801851	0.834348	0.861396	0.883852	0.902473	0.917912	0.930722
0.50	0.778009	0.816004	0.847522	0.873540	0.894957	0.912565	0.927041	0.938950	0.948761	0.956859
1.00	0.858331	0.883493	0.904055	0.920822	0.934493	0.945649	0.954766	0.962229	0.968354	0.973393
1.50	0.911324	0.927431	0.940465	0.951018	0.959574	0.966524	0.972182	0.976801	0.980583	0.983687
2.00	0.945158	0.955253	0.963375	0.969922	0.975211	0.979496	0.982976	0.985812	0.988131	0.990032
2.50	0.966331	0.972577	0.977585	0.981610	0.984856	0.987481	0.989610	0.991343	0.992759	0.993919
3.00	0.979419	0.983253	0.986321	0.988784	0.990766	0.992368	0.993666	0.994722	0.995584	0.996290
3.50	0.987450	0.989792	0.991665	0.993166	0.994373	0.995348	0.996138	0.996780	0.997304	0.997734
4.00	0.992355	0.993782	0.994922	0.995835	0.996569	0.997162	0.997642	0.998032	0.998351	0.998611
4.50	0.995342	0.996210	0.996903	0.997458	0.997904	0.998264	0.998556	0.998793	0.998986	0.999144
5.00	0.997159	0.997687	0.998107	0.998444	0.998715	0.998934	0.999111	0.999254	0.999372	0.999468
5.50	0.998263	0.998583	0.998838	0.999043	0.999207	0.999340	0.999447	0.999535	0.999606	0.999664
6.00	0.998933	0.999127	0.999282	0.999406	0.999506	0.999587	0.999652	0.999705	0.999748	0.999783
6.50	0.999340	0.999457	0.999552	0.999627	0.999687	0.999736	0.999776	0.999808	0.999834	0.999855
7.00	0.999586	0.999658	0.999715	0.999761	0.999797	0.999827	0.999851	0.999870	0.999886	0.999899
7.50	0.999736	0.999779	0.999814	0.999842	0.999864	0.999882	0.999896	0.999908	0.999918	0.999926
8.00	0.999827	0.999853	0.999874	0.999891	0.999904	0.999915	0.999924	0.999931	0.999937	0.999942

For $\beta_{\mathbb{S}} = 5.00$

$A = \dfrac{\theta_{\mathbb{S}}}{\theta_s}$ and $C = \dfrac{\mathbb{S}_0 - \delta_s}{\theta_s}$

$C \backslash A$	1.00	1.25	1.50	1.75	2.00	2.25	2.50	2.75	3.00	3.25
−1.50	0.172281	0.246375	0.325931	0.406031	0.482886	0.554014	0.618087	0.674652	0.723842	0.766144
−1.00	0.338711	0.420765	0.499093	0.571112	0.635507	0.691901	0.740535	0.782007	0.817084	0.846581
−0.50	0.515348	0.587763	0.652107	0.708064	0.755965	0.796500	0.830520	0.858908	0.882503	0.902066
0.00	0.667316	0.722672	0.769762	0.809345	0.842337	0.869673	0.892234	0.910808	0.926080	0.938632
0.50	0.781722	0.820414	0.852469	0.878863	0.900506	0.918209	0.932670	0.944477	0.954121	0.962005
1.00	0.860954	0.886545	0.907409	0.924374	0.938147	0.949325	0.958399	0.965770	0.971767	0.976652
1.50	0.913068	0.929430	0.942641	0.953301	0.961904	0.968852	0.974471	0.979022	0.982714	0.985717
2.00	0.946276	0.956524	0.964749	0.971356	0.976668	0.980946	0.984397	0.987187	0.989447	0.991283
2.50	0.967032	0.973370	0.978439	0.982499	0.985756	0.988374	0.990484	0.992187	0.993566	0.994685
3.00	0.979853	0.983743	0.986847	0.989329	0.991318	0.992915	0.994201	0.995238	0.996077	0.996758
3.50	0.987716	0.990093	0.991987	0.993499	0.994710	0.995682	0.996464	0.997095	0.997605	0.998018
4.00	0.992518	0.993965	0.995118	0.996038	0.996775	0.997365	0.997840	0.998223	0.998533	0.998784
4.50	0.995442	0.996322	0.997023	0.997582	0.998029	0.998388	0.998676	0.998909	0.999097	0.999249
5.00	0.997220	0.997754	0.998180	0.998519	0.998791	0.999009	0.999184	0.999325	0.999439	0.999531
5.50	0.998800	0.998624	0.998883	0.999089	0.999253	0.999386	0.999492	0.999577	0.999647	0.999703
6.00	0.998955	0.999152	0.999309	0.999434	0.999534	0.999614	0.999679	0.999731	0.999773	0.999807
6.50	0.999353	0.999473	0.999568	0.999644	0.999704	0.999753	0.999792	0.999823	0.999849	0.999870
7.00	0.999595	0.999667	0.999725	0.999771	0.999808	0.999837	0.999861	0.999880	0.999895	0.999908
7.50	0.999741	0.999785	0.999820	0.999848	0.999870	0.999888	0.999902	0.999914	0.999923	0.999931
8.00	0.999830	0.999856	0.999878	0.999895	0.999908	0.999919	0.999928	0.999935	0.999940	0.999945

appendix V percentage points of the chi-square distribution*

χ^2_α

ν	$\chi^2_{.995}$	$\chi^2_{.99}$	$\chi^2_{.975}$	$\chi^2_{.95}$	$\chi^2_{.90}$	$\chi^2_{.80}$	$\chi^2_{.75}$	$\chi^2_{.70}$	ν
1	.0000393	.000157	.000982	.00393	.0158	.0642	.102	.148	1
2	.0100	.0201	.0506	.103	.211	.446	.575	.713	2
3	.0717	.115	.216	.352	.584	1.005	1.213	1.424	3
4	.207	.297	.484	.711	1.064	1.649	1.923	2.195	4
5	.412	.554	.831	1.145	1.610	2.343	2.675	3.000	5
6	.676	.872	1.237	1.635	2.204	3.070	3.455	3.828	6
7	.989	1.239	1.690	2.167	2.833	3.822	4.255	4.671	7
8	1.344	1.646	2.180	2.733	3.490	4.594	5.071	5.527	8
9	1.735	2.088	2.700	3.325	4.168	5.380	5.899	6.393	9
10	2.156	2.558	3.247	3.940	4.865	6.179	6.737	7.267	10
11	2.603	3.053	3.816	4.575	5.578	6.989	7.584	8.148	11
12	3.074	3.571	4.404	5.226	6.304	7.807	8.438	9.034	12
13	3.565	4.107	5.009	5.892	7.042	8.634	9.299	9.926	13
14	4.075	4.660	5.629	6.571	7.790	9.467	10.165	10.821	14
15	4.601	5.229	6.262	7.261	8.574	10.307	11.306	11.721	15
16	5.142	5.812	6.908	7.962	9.312	11.152	11.192	12.624	16
17	5.697	6.408	7.564	8.672	10.085	12.002	12.792	13.531	17
18	6.265	7.015	8.231	9.390	10.865	12.857	13.675	14.440	18
19	6.844	7.633	8.907	10.117	11.651	13.716	14.562	15.352	19
20	7.434	8.260	9.591	10.851	12.443	14.578	15.452	16.266	20
21	8.034	8.897	10.283	11.591	13.240	15.445	16.344	17.182	21
22	8.643	9.542	10.982	12.338	14.041	16.314	17.240	18.101	22
23	9.260	10.196	11.688	13.091	14.848	17.187	18.137	19.021	23
24	9.886	10.856	12.401	13.848	15.659	18.062	19.037	19.943	24
25	10.520	11.524	13.120	14.611	16.473	18.940	19.939	20.867	25

χ^2_α (*Continued*)

ν	$\chi^2_{.995}$	$\chi^2_{.99}$	$\chi^2_{.975}$	$\chi^2_{.95}$	$\chi^2_{.90}$	$\chi^2_{.80}$	$\chi^2_{.75}$	$\chi^2_{.70}$	ν
26	11.160	12.198	13.844	15.379	17.292	19.820	20.843	21.792	26
27	11.808	12.879	14.573	16.151	18.114	20.703	21.749	22.719	27
28	12.461	13.565	15.308	16.928	18.939	21.588	22.657	23.647	28
29	13.121	14.256	16.047	17.708	19.768	22.475	23.567	24.577	29
30	13.787	14.953	16.791	18.493	20.599	23.364	24.478	25.508	30
35	17.156	18.484	20.558	22.462	24.812	27.820	29.058	30.181	35
40	20.674	22.142	24.423	26.507	29.067	32.326	33.664	34.874	40
45	24.281	25.880	28.356	30.610	33.367	36.863	38.294	39.586	45
50	27.962	29.687	32.348	34.762	37.706	41.426	42.944	44.314	50
55	31.708	33.552	36.390	38.956	42.078	46.011	47.612	49.055	55
60	35.510	37.467	40.474	43.186	46.478	50.614	52.295	53.808	60
65	39.360	41.427	44.595	47.448	50.902	55.233	56.991	58.572	65
70	43.253	45.426	48.750	51.737	55.349	59.868	61.698	63.344	70
75	47.186	49.460	52.935	56.052	59.815	64.515	66.416	68.125	75
80	51.153	53.526	57.146	60.390	64.299	69.174	71.144	72.913	80
85	55.151	57.621	61.382	64.748	68.799	73.843	75.880	77.707	85
90	59.179	61.741	65.640	69.124	73.313	78.522	80.623	82.508	90
95	63.963	65.886	69.919	73.518	77.841	83.210	85.374	87.314	95
100	67.312	70.053	74.216	77.928	82.381	87.906	90.131	92.125	100
105	71.414	74.241	78.530	82.352	86.933	92.610	94.894	96.941	105
110	75.536	78.448	82.861	86.790	91.495	97.321	99.663	101.761	110
115	79.679	82.672	87.207	91.240	96.067	102.038	104.437	106.585	115
120	83.839	86.913	91.567	95.703	100.648	106.762	109.216	111.413	120

*Taken from *Reliability Handbook*, AMCP 702-3, Headquarters, U. S. Army Materiel Command, Washington, D. C., October, 1968.

χ^2_α

ν	$\chi^2_{.50}$	$\chi^2_{.30}$	$\chi^2_{.25}$	$\chi^2_{.20}$	$\chi^2_{.10}$	$\chi^2_{.05}$	$\chi^2_{.025}$	$\chi^2_{.01}$	$\chi^2_{.005}$	ν
1	.455	1.074	1.323	1.642	2.706	3.841	5.024	6.635	7.879	1
2	1.386	2.408	2.773	3.219	4.605	5.991	7.378	9.210	10.597	2
3	2.366	3.665	4.108	4.642	6.251	7.815	9.348	11.345	12.838	3
4	3.357	4.878	5.385	5.989	7.779	9.488	11.143	13.277	14.860	4
5	4.351	6.064	6.626	7.289	9.236	11.070	12.832	15.086	16.750	5
6	5.348	7.231	7.841	8.558	10.645	12.592	14.449	16.812	18.548	6
7	6.346	8.383	9.037	9.803	12.017	14.067	16.013	18.475	20.278	7
8	7.344	9.524	10.219	11.030	13.362	15.507	17.535	20.090	21.955	8
9	8.343	10.656	11.389	12.242	14.684	16.919	19.023	21.666	23.589	9
10	9.342	11.781	12.549	13.442	15.987	18.307	20.483	23.209	25.188	10
11	10.341	12.899	13.701	14.631	17.275	19.675	21.920	24.725	26.757	11
12	11.340	14.011	14.845	15.812	18.549	21.920	23.337	26.217	28.300	12
13	12.340	15.119	15.984	16.985	19.812	22.362	24.736	27.688	29.819	13
14	13.339	16.222	17.117	18.151	21.064	23.685	26.119	29.141	31.319	14
15	14.339	17.322	18.245	19.311	22.307	24.996	27.488	30.578	32.801	15
16	15.338	18.418	19 369	20.465	23.542	26.296	28.845	32.000	34.267	16
17	16.338	19.511	20.489	21.615	24.769	27.587	30.191	33.409	35.718	17
18	17.338	20.601	21.605	22.760	25.989	28.869	31.526	34.805	37.156	18
19	18.338	21.689	22.718	23.900	27.204	30.144	32.852	36.191	38.582	19
20	19.337	22.775	23.828	25.038	28.412	31.410	34.170	37.566	39.997	20
21	20.337	23.858	24.935	26.171	29.615	32.671	35.479	38.932	41.401	21
22	21.337	24.939	26.039	27.301	30.813	33.924	36.781	40.289	42.796	22
23	22.337	26.018	27.141	28.429	32.007	35.172	38.076	41.638	44.181	23
24	23.337	27.096	28.241	29.553	33.196	36.415	39.364	42.980	45.558	24
25	24.337	28.172	29.339	30.675	34.382	37.652	40.646	44.314	46.928	25

χ^2_ν

ν	$\chi^2_{.50}$	$\chi^2_{.30}$	$\chi^2_{.25}$	$\chi^2_{.20}$	$\chi^2_{.10}$	$\chi^2_{.05}$	$\chi^2_{.025}$	$\chi^2_{.01}$	$\chi^2_{.005}$	ν
26	25.336	29.246	30.434	31.795	35.563	38.885	41.923	45.642	48.290	26
27	26.336	30.319	31.528	32.912	36.741	40.113	43.194	46.963	49.645	27
28	27.336	31.391	32.620	34.027	37.916	41.337	44.461	48.278	50.993	28
29	28.336	32.461	33.711	35.139	39.087	42.557	45.722	49.588	52.336	29
30	29.336	33.530	34.800	36.250	40.256	43.773	46.979	50.892	53.672	30
35	34.338	38.860	40.221	41.802	46.034	49.798	53.207	57.359	60.304	35
40	39.337	44.166	45.615	47.295	51.780	55.755	59.345	63.706	66.792	40
45	44.337	49.453	50.984	52.757	57.480	61.653	65.414	69.971	73.190	45
50	49.336	54.725	56.333	58.194	63.141	67.502	71.424	76.167	79.512	50
55	54.336	59.983	61.665	63.610	68.770	73.309	77.384	82.305	85.769	55
60	59.336	65.229	66.982	69.006	74.370	79.080	83.301	88.391	91.970	60
65	64.336	70.466	72.286	74.387	79.946	84.819	89.181	94.433	93.122	65
70	69.335	75.693	77.578	79.752	85.500	90.530	95.027	100.436	104.230	70
75	74.335	80.912	82.860	85.105	91.034	96.216	100.843	106.403	110.300	75
80	79.335	86.124	88.132	90.446	96.550	101.879	106.632	112.338	116.334	80
85	84.335	91.329	93.396	95.777	102.050	107.521	112.397	118.244	122.337	85
90	89.335	96.529	98.653	101.097	107.536	113.145	118.139	124.125	128.310	90
95	94.335	101.723	103.902	106.409	113.008	118.751	123.861	129.980	134.257	95
100	99.335	106.911	109.145	111.713	118.468	124.342	129.565	135.814	140.179	100
105	104.335	112.095	114.381	117.009	123.917	129.918	135.250	141.627	146.078	105
110	109.335	117.275	119.612	112.299	129.355	135.480	140.920	147.421	151.956	110
115	114.335	122.451	124.838	127.581	134.782	141.030	146.574	153.197	157.814	115
120	119.335	127.623	130.059	132.858	140.201	146.568	152.215	158.956	163.654	120

appendix VI percentage points of the *F*-distribution[a]

Upper

$\nu_2 \backslash \nu_1$	1	2	3	4	5	6	7	8	9
1	16211.	20000.	21615.	22500.	23056.	23437.	23715.	23925.	24091.
2	198.5	199.0	199.2	199.2	199.3	199.3	199.4	199.4	199.4
3	55.55	49.80	47.47	46.19	45.39	44.84	44.43	44.13	43.88
4	31.33	26.28	24.26	23.15	22.46	21.97	21.62	21.35	21.14
5	22.78	18.31	16.53	15.56	14.94	14.51	14.20	13.96	13.77
6	18.63	14.54	12.92	12.03	11.46	11.07	10.79	10.57	10.39
7	16.24	12.40	10.88	10.05	9.52	9.16	8.89	8.68	8.51
8	14.69	11.04	9.60	8.81	8.30	7.95	7.69	7.50	7.34
9	13.61	10.11	8.72	7.96	7.47	7.13	6.88	6.69	6.54
10	12.83	9.43	8.08	7.34	6.87	6.54	6.30	6.12	5.97
11	12.23	8.91	7.60	6.88	6.42	6.10	5.86	5.68	5.54
12	11.75	8.51	7.23	6.52	6.07	5.76	5.52	5.35	5.20
13	11.37	8.19	6.93	6.23	5.79	5.48	5.25	5.08	4.94
14	11.06	7.92	6.68	6.00	5.56	5.26	5.03	4.86	4.72
15	10.80	7.70	6.48	5.80	5.37	5.07	4.85	4.67	4.54
16	10.58	7.51	6.30	5.64	5.21	4.91	4.69	4.52	4.38
17	10.38	7.35	6.16	5.50	5.07	4.78	4.56	4.39	4.25
18	10.22	7.21	6.03	5.37	4.96	4.66	4.44	4.28	4.14
19	10.07	7.09	5.92	5.27	4.85	4.56	4.34	4.18	4.04
20	9.94	6.99	5.82	5.17	4.76	4.47	4.26	4.09	3.96
21	9.83	6.89	5.73	5.09	4.68	4.39	4.18	4.01	3.88
22	9.73	6.81	5.65	5.02	4.61	4.32	4.11	3.94	3.81
23	9.63	6.73	5.58	4.95	4.54	4.26	4.05	3.88	3.75
24	9.55	6.66	5.52	4.89	4.49	4.20	3.99	3.83	3.69
25	9.48	6.60	5.46	4.84	4.43	4.15	3.94	3.78	3.64
26	9.41	6.54	5.41	4.79	4.38	4.10	3.89	3.73	3.60
27	9.34	6.49	5.36	4.74	4.34	4.06	3.85	3.69	3.56
28	9.28	6.44	5.32	4.70	4.30	4.02	3.81	3.65	3.52
29	9.23	6.40	5.28	4.66	4.26	3.98	3.77	3.61	3.48
30	9.18	6.35	5.24	4.62	4.23	3.95	3.74	3.58	3.45
40	8.83	6.07	4.98	4.37	3.99	3.71	3.51	3.35	3.22
60	8.49	5.79	4.73	4.14	3.76	3.49	3.29	3.13	3.01
120	8.18	5.54	4.50	3.92	3.55	3.28	3.09	2.93	2.81
∞	7.88	5.30	4.28	3.72	3.35	3.09	2.90	2.74	2.62

$F_{0.005;\,\nu_1,\nu_2} = 1/F_{0.995,\nu_2,\nu_1}$; $P[F > F_{0.005;\,8,4}] = P[F > 21.35] = 0.005$

0.5% points

10	12	15	20	24	30	40	60	120	∞
24224.	24426.	24630.	24836.	24940.	25044.	25148.	25253.	25359.	25465.
199.4	199.4	199.4	199.4	199.5	199.5	199.5	199.5	199.5	199.5
43.69	43.39	43.08	42.78	42.62	42.47	42.31	42.15	41.99	41.83
20.97	20.70	20.44	20.17	20.03	19.89	19.75	19.61	19.47	19.32
13.62	13.38	13.15	12.90	12.78	12.66	12.53	12.40	12.27	12.14
10.25	10.03	9.81	9.59	9.47	9.36	9.24	9.12	9.00	8.88
8.38	8.18	7.97	7.75	7.65	7.53	7.42	7.31	7.10	7.08
7.21	7.01	6.81	6.61	6.50	6.40	6.29	6.18	6.06	5.95
6.42	6.23	6.03	5.83	5.73	5.62	5.52	5.41	5.30	5.19
5.85	5.66	5.47	5.27	5.17	5.07	4.97	4.86	4.75	4.61
5.42	5.24	5.05	4.86	4.76	4.65	4.55	4.44	4.34	4.23
5.09	4.91	4.72	4.53	4.43	4.33	4.23	4.12	4.01	3.90
4.82	4.64	4.46	4.27	4.17	4.07	3.97	3.87	3.76	3.65
4.60	4.43	4.25	4.06	3.96	3.86	3.76	3.66	3.55	3.44
4.42	4.25	4.07	3.88	3.79	3.69	3.58	3.48	3.37	3.26
4.27	4.10	3.92	3.73	3.64	3.54	3.44	3.33	3.22	3.11
4.14	3.97	3.79	3.61	3.51	3.41	3.31	3.21	3.10	2.98
4.03	3.86	3.68	3.50	3.40	3.30	3.20	3.10	2.09	2.87
3.93	3.76	3.59	3.40	3.31	3.21	3.11	3.00	2.89	2.78
3.85	3.68	3.50	3.32	3.22	3.12	3.02	2.92	2.81	2.69
3.77	3.60	3.43	3.24	3.15	3.05	2.95	2.84	2.73	2.61
3.70	3.54	3.36	3.18	3.08	2.98	2.88	2.77	2.66	2.55
3.64	3.47	3.30	3.12	3.02	2.92	2.82	2.71	2.60	2.48
3.59	3.42	3.25	3.06	2.97	2.87	2.77	2.66	2.55	2.43
3.54	3.37	3.20	3.01	2.92	2.82	2.72	2.61	2.50	2.38
3.49	3.33	3.15	2.97	2.87	2.77	2.67	2.56	2.45	2.33
3.45	3.28	3.11	2.93	2.83	2.73	2.63	2.52	2.41	2.29
3.41	3.25	3.07	2.89	2.79	2.69	2.59	2.48	2.37	2.25
3.38	3.21	3.04	2.86	2.76	2.66	2.56	2.45	2.33	2.21
3.34	3.18	3.01	2.82	2.73	2.63	2.52	2.42	2.30	2.18
3.12	2.95	2.78	2.60	2.50	2.40	2.30	2.18	2.06	1.93
2.90	2.74	2.57	2.39	2.29	2.19	2.08	1.96	1.83	1.69
2.71	2.54	2.37	2.19	2.09	1.98	1.87	1.75	1.61	1.43
2.52	2.36	2.19	2.00	1.90	1.79	1.67	1.53	1.36	1.00

Upper

$\nu_2 \backslash \nu_1$	1	2	3	4	5	6	7	8	9
1	4052.	4999.5	5403.	5625.	5764.	5859.	5928.	5982.	6022.
2	98.50	99.00	99.17	99.25	99.30	99.33	99.36	99.37	99.39
3	34.12	30.82	29.46	28.71	28.24	27.91	27.67	27.49	27.35
4	21.20	18.00	16.69	15.98	15.52	15.21	14.98	14.80	14.66
5	16.26	13.27	12.06	11.39	10.97	10.67	10.46	10.29	10.16
6	13.75	10.92	9.78	9.15	8.75	8.47	8.26	8.10	7.98
7	12.25	9.55	8.45	7.85	7.46	7.19	6.99	6.84	6.72
8	11.26	8.65	7.59	7.01	6.63	6.37	6.18	6.03	5.91
9	10.56	8.02	6.99	6.42	6.06	5.80	5.61	5.47	5.35
10	10.04	7.56	6.55	5.99	5.64	5.39	5.20	5.06	4.94
11	9.65	7.21	6.22	5.67	5.32	5.07	4.80	4.74	4.63
12	9.33	6.93	5.95	5.41	5.06	4.82	4.64	4.50	4.39
13	9.07	6.70	5.74	5.21	4.86	4.62	4.44	4.30	4.19
14	8.86	6.51	5.56	5.04	4.69	4.46	4.28	4.14	4.03
15	8.68	6.36	5.42	4.89	4.56	4.32	4.14	4.00	3.89
16	8.53	6.23	5.29	4.77	4.44	4.20	4.03	3.89	3.78
17	8.40	6.11	5.18	4.67	4.34	4.10	3.93	3.79	3.68
18	8.29	6.01	5.09	4.58	4.25	4.01	3.84	3.71	3.60
19	8.18	5.93	5.01	4.50	4.17	3.94	3.77	3.63	3.52
20	8.10	5.85	4.94	4.43	4.10	3.87	3.70	3.56	3.46
21	8.02	5.78	4.87	4.37	4.04	3.81	3.64	3.51	3.40
22	7.95	5.72	4.82	4.31	3.99	3.76	3.59	3.45	3.35
23	7.88	5.66	4.76	4.26	3.94	3.71	3.54	3.41	3.30
24	7.82	5.61	4.72	4.22	3.90	3.67	3.50	3.36	3.26
25	7.77	5.57	4.68	4.18	3.85	3.63	3.46	3.32	3.22
26	7.72	5.53	4.64	4.14	3.82	3.59	3.42	3.29	3.18
27	7.68	5.49	4.60	4.11	3.78	3.56	3.39	3.26	3.15
28	7.64	5.45	4.57	4.07	3.75	3.53	3.30	3.23	3.12
29	7.60	5.42	4.54	4.04	3.73	3.50	3.33	3.20	3.09
30	7.56	5.39	4.51	4.02	3.70	3.47	3.30	3.17	3.07
40	7.31	5.18	4.31	3.83	3.51	3.29	3.12	2.99	2.89
60	7.08	4.98	4.13	3.65	3.34	3.12	2.95	2.82	2.72
120	6.85	4.79	3.95	3.48	3.17	2.96	2.79	2.66	2.56
∞	6.63	4.61	3.78	3.32	3.02	2.80	2.64	2.51	2.41

1% points

10	12	15	20	24	30	40	60	120	∞
6056.	6106.	6157.	6209.	6235.	6261.	6287.	6313.	6339.	6366.
99.40	99.42	99.43	99.45	99.46	99.47	99.47	99.78	99.49	99.50
27.23	27.05	26.87	26.69	26.60	26.50	26.41	26.32	26.22	26.13
14.55	14.37	14.20	14.02	13.93	13.84	13.75	13.65	13.56	13.46
10.05	9.89	9.72	9.55	9.47	9.38	9.29	9.20	9.11	9.02
7.87	7.72	7.56	7.40	7.31	7.23	7.14	7.06	6.97	6.88
6.62	6.47	6.31	6.16	6.07	5.99	5.91	5.82	5.74	5.65
5.81	5.67	5.52	5.36	5.28	5.20	5.12	5.03	4.95	4.86
5.26	5.11	4.96	4.81	4.73	4.65	4.57	4.48	4.40	4.31
4.85	4.71	4.56	4.41	4.33	4.25	4.17	4.08	4.00	3.91
4.54	4.40	4.25	4.10	4.02	3.94	3.86	3.78	3.69	3.60
4.30	4.16	4.01	3.86	3.78	3.70	3.62	3.54	3.45	3.36
4.10	3.96	3.82	3.66	3.59	3.51	3.43	3.34	3.25	3.17
3.94	3.80	3.66	3.51	3.43	3.35	3.27	3.18	3.09	3.00
3.80	3.67	3.52	3.37	3.29	3.21	3.13	3.05	2.96	2.87
3.69	3.55	3.41	3.26	3.18	3.10	3.02	2.93	2.84	2.76
3.59	3.46	3.31	3.16	3.08	3.00	2.92	2.83	2.75	2.65
3.51	3.37	3.23	3.08	3.00	2.92	2.84	2.75	2.66	2.57
3.43	3.30	3.15	3.00	2.92	2.84	2.76	2.67	2.58	2.49
3.37	3.23	3.09	2.94	2.86	2.78	2.69	2.61	2.52	2.42
3.31	3.17	3.03	2.88	2.80	2.72	2.64	2.55	2.46	2.36
3.26	3.12	2.98	2.83	2.75	2.67	2.58	2.50	2.40	2.31
3.21	3.07	2.93	2.78	2.70	2.62	2.54	2.45	2.35	2.26
3.17	3.03	2.89	2.74	2.66	2.58	2.49	2.40	2.31	2.21
3.13	2.99	2.85	2.70	2.62	2.54	2.45	2.36	2.27	2.17
3.09	2.96	2.81	2.66	2.58	2.50	2.42	2.33	2.23	2.13
3.06	2.93	2.78	2.63	2.55	2.47	2.38	2.29	2.20	2.10
3.03	2.90	2.75	2.60	2.52	2.44	2.35	2.26	2.17	2.06
3.00	2.87	2.73	2.57	2.49	2.41	2.33	2.23	2.14	2.03
2.98	2.84	2.70	2.55	2.47	2.39	2.30	2.21	2.11	2.01
2.80	2.66	2.52	2.37	2.29	2.20	2.11	2.02	1.92	1.80
2.63	2.50	2.35	2.20	2.12	2.03	1.94	1.84	1.73	1.60
2.47	2.34	2.19	2.03	1.95	1.86	1.76	1.66	1.53	1.38
2.32	2.18	2.04	1.88	1.79	1.70	1.59	1.47	1.32	1.00

Upper

$\nu_2 \backslash \nu_1$	1	2	3	4	5	6	7	8	9
1	647.8	799.5	864.2	899.6	921.8	937.1	948.2	956.7	963.3
2	38.51	39.00	39.17	39.25	39.30	39.33	39.36	39.37	39.39
3	17.44	16.04	15.44	15.10	14.88	14.73	14.62	14.54	14.47
4	12.22	10.65	9.98	9.60	9.36	9.20	9.07	8.98	8.90
5	10.01	8.43	7.76	7.39	7.15	6.98	6.85	6.76	6.68
6	8.81	7.26	6.60	6.23	5.99	5.82	5.70	5.60	5.52
7	8.07	6.54	5.89	5.52	5.29	5.12	4.99	4.90	4.82
8	7.57	6.06	5.42	5.05	4.82	4.65	4.53	4.43	4.36
9	7.21	5.71	5.08	4.72	4.48	4.32	4.20	4.10	4.03
10	6.94	5.46	4.83	4.47	4.24	4.07	3.95	3.85	3.78
11	6.72	5.26	4.63	4.28	4.04	3.88	3.76	3.66	3.59
12	6.55	5.10	4.47	4.12	3.89	3.73	3.61	3.51	3.44
13	6.41	4.97	4.35	4.00	3.77	3.60	3.48	3.39	3.31
14	6.30	4.86	4.24	3.89	3.66	3.50	3.38	3.29	3.21
15	6.20	4.77	4.15	3.80	3.58	3.41	3.29	3.20	3.12
16	6.12	4.69	4.08	3.73	3.50	3.34	3.22	3.12	3.05
17	6.04	4.62	4.01	3.66	3.44	3.28	3.16	3.06	2.98
18	5.98	4.56	3.95	3.61	3.38	3.22	3.10	3.01	2.93
19	5.92	4.51	3.90	3.56	3.33	3.17	3.05	2.96	2.88
20	5.87	4.46	3.86	3.51	3.29	3.13	3.01	2.91	2.84
21	5.83	4.42	3.82	3.48	3.25	3.09	2.97	2.87	2.80
22	5.79	4.38	3.78	3.44	3.22	3.05	2.93	2.84	2.76
23	5.75	4.35	3.75	3.41	3.18	3.02	2.90	2.81	2.73
24	5.72	4.32	3.72	3.38	3.15	2.99	2.87	2.78	2.70
25	5.69	4.29	3.69	3.35	3.13	2.97	2.85	2.75	2.68
26	5.66	4.27	3.67	3.33	3.10	2.94	2.82	2.73	2.65
27	5.63	4.24	3.65	3.31	3.08	2.92	2.80	2.71	2.63
28	5.61	4.22	3.63	3.29	3.06	2.90	2.78	2.69	2.61
29	5.59	4.20	3.61	3.27	3.04	2.88	2.76	2.67	2.59
30	5.57	4.18	3.59	3.25	3.03	2.87	2.75	2.65	2.57
40	5.42	4.05	3.46	3.13	2.90	2.74	2.62	2.53	2.45
60	5.29	3.93	3.34	3.01	2.79	2.63	2.51	2.41	2.33
120	5.15	3.80	3.23	2.89	2.67	2.52	2.39	2.30	2.22
∞	5.02	3.69	3.12	2.79	2.57	2.41	2.29	2.19	2.11

2.5% points

10	12	15	20	24	30	40	60	120	∞
968.6	976.7	984.9	993.1	997.2	1001.	1006.	1010.	1014.	1018.
39.40	39.41	39.43	39.45	39.46	39.46	39.47	39.48	39.49	39.50
14.42	14.34	14.25	14.17	14.12	14.08	14.04	13.99	13.95	13.90
8.84	8.75	8.66	8.56	8.51	8.46	8.41	8.36	8.31	8.20
6.62	6.52	6.43	6.33	6.28	6.23	6.18	6.12	6.07	6.02
5.46	5.37	5.27	5.17	5.12	5.07	5.01	4.96	4.90	4.85
4.76	4.67	4.57	4.47	4.42	4.36	4.31	4.25	4.20	4.14
4.30	4.20	4.10	4.00	3.95	3.89	3.84	3.78	3.73	3.67
3.96	3.87	3.77	3.67	3.61	3.56	3.51	3.45	3.39	3.33
3.72	3.62	3.52	3.42	3.37	3.31	3.26	3.20	3.14	3.08
3.53	3.43	3.33	3.23	3.17	3.12	3.06	3.00	2.94	2.88
3.37	3.28	3.18	3.07	3.02	2.96	2.91	2.85	2.79	2.72
3.25	3.15	3.05	2.95	2.89	2.84	2.78	2.72	2.66	2.60
3.15	3.05	2.95	2.84	2.79	2.73	2.67	2.61	2.55	2.49
3.06	2.96	2.86	2.76	2.70	2.64	2.59	2.52	2.46	2.40
2.99	2.89	2.79	2.68	2.63	2.57	2.51	2.45	2.38	2.32
2.92	2.82	2.72	2.62	2.56	2.50	2.44	2.38	2.32	2.25
2.87	2.77	2.67	2.56	2.50	2.44	2.38	2.32	2.26	2.19
2.82	2.72	2.62	2.51	2.45	2.39	2.33	2.27	2.20	2.13
2.77	2.68	2.57	2.46	2.41	2.35	2.29	2.22	2.16	2.09
2.73	2.64	2.53	2.42	2.37	2.31	2.25	2.18	2.11	2.04
2.70	2.60	2.50	2.39	2.33	2.27	2.21	2.14	2.08	2.00
2.67	2.57	2.47	2.36	2.30	2.24	2.18	2.11	2.04	1.97
2.64	2.54	2.44	2.33	2.27	2.21	2.15	2.08	2.01	1.94
2.61	2.51	2.41	2.30	2.24	2.18	2.12	2.05	1.98	1.91
2.59	2.49	2.39	2.28	2.22	2.16	2.09	2.03	1.95	1.88
2.57	2.47	2.36	2.25	2.19	2.13	2.07	2.00	1.93	1.85
2.55	2.45	2.34	2.23	2.17	2.11	2.05	1.98	1.91	1.83
2.53	2.43	2.32	2.21	2.15	2.09	2.03	1.96	1.89	1.81
2.51	2.41	2.31	2.20	2.14	2.07	2.01	1.94	1.87	1.79
2.39	2.29	2.18	2.07	2.01	1.94	1.88	1.80	1.72	1.64
2.27	2.17	2.06	1.94	1.88	1.82	1.74	1.67	1.58	1.48
2.16	2.05	1.94	1.82	1.76	1.69	1.61	1.53	1.43	1.31
2.05	1.94	1.83	1.71	1.64	1.57	1.48	1.39	1.27	1.00

Upper

$\nu_2 \backslash \nu_1$	1	2	3	4	5	6	7	8	9
1	161.4	199.5	215.7	224.6	230.2	234.0	236.8	238.9	240.5
2	18.51	19.00	19.16	19.25	19.30	19.33	19.35	19.37	19.38
3	10.13	9.55	9.28	9.12	9.01	8.94	8.89	8.85	8.81
4	7.71	6.94	6.59	6.39	6.26	6.16	6.09	6.04	6.00
5	6.61	5.79	5.41	5.19	5.05	4.95	4.88	4.82	4.77
6	5.99	5.14	4.76	4.53	4.39	4.28	4.21	4.15	4.10
7	5.59	4.74	4.35	4.12	3.97	3.87	3.79	3.73	3.68
8	5.32	4.46	4.07	3.84	3.69	3.58	3.50	3.44	3.39
9	5.12	4.26	3.86	3.63	3.48	3.37	3.29	3.23	3.18
10	4.96	4.10	3.71	3.48	3.33	3.22	3.14	3.07	3.02
11	4.84	3.98	3.59	3.36	3.20	3.09	3.01	2.95	2.90
12	4.75	3.89	3.49	3.26	3.11	3.00	2.91	2.85	2.80
13	4.67	3.81	3.41	3.18	3.03	2.92	2.83	2.77	2.71
14	4.60	3.74	3.34	3.11	2.96	2.85	2.76	2.70	2.65
15	4.54	3.68	3.29	3.06	2.90	2.79	2.71	2.64	2.59
16	4.49	3.63	3.24	3.01	2.85	2.74	2.66	2.59	2.54
17	4.45	3.59	3.20	2.96	2.81	2.70	2.61	2.55	2.49
18	4.41	3.55	3.16	2.93	2.77	2.66	2.58	2.51	2.46
19	4.38	3.52	3.13	2.90	2.74	2.63	2.54	2.48	2.42
20	4.35	3.49	3.10	2.87	2.71	2.60	2.51	2.45	2.39
21	4.32	3.47	3.07	2.84	2.68	2.57	2.49	2.42	2.37
22	4.30	3.44	3.05	2.82	2.66	2.55	2.46	2.40	2.34
23	4.28	3.42	3.03	2.80	2.64	2.53	2.44	2.37	2.32
24	4.26	3.40	3.01	2.78	2.62	2.51	2.42	2.36	2.30
25	4.24	3.39	2.99	2.76	2.60	2.49	2.40	2.34	2.28
26	4.23	3.37	2.98	2.74	2.59	2.47	2.39	2.32	2.27
27	4.21	3.35	2.96	2.73	2.57	2.46	2.37	2.31	2.25
28	4.20	3.34	2.95	2.71	2.56	2.45	2.36	2.29	2.24
29	4.18	3.33	2.93	2.70	2.55	2.43	2.35	2.28	2.22
30	4.17	3.32	2.92	2.69	2.53	2.42	2.33	2.27	2.21
40	4.08	3.23	2.84	2.61	2.45	2.34	2.25	2.18	2.12
60	4.00	3.15	2.76	2.53	2.37	2.25	2.17	2.10	2.04
120	3.92	3.07	2.68	2.45	2.29	2.17	2.09	2.02	1.96
∞	3.84	3.00	2.60	2.37	2.21	2.10	2.01	1.94	1.88

5% points

10	12	15	20	24	30	40	60	120	∞
241.9	243.9	245.9	248.0	249.1	250.1	251.1	252.2	253.3	254.3
19.40	19.41	19.43	19.45	19.45	19.46	19.47	19.48	19.49	19.50
8.79	8.74	8.70	8.66	8.64	8.62	8.59	8.57	8.55	8.53
5.96	5.91	5.86	5.80	5.77	5.75	5.72	5.69	5.66	5.63
4.74	4.68	4.62	4.56	4.53	4.50	4.46	4.43	4.40	4.36
4.06	4.00	3.94	3.87	3.84	3.81	3.77	3.74	3.70	3.67
3.64	3.57	3.51	3.44	3.41	3.38	3.34	3.30	3.27	3.23
3.35	3.28	3.22	3.15	3.12	3.08	3.04	3.01	2.97	2.93
3.14	3.07	3.01	2.94	2.90	2.86	2.83	2.79	2.75	2.71
2.98	2.91	2.85	2.77	2.74	2.70	2.66	2.62	2.58	2.54
2.85	2.79	2.72	2.65	2.61	2.57	2.53	2.49	2.45	2.40
2.75	2.69	2.62	2.54	2.51	2.47	2.43	2.38	2.34	2.30
2.67	2.60	2.53	2.46	2.42	2.38	2.34	2.30	2.25	2.21
2.60	2.53	2.46	2.39	2.35	2.31	2.27	2.22	2.18	2.13
2.54	2.48	2.40	2.33	2.29	2.25	2.20	2.16	2.11	2.07
2.49	2.42	2.35	2.28	2.24	2.19	2.15	2.11	2.06	2.01
2.45	2.38	2.31	2.23	2.19	2.15	2.10	2.06	2.01	1.96
2.41	2.34	2.27	2.19	2.15	2.11	2.06	2.02	1.97	1.92
2.38	2.31	2.23	2.16	2.11	2.07	2.03	1.98	1.93	1.88
2.35	2.28	2.20	2.12	2.08	2.04	1.99	1.95	1.90	1.84
2.32	2.25	2.18	2.10	2.05	2.01	1.96	1.92	1.87	1.81
2.30	2.23	2.15	2.07	2.03	1.98	1.94	1.89	1.84	1.78
2.27	2.20	2.13	2.05	2.01	1.96	1.91	1.86	1.81	1.76
2.25	2.18	2.11	2.03	1.98	1.94	1.89	1.84	1.79	1.73
2.24	2.16	2.09	2.01	1.96	1.92	1.87	1.82	1.77	1.71
2.22	2.15	2.07	1.99	1.95	1.90	1.85	1.80	1.75	1.69
2.20	2.13	2.06	1.97	1.93	1.88	1.84	1.79	1.73	1.67
2.19	2.12	2.04	1.96	1.91	1.87	1.82	1.77	1.71	1.65
2.18	2.10	2.03	1.94	1.90	1.85	1.81	1.75	1.70	1.64
2.16	2.09	2.01	1.93	1.89	1.84	1.79	1.74	1.68	1.62
2.08	2.00	1.92	1.84	1.79	1.74	1.69	1.64	1.58	1.51
1.99	1.92	1.84	1.75	1.70	1.65	1.59	1.53	1.47	1.39
1.91	1.83	1.75	1.66	1.61	1.55	1.55	1.43	1.35	1.25
1.83	1.75	1.67	1.57	1.52	1.46	1.39	1.32	1.22	1.00

Upper

$\nu_2 \backslash \nu_1$	1	2	3	4	5	6	7	8	9
1	39.86	49.50	53.59	55.83	57.24	58.20	58.91	59.44	59.86
2	8.53	9.00	9.16	9.24	9.29	9.33	9.35	9.37	9.38
3	5.54	5.46	5.39	5.34	5.31	5.28	5.27	5.25	5.24
4	4.54	4.32	4.19	4.11	4.05	4.01	3.98	3.95	3.94
5	4.06	3.78	3.62	3.52	3.45	3.40	3.37	3.34	3.32
6	3.78	3.46	3.29	3.18	3.11	3.05	3.01	2.98	2.96
7	3.59	3.26	3.07	2.96	2.88	2.83	2.78	2.75	2.72
8	3.46	3.11	2.92	2.81	2.73	2.67	2.62	2.59	2.56
9	3.36	3.01	2.81	2.69	2.61	2.55	2.51	2.47	2.44
10	3.29	2.92	2.73	2.61	2.52	2.46	2.41	2.38	2.35
11	3.23	2.86	2.66	2.54	2.45	2.39	2.34	2.30	2.27
12	3.18	2.81	2.61	2.48	2.39	2.33	2.28	2.24	2.21
13	3.14	2.76	2.56	2.43	2.35	2.28	2.23	2.20	2.16
14	3.10	2.73	2.52	2.39	2.31	2.24	2.19	2.15	2.12
15	3.07	2.70	2.49	2.36	2.27	2.21	2.16	2.12	2.09
16	3.05	2.67	2.46	2.33	2.24	2.18	2.13	2.09	2.06
17	3.03	2.64	2.44	2.31	2.22	2.15	2.10	2.06	2.03
18	3.01	2.62	2.42	2.29	2.20	2.13	2.08	2.04	2.00
19	2.99	2.61	2.40	2.27	2.18	2.11	2.06	2.02	1.98
20	2.97	2.59	2.38	2.25	2.16	2.09	2.04	2.00	1.96
21	2.96	2.57	2.36	2.23	2.14	2.08	2.02	1.98	1.95
22	2.95	2.56	2.35	2.22	2.13	2.06	2.01	1.97	1.93
23	2.94	2.55	2.34	2.21	2.11	2.05	1.99	1.95	1.92
24	2.93	2.54	2.33	2.19	2.10	2.04	1.98	1.94	1.91
25	2.92	2.53	2.32	2.18	2.09	2.02	1.97	1.93	1.89
26	2.91	2.52	2.31	2.17	2.08	2.01	1.96	1.92	1.88
27	2.90	2.51	2.30	2.17	2.07	2.00	1.95	1.91	1.87
28	2.89	2.50	2.29	2.16	2.06	2.00	1.94	1.90	1.87
29	2.89	2.50	2.28	2.15	2.06	1.99	1.93	1.89	1.86
30	2.88	2.49	2.28	2.14	2.05	1.98	1.93	1.88	1.85
40	2.84	2.44	2.23	2.09	2.00	1.93	1.87	1.83	1.79
60	2.79	2.39	2.18	2.04	1.95	1.87	1.82	1.77	1.74
120	2.75	2.35	2.13	1.99	1.90	1.82	1.77	1.72	1.68
∞	2.71	2.30	2.08	1.94	1.85	1.77	1.72	1.67	1.63

10% points

10	12	15	20	24	30	40	60	120	∞
60.19	60.71	61.22	61.74	62.00	62.26	62.53	62.79	63.06	63.33
9.39	9.41	9.42	9.44	9.45	9.46	9.47	9.47	9.48	9.49
5.23	5.22	5.20	5.18	5.18	5.17	5.16	5.15	5.14	5.13
3.92	3.90	3.87	3.84	3.83	3.82	3.80	3.79	3.78	3.76
3.30	3.27	3.24	3.21	3.19	3.17	3.16	3.14	3.12	3.10
2.94	2.90	2.87	2.84	2.82	2.80	2.78	2.76	2.74	2.72
2.70	2.67	2.63	2.59	2.58	2.56	2.54	2.51	2.49	2.47
2.54	2.50	2.46	2.42	2.40	2.38	2.36	2.34	2.32	2.29
2.42	2.38	2.34	2.30	2.28	2.25	2.23	2.21	2.18	2.16
2.32	2.28	2.24	2.20	2.18	2.16	2.13	2.11	2.08	2.06
2.25	2.21	2.17	2.12	2.10	2.08	2.05	2.03	2.00	1.97
2.19	2.15	2.10	2.06	2.04	2.01	1.99	1.96	1.93	1.90
2.14	2.10	2.05	2.01	1.98	1.96	1.93	1.90	1.88	1.85
2.10	2.05	2.01	1.96	1.94	1.91	1.89	1.86	1.83	1.80
2.06	2.02	1.97	1.92	1.90	1.87	1.85	1.82	1.79	1.76
2.03	1.99	1.94	1.89	1.87	1.84	1.81	1.78	1.75	1.72
2.00	1.96	1.91	1.86	1.84	1.81	1.78	1.75	1.72	1.69
1.98	1.93	1.89	1.84	1.81	1.78	1.75	1.72	1.69	1.66
1.96	1.91	1.86	1.81	1.79	1.76	1.73	1.70	1.67	1.63
1.94	1.89	1.84	1.79	1.77	1.74	1.71	1.68	1.64	1.61
1.92	1.87	1.83	1.78	1.75	1.72	1.69	1.66	1.62	1.59
1.90	1.86	1.81	1.76	1.73	1.70	1.67	1.64	1.60	1.57
1.89	1.84	1.80	1.74	1.72	1.69	1.66	1.62	1.59	1.55
1.88	1.83	1.78	1.73	1.70	1.67	1.64	1.61	1.57	1.53
1.87	1.82	1.77	1.72	1.69	1.66	1.63	1.59	1.56	1.52
1.86	1.81	1.76	1.71	1.68	1.65	1.61	1.58	1.54	1.50
1.85	1.80	1.75	1.70	1.67	1.64	1.60	1.57	1.53	1.49
1.84	1.79	1.74	1.69	1.66	1.63	1.59	1.56	1.52	1.48
1.83	1.78	1.73	1.68	1.65	1.62	1.58	1.55	1.51	1.47
1.82	1.77	1.72	1.67	1.64	1.61	1.57	1.54	1.50	1.46
1.76	1.71	1.66	1.61	1.57	1.54	1.51	1.47	1.42	1.38
1.71	1.66	1.60	1.54	1.51	1.48	1.44	1.40	1.35	1.29
1.65	1.60	1.55	1.48	1.45	1.41	1.37	1.32	1.26	1.19
1.60	1.55	1.49	1.42	1.38	1.34	1.30	1.24	1.17	1.00

Upper

$\nu_2 \backslash \nu_1$	1	2	3	4	5	6	7	8	9
1	5.83	7.50	8.20	8.58	8.82	8.98	9.10	9.19	9.26
2	2.57	3.00	3.15	3.23	3.28	3.31	3.34	3.35	3.37
3	2.02	2.28	2.36	2.39	2.41	2.42	2.43	2.44	2.44
4	1.81	2.00	2.05	2.06	2.07	2.08	2.08	2.08	2.08
5	1.69	1.85	1.88	1.89	1.89	1.89	1.89	1.89	1.89
6	1.62	1.76	1.78	1.79	1.79	1.78	1.78	1.78	1.77
7	1.57	1.70	1.72	1.72	1.71	1.71	1.70	1.70	1.70
8	1.54	1.66	1.67	1.66	1.66	1.65	1.64	1.64	1.63
9	1.51	1.62	1.63	1.63	1.62	1.61	1.60	1.60	1.59
10	1.49	1.60	1.60	1.59	1.59	1.58	1.57	1.56	1.56
11	1.47	1.58	1.58	1.57	1.56	1.55	1.54	1.53	1.53
12	1.46	1.56	1.56	1.55	1.54	1.53	1.52	1.51	1.51
13	1.45	1.55	1.55	1.53	1.52	1.51	1.50	1.49	1.49
14	1.44	1.53	1.53	1.52	1.51	1.50	1.49	1.48	1.47
15	1.43	1.52	1.52	1.51	1.49	1.48	1.47	1.46	1.46
16	1.42	1.51	1.51	1.50	1.48	1.47	1.46	1.45	1.44
17	1.42	1.51	1.50	1.49	1.47	1.46	1.45	1.44	1.43
18	1.41	1.50	1.49	1.48	1.46	1.45	1.44	1.43	1.42
19	1.41	1.49	1.49	1.47	1.46	1.44	1.43	1.42	1.41
20	1.40	1.49	1.48	1.47	1.45	1.44	1.43	1.42	1.41
21	1.40	1.48	1.48	1.46	1.44	1.43	1.42	1.41	1.40
22	1.40	1.48	1.47	1.45	1.44	1.42	1.41	1.40	1.39
23	1.39	1.47	1.47	1.45	1.43	1.42	1.41	1.40	1.39
24	1.39	1.47	1.46	1.44	1.43	1.41	1.40	1.39	1.38
25	1.39	1.47	1.46	1.44	1.42	1.41	1.40	1.39	1.38
26	1.38	1.46	1.45	1.44	1.42	1.41	1.39	1.38	1.37
27	1.38	1.46	1.45	1.43	1.42	1.40	1.39	1.38	1.37
28	1.38	1.46	1.45	1.43	1.41	1.40	1.39	1.38	1.37
29	1.38	1.45	1.45	1.43	1.41	1.40	1.38	1.37	1.36
30	1.38	1.45	1.44	1.42	1.41	1.39	1.38	1.37	1.36
40	1.36	1.44	1.42	1.40	1.39	1.37	1.36	1.35	1.34
60	1.35	1.42	1.41	1.38	1.37	1.35	1.33	1.32	1.31
120	1.34	1.40	1.39	1.37	1.35	1.33	1.31	1.30	1.29
∞	1.32	1.39	1.37	1.35	1.33	1.31	1.29	1.28	1.27

[a]This table is reproduced from Table 18 of *Biometrika Tables for Statisticians*, Volume I, Second

25% points

10	12	15	20	24	30	40	60	120	∞
9.32	9.41	9.49	9.58	9.63	9.67	9.71	9.76	9.80	9.85
3.38	3.39	3.41	3.43	3.43	3.44	3.45	3.46	3.47	3.48
2.44	2.45	2.46	2.46	2.46	2.47	2.47	2.47	2.47	2.47
2.08	2.08	2.08	2.08	2.08	2.08	2.08	2.08	2.08	2.08
1.89	1.89	1.89	1.88	1.88	1.88	1.88	1.87	1.87	1.87
1.77	1.77	1.76	1.76	1.75	1.75	1.75	1.74	1.74	1.74
1.69	1.68	1.68	1.67	1.67	1.66	1.66	1.65	1.65	1.65
1.63	1.62	1.62	1.61	1.60	1.60	1.59	1.59	1.58	1.58
1.59	1.58	1.57	1.56	1.56	1.55	1.54	1.54	1.53	1.53
1.55	1.54	1.53	1.52	1.52	1.51	1.51	1.50	1.49	1.48
1.52	1.51	1.50	1.49	1.49	1.48	1.47	1.47	1.46	1.45
1.50	1.49	1.48	1.47	1.46	1.45	1.45	1.44	1.43	1.42
1.48	1.47	1.46	1.45	1.44	1.43	1.42	1.42	1.41	1.40
1.46	1.45	1.44	1.43	1.42	1.41	1.41	1.40	1.39	1.38
1.45	1.44	1.43	1.41	1.41	1.40	1.39	1.38	1.37	1.36
1.44	1.43	1.41	1.40	1.39	1.38	1.37	1.36	1.35	1.34
1.43	1.41	1.40	1.39	1.38	1.37	1.36	1.35	1.34	1.33
1.42	1.40	1.39	1.38	1.37	1.36	1.35	1.34	1.33	1.32
1.41	1.40	1.38	1.37	1.36	1.35	1.34	1.33	1.32	1.30
1.40	1.39	1.37	1.36	1.35	1.34	1.33	1.32	1.31	1.29
1.39	1.38	1.37	1.35	1.34	1.33	1.32	1.31	1.30	1.28
1.39	1.37	1.36	1.34	1.33	1.32	1.31	1.30	1.29	1.28
1.38	1.37	1.35	1.34	1.33	1.32	1.31	1.30	1.28	1.27
1.38	1.36	1.35	1.33	1.32	1.31	1.30	1.29	1.28	1.26
1.37	1.36	1.34	1.33	1.32	1.31	1.29	1.28	1.27	1.25
1.37	1.35	1.34	1.32	1.31	1.30	1.29	1.28	1.26	1.25
1.36	1.35	1.33	1.32	1.31	1.30	1.28	1.27	1.26	1.24
1.36	1.34	1.33	1.31	1.30	1.29	1.28	1.27	1.25	1.24
1.35	1.34	1.32	1.31	1.30	1.29	1.27	1.26	1.25	1.23
1.35	1.34	1.32	1.30	1.29	1.28	1.27	1.26	1.24	1.23
1.33	1.31	1.30	1.28	1.26	1.25	1.24	1.22	1.21	1.19
1.30	1.29	1.27	1.25	1.24	1.22	1.21	1.19	1.17	1.15
1.28	1.26	1.24	1.22	1.21	1.19	1.18	1.16	1.13	1.10
1.25	1.24	1.22	1.19	1.18	1.16	1.14	1.12	1.08	1.00

Edition, 1958, by permission of the Biometrika Trustees.

appendix VII operating characteristic curves for life testing -exponential distribution

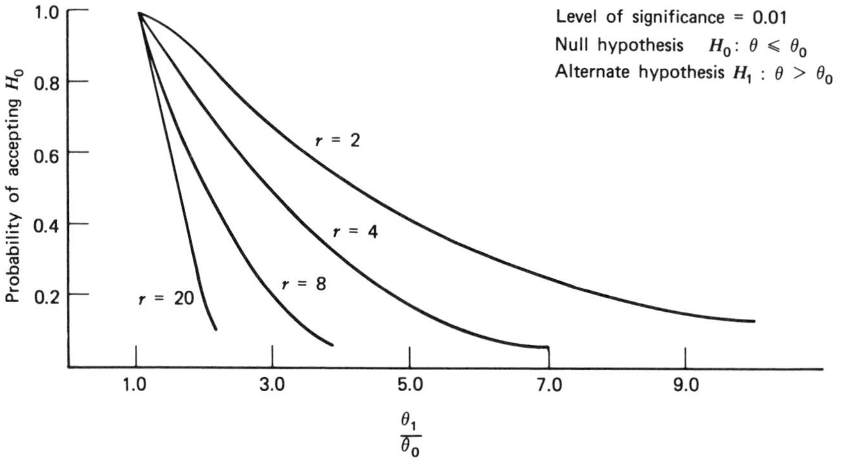

Figure A VII.1 Operating characteristic curves for life testing.

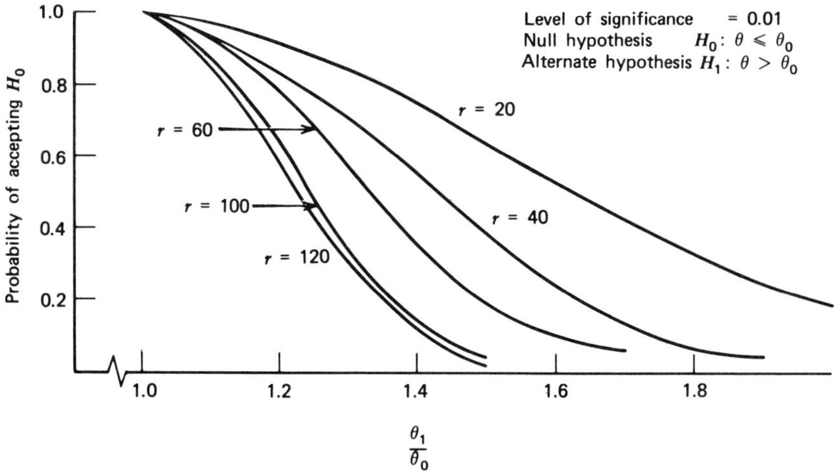

Figure A VII.2 Operating characteristic curves for life testing.

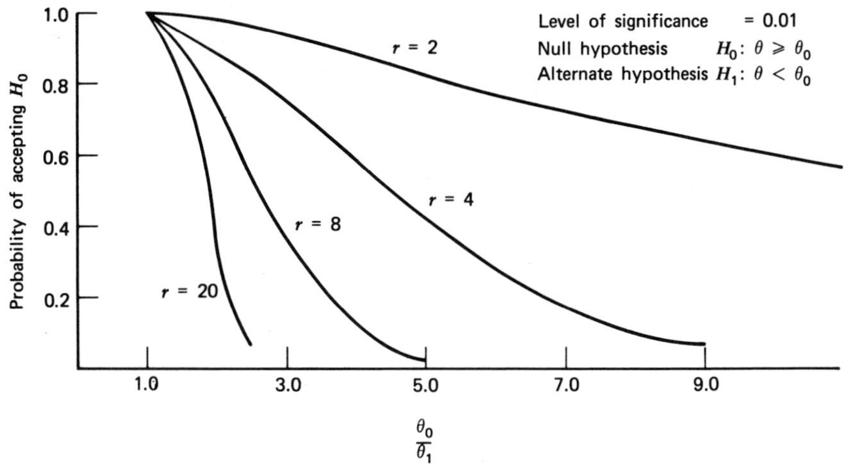

Figure A VII.3 Operating characteristic curves for life testing.

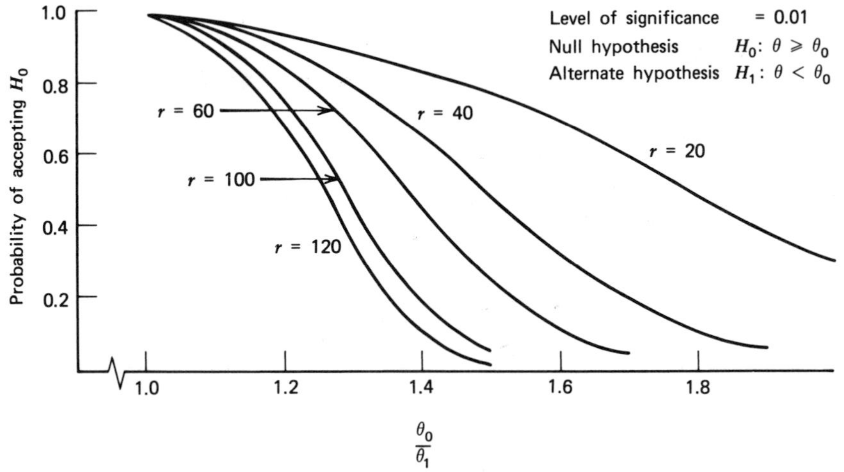

Figure A VII.4 Operating characteristic curves for life testing.

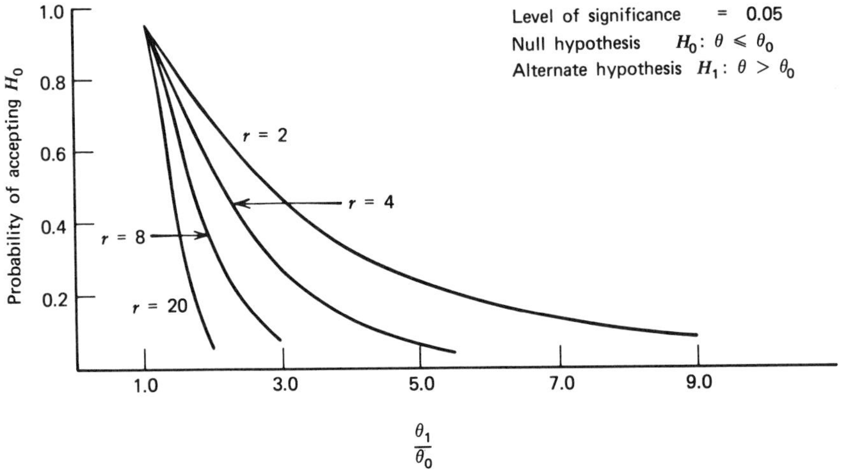

Figure A VII.5 Operating characteristic curves for life testing.

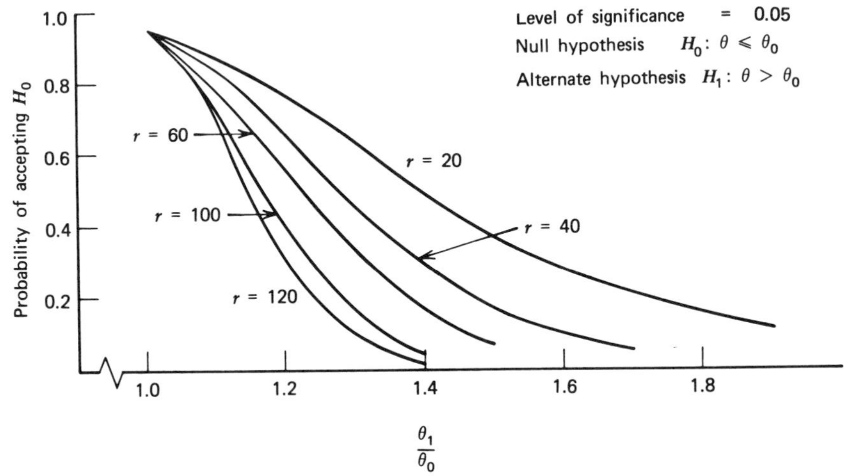

Figure A VII.6 Operating characteristic curves for life testing.

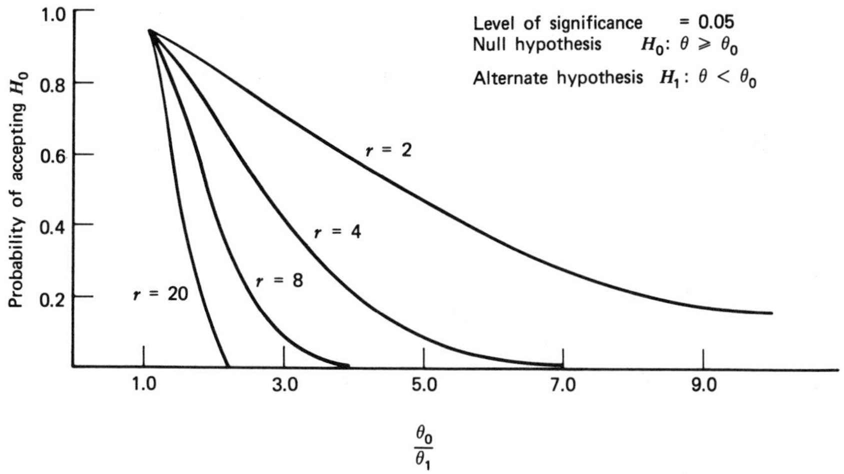

Figure A VII.7 Operating characteristic curves for life testing.

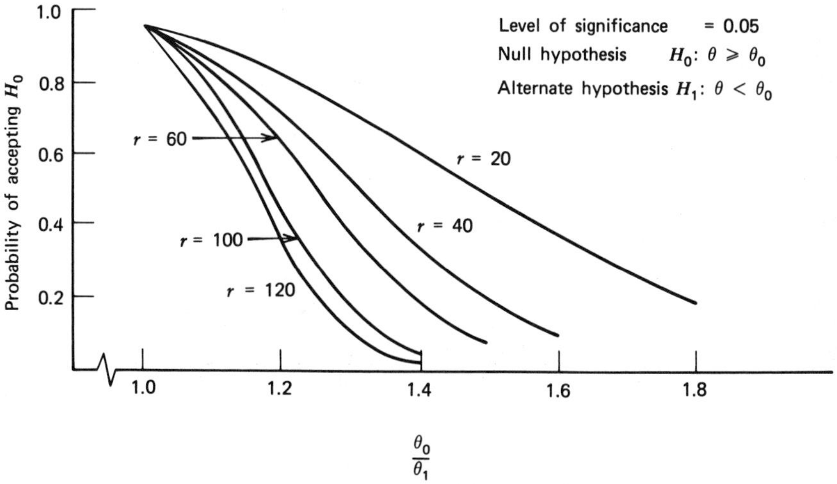

Figure A VII.8 Operating characteristic curves for life testing.

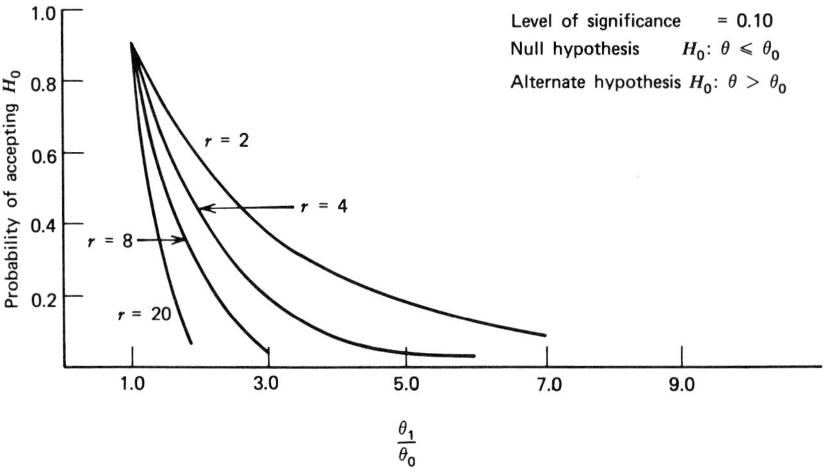

Figure A VII.9 Operating characteristic curves for life testing.

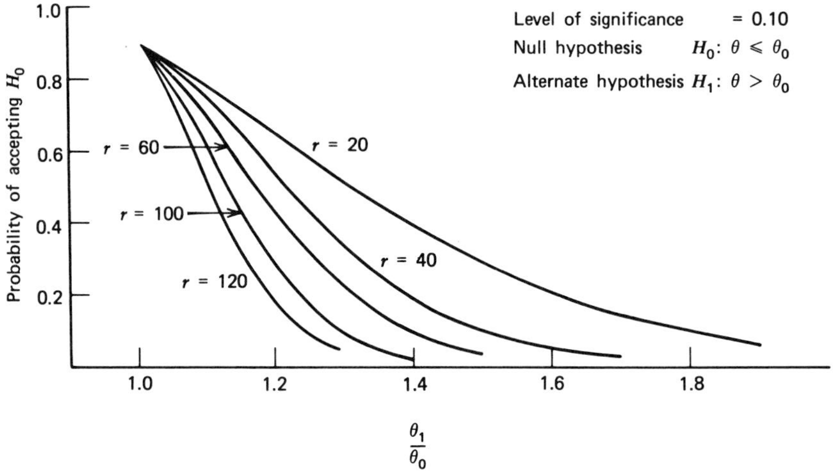

Figure A VII.10 Operating characteristic curves for life testing.

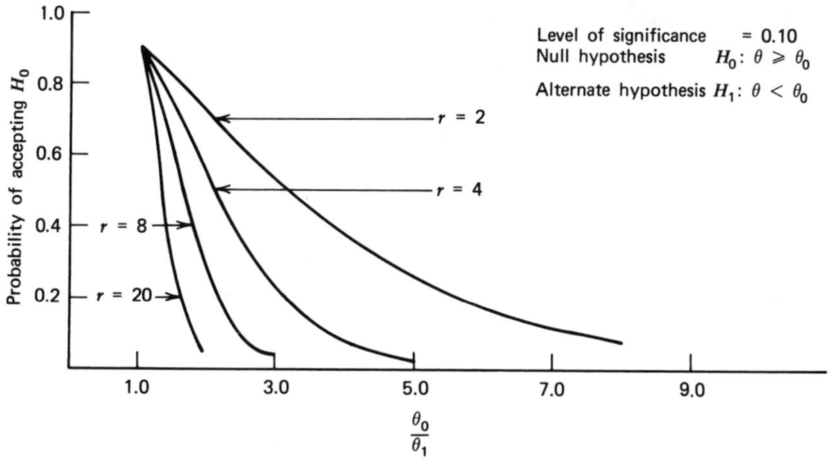

Figure A VII.11 Operating characteristic curves for life testing.

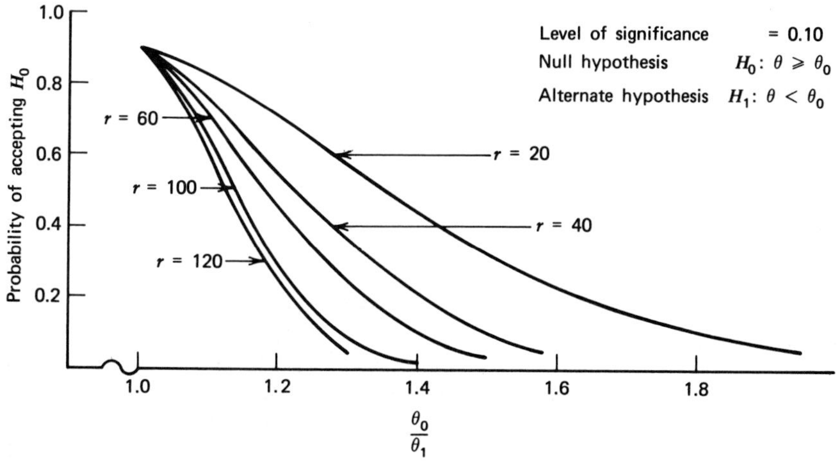

Figure A VII.12 Operating characteristic curves for life testing.

appendix VIII rank tables-(median, 5%, 95%)[a]

MEDIAN RANKS

	SAMPLE SIZE									
$j \backslash n$	1	2	3	4	5	6	7	8	9	10
1	50.000	29.289	20.630	15.910	12.945	10.910	9.428	8.300	7.412	6.697
2		70.711	50.000	38.573	31.381	26.445	22.849	20.113	17.962	16.226
3			79.370	61.427	50.000	42.141	36.412	32.052	28.624	25.857
4				84.090	68.619	57.859	50.000	44.015	39.308	35.510
5					87.055	73.555	63.588	55.984	50.000	45.169
6						89.090	77.151	67.948	60.691	54.831
7							90.572	79.887	71.376	64.490
8								91.700	82.038	74.142
9									92.587	83.774
10										93.303

MEDIAN RANKS

	SAMPLE SIZE									
$j \backslash n$	11	12	13	14	15	16	17	18	19	20
1	6.107	5.613	5.192	4.830	4.516	4.240	3.995	3.778	3.582	3.406
2	14.796	13.598	12.579	11.702	10.940	10.270	9.678	9.151	8.677	8.251
3	23.578	21.669	20.045	18.647	17.432	16.365	15.422	14.581	13.827	13.147
4	32.380	29.758	27.528	25.608	23.939	22.474	21.178	20.024	18.988	18.055
5	41.189	37.853	35.016	32.575	30.452	28.589	26.940	25.471	24.154	22.967
6	50.000	45.951	42.508	39.544	36.967	34.705	32.704	30.921	29.322	27.880
7	58.811	54.049	50.000	46.515	43.483	40.823	38.469	36.371	34.491	32.795
8	67.620	62.147	57.492	53.485	50.000	46.941	44.234	41.823	39.660	37.710
9	76.421	70.242	64.984	60.456	56.517	53.059	50.000	47.274	44.830	42.626
10	85.204	78.331	72.472	67.425	63.033	59.177	55.766	52.726	50.000	47.542
11	93.893	86.402	79.955	74.392	69.548	65.295	61.531	58.177	55.170	52.458
12		94.387	87.421	81.353	76.061	71.411	67.296	63.629	60.340	57.374
13			94.808	88.298	82.568	77.525	73.060	69.079	65.509	62.289
14				95.169	89.060	83.635	78.821	74.529	70.678	67.205
15					95.484	89.730	84.578	79.976	75.846	72.119
16						95.760	90.322	85.419	81.011	77.033
17							96.005	90.849	86.173	81.945
18								96.222	91.322	86.853
19									96.418	91.749
20										96.594

Appendix VIII (*Continued*)

MEDIAN RANKS

	SAMPLE SIZE									
j\n	21	22	23	24	25	26	27	28	29	30
1	3.247	3.101	2.969	2.847	2.734	2.631	2.534	2.445	2.362	2.284
2	7.864	7.512	7.191	6.895	6.623	6.372	6.139	5.922	5.720	5.532
3	12.531	11.970	11.458	10.987	10.553	10.153	9.781	9.436	9.114	8.814
4	17.209	16.439	15.734	15.088	14.492	13.942	13.432	12.958	12.517	12.104
5	21.890	20.911	20.015	19.192	18.435	17.735	17.086	16.483	15.922	15.397
6	26.574	25.384	24.297	23.299	22.379	21.529	20.742	20.010	19.328	18.691
7	31.258	29.859	28.580	27.406	26.324	25.325	24.398	23.537	22.735	21.986
8	35.943	34.334	32.863	31.513	30.269	29.120	28.055	27.065	26.143	25.281
9	40.629	38.810	37.147	35.621	34.215	32.916	31.712	30.593	29.550	28.576
10	45.314	43.286	41.431	39.729	38.161	36.712	35.370	34.121	32.958	31.872
11	50.000	47.762	45.716	43.837	42.107	40.509	39.027	37.650	36.367	35.168
12	54.686	52.238	50.000	47.946	46.054	44.305	42.685	41.178	39.775	38.464
13	59.371	56.714	54.284	52.054	50.000	48.102	46.342	44.707	43.183	41.760
14	64.057	61.190	58.568	56.162	53.946	51.898	50.000	48.236	46.592	45.056
15	68.742	65.665	62.853	60.271	57.892	55.695	53.658	51.764	50.000	48.352
16	73.426	70.141	67.137	64.379	61.839	59.491	57.315	55.293	53.408	51.648
17	78.109	74.616	71.420	68.487	65.785	63.287	60.973	58.821	56.817	54.944
18	82.791	79.089	75.703	72.594	69.730	67.084	64.630	62.350	60.225	58.240
19	87.469	83.561	79.985	76.701	73.676	70.880	68.288	65.878	63.633	61.536
20	92.136	88.030	84.266	80.808	77.621	74.675	71.945	69.407	67.041	64.852
21	96.753	92.488	88.542	84.912	81.565	78.471	75.602	72.935	70.450	68.128
22		96.898	92.809	89.013	85.507	82.265	79.258	76.463	73.857	71.424
23			97.031	93.105	89.447	86.058	82.914	79.990	77.265	74.719
24				97.153	93.377	89.847	86.568	83.517	80.672	78.014
25					97.265	93.628	90.219	87.042	84.078	81.309
26						97.369	93.861	90.564	87.483	84.603
27							97.465	94.078	90.865	87.896
28								97.555	94.280	91.186
29									97.638	94.468
30										97.716

Appendix VIII (*Continued*)

MEDIAN RANKS

	SAMPLE SIZE									
$j \backslash n$	31	32	33	34	35	36	37	38	39	40
1	2.211	2.143	2.078	2.018	1.961	1.907	1.856	1.807	1.762	1.718
2	5.355	5.190	5.034	4.887	4.749	4.618	4.495	4.377	4.266	4.160
3	8.533	8.269	8.021	7.787	7.567	7.359	7.162	6.975	6.798	6.629
4	11.716	11.355	11.015	10.694	10.391	10.105	9.835	9.578	9.335	9.103
5	14.905	14.445	14.011	13.603	13.218	12.855	12.510	12.184	11.874	11.580
6	18.094	17.535	17.009	16.514	16.046	15.605	15.187	14.791	14.415	14.057
7	21.284	20.626	20.007	19.425	18.875	18.355	17.864	17.398	16.956	16.535
8	24.474	23.717	23.006	22.336	21.704	21.107	20.541	20.005	19.497	19.013
9	27.664	26.809	26.005	25.247	24.533	23.858	23.219	22.613	22.038	21.492
10	30.855	29.901	29.004	28.159	27.362	26.609	25.897	25.221	24.580	23.971
11	34.046	32.993	32.003	31.071	30.192	29.361	28.575	27.829	27.122	26.449
12	37.236	36.085	35.003	33.983	33.022	32.113	31.253	30.437	29.664	28.928
13	40.427	39.177	38.002	36.895	35.851	34.865	33.931	33.046	32.206	31.407
14	43.618	42.269	41.001	39.807	38.681	37.616	36.609	35.654	34.748	33.886
15	46.809	45.362	44.001	42.720	41.511	40.368	39.287	38.262	37.290	36.365
16	50.000	48.454	47.000	45.632	44.340	43.120	41.965	40.871	39.832	38.844
17	53.191	51.546	50.000	48.544	47.170	45.872	44.644	43.479	42.374	41.323
18	56.382	54.638	52.999	51.456	50.000	48.624	47.322	46.087	44.916	43.802
19	59.573	57.731	55.999	54.368	52.830	51.376	50.000	48.696	47.458	46.281
20	62.763	60.823	58.998	57.280	55.660	54.128	52.678	51.304	50.000	48.760
21	65.954	63.915	61.998	60.193	58.489	56.830	55.356	53.913	52.542	51.239
22	69.145	67.007	64.997	63.105	61.319	59.632	58.035	56.521	55.084	53.719
23	72.335	70.099	67.997	66.017	64.149	62.383	60.713	59.129	57.626	56.198
24	75.526	73.191	70.996	68.929	66.973	65.135	63.391	61.738	60.168	58.677
25	78.716	76.283	73.995	71.841	69.808	67.837	66.069	64.346	62.710	61.156
26	81.906	79.374	76.994	74.752	72.637	70.639	68.747	66.954	65.252	63.635
27	85.094	82.465	79.993	77.664	75.467	73.391	71.425	69.562	67.794	66.114
28	88.282	85.555	82.991	80.575	78.296	76.142	74.103	72.171	70.336	68.593
29	91.467	88.644	85.989	83.486	81.125	78.899	76.781	74.779	72.878	71.072
30	94.645	91.731	88.985	86.397	83.954	81.645	79.459	77.387	75.420	73.550
31	97.789	94.810	91.979	89.306	86.782	84.395	82.136	79.994	77.962	76.029
32		97.857	94.966	92.213	89.608	87.145	84.813	82.602	80.503	78.508
33			97.921	95.113	92.433	89.894	87.490	85.209	83.044	80.986
34				97.982	95.251	92.641	90.165	87.816	85.585	83.465
35					98.039	95.382	92.838	90.422	88.126	85.943
36						98.093	95.505	93.025	90.665	88.420
37							98.144	95.622	93.202	90.897
38								98.192	95.734	93.371
39									98.238	95.839
40										98.282

Appendix VIII (*Continued*)

MEDIAN RANKS

	SAMPLE SIZE									
*j**n*	41	42	43	44	45	46	47	48	49	50
1	1.676	1.637	1.599	1.563	1.528	1.495	1.464	1.434	1.405	1.377
2	4.060	3.964	3.872	3.785	3.702	3.622	3.545	3.472	3.402	3.334
3	6.469	6.316	6.170	6.031	5.898	5.771	5.649	5.532	5.420	5.312
4	8.883	8.673	8.473	8.282	8.099	7.925	7.757	7.597	7.443	7.295
5	11.300	11.033	10.778	10.535	10.303	10.080	9.867	9.663	9.467	9.279
6	13.717	13.393	13.084	12.789	12.507	12.237	11.979	11.731	11.493	11.265
7	16.135	15.754	15.391	15.043	14.712	14.394	14.090	13.799	13.519	13.250
8	18.554	18.115	17.697	17.298	16.917	16.551	16.202	15.867	15.545	15.236
9	20.972	50.477	20.004	19.553	19.122	18.709	18.314	17.935	17.571	17.222
10	23.391	22.838	22.311	21.808	21.327	20.867	20.426	20.003	19.598	19.209
11	25.810	25.200	24.618	24.063	23.532	23.025	22.538	22.072	21.625	21.195
12	28.228	27.562	26.926	26.318	25.738	25.182	24.650	24.140	23.651	23.181
13	30.647	29.924	29.233	28.574	27.943	27.340	26.763	26.209	25.678	25.168
14	33.066	32.285	31.540	30.829	30.149	29.498	28.875	28.278	27.705	27.154
15	35.485	34.647	33.848	33.084	32.355	31.656	30.988	30.347	29.731	29.141
16	37.905	37.009	36.155	35.340	34.560	33.814	33.100	32.415	31.758	31.127
17	40.324	39.371	38.463	37.595	36.766	35.972	35.212	34.484	33.785	33.114
18	42.743	41.733	40.770	39.851	38.972	38.130	37.325	36.553	35.822	35.100
19	45.162	44.095	43.078	42.106	41.177	40.289	39.437	38.622	37.839	37.087
20	47.581	46.457	45.385	44.361	43.383	42.447	41.550	40.690	39.866	39.074
21	50.000	48.819	47.692	46.617	45.589	44.605	43.662	42.759	41.892	41.060
22	52.419	51.181	50.000	48.872	47.794	46.763	45.775	44.825	43.919	43.047
23	54.838	53.543	52.307	51.128	50.000	48.921	47.887	46.897	45.946	45.033
24	57.257	55.905	54.615	53.383	52.206	51.079	50.000	48.966	47.973	47.020
25	59.676	58.267	56.922	55.639	54.411	53.237	52.112	51.034	50.000	49.007
26	62.095	60.629	59.230	57.894	56.617	55.395	54.225	53.103	52.027	50.993
27	64.514	62.991	61.537	60.149	58.823	57.553	56.337	55.172	54.054	52.980
28	66.933	65.353	63.845	62.405	61.028	59.711	58.450	57.241	56.081	54.966
29	69.352	67.714	66.152	64.660	63.234	61.869	60.562	59.310	58.107	56.953
30	71.771	70.076	68.459	66.916	65.440	64.027	62.675	61.378	60.134	58.940
31	74.190	72.438	70.767	69.171	67.645	66.186	64.767	63.447	62.161	60.926
32	76.609	74.800	73.074	71.426	69.851	68.344	66.900	65.516	64.188	62.913
33	79.028	77.162	75.381	73.681	72.056	70.502	69.012	67.585	66.215	64.899
34	81.446	79.523	77.689	75.937	74.262	72.660	71.125	69.653	68.242	66.886
35	83.865	81.885	79.996	78.192	76.467	74.817	73.237	71.722	70.268	68.873
36	86.283	84.246	82.303	80.447	78.673	76.975	75.349	73.791	72.295	70.859
37	88.700	86.607	84.609	82.702	80.878	79.133	77.462	75.859	74.322	72.646
38	91.117	86.967	86.916	84.956	83.083	81.291	79.574	77.928	76.349	74.832
39	93.531	91.327	89.222	87.211	85.283	83.448	81.686	79.997	78.375	76.819
40	95.940	93.684	91.527	89.465	87.493	85.606	83.798	82.065	80.402	78.805
41	98.324	96.036	93.830	91.718	89.697	87.763	85.910	84.133	82.428	80.791
42		95.363	96.127	93.969	91.900	89.920	88.021	86.201	84.455	82.778
43			98.401	96.215	94.102	92.075	90.132	88.269	86.481	84.764
44				98.437	96.298	94.229	92.243	90.337	88.507	86.750
45					98.471	96.378	94.351	92.403	90.532	88.735
46						98.504	96.455	94.468	92.557	90.721
47							98.536	96.528	94.580	92.705
48								98.566	96.598	94.688
49									98.595	96.666
50										98.623

Appendix VIII (*Continued*)

5 PERCENT RANKS

	SAMPLE SIZE									
$j \backslash n$	1	2	3	4	5	6	7	8	9	10
1	5.000	2.532	1.695	1.274	1.021	0.851	0.730	0.639	0.568	0.512
2		22.361	13.535	9.761	7.644	6.285	5.337	4.639	4.102	3.677
3			36.840	24.860	18.925	15.316	12.876	11.111	9.775	8.726
4				47.237	34.259	27.134	22.532	19.290	16.875	15.003
5					54.928	41.820	34.126	28.924	25.137	22.244
6						60.696	47.930	40.031	34.494	30.354
7							65.184	52.932	45.036	39.338
8								68.766	57.086	49.310
9									71.687	60.584
10										74.113

5 PERCENT RANKS

	SAMPLE SIZE									
$j \backslash n$	11	12	13	14	15	16	17	18	19	20
1	0.465	0.426	0.394	0.366	0.341	0.320	0.301	0.285	0.270	0.256
2	3.332	3.046	2.805	2.600	2.423	2.268	2.132	2.011	1.903	1.806
3	7.882	7.187	6.605	6.110	5.685	5.315	4.990	4.702	4.446	4.217
4	13.507	12.285	11.267	10.405	9.666	9.025	8.464	7.969	7.529	7.135
5	19.958	18.102	16.566	15.272	14.166	13.211	12.377	11.643	10.991	10.408
6	27.125	24.530	22.395	20.607	19.086	17.777	16.636	15.634	14.747	13.955
7	34.981	31.524	28.705	26.358	24.373	22.669	21.191	19.895	18.750	17.731
8	43.563	39.086	35.480	32.503	29.999	27.860	26.011	24.396	22.972	21.707
9	52.991	47.267	42.738	39.041	35.956	33.337	31.083	29.120	27.395	25.865
10	63.564	56.189	50.535	45.999	42.256	39.101	36.401	34.060	32.009	30.195
11	76.160	66.132	58.990	53.434	48.925	45.165	41.970	39.215	36.811	34.693
12		77.908	68.366	61.461	56.022	51.560	47.808	44.595	41.806	39.358
13			79.418	70.327	63.656	58.343	53.945	50.217	47.003	44.197
14				80.736	72.060	65.617	60.436	56.112	52.420	49.218
15					81.896	73.604	67.381	62.332	58.088	54.442
16						82.925	74.988	68.974	64.057	59.897
17							83.843	76.234	70.420	65.634
18								84.668	77.363	71.738
19									85.413	78.389
20										86.089

Appendix VIII (*Continued*)

5 PERCENT RANKS

	SAMPLE SIZE									
j\n	21	22	23	24	25	26	27	28	29	30
1	0.244	0.233	0.223	0.213	0.205	0.197	0.190	0.183	0.177	0.171
2	1.719	1.640	1.567	1.501	1.440	1.384	1.332	1.284	1.239	1.198
3	4.010	3.822	3.651	3.495	3.352	3.220	3.098	2.985	2.879	2.781
4	6.781	6.460	6.167	5.901	5.656	5.431	5.223	5.031	4.852	4.685
5	9.884	9.411	8.981	8.588	8.229	7.899	7.594	7.311	7.049	6.806
6	13.245	12.603	12.021	11.491	11.006	10.560	10.148	9.768	9.415	9.087
7	16.818	15.994	15.248	14.569	13.947	13.377	12.852	12.367	11.917	11.499
8	20.575	19.556	18.634	17.796	17.030	16.328	15.682	15.085	14.532	14.018
9	24.499	23.272	22.164	21.157	20.238	19.396	18.622	17.908	17.246	16.633
10	28.580	27.131	25.824	24.639	23.559	22.570	21.662	20.824	20.050	19.331
11	32.811	31.126	29.609	28.236	26.985	25.842	24.793	23.827	22.934	22.106
12	37.190	35.254	33.515	31.942	30.513	29.508	28.012	26.911	25.894	24.953
13	41.720	39.516	37.539	35.756	34.139	32.664	31.314	30.072	28.927	27.867
14	46.406	43.913	41.684	39.678	37.862	36.209	34.697	33.309	32.030	30.846
15	51.261	48.454	45.954	43.711	41.684	39.842	38.161	36.620	35.200	33.889
16	56.302	53.151	50.356	47.858	45.607	43.566	41.707	40.004	38.439	36.995
17	61.559	58.020	54.902	52.127	49.636	47.384	45.336	43.464	41.746	40.163
18	67.079	63.091	59.610	56.531	53.779	51.300	49.052	47.002	45.123	43.394
19	72.945	68.409	64.507	61.086	58.048	55.323	52.861	50.621	48.573	46.691
20	79.327	74.053	69.636	65.819	62.459	59.465	56.770	54.327	52.099	50.056
21	86.705	80.188	75.075	70.773	67.039	63.740	60.790	58.127	55.706	53.493
22		87.269	80.980	76.020	71.828	68.176	64.936	62.033	59.403	57.007
23			87.788	81.711	76.896	72.810	69.237	66.060	63.200	60.605
24				88.265	82.388	77.711	73.726	70.231	67.113	64.299
25					88.707	83.017	78.470	74.583	71.168	68.103
26						89.117	83.603	79.179	75.386	72.038
27							89.498	84.149	79.844	76.140
28								89.853	84.661	80.467
29									90.185	85.140
30										90.497

Appendix VIII (*Continued*)

5 PERCENT RANKS

	SAMPLE SIZE									
j\n	31	32	33	34	35	36	37	38	39	40
1	0.165	0.160	0.155	0.151	0.146	0.142	0.138	0.135	0.131	0.128
2	1.158	1.122	1.086	1.055	1.025	0.996	0.969	0.943	0.919	0.896
3	2.690	2.604	2.524	2.448	2.377	2.310	2.246	2.186	2.129	2.075
4	4.530	4.384	4.246	4.120	3.999	3.885	3.778	3.676	3.580	3.488
5	6.578	6.365	3.166	5.978	5.802	5.636	5.479	5.331	5.190	5.057
6	8.781	8.495	8.227	7.976	7.739	7.516	7.306	7.107	6.919	6.740
7	11.109	10.745	10.404	10.084	9.783	9.499	9.232	8.979	8.740	8.513
8	13.540	13.093	12.675	12.283	11.914	11.567	11.240	10.931	10.638	10.361
9	16.061	15.528	15.029	14.561	14.122	13.708	13.318	12.950	12.601	12.271
10	18.662	18.038	17.455	16.909	16.396	15.913	15.458	15.028	14.622	14.237
11	21.336	20.618	19.948	19.319	18.730	18.175	17.653	17.160	16.694	16.252
12	24.077	23.262	22.501	21.788	21.119	20.491	19.898	19.340	18.812	18.312
13	26.883	25.966	25.111	24.310	23.560	22.855	22.191	21.565	20.973	20.413
14	29.749	28.727	27.775	26.884	26.049	25.265	24.527	23.832	23.175	22.553
15	32.674	31.544	30.491	29.507	28.585	27.719	26.905	26.138	25.414	24.729
16	35.657	34.415	33.258	32.177	31.165	30.216	29.324	28.483	27.690	26.940
17	38.698	37.339	36.074	34.894	33.789	32.754	31.781	30.865	30.001	29.185
18	41.797	40.317	38.940	37.657	36.457	35.332	34.276	33.283	32.346	31.461
19	44.956	43.349	41.656	40.466	39.167	37.951	36.809	35.736	34.725	33.770
20	48.175	46.436	44.823	43.321	41.920	40.609	39.380	38.224	37.136	36.109
21	51.458	49.581	47.841	46.225	44.717	43.309	41.988	40.748	39.581	38.480
22	54.810	52.786	50.914	49.177	47.560	46.049	44.634	43.307	42.058	40.881
23	58.234	56.055	54.344	52.181	50.448	48.832	47.320	45.902	44.569	43.314
24	64.739	59.314	57.235	55.239	53.385	51.658	50.045	48.534	47.114	45.778
25	65.336	62.810	60.493	58.355	56.374	54.532	52.812	51.204	49.694	48.275
26	69.036	66.313	63.824	61.534	59.416	57.454	55.624	53.914	52.311	50.805
27	72.563	69.916	67.237	64.754	62.523	60.429	58.483	56.666	54.966	53.370
28	76.650	73.640	70.748	68.113	65.695	63.483	61.392	59.463	57.661	55.972
29	81.054	77.518	74.375	71.535	68.944	66.561	64.357	62.309	60.399	58.612
30	85.591	81.606	76.150	75.069	72.282	69.732	67.384	65.209	63.185	61.294
31	90.789	86.015	82.127	78.747	75.728	72.990	70.482	68.168	66.021	64.021
32		91.063	86.415	82.619	79.312	76.352	73.663	71.196	68.916	66.797
33			91.322	86.793	83.085	79.848	76.946	74.304	71.876	69.629
34				91.566	87.150	83.526	80.357	77.510	74.915	72.525
35					91.797	87.488	83.946	80.841	78.048	75.497
36						92.015	87.809	84.344	81.302	78.560
37							92.222	88.115	84.723	81.741
38								92.419	88.405	85.085
39									92.606	88.681
40										92.784

Appendix VIII (*Continued*)

5 PERCENT RANKS

	SAMPLE SIZE									
$j\backslash n$	41	42	43	44	45	46	47	48	49	50
1	0.125	0.122	0.119	0.116	0.114	0.111	0.109	0.107	0.105	0.102
2	0.874	0.853	0.833	0.814	0.795	0.778	0.761	0.745	0.730	0.715
3	2.024	1.975	1.928	1.884	1.842	1.801	1.762	1.725	1.689	1.655
4	3.402	3.319	3.240	3.165	3.093	3.025	2.959	2.897	2.836	2.779
5	4.930	4.810	4.695	4.586	4.481	4.382	4.286	4.195	4.108	4.024
6	6.570	6.409	6.256	6.109	5.969	5.836	5.708	5.586	5.469	5.357
7	8.298	8.093	7.898	7.713	7.536	7.366	7.205	7.050	6.902	6.760
8	10.097	9.847	9.609	9.382	9.166	8.959	8.762	8.573	8.392	8.218
9	11.958	11.660	11.377	11.107	10.850	10.605	10.370	10.146	9.931	9.725
10	13.872	13.525	13.195	12.881	12.582	12.296	12.023	11.762	11.512	11.272
11	15.833	15.436	15.058	14.698	14.355	14.028	13.715	13.416	13.130	12.856
12	17.838	17.389	16.961	16.554	16.166	15.796	15.443	15.105	14.782	14.472
13	19.883	19.379	18.901	18.445	18.012	17.598	17.203	16.825	16.464	16.117
14	21.964	21.406	20.875	20.370	19.889	19.430	18.993	18.574	18.174	17.790
15	24.081	23.466	22.881	22.326	21.796	21.292	20.810	20.350	19.910	19.488
16	26.230	25.557	24.918	24.311	23.732	23.180	22.654	22.151	21.671	21.210
17	28.412	27.679	26.984	26.323	25.694	25.095	24.523	23.977	23.455	22.955
18	30.624	29.831	29.078	28.363	27.683	27.034	26.416	25.825	25.261	24.721
19	32.867	32.011	31.200	30.429	29.696	28.997	28.331	27.696	27.088	26.507
20	35.138	34.219	33.348	32.520	31.733	30.984	30.269	29.588	28.936	28.313
21	37.440	36.455	35.522	34.636	33.794	32.993	32.229	31.500	30.804	30.138
22	39.770	38.719	37.722	36.777	35.879	35.025	34.210	33.434	32.692	31.980
23	42.129	41.009	39.949	38.943	37.987	37.078	36.212	35.387	34.599	33.845
24	44.518	43.328	42.201	41.133	40.118	39.154	38.235	37.360	36.524	35.726
25	46.937	45.674	44.480	43.347	42.273	41.251	40.279	39.353	38.469	37.625
26	49.388	48.050	46.785	45.587	44.451	43.371	42.344	41.366	10.432	39.541
27	51.869	50.454	49.117	47.852	46.652	45.513	44.430	43.398	42.415	41.476
28	54.385	52.889	51.478	50.143	48.878	47.678	46.537	45.451	44.416	43.428
29	56.935	55.356	53.868	52.461	51.129	49.866	48.666	47.524	46.436	45.399
30	59.522	57.857	56.288	54.807	53.406	52.078	50.817	49.618	48.477	47.388
31	62.149	60.393	58.741	57.183	55.710	54.315	52.991	51.734	50.537	49.396
32	64.820	62.968	61.228	59.590	58.042	56.578	55.190	53.871	52.617	51.423
33	67.539	56.585	63.753	62.029	60.404	58.868	57.413	56.032	54.720	53.470
34	70.311	68.248	66.318	64.505	62.798	61.187	59.662	58.217	56.844	55.538
35	73.146	70.963	68.927	67.020	65.227	63.537	61.940	60.427	58.991	57.627
36	76.053	73.738	71.587	69.578	67.694	65.921	64.247	62.664	61.164	59.738
37	79.049	76.584	74.306	72.185	70.203	68.341	66.587	64.931	63.362	61.874
38	82.160	79.517	77.093	74.849	72.759	70.805	68.963	67.228	65.589	64.034
39	85.429	82.561	79.964	77.580	75.370	73.309	71.378	69.561	67.846	66.222
40	88.945	85.759	82.944	80.392	78.046	75.870	73.838	71.932	70.136	68.440
41	92.954	89.196	86.073	83.310	80.802	78.494	76.350	74.347	72.465	70.691
42		93.116	89.437	86.374	83.661	81.196	78.924	76.812	74.836	72.978
43			93.270	89.666	86.662	83.998	81.573	79.337	77.256	75.306
44				93.418	89.887	86.939	84.321	81.936	79.734	77.683
45					93.560	90.098	87.204	84.631	82.285	80.117
46						93.695	90.300	87.459	84.929	82.621
47							93.825	90.494	87.703	85.216
48								93.950	90.681	87.939
49									94.069	90.860
50										94.184

Appendix VIII (*Continued*)

95 PERCENT RANKS

	SAMPLE SIZE									
$j \backslash n$	1	2	3	4	5	6	7	8	9	10
1	95.000	77.639	63.160	52.713	45.072	39.304	34.816	31.234	28.313	25.887
2		97.468	86.465	75.139	65.741	58.180	52.070	47.068	42.914	39.416
3			98.305	90.239	81.075	72.866	65.874	59.969	54.964	50.690
4				98.726	92.356	84.684	77.468	71.076	65.506	60.662
5					98.979	93.715	87.124	80.710	74.863	69.646
6						99.149	94.662	88.889	83.125	77.756
7							99.270	95.361	90.225	84.997
8								99.361	95.898	91.274
9									99.432	96.323
10										99.488

95 PERCENT RANKS

	SAMPLE SIZE									
$j \backslash n$	11	12	13	14	15	16	17	18	19	20
1	23.840	22.092	20.582	19.264	18.104	17.075	16.157	15.332	14.587	13.911
2	36.436	33.868	31.634	29.673	27.940	26.396	25.012	23.766	22.637	21.611
3	47.009	43.811	41.010	38.539	36.344	34.383	32.619	31.026	29.580	28.262
4	56.437	52.733	49.465	46.566	43.978	41.657	39.564	37.668	35.943	34.366
5	65.019	60.914	57.262	54.000	51.075	48.440	46.055	43.888	41.912	40.103
6	72.875	68.476	64.520	60.928	57.744	54.835	52.192	49.783	47.580	45.558
7	80.042	75.470	71.295	67.497	64.043	60.899	58.029	55.404	52.997	50.782
8	86.492	81.898	77.604	73.641	70.001	66.663	63.599	60.784	58.194	55.803
9	92.118	87.715	83.434	79.393	75.627	72.140	68.917	65.940	63.188	60.641
10	96.668	92.813	88.733	84.728	80.913	77.331	73.989	70.880	67.991	65.307
11	99.535	96.954	93.395	89.595	85.834	82.223	78.809	75.604	72.605	69.805
12		99.573	97.195	93.890	90.334	86.789	83.364	80.105	77.028	74.135
13			99.606	97.400	94.315	90.975	87.623	84.366	81.250	78.293
14				99.634	97.577	94.685	91.535	88.357	85.253	82.269
15					99.659	97.732	95.010	92.030	89.009	86.045
16						99.680	97.868	95.297	92.471	89.592
17							99.699	97.989	95.553	92.865
18								99.715	98.097	95.783
19									99.730	98.193
20										99.744

Appendix VIII (*Continued*)

95 PERCENT RANKS

	SAMPLE SIZE									
$j \backslash n$	21	22	23	24	25	26	27	28	29	30
1	13.295	12.731	12.212	11.735	11.293	10.883	10.502	10.147	9.814	9.503
2	20.673	19.812	19.020	18.289	17.612	16.983	16.397	15.851	15.339	14.860
3	27.055	25.947	24.925	23.980	23.104	22.289	21.530	20.821	20.156	19.533
4	32.921	31.591	30.364	29.227	28.172	27.190	26.274	25.417	24.614	23.860
5	38.441	36.909	35.193	34.181	32.961	31.824	30.763	29.769	28.837	27.962
6	43.698	41.980	40.390	38.914	37.541	36.260	35.062	33.940	32.887	31.897
7	48.739	46.849	45.097	43.469	41.952	40.535	39.210	37.967	36.800	35.701
8	53.594	51.546	49.643	47.873	46.221	44.677	43.230	41.873	40.597	39.395
9	58.280	56.087	54.046	52.142	50.364	48.700	47.139	45.673	44.294	42.993
10	62.810	60.484	58.315	56.289	54.393	52.616	50.948	49.379	47.901	46.507
11	67.189	64.746	62.461	60.321	58.316	56.434	54.664	52.998	51.427	49.944
12	71.420	68.874	66.485	64.244	62.138	60.158	58.293	56.536	54.877	53.309
13	75.501	72.869	70.391	68.058	65.861	63.791	61.839	59.996	58.254	56.605
14	79.425	76.728	74.176	71.764	69.487	67.336	65.303	63.380	61.561	59.837
15	83.182	80.444	77.836	75.361	73.015	70.792	68.686	66.691	64.799	63.005
16	86.755	84.006	81.366	78.843	76.441	74.158	71.988	69.927	67.970	66.111
17	90.116	87.397	84.752	82.204	79.762	77.430	75.207	73.089	71.073	69.154
18	93.219	90.589	87.978	85.431	82.970	80.604	78.338	46.173	74.106	72.133
19	95.990	93.540	91.019	88.509	86.052	83.672	81.378	79.176	77.066	75.047
20	98.281	96.178	93.832	91.411	88.994	86.623	84.318	82.092	79.950	77.894
21	99.756	98.360	96.348	94.099	91.771	89.440	87.148	84.915	82.753	80.669
22		99.767	98.433	96.505	94.344	92.101	89.851	87.633	85.468	83.367
23			99.777	98.499	96.648	94.569	92.406	90.232	88.083	85.981
24				99.786	98.560	96.780	94.777	92.689	90.584	88.501
25					99.795	98.616	96.902	94.969	92.951	90.913
26						99.803	98.668	97.015	95.148	93.194
27							99.810	98.716	97.120	95.314
28								99.817	98.761	97.218
29									99.823	98.802
30										99.829

Appendix VIII (*Continued*)

95 PERCENT RANKS

	SAMPLE SIZE									
$j\backslash n$	31	32	33	34	35	36	37	38	39	40
1	9.211	8.937	8.678	8.434	8.203	7.985	7.778	7.581	7.394	7.216
2	14.409	13.985	13.585	13.207	12.850	12.512	12.191	11.885	11.595	11.319
3	18.946	18.394	17.873	17.381	16.915	16.474	16.054	15.656	15.277	14.915
4	23.150	22.482	21.850	21.253	20.688	20.152	19.643	19.159	18.698	18.259
5	27.137	26.360	25.625	24.931	24.272	23.648	23.054	22.490	21.952	21.440
6	30.964	30.084	29.252	28.465	27.718	27.010	26.337	25.696	25.085	24.503
7	34.665	33.687	32.763	31.887	31.056	30.268	29.518	28.804	28.124	27.475
8	38.261	37.190	36.176	35.216	34.305	33.439	32.616	31.832	31.084	30.371
9	41.766	40.606	39.507	68.466	37.477	36.537	35.643	34.791	33.979	33.203
10	45.190	43.945	42.765	41.645	40.582	39.571	38.608	37.691	36.815	35.979
11	48.542	57.214	45.956	44.761	43.626	42.546	41.517	40.537	39.601	38.706
12	51.825	50.419	49.086	47.819	46.615	45.468	44.376	43.334	42.339	41.388
13	55.044	53.564	52.159	50.823	49.552	48.341	47.187	46.086	45.034	44.028
14	58.203	56.651	55.177	53.775	52.440	51.168	49.955	48.796	47.689	46.630
15	61.302	59.683	58.144	56.678	55.282	53.951	52.680	51.466	50.305	49.195
16	64.343	62.661	61.060	59.534	58.080	56.691	55.366	54.098	52.886	51.725
17	67.326	65.585	63.926	62.343	60.833	59.391	58.012	56.693	55.431	54.222
18	70.251	68.456	66.742	65.106	63.543	62.049	60.620	59.252	57.942	56.686
19	73.117	71.272	69.509	67.823	66.210	64.668	63.190	61.776	60.419	59.119
20	75.922	74.034	72.225	70.493	68.835	67.246	65.723	64.264	62.864	61.520
21	78.664	76.738	74.889	73.116	71.415	69.784	68.219	66.717	65.275	63.891
22	81.338	79.382	77.499	75.689	73.951	72.280	70.676	69.135	67.654	66.230
23	83.939	81.961	80.052	78.212	76.440	74.735	73.094	71.517	69.999	68.539
24	86.460	84.472	82.545	80.680	78.881	77.145	75.473	73.862	72.310	70.815
25	88.891	86.907	84.971	83.091	81.270	79.509	77.809	76.168	74.586	73.060
26	91.219	89.255	87.325	85.439	83.604	81.825	80.101	78.435	76.825	75.270
27	93.422	91.505	89.596	87.717	85.878	84.087	82.347	80.660	79.027	77.447
28	95.470	93.635	91.772	89.916	88.086	86.292	84.542	82.840	81.188	79.587
29	97.310	95.615	93.834	92.024	90.217	88.433	86.682	84.972	83.306	81.688
30	98.841	97.396	95.752	94.021	92.261	90.501	88.760	87.050	85.378	83.746
31	99.835	98.878	97.476	95.880	94.198	92.483	90.768	89.069	87.399	85.763
32		99.840	98.912	97.552	96.001	94.364	92.694	91.021	89.362	87.729
33			99.845	98.945	97.623	96.114	94.521	92.893	91.260	89.639
34				99.849	98.975	97.690	96.222	94.669	93.081	91.487
35					99.854	99.004	97.754	96.324	94.810	93.260
36						99.858	99.031	97.814	96.420	94.943
37							99.861	99.057	97.871	96.511
38								99.865	99.081	97.925
39									99.869	99.104
40										99.872

95 PERCENT RANKS

	SAMPLE SIZE									
$j \backslash n$	41	42	43	44	45	46	47	48	49	50
1	7.046	6.884	6.730	6.582	6.440	6.305	6.175	6.050	5.931	5.816
2	11.055	10.804	10.563	10.334	10.113	9.902	9.700	9.506	9.319	9.140
3	14.571	14.241	13.927	13.626	13.338	13.061	12.796	12.541	12.297	12.061
4	17.840	17.439	17.056	16.690	16.339	16.002	15.679	15.369	15.071	14.784
5	20.951	20.483	20.036	19.608	19.198	18.804	18.427	18.064	17.715	17.379
6	23.947	23.416	22.907	22.420	21.954	21.506	21.076	20.663	20.266	19.883
7	26.854	26.262	25.694	25.151	24.630	24.130	23.650	23.188	22.744	22.317
8	29.689	29.037	28.413	27.814	27.241	26.691	26.162	25.623	25.164	24.694
9	32.461	31.752	31.073	30.422	29.797	29.198	28.622	28.068	27.535	27.022
10	35.180	34.415	33.682	32.980	32.306	31.659	31.037	30.439	29.864	29.309
11	37.851	37.032	36.247	35.495	34.773	34.079	33.413	32.772	32.154	31.560
12	40.478	39.607	38.772	37.971	37.202	36.463	35.753	35.069	34.411	33.778
13	43.065	42.143	41.259	40.410	39.596	38.813	38.060	37.336	36.698	35.933
14	45.615	44.644	43.712	42.817	41.958	41.132	40.338	39.573	38.836	38.126
15	48.131	47.110	46.132	45.193	44.290	43.422	42.587	41.783	41.008	40.262
16	50.612	49.546	48.522	47.539	46.594	45.665	44.810	43.968	43.156	42.373
17	53.062	51.950	50.883	49.857	48.871	47.922	47.009	46.129	45.280	44.462
18	55.482	54.326	53.215	52.148	51.122	50.134	49.183	48.266	47.382	46.530
19	57.871	56.672	55.520	54.413	53.348	52.322	51.334	50.382	49.463	48.577
20	60.230	58.991	57.799	56.653	55.549	54.487	53.463	52.476	51.523	50.604
21	62.560	61.281	60.051	58.867	57.727	56.629	55.570	54.549	53.563	52.612
22	64.861	63.545	62.273	61.057	59.882	58.749	57.556	56.602	55.584	54.801
23	67.133	65.781	64.478	63.223	62.013	60.846	59.721	58.634	57.585	56.572
24	69.376	67.989	66.652	65.363	64.121	62.922	61.765	60.647	59.568	58.524
25	71.588	70.169	68.800	67.480	66.205	64.975	63.787	62.640	61.531	60.459
26	73.769	72.320	70.922	69.571	68.267	67.007	65.790	64.613	63.476	62.375
27	75.919	74.443	73.016	71.637	70.304	69.016	67.771	66.566	65.401	64.274
28	78.035	76.534	75.082	73.677	72.317	71.002	69.730	68.500	67.308	66.155
29	80.117	78.594	77.119	75.689	74.306	72.966	71.668	70.412	69.196	68.017
30	82.162	80.621	79.125	77.674	76.268	74.905	73.584	72.304	71.064	69.862
31	84.166	82.611	81.099	79.630	78.203	76.819	75.477	74.175	72.912	71.687
32	86.128	84.564	83.039	81.554	80.111	78.708	77.346	76.023	74.739	73.493
33	88.042	86.475	84.942	83.446	81.988	80.569	79.190	77.848	76.545	75.279
34	89.903	88.340	86.805	85.302	83.834	82.402	81.007	79.650	78.329	77.045
35	91.702	90.153	88.623	87.119	85.645	84.204	82.797	81.426	80.090	78.790
36	93.430	91.907	90.391	88.892	87.418	85.972	84.557	83.175	81.826	80.511
37	95.070	93.591	92.102	90.618	89.150	87.704	86.285	84.895	83.536	82.210
38	96.598	95.190	93.744	92.287	90.834	89.395	87.977	86.584	85.218	83.882
39	97.976	96.681	95.305	93.891	92.464	91.041	89.630	88.238	86.870	85.528
40	99.126	98.025	96.760	95.414	94.030	92.633	91.238	89.854	88.488	87.144
41	99.875	99.147	98.071	96.835	95.518	94.164	92.795	91.427	90.069	88.728
42		99.878	99.167	98.116	96.907	95.618	94.291	92.950	91.608	90.275
43			99.881	99.186	98.158	96.975	95.714	94.414	93.098	91.781
44				99.883	99.205	98.199	97.041	95.805	94.531	93.240
45					99.886	99.222	98.238	97.103	95.892	94.643
46						99.889	99.239	98.275	97.163	95.976
47							99.891	99.255	98.311	97.221
48								99.893	99.270	98.345
49									99.895	99.285
50										99.897

[a]Courtesy of General Motors Research Laboratories, Warren, Michigan.

appendix IX weights for obtaining estimates of the parameters of the weibull distribution[a]

[a] Taken from: Mann, N. R.; *Results on Location and Scale Parameter Estimation with Application to the Extreme-Value Distribution*; Aerospace Research Laboratories, Wright–Patterson Air Force Base, Ohio, ARL 67-0023, Contract No. AF 33(615)-2818, February, 1967.

Appendix IX

	n	r	i	a_i	c_i
	2	2	1	0.110731	−0.421383
			2	0.889269	0.421383
E(LU)[b]				0.65712995	
E(CP)				0.03757418	
E(LB)					0.41583918
	3	2	1	−0.166001	−0.452110
			2	1.166001	0.452110
E(LU)				0.79546061	
E(CP)				0.25750956	
E(LB)					0.45005549
	3	3	1	0.081063	−0.278666
			2	0.251001	−0.190239
			3	0.667936	0.468904
E(LU)				0.40240741	
E(CP)				−0.01842169	
E(LB)					0.25634620
	4	2	1	−0.346974	−0.465455
			2	1.346974	0.465455
E(LU)				1.01477788	
E(CP)				0.41350875	
E(LB)					0.46438768

	n	r	i	a_i	c_i
	4	3	1	−0.044975	−0.297651
			2	0.088057	−0.234054
			3	0.956918	0.531705
E(LU)				0.42315147	
E(CP)				0.08477554	
E(LB)					0.28172930
	4	4	1	0.064336	−0.203052
			2	0.147340	−0.182749
			3	0.261510	−0.070109
			4	0.526813	0.455910
E(LU)				0.29247651	
E(CP)				−0.02831210	
E(LB)					0.18386193
	5	2	1	−0.481434	−0.472962
			2	1.481434	0.472962
E(LU)				1.24921018	
E(CP)				0.53379141	
E(LB)					0.47230837
	5	3	1	−0.137958	−0.306562
			2	−0.025510	−0.257087
			3	1.163468	0.563650

[b] E(LU) = Expected loss for estimate of u.
E(CP) = Expected cross product.
E(LB) = Expected loss for estimate of b.

Appendix IX *(Continued)*

	n	r	i	a_i	c_i		n	r	i	a_i	c_i
E(LU)				0.49029288		E(LU)				0.57539484	
E(CP)				0.16612899		E(CP)				0.23269670	
E(LB)					0.29419192	E(LB)					0.30173252
	5	4	1	−0.006983	−0.217766		6	4	1	−0.063569	−0.225141
			2	0.059652	−0.199351				2	−0.006726	−0.209083
			3	0.156664	−0.118927				3	0.079882	−0.146386
			4	0.790668	0.536044				4	0.990412	0.580610
E(LU)				0.29062766		E(LU)				0.31552097	
E(CP)				0.03076329		E(CP)				0.08035062	
E(LB)					0.20241894	E(LB)					0.21242254
	5	5	1	0.052975	−0.158131		6	5	1	0.007521	−0.169920
			2	0.103531	−0.155707				2	0.048328	−0.166319
			3	0.163808	−0.111820				3	0.101608	−0.129510
			4	0.246092	−0.005600				4	0.172859	−0.054453
			5	0.433593	0.431259				5	0.669685	0.520201
E(LU)				0.23040495		E(LU)				0.22351297	
E(CP)				−0.02913523		E(CP)				0.00888019	
E(LB)					0.14284288	E(LB)					0.15690540
	6	2	1	−0.588298	−0.477782		6	6	1	0.044826	−0.128810
			2	1.588298	0.477782				2	0.079377	−0.132102
E(LU)				1.48102383					3	0.117541	−0.111951
E(CP)				0.63148980					4	0.163591	−0.064666
E(LB)					0.47734078				5	0.226486	0.031796
									6	0.368179	0.405733
	6	3	1	−0.211474	−0.311847	E(LU)				0.19030430	
			2	−0.112994	−0.271381	E(CP)				−0.02771574	
			3	1.324468	0.583229	E(LB)					0.11657671

	7	2	1	−0.676894		−0.481140
			2	1.676894		0.481140
E(LU)				1.70468001		
E(CP)					0.71366553	
E(LB)						0.48082310
	7	3	1	−0.272195		−0.315369
			2	−0.184061		−0.281139
			3	1.456255		0.596507
E(LU)				0.66758707		
E(CP)					0.28885432	
E(LB)						0.30681307
	7	4	1	−0.110274		−0.229691
			2	−0.060226		−0.215613
			3	0.018671		−0.164168
			4	1.151829		0.609472
E(LU)				0.35340223		
E(CP)					0.12260834	
E(LB)						0.21884662
	7	5	1	−0.030368		−0.176203
			2	0.004333		−0.172399
			3	0.052957		−0.141218
			4	0.117599		−0.082820
			5	0.855480		0.572640
E(LU)				0.23316740		
E(CP)					0.04212562	
E(LB)						0.16497315
	7	6	1	0.013524		−0.138436
			2	0.041588		−0.140342
			3	0.075499		−0.121821
			4	0.117461		−0.082938
			5	0.172092		−0.015394
			6	0.579835		0.498931
E(LU)				0.18269947		
E(CP)					−0.00130057	
E(LB)						0.12760617
	7	7	1	0.038743		−0.108323
			2	0.064086		−0.113479
			3	0.090785		−0.103569
			4	0.120971		−0.078748
			5	0.157657		−0.032632
			6	0.207825		0.054727
			7	0.319934		0.382022
E(LU)				0.16219070		
E(CP)					−0.02578937	
E(LB)						0.09836496
	8	2	1	−0.752513		−0.483616
			2	1.752513		0.483616
E(LU)				1.91861540		
E(CP)					0.78453314	
E(LB)						0.48337662
	8	3	1	−0.323875		−0.317890
			2	−0.243808		−0.288231
			3	1.567683		0.606120

Appendix IX (*Continued*)

	n	r	i	a_i	c_i
E(LU)				0.76198737	
E(CP)				0.33734068	
E(LB)					0.31047652
	8	4	1	−0.149973	−0.232805
			2	−0.105015	−0.220324
			3	−0.032257	−0.176675
			4	1.287245	0.629805
E(LU)				0.39805551	
E(CP)				0.15928131	
E(LB)					0.22335819
	8	5	1	−0.062656	−0.180231
			2	−0.032248	−0.176510
			3	0.012767	−0.149566
			4	0.072446	−0.101642
			5	1.009691	0.607948
E(LU)				0.25192092	
E(CP)				0.07129172	
E(LB)					0.17037848
	8	6	1	−0.013509	−0.143834
			2	0.010292	−0.145006
			3	0.041357	−0.128393
			4	0.080475	−0.095696
			5	0.130327	−0.043280
			6	0.751058	0.556209
E(LU)				0.18599844	
E(CP)				0.02247163	
E(LB)					0.13422386

	n	r	i	a_i	c_i
	8	7	1	0.015973	−0.116317
			2	0.036729	−0.120331
			3	0.060439	−0.110582
			4	0.088239	−0.088450
			5	0.122062	−0.050995
			6	0.165529	0.009700
			7	0.511030	0.476975
E(LU)				0.15505149	
E(CP)				−0.00641304	
E(LB)					0.10726405
	8	8	1	0.034052	−0.093270
			2	0.053552	−0.098886
			3	0.073452	−0.093994
			4	0.095062	−0.079752
			5	0.119768	−0.053918
			6	0.149934	−0.010179
			7	0.191236	0.069325
			8	0.282943	0.360675
E(LU)				0.14136026	
E(CP)				−0.02386561	
E(LB)					0.08501680
	9	2	1	−0.818444	−0.485517
			2	1.818444	0.485517
E(LU)				2.12272209	
E(CP)				0.84680378	
E(LB)					0.48532951

	9	3	1	−0.368833	−0.319786
			2	−0.295280	−0.293621
			3	1.664113	0.613407
E(LU)				0.85621748	
E(CP)				0.37995861	
E(LB)					0.31324611
	9	4	1	−0.184461	−0.235080
			2	−0.143505	−0.223891
			3	−0.075815	−0.185970
			4	1.403781	0.644941
E(LU)				0.44625568	
E(CP)				0.19160927	
E(LB)					0.22671251
	9	5	1	−0.090726	−0.183061
			2	−0.063541	−0.179515
			3	−0.021495	−0.155825
			4	0.034159	−0.115133
			5	1.141604	0.633534
E(LU)				0.27605014	
E(CP)				0.09715351	
E(LB)					0.17429417
	9	6	1	−0.037118	−0.147411
			2	−0.016377	−0.148150
			3	0.012499	−0.133219
			4	0.049305	−0.105060
			5	0.095614	−0.062073
			6	0.896078	0.595913
E(LU)				0.19579592	
E(CP)				0.04378261	
E(LB)					0.13880129
	9	7	1	−0.004220	−0.120988
			2	0.013386	−0.124245
			3	0.035068	−0.115091
			4	0.061198	−0.095508
			5	0.093013	−0.064162
			6	0.132740	−0.017187
			7	0.668815	0.537180
E(LU)				0.15547192	
E(CP)				0.01139509	
E(LB)					0.11278822
	9	8	1	0.016797	−0.100011
			2	0.032919	−0.104750
			3	0.050582	−0.099608
			4	0.070497	−0.086226
			5	0.093635	−0.063541
			6	0.121560	−0.028346
			7	0.157175	0.026525
			8	0.456836	0.455956
E(LU)				0.13496842	
E(CP)				−0.00906894	
E(LB)					0.09236358
	9	9	1	0.030338	−0.081777
			2	0.045872	−0.087308
			3	0.061368	−0.085084
			4	0.077742	−0.076470
			5	0.095769	−0.060667
			6	0.116517	−0.035136
			7	0.141932	0.006001
			8	0.176764	0.078828
			9	0.253697	0.341614

Appendix IX (*Continued*)

	n	r	i	a_i	c_i
E(LU)				0.12529518	
E(CP)				−0.02209438	
E(LB)					0.07482425
	10	2	1	−0.876869	−0.487022
			2	1.876869	0.487022
E(LU)				2.31744054	
E(CP)				0.90232208	
E(LB)					0.48687150
	10	3	1	−0.408602	−0.321265
			2	−0.340443	−0.297858
			3	1.749045	0.619124
E(LU)				0.94907551	
E(CP)				0.41795081	
E(LB)					0.31541467
	10	4	1	−0.214930	−0.236817
			2	−0.177223	−0.226688
			3	−0.113820	−0.193159
			4	1.505973	0.656663
E(LU)				0.49619736	
E(CP)				0.22047816	
E(LB)					0.22930885
	10	5	1	−0.115524	−0.185169
			2	−0.090868	−0.181821
			3	−0.051341	−0.160697
			4	0.000925	−0.125311
			5	1.256809	0.652997
E(LU)				0.30344549	
E(CP)				0.12033056	
E(LB)					0.17727542
	10	6	1	−0.058017	−0.149985
			2	−0.039595	−0.150451
			3	−0.012513	−0.136941
			4	0.022314	−0.112224
			5	0.065750	−0.075721
			6	1.022062	0.625321
E(LU)				0.20973843	
E(CP)				0.06299841	
E(LB)					0.14219828
	10	7	1	−0.022198	−0.124170
			2	−0.006909	−0.126894
			3	0.013224	−0.118392
			4	0.037994	−0.100924
			5	0.068153	−0.073988
			6	0.105164	−0.035501
			7	0.804572	0.579868
E(LU)				0.16066059	
E(CP)				0.02762724	
E(LB)					0.11670571
	10	8	1	0.001179	−0.104082
			2	0.014889	−0.108163
			3	0.030998	−0.103119
			4	0.049734	−0.090835
			5	0.071745	−0.070902
			6	0.098114	−0.041560
			7	0.130649	0.000799
			8	0.602692	0.517864
E(LU)				0.13403554	
E(CP)				0.00474963	
E(LB)					0.09704810

	10	9	1	0.016841	−0.087538
			2	0.029807	−0.092405
			3	0.043570	−0.089839
			4	0.058640	−0.081428
			5	0.075576	−0.066855
			6	0.095169	−0.044670
			7	0.118707	−0.011816
			8	0.148575	0.038159
			9	0.413116	0.436394
E(LU)				0.11965747	
E(CP)				−0.01043859	
E(LB)				0.08100409	
	10	10	1	0.027331	−0.072734
			2	0.040034	−0.077971
			3	0.052496	−0.077242
			4	0.065408	−0.071876
			5	0.079263	−0.061652
			6	0.094638	−0.045420
			7	0.112414	−0.020698
			8	0.134239	0.017927
			9	0.164178	0.085070
			10	0.230001	0.324597
E(LU)				0.11252220	
E(CP)				−0.02050852	
E(LB)				0.06679250	
	11	2	1	−0.929310	−0.488243
			2	1.929310	0.488243
E(LU)				2.50340024	
E(CP)				0.95239887	
E(LB)				0.48812000	
	11	3	1	−0.444245	−0.322452
			2	−0.380642	−0.301277
			3	1.824887	0.623729
E(LU)				1.03995578	
E(CP)				0.45220741	
E(LB)				0.31715930	
	11	4	1	−0.242206	−0.238188
			2	−0.207204	−0.228941
			3	−0.147490	−0.198888
			4	1.596900	0.666017
E(LU)				0.54681985	
E(CP)				0.24653583	
E(LB)				0.23138012	
	11	5	1	−0.137718	−0.186803
			2	−0.115110	−0.183651
			3	−0.077762	−0.164597
			4	−0.028411	−0.133278
			5	1.359000	0.668329
E(LU)				0.33282848	
E(CP)				0.14129911	
E(LB)				0.17962678	
	11	6	1	−0.076739	−0.151936
			2	−0.060142	−0.152221
			3	−0.034581	−0.139907
			4	−0.001490	−0.117886
			5	0.039518	−0.086131
			6	1.133434	0.648081
E(LU)				0.22640907	
E(CP)				0.08045010	
E(LB)				0.14483423	

Appendix IX (*Continued*)

	n	r	i	a_i	c_i
	11	7	1	−0.038349	−0.126507
			2	−0.024842	−0.128838
			3	−0.005964	−0.120951
			4	0.017632	−0.105219
			5	0.046354	−0.081602
			6	0.081182	−0.048929
			7	0.923987	0.612047
E(LU)				0.16905710	
E(CP)				0.04246025	
E(LB)					0.11966982
	11	8	1	−0.012943	−0.106922
			2	−0.001050	−0.110498
			3	0.013869	−0.105662
			4	0.031661	−0.094405
			5	0.052723	−0.076693
			6	0.077815	−0.051525
			7	0.108161	−0.016860
			8	0.729765	0.562564
E(LU)				0.13669382	
E(CP)				0.01751192	
E(LB)					0.10043756
	11	9	1	0.004425	−0.091115
			2	0.015498	−0.095437
			3	0.028023	−0.092780
			4	0.042178	−0.084833
			5	0.058340	−0.071581
			6	0.077093	−0.052182
			7	0.099349	−0.024880
			8	0.126592	0.013606
			9	0.548502	0.499201
E(LU)				0.11809425	
E(CP)				0.00058414	
E(LB)					0.08503131
	11	10	1	0.016502	−0.077717
			2	0.027205	−0.082449
			3	0.038291	−0.081388
			4	0.050160	−0.075977
			5	0.063170	−0.066222
			6	0.077772	−0.051429
			7	0.094625	−0.030120
			8	0.114811	0.000537
			9	0.140333	0.046381
			10	0.377130	0.418384
E(LU)				0.10756449	
E(CP)				−0.01109747	
E(LB)					0.07207183
	11	11	1	0.024850	−0.065444
			2	0.035456	−0.070318
			3	0.045727	−0.070456
			4	0.056215	−0.067076
			5	0.067261	−0.060207
			6	0.079220	−0.049300
			7	0.092560	−0.033156
			8	0.108034	−0.009427
			9	0.127068	0.026879
			10	0.153197	0.089148
			11	0.210412	0.309357
E(LU)				0.10212039	
E(CP)				−0.01910164	
E(LB)					0.06030372

	12	2	1	−0.976872		−0.489254
			2	1.976872		0.489254
E(LU)				2.68127021		
E(CP)					0.99799849	
E(LB)						0.48915157
	12	3	1	−0.476530		−0.323426
			2	−0.416836		−0.304093
			3	1.893367		0.627519
E(LU)				1.12857097		
E(CP)					0.48338667	
E(LB)						0.31859354
	12	4	1	−0.266888		−0.239300
			2	−0.234180		−0.230796
			3	−0.177681		−0.203562
			4	1.678749		0.673657
E(LU)				0.59748043		
E(CP)					0.27026774	
E(LB)						0.23307201
	12	5	1	−0.157792		−0.188109
			2	−0.136884		−0.185142
			3	−0.101445		−0.167790
			4	−0.054640		−0.139693
			5	1.450761		0.680734
E(LU)				0.36338878		
E(CP)					0.16042600	
E(LB)						0.18153147
	12	6	1	−0.093679		−0.153471
			2	−0.078561		−0.153632
			3	−0.054320		−0.142329
			4	−0.022769		−0.122474
			5	0.016136		−0.094355
			6	1.233193		0.666261
E(LU)				0.24490094		
E(CP)					0.09641022	
E(LB)						0.14694548
	12	7	1	−0.052987		−0.128308
			2	−0.040893		−0.130339
			3	−0.023072		−0.123007
			4	−0.000515		−0.108712
			5	0.026930		−0.087681
			6	0.059918		−0.059256
			7	1.030620		0.637304
E(LU)				0.17967935		
E(CP)					0.05607919	
E(LB)						0.12200601
	12	8	1	−0.025785		−0.109045
			2	−0.015312		−0.112224
			3	−0.001353		−0.107627
			4	0.015634		−0.097276
			5	0.035853		−0.081361
			6	0.059835		−0.059315
			7	0.088444		−0.029900
			8	0.842684		0.596748

Appendix IX (*Continued*)

	n	r	i	a_i	c_i
E(LU)				0.14186580	
E(CP)				0.02930146	
E(LB)					0.10304331
	12	9	1	−0.006944	−0.093658
			2	0.002669	−0.097540
			3	0.014239	−0.094893
			4	0.027669	−0.087448
			5	0.043189	−0.075371
			6	0.061225	−0.058180
			7	0.082441	−0.034802
			8	0.107856	−0.003342
			9	0.667655	0.545234
E(LU)				0.11929957	
E(CP)				0.01087297	
E(LB)					0.08799386
	12	10	1	0.006411	−0.080881
			2	0.015598	−0.085171
			3	0.025675	−0.083952
			4	0.036799	−0.078714
			5	0.049211	−0.069610
			6	0.063256	−0.056237
			7	0.079438	−0.037675
			8	0.098522	−0.012272
			9	0.121752	0.022956
			10	0.503338	0.481555
E(LU)				0.10573191	
E(CP)				−0.00210755	
E(LB)					0.07557509
	12	11	1	0.015982	−0.069798
			2	0.024997	−0.074285
			3	0.034156	−0.074131
			4	0.043790	−0.070617
			5	0.054149	−0.063891
			6	0.065515	−0.053621
			7	0.078264	−0.039034
			8	0.092958	−0.018715
			9	0.110521	0.009948
			10	0.132666	0.052280
			11	0.347003	0.401864
E(LU)				0.09775217	
E(CP)				−0.01134890	
E(LB)					0.06487266
	12	12	1	0.022771	−0.059449
			2	0.031776	−0.063952
			3	0.040408	−0.064601
			4	0.049122	−0.062489
			5	0.058175	−0.057754
			6	0.067800	−0.050137
			7	0.078281	−0.039010
			8	0.090017	−0.023199
			9	0.103664	−0.000505
			10	0.120475	0.033696
			11	0.143566	0.091751
			12	0.193947	0.295648
E(LU)				0.09348388	
E(CP)				−0.01785537	
E(LB)					0.05495436

	13	2	1	−1.020378	−0.490105	
			2	2.020377	0.490105	
E(LU)				2.85169694		
E(CP)					1.03985071	
E(LB)						0.49001823
	13	3	1	−0.506031	−0.324239	
			2	−0.449735	−0.306454	
			3	1.955765	0.630694	
E(LU)				1.21480934		
E(CP)					0.51198847	
E(LB)						0.31979363
	13	4	1	−0.289420	−0.240219	
			2	−0.258687	−0.232349	
			3	−0.205024	−0.207450	
			4	1.753131	0.680018	
E(LU)				0.64778295		
E(CP)					0.29204583	
E(LB)						0.23448055
	13	5	1	−0.176109	−0.189177	
			2	−0.156637	−0.186381	
			3	−0.122893	−0.170454	
			4	−0.078337	−0.144971	
			5	1.533976	0.690983	
E(LU)				0.39459617		
E(CP)					0.17799724	
E(LB)						0.18310709
	13	6	1	−0.109140	−0.154711	
			2	−0.095246	−0.154785	
			3	−0.072165	−0.144347	
			4	−0.041997	−0.126268	
			5	−0.004940	−0.101028	
			6	1.323488	0.681140	
E(LU)				0.26460952		
E(CP)					0.11109896	
E(LB)						0.14867755
	13	7	1	−0.066358	−0.129743	
			2	−0.055414	−0.131538	
			3	−0.038503	−0.124701	
			4	−0.016879	−0.111609	
			5	0.009416	−0.092649	
			6	0.040810	−0.067475	
			7	1.126930	0.657714	
E(LU)				0.19187273		
E(CP)					0.06864731	
E(LB)						0.12390133
	13	8	1	−0.037540	−0.110704	
			2	−0.028206	−0.113563	
			3	−0.015049	−0.109206	
			4	0.001231	−0.099644	
			5	0.020686	−0.085204	
			6	0.043677	−0.065581	
			7	0.070830	−0.039995	
			8	0.944372	0.623896	

Appendix IX (*Continued*)

	n	r	i	a_i	c_i
E(LU)				0.14885020	
E(CP)				0.04022462	
E(LB)					0.10512398
	13	9	1	−0.017389	−0.095590
			2	−0.008934	−0.099109
			3	0.001863	−0.096521
			4	0.014684	−0.089554
			5	0.029637	−0.078490
			6	0.047027	−0.063068
			7	0.067346	−0.042607
			8	0.091328	−0.015928
			9	0.774437	0.580865
E(LU)				0.12250342	
E(CP)				0.02046326	
E(LB)					0.09030201
	13	10	1	−0.002927	−0.083170
			2	0.005067	−0.087085
			3	0.014356	−0.085792
			4	0.024891	−0.080789
			5	0.036816	−0.072325
			6	0.050389	−0.060181
			7	0.065995	−0.043768
			8	0.084201	−0.022048
			9	0.105863	0.006715
			10	0.615348	0.528441
E(LU)				0.10607774	
E(CP)				0.00635741	
E(LB)					0.07818835

	n	r	i	a_i	c_i
	13	11	1	0.007628	−0.072617
			2	0.015408	−0.076746
			3	0.023732	−0.076418
			4	0.032743	−0.072938
			5	0.042611	−0.066531
			6	0.053556	−0.057014
			7	0.065876	−0.043886
			8	0.080005	−0.026244
			9	0.096594	−0.002552
			10	0.116703	0.029910
			11	0.465143	0.465037
E(LU)				0.09583611	
E(CP)				−0.00388188	
E(LB)					0.06795140
	13	12	1	0.015382	−0.063288
			2	0.023100	−0.067492
			3	0.030818	−0.067892
			4	0.038824	−0.065622
			5	0.047302	−0.060887
			6	0.056444	−0.053540
			7	0.066482	−0.043158
			8	0.077739	−0.028970
			9	0.090699	−0.009644
			10	0.106166	0.017233
			11	0.125627	0.056547
			12	0.321416	0.386713
E(LU)				0.08961947	
E(CP)				−0.01136145	
E(LB)					0.05895232

	13	13	1	0.021005	−0.054436	
			2	0.028757	−0.058585	
			3	0.036127	−0.059535	
			4	0.043501	−0.058259	
			5	0.051078	−0.054942	
			6	0.059028	−0.049472	
			7	0.067533	−0.041504	
			8	0.076831	−0.030398	
			9	0.087274	−0.015037	
			10	0.099441	0.006644	
			11	0.114446	0.038943	
			12	0.135068	0.093324	
			13	0.179913	0.283257	
E(LU)				0.08619744		
E(CP)					−0.01674914	
E(LB)						0.05046988
	14	2	1	−1.060461	−0.490831	
			2	2.060461	0.490831	
E(LU)				3.01527998		
E(CP)					1.07852097	
E(LB)						0.49075663
	14	3	1	−0.533185	−0.324929	
			2	−0.479874	−0.308462	
			3	2.013059	0.633391	
E(LU)				1.29865775		
E(CP)					0.53840104	
E(LB)						0.32081269
	14	4	1	−0.310144	−0.240992	
			2	−0.281132	−0.233670	
			3	−0.229990	−0.210735	
			4	1.821266	0.685397	
E(LU)				0.69748231		
E(CP)					0.31216081	
E(LB)						0.23567174
	14	5	1	−0.192947	−0.190068	
			2	−0.174709	−0.187427	
			3	−0.142478	−0.172710	
			4	−0.099930	−0.149393	
			5	1.610065	0.699598	
E(LU)				0.42609561		
E(CP)					0.19423903	
E(LB)						0.18443288
	14	6	1	−0.123352	−0.155736	
			2	−0.110490	−0.155747	
			3	−0.088443	−0.146054	
			4	−0.059523	−0.129460	
			5	−0.024111	−0.106556	
			6	1.405919	0.693553	
E(LU)				0.28511973		
E(CP)					0.12469427	
E(LB)						0.15012578
	14	7	1	−0.078656	−0.130915	
			2	−0.068666	−0.132521	
			3	−0.052554	−0.126123	
			4	−0.031776	−0.114051	
			5	−0.006522	−0.096788	
			6	0.023467	−0.074184	
			7	1.214708	0.674581	
E(LU)				0.20518434		
E(CP)					0.08030259	
E(LB)						0.12547311

Appendix IX (*Continued*)

	n	r	i	a_i		c_i
	14	8	1	−0.048365		−0.112041
			2	−0.039964		−0.114637
			3	−0.027495		−0.110509
			4	−0.011849		−0.101635
			5	0.006905		−0.088422
			6	0.029002		−0.070735
			7	0.054897		−0.048074
			8	1.036868		0.646052
E(LU)				0.15716466		
E(CP)					0.05038249	
E(LB)						0.10683049
	14	9	1	−0.027030		−0.097117
			2	−0.019516		−0.100334
			3	−0.009363		−0.097827
			4	0.002928		−0.091298
			5	0.017368		−0.081103
			6	0.034165		−0.067124
			7	0.053685		−0.048921
			8	0.076476		−0.025720
			9	0.871287		0.609445
E(LU)				0.12719148		
E(CP)					0.02941694	
E(LB)						0.09216556
	14	10	1	−0.011580		−0.084931
			2	−0.004548		−0.088528
			3	0.004100		−0.087207
			4	0.014144		−0.082451
			5	0.025647		−0.074573
			6	0.038794		−0.063473
			7	0.053879		−0.048768
			8	0.071335		−0.029776
			9	0.091783		−0.005398
			10	0.716445		0.565105
E(LU)				0.10803536		
E(CP)					0.01430729	
E(LB)						0.08024763
	14	11	1	−0.000170		−0.074686
			2	0.006622		−0.078499
			3	0.014283		−0.078064
			4	0.022800		−0.074680
			5	0.032273		−0.068624
			6	0.042866		−0.059816
			7	0.054817		−0.047926
			8	0.068463		−0.032355
			9	0.084290		−0.012126
			10	0.103025		0.014349
			11	0.570731		0.512429

E(LU)				0.09566494	
E(CP)				0.00320055	
E(LB)					0.07027548
	14	12	1	0.008361	−0.065816
			2	0.015058	−0.069728
			3	0.022076	−0.069962
			4	0.029552	−0.067659
			5	0.037615	−0.063070
			6	0.046411	−0.056130
			7	0.056132	−0.046558
			8	0.067039	−0.033834
			9	0.079506	−0.017101
			10	0.094096	0.005064
			11	0.111723	0.035156
			12	0.432431	0.449638
E(LU)				0.08771669	
E(CP)				−0.00506397	
E(LB)					0.06168210
	14	13	1	0.014760	−0.057849
			2	0.021453	−0.061764
			3	0.028064	−0.062506
			4	0.034842	−0.061074
			5	0.041933	−0.057693
			6	0.049474	−0.052317
			7	0.057619	−0.044707
			8	0.066569	−0.034420
			9	0.076605	−0.020713
			10	0.088151	−0.002338
			11	0.101914	0.022943
			12	0.119200	0.059643
			13	0.299416	0.372795
E(LU)				0.08276211	
E(CP)				−0.01123278	
E(LB)					0.05400148
	14	14	1	0.019487	−0.050186
			2	0.026238	−0.054008
			3	0.032614	−0.055130
			4	0.038947	−0.054419
			5	0.045399	−0.052075
			6	0.052097	−0.048066
			7	0.059168	−0.042197
			8	0.066767	−0.034099
			9	0.075102	−0.023149
			10	0.084482	−0.008285
			11	0.095428	0.012430
			12	0.108942	0.043015
			13	0.127523	0.094166
			14	0.167807	0.272004
E(LU)				0.07996685	
E(CP)				−0.01576372	
E(LB)					0.04665712
	15	2	1	−1.097617	−0.491458
			2	2.097617	0.491458
E(LU)				3.17256460	
E(CP)				1.11445612	
E(LB)					0.49139327
	15	3	1	−0.558336	−0.325521
			2	−0.507671	−0.310191
			3	2.066007	0.635712
E(LU)				1.38015851	
E(CP)				0.56293169	
E(LB)					0.32168886

Appendix IX (*Continued*)

	n	r	i	a_i	c_i
	15	4	1	−0.329324	−0.241651
			2	−0.301829	−0.234806
			3	−0.252948	−0.213548
			4	1.884101	0.690005
E(LU)				0.74642859	
E(CP)				0.33084387	
E(LB)					0.23669248
	15	5	1	−0.208525	−0.190823
			2	−0.191357	−0.188323
			3	−0.160491	−0.174645
			4	−0.119748	−0.153153
			5	1.680121	0.706944
E(LU)				0.45764555	
E(CP)				0.20933279	
E(LB)					0.18556433
	15	6	1	−0.136498	−0.156597
			2	−0.124518	−0.156563
			3	−0.103401	−0.147517
			4	−0.075614	−0.132182
			5	−0.041680	−0.111215
			6	1.481712	0.704074
E(LU)				0.30614004	
E(CP)				0.13734100	
E(LB)					0.15135556
	15	7	1	−0.090036	−0.131891
			2	−0.080850	−0.133342
			3	−0.065446	−0.127335
			4	−0.045441	−0.116138
			5	−0.021137	−0.100291
			6	0.007597	−0.079774
			7	1.295312	0.688771
E(LU)				0.21929214	
E(CP)				0.09116039	
E(LB)					0.12679942
	15	8	1	−0.058390	−0.113143
			2	−0.050767	−0.115520
			3	−0.038897	−0.111607
			4	−0.023825	−0.103332
			5	−0.005717	−0.091156
			6	0.015565	−0.075053
			7	0.040351	−0.054703
			8	1.121680	0.664514
E(LU)				0.16646559	
E(CP)				0.05986446	
E(LB)					0.10825884

	15	9	1	−0.035972		−0.098361
			2	−0.029235		−0.101322
			3	−0.019633		−0.098904
			4	−0.007812		−0.092773
			5	0.006156		−0.083327
			6	0.022403		−0.070544
			7	0.041203		−0.054142
			8	0.062969		−0.033595
			9	0.959920		0.632967
E(LU)				0.13300106		
E(CP)					0.03779810	
E(LB)						0.09370837
	15	10	1	−0.019626		−0.086339
			2	−0.013383		−0.089664
			3	−0.005271		−0.088341
			4	0.004351		−0.083828
			5	0.015475		−0.076474
			6	0.028227		−0.066261
			7	0.042832		−0.052943
			8	0.059624		−0.036054
			9	0.079072		−0.014863
			10	0.808700		0.594768
E(LU)				0.11121862		
E(CP)					0.02177795	
E(LB)						0.08192616
	15	11	1	−0.007450		−0.076297
			2	−0.001467		−0.079835
			3	0.005652		−0.079332
			4	0.013759		−0.076068
			5	0.022893		−0.070355
			6	0.033174		−0.062181
			7	0.044787		−0.051331
			8	0.057997		−0.037396
			9	0.073180		−0.019723
			10	0.090865		0.002701
			11	0.666610		0.549817
E(LU)				0.09681113		
E(CP)					0.00989471	
E(LB)						0.07212492
	15	12	1	0.001756		−0.067695
			2	0.007624		−0.071342
			3	0.014079		−0.071459
			4	0.021133		−0.069178
			5	0.028861		−0.064779
			6	0.037374		−0.058256
			7	0.046827		−0.049425
			8	0.057431		−0.037926
			9	0.069479		−0.023180
			10	0.083393		−0.004280
			11	0.099799		0.020236
			12	0.532243		0.497284
E(LU)				0.08723346		
E(CP)					0.00094612	
E(LB)						0.06376409

Appendix IX (*Continued*)

	n	r	i	a_i	c_i
	15	13	1	0.008779	−0.060130
			2	0.014620	−0.063805
			3	0.020637	−0.064394
			4	0.026961	−0.062900
			5	0.033693	−0.059574
			6	0.040939	−0.054417
			7	0.048828	−0.047269
			8	0.057528	−0.037821
			9	0.067265	−0.025565
			10	0.078368	−0.009694
			11	0.091330	0.011113
			12	0.106947	0.039155
			13	0.404106	0.435302
E(LU)				0.08092217	
E(CP)				−0.00585240	
E(LB)					0.05644073
	15	14	1	0.014143	−0.053241
			2	0.020013	−0.056879
			3	0.025750	−0.057827
			4	0.031576	−0.056973
			5	0.037611	−0.054542
			6	0.043958	−0.050539
			7	0.050725	−0.044833
			8	0.058045	−0.037157
			9	0.066092	−0.027072
			10	0.075114	−0.013872
			11	0.085490	0.003612
			12	0.097844	0.027465
			13	0.113340	0.061879
			14	0.280298	0.359980
E(LU)				0.07689745	
E(CP)				−0.01102126	
E(LB)					0.04980248
	15	15	1	0.018170	−0.046538
			2	0.024108	−0.050064
			3	0.029685	−0.051279
			4	0.035191	−0.050957
			5	0.040762	−0.049298
			6	0.046496	−0.046315
			7	0.052488	−0.041899
			8	0.058844	−0.035827
			9	0.065696	−0.027731
			10	0.073230	−0.017008
			11	0.081725	−0.002653
			12	0.091651	0.017156
			13	0.103914	0.046191
			14	0.120784	0.094483
			15	0.157255	0.261738
E(LU)				0.07457775	
E(CP)				−0.01488220	
E(LB)					0.04337628
	16	2	1	−1.132243	−0.492005
			2	2.132243	0.492005
E(LU)				3.32404220	
E(CP)				1.14801534	
E(LB)					0.49194784

	16	3	1	−0.581757	−0.326035
			2	−0.533457	−0.311694
			3	2.115214	0.637730
E(LU)				1.45938438	
E(CP)				0.58582769	
E(LB)					0.32245028
	16	4	1	−0.347172	−0.242220
			2	−0.321026	−0.235794
			3	−0.274186	−0.215984
			4	1.942384	0.693998
E(LU)				0.79453329	
E(CP)				0.34828173	
E(LB)					0.23757701
	16	5	1	−0.223015	−0.191470
			2	−0.206788	−0.189099
			3	−0.177158	−0.176323
			4	−0.138048	−0.156390
			5	1.745009	0.713282
E(LU)				0.48908000	
E(CP)				0.22342597	
E(LB)					0.18654151
	16	6	1	−0.148725	−0.157331
			2	−0.137508	−0.157263
			3	−0.117232	−0.148785
			4	−0.090481	−0.134532
			5	−0.057883	−0.115196
			6	1.551828	0.713108
E(LU)				0.32746210	
E(CP)				0.14915808	
E(LB)					0.15241337
	16	7	1	−0.100621	−0.132718
			2	−0.092121	−0.134040
			3	−0.077354	−0.128381
			4	−0.058057	−0.117942
			5	−0.034624	−0.103296
			6	−0.007020	−0.084506
			7	1.369798	0.700883
E(LU)				0.23396225	
E(CP)				0.10131710	
E(LB)					0.12793461
	16	8	1	−0.067719	−0.114069
			2	−0.060754	−0.116260
			3	−0.049415	−0.112545
			4	−0.034868	−0.104798
			5	−0.017357	−0.093508
			6	0.003178	−0.078726
			7	0.026973	−0.060251
			8	1.199963	0.680158
E(LU)				0.17650200	
E(CP)				0.06874770	
E(LB)					0.10947376
	16	9	1	−0.044303	−0.099396
			2	−0.038218	−0.102138
			3	−0.029094	−0.099811
			4	−0.017697	−0.094037
			5	−0.004166	−0.085242
			6	0.011570	−0.073467
			7	0.029712	−0.058535
			8	0.050576	−0.040084
			9	1.041619	0.652711
E(LU)				0.13966768	
E(CP)				0.04566615	
E(LB)					0.09501012

Appendix IX (*Continued*)

	n	r	i	a_i	c_i
	16	10	1	−0.027135	−0.087496
			2	−0.021550	−0.090585
			3	−0.013895	−0.089277
			4	−0.004646	−0.084992
			5	0.006132	−0.078105
			6	0.018515	−0.068653
			7	0.032675	−0.056482
			8	0.048869	−0.041268
			9	0.067459	−0.022503
			10	0.893576	0.619360
E(LU)				0.11534960	
E(CP)				0.02881067	
E(LB)					0.08332716
	16	11	1	−0.014263	−0.077597
			2	−0.008950	−0.080895
			3	−0.002286	−0.080349
			4	0.005469	−0.077213
			5	0.014303	−0.071820
			6	0.024297	−0.064207
			7	0.035593	−0.054237
			8	0.048404	−0.041625
			9	0.063020	−0.025917
			10	0.079847	−0.006432
			11	0.754566	0.580293
E(LU)				0.09897866	
E(CP)				0.01622073	
E(LB)					0.07364497

	n	r	i	a_i	c_i
	16	12	1	−0.004450	−0.069172
			2	0.000732	−0.072584
			3	0.006721	−0.072615
			4	0.013424	−0.070383
			5	0.020868	−0.066184
			6	0.029134	−0.060054
			7	0.038344	−0.051876
			8	0.048668	−0.041398
			9	0.060342	−0.028216
			10	0.073692	−0.011716
			11	0.089173	0.009035
			12	0.623351	0.535164
E(LU)				0.08784015	
E(CP)				0.00665801	
E(LB)					0.06543511
	16	13	1	0.003118	−0.061843
			2	0.008256	−0.065297
			3	0.013789	−0.065770
			4	0.019747	−0.064259
			5	0.026189	−0.061031
			6	0.033196	−0.056120
			7	0.040872	−0.049427
			8	0.049357	−0.040731
			9	0.058836	−0.029675
			10	0.069568	−0.015710
			11	0.081920	0.002010
			12	0.096438	0.024833
			13	0.498713	0.483018

E(LU)				0.08025299	
E(CP)					−0.00069037
E(LB)					0.05831799
	16	14	1	0.008992	−0.055309
			2	0.014141	−0.058750
			3	0.019370	−0.059563
			4	0.024804	−0.058635
			5	0.030525	−0.056208
			6	0.036615	−0.052317
			7	0.043164	−0.046878
			8	0.050284	−0.039699
			9	0.058124	−0.030467
			10	0.066884	−0.018695
			11	0.076854	−0.003625
			12	0.088469	0.015969
			13	0.102433	0.042224
			14	0.379341	0.421953
E(LU)				0.07514429	
E(CP)					−0.00637294
E(LB)					0.05199709
	16	15	1	0.013547	−0.049291
			2	0.018743	−0.052670
			3	0.023778	−0.053739
			4	0.028849	−0.053290
			5	0.034060	−0.051538
			6	0.039489	−0.048520
			7	0.045218	−0.044164
			8	0.051338	−0.038307
			9	0.057965	−0.030678
			10	0.065253	−0.020850
			11	0.073425	−0.008156
			12	0.082818	0.008503
			13	0.093994	0.031075
			14	0.107995	0.063476
			15	0.263528	0.348149
E(LU)				0.07182155	
E(CP)					−0.01076262
E(LB)					0.04619787

Appendix IX (*Continued*)

	n	r	i	a_i	c_i
	16	16	1	0.017016	−0.043375
			2	0.022284	−0.046633
			3	0.027208	−0.047890
			4	0.032046	−0.047839
			5	0.036912	−0.046675
			6	0.041887	−0.044432
			7	0.047042	−0.041053
			8	0.052455	−0.036402
			9	0.058216	−0.030249
			10	0.064444	−0.022230
			11	0.071304	−0.011772
			12	0.079051	0.002079
			13	0.088111	0.021044
			14	0.099315	0.048675
			15	0.114733	0.094419
			16	0.147977	0.252333
E(LU)				0.06987019	
E(CP)				−0.01409012	
E(LB)					0.04052374
	17	2	1	−1.164659	−0.492486
			2	2.164659	0.492486
E(LU)				3.47015408	
E(CP)				1.17949167	
E(LB)					0.49243526
	17	3	1	−0.603668	−0.326486
			2	−0.557497	−0.313014
			3	2.161166	0.639500
E(LU)				1.53642388	
E(CP)				0.60729095	
E(LB)					0.32311812
	17	4	1	−0.363861	−0.242716
			2	−0.338922	−0.236662
			3	−0.293934	−0.218114
			4	1.996717	0.697492
E(LU)				0.84174810	
E(CP)				0.36462724	
E(LB)					0.23835098
	17	5	1	−0.236557	−0.192031
			2	−0.221164	−0.189778
			3	−0.192661	−0.177793
			4	−0.155037	−0.159206
			5	1.805419	0.718809
E(LU)				0.52028442	
E(CP)				0.23663986	
E(LB)					0.18739415
	17	6	1	−0.160149	−0.157965
			2	−0.149601	−0.157871
			3	−0.130090	−0.149896
			4	−0.104290	−0.136581
			5	−0.072907	−0.118639
			6	1.617037	0.720952
E(LU)				0.34893506	
E(CP)				0.16024410	
E(LB)					0.15333326

	17	7	1	−0.110512	−0.133428
			2	−0.102606	−0.134640
			3	−0.088415	−0.129294
			4	−0.069771	−0.119517
			5	−0.047139	−0.105901
			6	−0.020560	−0.088568
			7	1.439003	0.711349
E(LU)				0.24902198	
E(CP)				0.11085361	
E(LB)					0.12891783
	17	8	1	−0.076441	−0.114859
			2	−0.070039	−0.116891
			3	−0.059173	−0.113357
			4	−0.045110	−0.106076
			5	−0.028154	−0.095554
			6	−0.008307	−0.081890
			7	0.014595	−0.064968
			8	1.272628	0.693595
E(LU)				0.18708688	
E(CP)				0.07709833	
E(LB)					0.11052085
	17	9	1	−0.052096	−0.100271
			2	−0.046565	−0.102825
			3	−0.037862	−0.100587
			4	−0.026851	−0.095136
			5	−0.013728	−0.086910
			6	0.001531	−0.075995
			7	0.019069	−0.062288
			8	0.039129	−0.045535
			9	1.117373	0.669546
E(LU)				0.14699387	
E(CP)				0.05307401	
E(LB)					0.09612512
	17	10	1	−0.034167	−0.088465
			2	−0.029139	−0.091350
			3	−0.021881	−0.090064
			4	−0.012965	−0.085992
			5	−0.002507	−0.079521
			6	0.009531	−0.070728
			7	0.023273	−0.059520
			8	0.038922	−0.045671
			9	0.056761	−0.028822
			10	0.972172	0.640135
E(LU)				0.12022174	
E(CP)				0.03544569	
E(LB)					0.08451762
	17	11	1	−0.020654	−0.078673
			2	−0.015906	−0.081761
			3	−0.009632	−0.081188
			4	−0.002186	−0.078180
			5	0.006378	−0.073083
			6	0.016104	−0.065964
			7	0.027102	−0.056744
			8	0.039540	−0.045224
			9	0.053648	−0.031078
			10	0.069744	−0.013827
			11	0.835861	0.605723
E(LU)				0.10195092	
E(CP)				0.02220540	
E(LB)					0.07492279

Appendix IX (*Continued*)

	n	r	i	a_i	c_i
	17	12	1	−0.010288	−0.070375
			2	−0.005683	−0.073577
			3	−0.000086	−0.073546
			4	0.006316	−0.071375
			5	0.013511	−0.067372
			6	0.021553	−0.061602
			7	0.030535	−0.053996
			8	0.040597	−0.044377
			9	0.051928	−0.032455
			10	0.064785	−0.017797
			11	0.079517	0.000228
			12	0.707314	0.566244
E(LU)				0.08930564	
E(CP)				0.01208216	
E(LB)					0.06681858
	17	13	1	−0.002231	−0.063202
			2	0.002318	−0.066454
			3	0.007448	−0.066839
			4	0.013101	−0.065335
			5	0.019298	−0.062220
			6	0.026098	−0.057556
			7	0.033584	−0.051282
			8	0.041872	−0.043242
			9	0.051113	−0.033181
			10	0.061516	−0.020708
			11	0.073364	−0.005250
			12	0.087058	0.014056
			13	0.585461	0.521211
E(LU)				0.07436842	
E(CP)				−0.00189289	
E(LB)					0.05369960

	n	r	i	a_i	c_i
E(LU)				0.08049558	
E(CP)				0.00423893	
E(LB)					0.05983608
	17	14	1	0.004088	−0.056878
			2	0.008636	−0.060131
			3	0.013446	−0.060836
			4	0.018560	−0.059871
			5	0.024028	−0.057487
			6	0.029909	−0.053742
			7	0.036278	−0.048586
			8	0.043231	−0.041881
			9	0.050892	−0.033402
			10	0.059426	−0.022799
			11	0.069060	−0.009558
			12	0.080119	0.007112
			13	0.093083	0.028459
			14	0.469244	0.469601

	17	15	1	0.009066	−0.051176			
			2	0.013648	−0.054390			
			3	0.018244	−0.055341			
			4	0.022974	−0.054815			
			5	0.027908	−0.053042			
			6	0.033111	−0.050075			
			7	0.038648	−0.045871			
			8	0.044600	−0.040314			
			9	0.051065	−0.033203			
			10	0.058176	−0.024231			
			11	0.066111	−0.012936			
			12	0.075128	0.001394			
			13	0.085616	0.019905			
			14	0.098200	0.044587			
			15	0.357506	0.409507			
E(LU)						0.07016498		
E(CP)							−0.00670775	
E(LB)								0.04818440
	17	16	1	0.012979	−0.045870			
			2	0.017617	−0.049009			
			3	0.022076	−0.050145			
			4	0.026538	−0.049982			
			5	0.031091	−0.048727			
			6	0.035799	−0.046430			
			7	0.040724	−0.043057			
			8	0.045932	−0.038508			
			9	0.051504	−0.032609			
			10	0.057542	−0.025090			
			11	0.064186	−0.015545			
			12	0.071635	−0.003341			
			13	0.080195	0.012556			
			14	0.090373	0.033974			
			15	0.103110	0.064588			
			16	0.248699	0.337194			
E(LU)						0.06738336		
E(CP)							−0.01047916	
E(LB)								0.04307100
	17	17	1	0.015998	−0.040607			
			2	0.020706	−0.043624			
			3	0.025089	−0.044891			
			4	0.029378	−0.045031			
			5	0.033671	−0.044229			
			6	0.038035	−0.042531			
			7	0.042527	−0.039913			
			8	0.047204	−0.036289			
			9	0.052133	−0.031512			
			10	0.057392	−0.025352			
			11	0.063089	−0.017458			
			12	0.069375	−0.007282			
			13	0.076482	0.006082			
			14	0.084803	0.024262			
			15	0.095098	0.050618			
			16	0.109270	0.094076			
			17	0.139752	0.243681			
E(LU)						0.06572241		
E(CP)							−0.01337530	
E(LB)								0.03802109
	18	2	1	−1.195128	−0.492912			
			2	2.195128	0.492912			
E(LU)						3.61129585		
E(CP)							1.20912723	
E(LB)								0.49286703
	18	3	1	−0.624252	−0.326884			
			2	−0.580008	−0.314183			
			3	2.204260	0.641066			

Appendix IX (*Continued*)

	n	r	i	a_i	c_i
E(LU)				1.61137253	
E(CP)				0.62748837	
E(LB)					0.32370865
	18	4	1	−0.379529	−0.243153
			2	−0.355679	−0.237429
			3	−0.312382	−0.219992
			4	2.047590	0.700574
E(LU)				0.88805128	
E(CP)				0.38000703	
E(LB)					0.23903395
	18	5	1	−0.249266	−0.192523
			2	−0.234618	−0.190376
			3	−0.207148	−0.179091
			4	−0.170883	−0.161679
			5	1.861914	0.723670
E(LU)				0.55118001	
E(CP)				0.24907530	
E(LB)					0.18814472
	18	6	1	−0.170868	−0.158518
			2	−0.160910	−0.158405
			3	−0.142100	−0.150876
			4	−0.117175	−0.138383
			5	−0.086906	−0.121647
			6	1.677960	0.727829

	n	r	i	a_i	c_i
E(LU)				0.37044855	
E(CP)				0.17068152	
E(LB)					0.15414076
	18	7	1	−0.119793	−0.134044
			2	−0.112406	−0.135163
			3	−0.098738	−0.130098
			4	−0.080698	−0.120904
			5	−0.058807	−0.108183
			6	−0.033165	−0.092095
			7	1.503605	0.720486
E(LU)				0.26434202	
E(CP)				0.11983820	
E(LB)					0.12977806
	18	8	1	−0.084626	−0.115541
			2	−0.078711	−0.117434
			3	−0.068272	−0.114068
			4	−0.054656	−0.107202
			5	−0.038217	−0.097349
			6	−0.019006	−0.084645
			7	0.003084	−0.069031
			8	1.340405	0.705270
E(LU)				0.19807869	
E(CP)				0.08497296	
E(LB)					0.11143330

	18	9	1	−0.059414		−0.101022
			2	−0.054359		−0.103411
			3	−0.046030		−0.101260
			4	−0.035375		−0.096099
			5	−0.022631		−0.088374
			6	−0.007819		−0.078203
			7	0.009161		−0.065532
			8	0.028495		−0.050186
			9	1.187973		0.684087
E(LU)				0.15482946		
E(CP)					0.06006815	
E(LB)						0.09709201
	18	10	1	−0.040776		−0.089291
			2	−0.036223		−0.091997
			3	−0.029314		−0.090739
			4	−0.020701		−0.086863
			5	−0.010540		−0.080764
			6	0.001172		−0.072544
			7	0.014523		−0.062157
			8	0.029671		−0.049445
			9	0.046841		−0.034147
			10	1.045347		0.657947
E(LU)				0.12567798		
E(CP)					0.04172006	
E(LB)						0.08554362
	18	11	1	−0.026669		−0.079582
			2	−0.022402		−0.082484
			3	−0.016466		−0.081896
			4	−0.009294		−0.079012
			5	−0.000979		−0.074183
			6	0.008496		−0.067503
			7	0.019212		−0.058930
			8	0.031300		−0.048324
			9	0.044947		−0.035451
			10	0.060404		−0.019962
			11	0.911449		0.627325
E(LU)				0.10556433		
E(CP)					0.02787638	
E(LB)						0.07601539
	18	12	1	−0.015793		−0.071378
			2	−0.011677		−0.074393
			3	−0.006416		−0.074315
			4	−0.000278		−0.072211
			5	0.006695		−0.068395
			6	0.014529		−0.062952
			7	0.023297		−0.055848
			8	0.033110		−0.046959
			9	0.044122		−0.036073
			10	0.056540		−0.022877
			11	0.070637		−0.006924
			12	0.785235		0.592326
E(LU)				0.09145851		
E(CP)					0.01723593	
E(LB)						0.06798899

Appendix IX (*Continued*)

	n	r	i	a_i	c_i
	18	13	1	−0.007289	−0.064317
			2	−0.003238	−0.067387
			3	0.001550	−0.067701
			4	0.006940	−0.066218
			5	0.012925	−0.063222
			6	0.019540	−0.058792
			7	0.026851	−0.052898
			8	0.034951	−0.045430
			9	0.043969	−0.036200
			10	0.054072	−0.024926
			11	0.065486	−0.011201
			12	0.078516	0.005561
			13	0.665728	0.552731
E(LU)				0.08146655	
E(CP)					0.00893995
E(LB)					0.06110111
	18	14	1	−0.000568	−0.058133
			2	0.003471	−0.061213
			3	0.007930	−0.061830
			4	0.012775	−0.060849
			5	0.018027	−0.058527
			6	0.023730	−0.054936
			7	0.029942	−0.050053
			8	0.036744	−0.043781
			9	0.044239	−0.035952
			10	0.052564	−0.026314
			11	0.061904	−0.014497
			12	0.072509	0.000034
			13	0.084730	0.018080
			14	0.552004	0.507970
E(LU)				0.07436294	
E(CP)					0.00240300
E(LB)					0.05508562
	18	15	1	0.004780	−0.052617
			2	0.008843	−0.055674
			3	0.013074	−0.056526
			4	0.017522	−0.055953
			5	0.022232	−0.054191
			6	0.027249	−0.051307
			7	0.032630	−0.047281
			8	0.038443	−0.042029
			9	0.044772	−0.035403
			10	0.051728	−0.027176
			11	0.059460	−0.017018
			12	0.068169	−0.004442
			13	0.078145	0.011289
			14	0.089813	0.031340
			15	0.443142	0.456986
E(LU)				0.06933298	
E(CP)					−0.00278409
E(LB)					0.04973648

18	16	1	0.009048	−0.047594
		2	0.013157	−0.050597
		3	0.017235	−0.051629
		4	0.021397	−0.051393
		5	0.025706	−0.050102
		6	0.030212	−0.047820
		7	0.034966	−0.044532
		8	0.040027	−0.040165
		9	0.045465	−0.034587
		10	0.051368	−0.027599
		11	0.057855	−0.018906
		12	0.065087	−0.008069
		13	0.073294	0.005581
		14	0.082827	0.023119
		15	0.094248	0.046408
		16	0.338109	0.397887
E(LU)			0.06582537	
E(CP)				−0.00691185
E(LB)				0.04487895
18	17	1	0.012444	−0.042879
		2	0.016611	−0.045800
		3	0.020593	−0.046965
		4	0.024555	−0.047008
		5	0.028573	−0.046121
		6	0.032702	−0.044362
		7	0.036990	−0.041722
		8	0.041487	−0.038137
		9	0.046252	−0.033494
		10	0.051355	−0.027618
		11	0.056889	−0.020248
		12	0.062980	−0.010994
		13	0.069810	0.000742
		14	0.077655	0.015937
		15	0.086979	0.036313
		16	0.098636	0.065331
		17	0.235490	0.327023
E(LU)			0.06346845	
E(CP)				−0.01018489
E(LB)				0.04033369

Appendix IX (*Continued*)

	n	r	i	a_i	c_i
	18	18	1	0.015092	−0.038165
			2	0.019328	−0.040965
			3	0.023258	−0.042221
			4	0.027089	−0.042497
			5	0.030909	−0.041963
			6	0.034773	−0.040676
			7	0.038728	−0.038627
			8	0.042820	−0.035765
			9	0.047095	−0.031992
			10	0.051612	−0.027160
			11	0.056443	−0.021041
			12	0.061685	−0.013300
			13	0.067477	−0.003410
			14	0.074032	0.009488
			15	0.081713	0.026940
			16	0.091221	0.052132
			17	0.104314	0.093529
			18	0.132411	0.235693
E(LU)				0.06204005	
E(CP)					−0.01272745
E(LB)					0.03580789
	19	2	1	−1.223869	−0.493292
			2	2.223869	0.493292
E(LU)				3.74782267	
E(CP)					1.23712437
E(LB)					0.49325215
	19	3	1	−0.643659	−0.327238
			2	−0.601169	−0.315224
			3	2.244827	0.642462

	n	r	i	a_i	c_i
E(LU)				1.68432765	
E(CP)					0.64655945
E(LB)					0.32423458
	19	4	1	−0.394294	−0.243540
			2	−0.371431	−0.238113
			3	−0.329685	−0.221662
			4	2.095409	0.703314
E(LU)				0.93343886	
E(CP)					0.39452713
E(LB)					0.23964110
	19	5	1	−0.261237	−0.192958
			2	−0.247259	−0.190909
			3	−0.220739	−0.180245
			4	−0.185724	−0.163869
			5	1.914959	0.727980
E(LU)				0.58171310	
E(CP)					0.26081700
E(LB)					0.18881059
	19	6	1	−0.180964	−0.159004
			2	−0.171530	−0.158877
			3	−0.153365	−0.151748
			4	−0.129250	−0.139981
			5	−0.100003	−0.124297
			6	1.735111	0.733908
E(LU)				0.39192137	
E(CP)					0.18053993
E(LB)					0.15485543

	19	7	1	−0.128533		−0.134583
			2	−0.121602		−0.135622
			3	−0.108414		−0.130811
			4	−0.090935		−0.122136
			5	−0.069730		−0.110197
			6	−0.044949		−0.095185
			7	1.564162		0.728535
E(LU)				0.27982455		
E(CP)					0.12832890	
E(LB)						0.13053726
	19	8	1	−0.092336		−0.116135
			2	−0.086846		−0.117908
			3	−0.076795		−0.114696
			4	−0.063593		−0.108201
			5	−0.047637		−0.098937
			6	−0.029016		−0.087065
			7	−0.007670		−0.072570
			8	1.403893		0.715513
E(LU)				0.20936888		
E(CP)					0.09242024	
E(LB)						0.11223593
	19	9	1	−0.066309		−0.101674
			2	−0.061667		−0.103918
			3	−0.053675		−0.101850
			4	−0.043347		−0.096952
			5	−0.030960		−0.089671
			6	−0.016565		−0.080147
			7	−0.000103		−0.068365
			8	0.018570		−0.054204
			9	1.254056		0.696782
E(LU)				0.16305851		
E(CP)					0.06668914	
E(LB)						0.09793914
	19	10	1	−0.047007		−0.090004
			2	−0.042865		−0.092551
			3	−0.036265		−0.091324
			4	−0.027929		−0.087629
			5	−0.018045		−0.081863
			6	−0.006641		−0.074147
			7	0.006343		−0.064468
			8	0.021029		−0.052718
			9	0.037595		−0.038703
			10	1.113786		0.673407
E(LU)				0.13159684		
E(CP)					0.04766707	
E(LB)						0.08643819
	19	11	1	−0.032345		−0.080360
			2	−0.028492		−0.083097
			3	−0.022852		−0.082502
			4	−0.015928		−0.079736
			5	−0.007843		−0.075152
			6	0.001395		−0.068861
			7	0.011842		−0.060851
			8	0.023603		−0.051025
			9	0.036827		−0.039209
			10	0.051716		−0.025147
			11	0.982076		0.645940
E(LU)				0.10969257		
E(CP)					0.03326000	
E(LB)						0.07696225

Appendix IX (*Continued*)

	n	r	i	a_i	c_i
	19	12	1	−0.020995	−0.072230
			2	−0.017298	−0.075078
			3	−0.012331	−0.074965
			4	−0.006425	−0.072929
			5	0.000344	−0.069288
			6	0.007984	−0.064141
			7	0.016548	−0.057480
			8	0.026124	−0.049218
			9	0.036839	−0.039200
			10	0.048859	−0.027194
			11	0.062404	−0.012872
			12	0.857947	0.614595
E(LU)				0.09416748	
E(CP)				0.02213853	
E(LB)					0.06899532
	19	13	1	−0.012077	−0.065253
			2	−0.008453	−0.068158
			3	−0.003960	−0.068416
			4	0.001201	−0.066962
			5	0.006996	−0.064084
			6	0.013442	−0.059872
			7	0.020588	−0.054320
			8	0.028511	−0.047351
			9	0.037316	−0.038827
			10	0.047141	−0.028538
			11	0.058169	−0.016185
			12	0.070640	−0.001353
			13	0.740487	0.579318
E(LU)				0.08302831	
E(CP)				0.01342381	
E(LB)					0.06217738
	19	14	1	−0.004989	−0.059169
			2	−0.001384	−0.062091
			3	0.002773	−0.062636
			4	0.007387	−0.061652
			5	0.012453	−0.059399
			6	0.017997	−0.055961
			7	0.024066	−0.051334
			8	0.030726	−0.045450
			9	0.038064	−0.038185
			10	0.046194	−0.029351
			11	0.055267	−0.018676
			12	0.065482	−0.005781
			13	0.077109	0.009882
			14	0.628854	0.539802
E(LU)				0.07498023	
E(CP)				0.00651543	
E(LB)					0.05624730

	19	15	1	0.000692	−0.053779
			2	0.004313	−0.056686
			3	0.008234	−0.057456
			4	0.012444	−0.056855
			5	0.016963	−0.055121
			6	0.021823	−0.052332
			7	0.027068	−0.048486
			8	0.032757	−0.043523
			9	0.038961	−0.037334
			10	0.045774	−0.029749
			11	0.053320	−0.020523
			12	0.061762	−0.009310
			13	0.071323	0.004394
			14	0.082316	0.021334
			15	0.522250	0.495426
E(LU)				0.06915784	
E(CP)				0.00099208	
E(LB)					0.05100764

	19	16	1	0.005271	−0.048924
			2	0.008929	−0.051792
			3	0.012687	−0.052735
			4	0.016601	−0.052449
			5	0.020708	−0.051151
			6	0.025048	−0.048913
			7	0.029663	−0.045735
			8	0.034603	−0.041566
			9	0.039929	−0.036308
			10	0.045717	−0.029809
			11	0.052068	−0.021850
			12	0.059114	−0.012118
			13	0.067036	−0.000151
			14	0.076094	0.014738
			15	0.086669	0.033642
			16	0.419861	0.445119
E(LU)				0.06496979	
E(CP)				−0.00344791	
E(LB)					0.04630055

Appendix IX (*Continued*)

	n	r	i	a_i	c_i
	19	17	1	0.008968	−0.044464
			2	0.012677	−0.047270
			3	0.016326	−0.048344
			4	0.020023	−0.048319
			5	0.023825	−0.047390
			6	0.027772	−0.045626
			7	0.031906	−0.043028
			8	0.036272	−0.039552
			9	0.040920	−0.035112
			10	0.045913	−0.029573
			11	0.051331	−0.022740
			12	0.057280	−0.014330
			13	0.063906	−0.003928
			14	0.071419	0.009098
			15	0.080136	0.025759
			16	0.090563	0.047806
			17	0.320763	0.387016
E(LU)				0.06200679	
E(CP)				−0.00702291	
E(LB)					0.04198714

	n	r	i	a_i	c_i
	19	18	1	0.011941	−0.040244
			2	0.015709	−0.042966
			3	0.019289	−0.044136
			4	0.022835	−0.044327
			5	0.026412	−0.043716
			6	0.030068	−0.042366
			7	0.033841	−0.040281
			8	0.037772	−0.037423
			9	0.041903	−0.033716
			10	0.046286	−0.029044
			11	0.050984	−0.023232
			12	0.056083	−0.016030
			13	0.061695	−0.007067
			14	0.067989	0.004227
			15	0.075216	0.018774
			16	0.083802	0.038205
			17	0.094528	0.065788
			18	0.223648	0.317554
E(LU)				0.05998848	
E(CP)				−0.00988868	
E(LB)					0.03791809

	19	19	1	0.014282		−0.035995
			2	0.018115		−0.038600
			3	0.021661		−0.039833
			4	0.025107		−0.040204
			5	0.028531		−0.039873
			6	0.031980		−0.038897
			7	0.035494		−0.037282
			8	0.039109		−0.034997
			9	0.042861		−0.031977
			10	0.046794		−0.028121
			11	0.050958		−0.023280
			12	0.055419		−0.017234
			13	0.060267		−0.009660
			14	0.065629		−0.000055
			15	0.071704		0.012400
			16	0.078826		0.029177
			17	0.087648		0.053305
			18	0.099799		0.092832
			19	0.125817		0.228292
E(LU)				0.05874886		
E(CP)					−0.01213794	
E(LB)						0.03383684
	20	2	1	−1.251068		−0.493634
			2	2.251068		0.493634
E(LU)				3.88005370		
E(CP)					1.26365389	
E(LB)						0.49359782
	20	3	1	−0.662014		−0.327555
			2	−0.621129		−0.316157
			3	2.283144		0.643713
E(LU)				1.75538518		
E(CP)					0.66462201	
E(LB)						0.32470597
	20	4	1	−0.408252		−0.243885
			2	−0.386289		−0.238726
			3	−0.345972		−0.223154
			4	2.140513		0.705766
E(LU)				0.97791855		
E(CP)					0.40827717	
E(LB)						0.24018443
	20	5	1	−0.272551		−0.193344
			2	−0.259179		−0.191385
			3	−0.233536		−0.181278
			4	−0.199675		−0.165821
			5	1.964941		0.731828
E(LU)				0.61184794		
E(CP)					0.27193675	
E(LB)						0.18940540
	20	6	1	−0.190502		−0.159435
			2	−0.181539		−0.159298
			3	−0.163969		−0.152528
			4	−0.140605		−0.141408
			5	−0.112303		−0.126651
			6	1.788917		0.739321
E(LU)				0.41329354		
E(CP)					0.18987852	
E(LB)						0.15549251

Appendix IX (*Continued*)

	n	r	i	a_i	c_i
	20	7	1	−0.136790	−0.135060
			2	−0.130264	−0.136029
			3	−0.117518	−0.131448
			4	−0.100561	−0.123236
			5	−0.079995	−0.111990
			6	−0.056007	−0.097918
			7	1.621135	0.735681
E(LU)				0.29539488	
E(CP)				0.13637533	
E(LB)					0.13121241
	20	8	1	−0.099621	−0.116659
			2	−0.094504	−0.118326
			3	−0.084808	−0.115255
			4	−0.071993	−0.109093
			5	−0.056488	−0.100352
			6	−0.038416	−0.089210
			7	−0.017755	−0.075681
			8	1.463585	0.724575
E(LU)				0.22087332	
E(CP)				0.09948206	
E(LB)					0.11294771
	20	9	1	−0.072826	−0.102246
			2	−0.068544	−0.104362
			3	−0.060858	−0.102371
			4	−0.050834	−0.097711
			5	−0.038781	−0.090828
			6	−0.024779	−0.081874
			7	−0.008798	−0.070863
			8	0.009270	−0.057714
			9	1.316151	0.707969
E(LU)				0.17159045	
E(CP)				0.07297238	
E(LB)					0.09868793
	20	10	1	−0.052900	−0.090626
			2	−0.049115	−0.093031
			3	−0.042792	−0.091837
			4	−0.034710	−0.088309
			5	−0.025087	−0.082842
			6	−0.013973	−0.075573
			7	−0.001335	−0.066511
			8	0.012921	−0.055584
			9	0.028939	−0.042651
			10	1.178052	0.686964
E(LU)				0.13788300	
E(CP)				0.05331635	
E(LB)					0.08722579
	20	11	1	−0.037716	−0.081036
			2	−0.034222	−0.083625
			3	−0.028845	−0.083028
			4	−0.022146	−0.080373
			5	−0.014276	−0.076014
			6	−0.005262	−0.070070
			7	0.004930	−0.062554
			8	0.016382	−0.053398
			9	0.029216	−0.042476
			10	0.043593	−0.029594
			11	1.048347	0.662168
E(LU)				0.11423656	
E(CP)				0.03838053	
E(LB)					0.07779186

	n	r	i		
	20	12	1	−0.025922	−0.072964
			2	−0.022589	−0.075662
			3	−0.017879	−0.075522
			4	−0.012183	−0.073554
			5	−0.005600	−0.070076
			6	0.001858	−0.065197
			7	0.010227	−0.058928
			8	0.019578	−0.051211
			9	0.030012	−0.041931
			10	0.041668	−0.030912
			11	0.054724	−0.017911
			12	0.926107	0.633868
E(LU)				0.09732994	
E(CP)				0.02680890	
E(LB)					0.06987174
	20	13	1	−0.016619	−0.066052
			2	−0.013364	−0.068809
			3	−0.009129	−0.069021
			4	−0.004170	−0.067601
			5	0.001453	−0.064835
			6	0.007742	−0.060825
			7	0.014732	−0.055581
			8	0.022485	−0.049051
			9	0.031087	−0.041132
			10	0.040654	−0.031666
			11	0.051333	−0.020429
			12	0.063321	−0.007116
			13	0.810474	0.602120
E(LU)				0.08507430	
E(CP)				0.01770390	
E(LB)					0.06310742
	20	14	1	−0.009191	−0.060043
			2	−0.005961	−0.062821
			3	−0.002065	−0.063307
			4	0.002348	−0.062329
			5	0.007248	−0.060148
			6	0.012649	−0.056856
			7	0.018585	−0.052465
			8	0.025109	−0.046929
			9	0.032297	−0.040154
			10	0.040241	−0.031999
			11	0.049069	−0.022261
			12	0.058941	−0.010659
			13	0.070069	0.003196
			14	0.700660	0.566775
E(LU)				0.07610847	
E(CP)				0.01045130	
E(LB)					0.05724068
	20	15	1	−0.003203	−0.054744
			2	0.000035	−0.057513
			3	0.003690	−0.058213
			4	0.007695	−0.057596
			5	0.012048	−0.055899
			6	0.016769	−0.053209
			7	0.021892	−0.049537
			8	0.027467	−0.044842
			9	0.033552	−0.039043
			10	0.040230	−0.032010
			11	0.047602	−0.023560
			12	0.055804	−0.013436
			13	0.065012	−0.001280
			14	0.075467	0.013416
			15	0.595940	0.527466

Appendix IX (*Continued*)

	n	r	i	a_i	c_i		n	r	i	a_i	c_i
E(LU)				0.0695 1916			20	17	1	0.005617	−0.045695
E(CP)				0.00461910					2	0.008931	−0.048385
E(LB)					0.05207861				3	0.012297	−0.049380
									4	0.015773	−0.049304
	20	16	1	0.001656	−0.050002				5	0.019394	−0.048357
			2	0.004926	−0.052742				6	0.023192	−0.046613
			3	0.008410	−0.053608				7	0.027201	−0.044083
			4	0.012111	−0.053287				8	0.031459	−0.040736
			5	0.016048	−0.051997				9	0.036010	−0.036510
			6	0.020247	−0.049816				10	0.040910	−0.031298
			7	0.024742	−0.046757				11	0.046228	−0.024954
			8	0.029575	−0.042785				12	0.052055	−0.017265
			9	0.034801	−0.037825				13	0.058509	−0.007934
			10	0.040482	−0.031763				14	0.065756	0.003474
			11	0.046707	−0.024432				15	0.074029	0.017606
			12	0.053585	−0.015601				16	0.083671	0.035486
			13	0.061262	−0.004939				17	0.398968	0.433947
			14	0.069938	0.008023	E(LU)				0.06114864	
			15	0.079893	0.023984	E(CP)				−0.00394307	
			16	0.495616	0.483548	E(LB)					0.04329477
E(LU)				0.06467887							
E(CP)				−0.00010333							
E(LB)					0.04747116						

20	18	1	0.008847	−0.041706
		2	0.012215	−0.044331
		3	0.015502	−0.045422
		4	0.018813	−0.045550
		5	0.022197	−0.044896
		6	0.025690	−0.043529
		7	0.029324	−0.041460
		8	0.033136	−0.038666
		9	0.037162	−0.035086
		10	0.041450	−0.030632
		11	0.046055	−0.025168
		12	0.051050	−0.018506
		13	0.056532	−0.010374
		14	0.062634	−0.000381
		15	0.069547	0.012071
		16	0.077558	0.027938
		17	0.087131	0.048871
		18	0.305157	0.376826
E(LU)			0.05861867	
E(CP)			−0.00706723	
E(LB)				0.03943688

20	19	1	0.011469	−0.037905
		2	0.014895	−0.040446
		3	0.018135	−0.041607
		4	0.021329	−0.041903
		5	0.024538	−0.041503
		6	0.027802	−0.040467
		7	0.031153	−0.038810
		8	0.034624	−0.036509
		9	0.038247	−0.033514
		10	0.042061	−0.029746
		11	0.046112	−0.025085
		12	0.050458	−0.019364
		13	0.055176	−0.012340
		14	0.060372	−0.003660
		15	0.066198	0.007217
		16	0.072887	0.021167
		17	0.080830	0.039737
		18	0.090746	0.066024
		19	0.212971	0.308714
E(LU)			0.05687410	
E(CP)			−0.00959610	
E(LB)				0.03577112

Appendix IX (*Continued*)

n	r	i	a_i	c_i	n	r	i	a_i	c_i
20	20	1	0.013553	−0.034055	E(LU)			0.05578958	
		2	0.017039	−0.036484	E(CP)				−0.01159947
		3	0.020257	−0.037686	E(LB)				0.03207039
		4	0.023376	−0.038123					
		5	0.026464	−0.037945					
		6	0.029565	−0.037211					
		7	0.032711	−0.035932					
		8	0.035932	−0.034091					
		9	0.039258	−0.031646					
		10	0.042720	−0.028527					
		11	0.046357	−0.024632					
		12	0.050215	−0.019814					
		13	0.054354	−0.013860					
		14	0.058856	−0.006460					
		15	0.063842	0.002866					
		16	0.069496	0.014902					
		17	0.076128	0.031052					
		18	0.084346	0.054203					
		19	0.095669	0.092028					
		20	0.119862	0.221415					

appendix X percentiles of the distribution of the statistic[a] $W=\tilde{b}/b$

Percentiles of the distribution of W

		$1-\alpha$										
n	r	0.02	0.05	0.10	0.25	0.40	0.50	0.60	0.75	0.90	0.95	0.98
3	3	0.11	0.17	0.25	0.42	0.57	0.67	0.78	0.99	1.33	1.56	1.86
4	3	0.10	0.15	0.22	0.39	0.53	0.64	0.75	0.96	1.32	1.56	1.90
	4	0.20	0.28	0.37	0.54	0.68	0.77	0.86	1.05	1.33	1.53	1.77
5	3	0.09	0.14	0.21	0.37	0.51	0.61	0.73	0.94	1.32	1.59	1.93
	4	0.18	0.26	0.34	0.50	0.64	0.74	0.84	1.03	1.35	1.55	1.82
	5	0.28	0.36	0.44	0.60	0.73	0.82	0.91	1.07	1.33	1.50	1.70
6	3	0.09	0.14	0.21	0.36	0.50	0.61	0.72	0.93	1.32	1.59	1.92
	4	0.18	0.25	0.32	0.49	0.62	0.72	0.82	1.01	1.33	1.55	1.84
	5	0.25	0.33	0.42	0.58	0.71	0.79	0.89	1.05	1.33	1.51	1.73
	6	0.33	0.41	0.50	0.65	0.77	0.85	0.93	1.07	1.31	1.46	1.64
7	3	0.08	0.14	0.20	0.35	0.49	0.59	0.71	0.92	1.30	1.56	1.92
	4	0.17	0.24	0.31	0.48	0.62	0.71	0.81	1.01	1.32	1.54	1.82
	5	0.25	0.32	0.40	0.56	0.70	0.78	0.88	1.05	1.33	1.52	1.75
	6	0.32	0.39	0.47	0.63	0.75	0.84	0.92	1.07	1.32	1.48	1.67
	7	0.38	0.46	0.54	0.69	0.80	0.87	0.95	1.08	1.30	1.43	1.60

Example: For $n=5$, $r=4$, we have $W_{0.05}=1.55$.

[a]Taken from: Mann, N. R.; K. W. Fertig and E. M. Scheuer; *Confidence and Tolerance Bounds and a New Goodness-of-Fit Test for Two-Parameter Weibull or Extreme-Value Distribution (With Tables for Censored Samples of Size* 3(1)25); Aerospace Research Laboratories, Wright-Patterson Air Force Base, Ohio, ARL 71-0077, Contract No. F33(615)-70-C-1216, May 1971.

Percentiles of the distribution of W

		$1-\alpha$										
n	r	0.02	0.05	0.10	0.25	0.40	0.50	0.60	0.75	0.90	0.95	0.98
8	3	0.08	0.13	0.19	0.35	0.49	0.59	0.70	0.92	1.31	1.58	1.95
	4	0.16	0.23	0.31	0.47	0.61	0.70	0.81	1.00	1.33	1.55	1.83
	5	0.23	0.31	0.39	0.55	0.68	0.77	0.87	1.05	1.33	1.52	1.76
	6	0.30	0.38	0.46	0.62	0.74	0.82	0.91	1.06	1.32	1.49	1.69
	7	0.36	0.44	0.52	0.67	0.78	0.86	0.94	1.08	1.30	1.45	1.62
	8	0.42	0.50	0.58	0.71	0.82	0.89	0.96	1.09	1.28	1.41	1.56
9	3	0.08	0.13	0.19	0.34	0.49	0.59	0.70	0.92	1.31	1.58	1.92
	4	0.16	0.23	0.31	0.47	0.60	0.70	0.80	1.00	1.33	1.55	1.84
	5	0.23	0.31	0.39	0.54	0.68	0.77	0.86	1.04	1.33	1.52	1.76
	6	0.30	0.38	0.45	0.60	0.73	0.81	0.90	1.06	1.31	1.48	1.70
	7	0.35	0.43	0.50	0.66	0.77	0.85	0.93	1.07	1.30	1.46	1.65
	8	0.40	0.48	0.55	0.70	0.81	0.88	0.95	1.08	1.28	1.42	1.59
	9	0.45	0.53	0.60	0.74	0.84	0.90	0.97	1.08	1.27	1.39	1.53
10	3	0.08	0.13	0.19	0.34	0.48	0.59	0.71	0.93	1.31	1.59	1.92
	4	0.16	0.23	0.30	0.46	0.60	0.70	0.80	1.00	1.33	1.57	1.86
	5	0.23	0.30	0.38	0.54	0.68	0.77	0.86	1.04	1.33	1.53	1.77
	6	0.29	0.37	0.45	0.60	0.73	0.81	0.90	1.06	1.32	1.49	1.71
	7	0.34	0.42	0.50	0.65	0.77	0.84	0.92	1.07	1.31	1.46	1.66
	8	0.39	0.47	0.54	0.69	0.80	0.87	0.95	1.08	1.29	1.43	1.60
	9	0.43	0.51	0.59	0.73	0.83	0.89	0.96	1.08	1.28	1.40	1.55
	10	0.48	0.55	0.62	0.76	0.85	0.91	0.98	1.09	1.26	1.38	1.51
11	3	0.08	0.13	0.19	0.34	0.48	0.59	0.71	0.92	1.31	1.60	1.97
	4	0.15	0.22	0.30	0.46	0.60	0.70	0.80	1.00	1.34	1.58	1.87
	5	0.22	0.30	0.38	0.54	0.67	0.76	0.86	1.04	1.34	1.54	1.82
	6	0.28	0.36	0.44	0.60	0.73	0.81	0.90	1.07	1.33	1.52	1.73
	7	0.33	0.41	0.49	0.65	0.76	0.84	0.92	1.08	1.32	1.48	1.67
	8	0.38	0.46	0.54	0.68	0.80	0.87	0.95	1.08	1.31	1.45	1.62
	9	0.42	0.50	0.57	0.71	0.82	0.89	0.96	1.09	1.29	1.42	1.58
	10	0.46	0.54	0.61	0.74	0.85	0.91	0.98	1.09	1.27	1.38	1.53
	11	0.50	0.57	0.64	0.77	0.87	0.93	0.99	1.09	1.25	1.36	1.49
12	3	0.08	0.13	0.19	0.34	0.48	0.58	0.70	0.92	1.30	1.56	1.87
	4	0.16	0.22	0.30	0.46	0.60	0.70	0.80	1.00	1.33	1.55	1.82
	5	0.23	0.30	0.38	0.54	0.67	0.76	0.86	1.04	1.33	1.53	1.78
	6	0.29	0.36	0.44	0.60	0.72	0.81	0.90	1.06	1.33	1.49	1.72
	7	0.34	0.41	0.50	0.65	0.76	0.84	0.93	1.08	1.31	1.47	1.66
	8	0.38	0.46	0.54	0.68	0.79	0.87	0.95	1.08	1.30	1.45	1.61
	9	0.42	0.50	0.57	0.71	0.82	0.89	0.96	1.09	1.29	1.43	1.58
	10	0.45	0.53	0.61	0.74	0.84	0.90	0.97	1.09	1.28	1.40	1.55
	11	0.49	0.56	0.64	0.76	0.86	0.92	0.98	1.09	1.27	1.37	1.51
	12	0.53	0.60	0.66	0.78	0.87	0.93	0.99	1.09	1.24	1.35	1.46

		$1-\alpha$										
n	r	0.02	0.05	0.10	0.25	0.40	0.50	0.60	0.75	0.90	0.95	0.98
13	3	0.08	0.13	0.19	0.33	0.48	0.58	0.69	0.91	1.30	1.58	1.95
	4	0.15	0.22	0.29	0.45	0.59	0.69	0.79	0.99	1.33	1.57	1.86
	5	0.22	0.30	0.37	0.53	0.67	0.75	0.85	1.03	1.34	1.55	1.79
	6	0.28	0.36	0.43	0.59	0.72	0.80	0.89	1.06	1.33	1.51	1.72
	7	0.33	0.40	0.48	0.64	0.76	0.84	0.92	1.07	1.32	1.48	1.67
	8	0.37	0.45	0.53	0.67	0.79	0.86	0.94	1.08	1.30	1.45	1.62
	9	0.42	0.49	0.56	0.70	0.81	0.88	0.95	1.08	1.29	1.42	1.58
	10	0.44	0.52	0.60	0.73	0.83	0.89	0.96	1.08	1.28	1.40	1.55
	11	0.48	0.55	0.62	0.75	0.85	0.91	0.97	1.08	1.26	1.38	1.51
	12	0.51	0.58	0.65	0.77	0.86	0.92	0.98	1.08	1.25	1.36	1.47
	13	0.54	0.61	0.68	0.79	0.88	0.93	0.99	1.09	1.24	1.33	1.44
14	3	0.08	0.13	0.19	0.34	0.48	0.58	0.69	0.91	1.31	1.58	1.94
	4	0.16	0.22	0.30	0.45	0.59	0.69	0.80	0.99	1.33	1.57	1.86
	5	0.22	0.30	0.37	0.53	0.67	0.76	0.86	1.04	1.34	1.54	1.77
	6	0.28	0.35	0.43	0.59	0.72	0.81	0.90	1.06	1.33	1.51	1.71
	7	0.33	0.40	0.48	0.64	0.76	0.84	0.93	1.07	1.32	1.48	1.67
	8	0.38	0.45	0.53	0.67	0.79	0.86	0.94	1.08	1.30	1.45	1.63
	9	0.41	0.49	0.56	0.70	0.82	0.88	0.96	1.08	1.29	1.42	1.59
	10	0.45	0.52	0.59	0.72	0.83	0.90	0.96	1.09	1.28	1.40	1.56
	11	0.48	0.55	0.62	0.75	0.85	0.91	0.97	1.09	1.26	1.38	1.52
	12	0.50	0.57	0.64	0.77	0.86	0.92	0.98	1.09	1.25	1.36	1.49
	13	0.53	0.60	0.67	0.79	0.88	0.93	0.99	1.09	1.24	1.34	1.46
	14	0.57	0.63	0.69	0.81	0.89	0.94	0.99	1.09	1.23	1.32	1.43
15	3	0.08	0.13	0.19	0.33	0.47	0.57	0.68	0.90	1.29	1.57	1.92
	4	0.16	0.22	0.29	0.45	0.59	0.68	0.79	0.99	1.33	1.56	1.85
	5	0.22	0.29	0.37	0.53	0.66	0.75	0.85	1.04	1.33	1.53	1.79
	6	0.28	0.35	0.44	0.59	0.71	0.80	0.89	1.06	1.32	1.50	1.71
	7	0.33	0.41	0.49	0.63	0.75	0.83	0.92	1.07	1.32	1.48	1.67
	8	0.37	0.45	0.52	0.67	0.78	0.86	0.94	1.08	1.30	1.45	1.63
	9	0.41	0.49	0.56	0.69	0.81	0.88	0.95	1.08	1.29	1.43	1.59
	10	0.45	0.52	0.59	0.72	0.82	0.89	0.96	1.09	1.28	1.41	1.56
	11	0.48	0.54	0.61	0.74	0.84	0.90	0.97	1.09	1.27	1.39	1.54
	12	0.50	0.57	0.63	0.76	0.86	0.91	0.98	1.09	1.26	1.37	1.50
	13	0.52	0.59	0.66	0.78	0.87	0.92	0.98	1.08	1.25	1.35	1.47
	14	0.55	0.62	0.68	0.80	0.88	0.94	0.99	1.08	1.24	1.33	1.45
	15	0.58	0.64	0.70	0.81	0.89	0.94	0.99	1.09	1.23	1.32	1.42

Percentiles of the distribution of W

						$1-\alpha$						
n	r	0.02	0.05	0.10	0.25	0.40	0.50	0.60	0.75	0.90	0.95	0.98
16	3	0.08	0.13	0.19	0.33	0.47	0.56	0.68	0.90	1.29	1.58	1.94
	4	0.15	0.22	0.29	0.45	0.59	0.68	0.79	0.99	1.33	1.56	1.86
	5	0.22	0.29	0.36	0.53	0.66	0.75	0.85	1.03	1.33	1.54	1.78
	6	0.27	0.35	0.43	0.58	0.71	0.79	0.88	1.05	1.33	1.51	1.74
	7	0.31	0.40	0.48	0.63	0.75	0.83	0.91	1.07	1.31	1.47	1.69
	8	0.36	0.44	0.52	0.66	0.78	0.85	0.93	1.07	1.30	1.45	1.64
	9	0.40	0.48	0.55	0.69	0.80	0.87	0.94	1.08	1.29	1.43	1.60
	10	0.43	0.51	0.58	0.72	0.82	0.89	0.96	1.08	1.28	1.41	1.57
	11	0.46	0.53	0.61	0.74	0.84	0.90	0.97	1.09	1.27	1.39	1.54
	12	0.49	0.56	0.63	0.76	0.85	0.91	0.97	1.09	1.26	1.38	1.50
	13	0.51	0.59	0.65	0.77	0.86	0.92	0.98	1.09	1.25	1.36	1.48
	14	0.54	0.61	0.67	0.76	0.88	0.93	0.99	1.09	1.24	1.34	1.46
	15	0.56	0.63	0.69	0.80	0.89	0.94	0.99	1.09	1.23	1.32	1.43
	16	0.59	0.65	0.71	0.82	0.90	0.95	1.00	1.09	1.21	1.29	1.40
17	3	0.08	0.13	0.18	0.33	0.48	0.58	0.69	0.92	1.33	1.59	1.95
	4	0.15	0.22	0.30	0.45	0.59	0.69	0.80	1.00	1.35	1.58	1.87
	5	0.22	0.30	0.37	0.53	0.67	0.76	0.86	1.04	1.34	1.55	1.79
	6	0.28	0.35	0.43	0.59	0.71	0.80	0.89	1.06	1.33	1.52	1.73
	7	0.33	0.40	0.48	0.63	0.75	0.83	0.92	1.07	1.32	1.48	1.68
	8	0.37	0.44	0.52	0.67	0.78	0.86	0.94	1.08	1.31	1.47	1.63
	9	0.40	0.48	0.55	0.70	0.80	0.88	0.95	1.09	1.30	1.44	1.60
	10	0.44	0.51	0.58	0.72	0.82	0.89	0.96	1.09	1.29	1.42	1.58
	11	0.46	0.54	0.61	0.74	0.84	0.90	0.97	1.09	1.28	1.39	1.55
	12	0.49	0.56	0.63	0.76	0.85	0.91	0.98	1.09	1.27	1.38	1.50
	13	0.51	0.58	0.65	0.78	0.87	0.92	0.99	1.09	1.26	1.36	1.48
	14	0.53	0.61	0.67	0.79	0.88	0.93	0.99	1.09	1.24	1.34	1.46
	15	0.56	0.63	0.69	0.80	0.89	0.94	1.00	1.09	1.23	1.33	1.44
	16	0.58	0.65	0.71	0.82	0.90	0.95	1.00	1.09	1.22	1.31	1.41
	17	0.61	0.67	0.73	0.83	0.91	0.95	1.00	1.09	1.21	1.29	1.39

		$1-\alpha$										
n	r	0.02	0.05	0.10	0.25	0.40	0.50	0.60	0.75	0.90	0.95	0.98
18	3	0.07	0.12	0.19	0.33	0.48	0.58	0.69	0.92	1.31	1.59	1.93
	4	0.15	0.22	0.29	0.45	0.59	0.69	0.79	0.99	1.34	1.58	1.87
	5	0.22	0.29	0.37	0.53	0.66	0.75	0.85	1.03	1.34	1.54	1.79
	6	0.27	0.35	0.43	0.58	0.71	0.80	0.89	1.05	1.32	1.51	1.73
	7	0.33	0.40	0.47	0.62	0.75	0.83	0.91	1.07	1.31	1.48	1.68
	8	0.37	0.44	0.51	0.66	0.77	0.85	0.93	1.08	1.30	1.46	1.63
	9	0.40	0.47	0.55	0.69	0.80	0.87	0.95	1.08	1.30	1.43	1.69
	10	0.43	0.51	0.58	0.71	0.81	0.88	0.96	1.08	1.28	1.41	1.56
	11	0.46	0.53	0.60	0.73	0.83	0.90	0.97	1.08	1.27	1.39	1.54
	12	0.49	0.56	0.63	0.75	0.85	0.91	0.97	1.09	1.26	1.38	1.52
	13	0.51	0.58	0.65	0.77	0.86	0.92	0.98	1.09	1.25	1.36	1.48
	14	0.54	0.60	0.66	0.78	0.87	0.93	0.99	1.09	1.25	1.35	1.46
	15	0.56	0.62	0.68	0.79	0.88	0.93	0.99	1.08	1.23	1.33	1.44
	16	0.57	0.64	0.70	0.81	0.89	0.94	0.99	1.09	1.22	1.31	1.42
	17	0.59	0.65	0.71	0.82	0.90	0.95	1.00	1.09	1.22	1.30	1.39
	18	0.61	0.67	0.73	0.83	0.91	0.95	1.00	1.08	1.20	1.28	1.37
19	3	0.08	0.12	0.18	0.33	0.47	0.57	0.69	0.91	1.30	1.57	1.91
	4	0.15	0.22	0.29	0.45	0.59	0.68	0.79	0.99	1.33	1.56	1.84
	5	0.22	0.29	0.37	0.52	0.66	0.76	0.85	1.03	1.33	1.54	1.78
	6	0.28	0.35	0.43	0.58	0.71	0.80	0.89	1.05	1.33	1.51	1.74
	7	0.32	0.40	0.48	0.63	0.75	0.83	0.91	1.06	1.32	1.48	1.69
	8	0.36	0.44	0.51	0.66	0.78	0.85	0.93	1.08	1.31	1.46	1.65
	9	0.40	0.47	0.55	0.69	0.80	0.87	0.95	1.08	1.30	1.43	1.60
	10	0.43	0.50	0.57	0.71	0.82	0.89	0.96	1.09	1.28	1.42	1.57
	11	0.46	0.53	0.60	0.73	0.83	0.90	0.97	1.09	1.27	1.40	1.55
	12	0.48	0.55	0.62	0.75	0.85	0.91	0.97	1.09	1.26	1.38	1.51
	13	0.50	0.58	0.64	0.77	0.86	0.92	0.98	1.09	1.26	1.37	1.49
	14	0.53	0.60	0.67	0.78	0.87	0.93	0.98	1.09	1.25	1.35	1.47
	15	0.54	0.62	0.68	0.79	0.88	0.93	0.99	1.09	1.24	1.34	1.45
	16	0.56	0.63	0.69	0.81	0.89	0.94	0.99	1.09	1.23	1.32	1.42
	17	0.58	0.65	0.71	0.82	0.90	0.95	1.00	1.09	1.22	1.31	1.41
	18	0.60	0.67	0.73	0.83	0.90	0.95	1.00	1.08	1.21	1.29	1.39
	19	0.62	0.68	0.74	0.84	0.91	0.96	1.00	1.08	1.20	1.28	1.36

Percentiles of the distribution of W

		$1-\alpha$										
n	r	0.02	0.05	0.10	0.25	0.40	0.50	0.60	0.75	0.90	0.95	0.98
20	3	0.07	0.12	0.18	0.33	0.47	0.58	0.69	0.91	1.31	1.60	1.97
	4	0.15	0.22	0.29	0.45	0.59	0.69	0.79	0.99	1.34	1.57	1.89
	5	0.22	0.29	0.37	0.52	0.66	0.75	0.85	1.03	1.34	1.55	1.81
	6	0.27	0.35	0.43	0.58	0.71	0.79	0.89	1.05	1.33	1.52	1.75
	7	0.32	0.40	0.47	0.62	0.75	0.83	0.91	1.07	1.32	1.49	1.70
	8	0.35	0.43	0.51	0.66	0.78	0.85	0.93	1.08	1.30	1.46	1.65
	9	0.39	0.47	0.55	0.69	0.80	0.87	0.95	1.08	1.30	1.44	1.61
	10	0.43	0.50	0.57	0.71	0.82	0.89	0.96	1.09	1.29	1.42	1.58
	11	0.45	0.53	0.60	0.73	0.83	0.90	0.97	1.09	1.28	1.40	1.54
	12	0.48	0.55	0.62	0.75	0.85	0.91	0.98	1.09	1.27	1.38	1.52
	13	0.50	0.57	0.64	0.77	0.86	0.92	0.98	1.09	1.26	1.36	1.50
	14	0.52	0.60	0.66	0.78	0.87	0.93	0.99	1.09	1.25	1.35	1.48
	15	0.54	0.61	0.68	0.79	0.88	0.93	0.99	1.09	1.24	1.34	1.46
	16	0.56	0.63	0.69	0.81	0.89	0.94	1.00	1.09	1.23	1.33	1.44
	17	0.58	0.65	0.71	0.82	0.90	0.95	1.00	1.09	1.22	1.31	1.42
	18	0.60	0.66	0.72	0.83	0.90	0.95	1.00	1.08	1.22	1.30	1.40
	19	0.62	0.68	0.74	0.84	0.91	0.96	1.00	1.08	1.21	1.28	1.37
	20	0.64	0.70	0.75	0.85	0.92	0.96	1.01	1.08	1.20	1.27	1.36

appendix XI percentiles of the distribution of the statistic[a] $(\tilde{u}-u)/\tilde{b}$

Percentiles of the distribution of $(\tilde{u}-u)/\tilde{b}$

		$1-\alpha$										
n	r	0.02	0.05	0.10	0.25	0.40	0.50	0.60	0.75	0.90	0.95	0.98
3	3	−4.47	−2.54	−1.49	−0.52	−0.10	0.10	0.31	0.69	1.46	2.12	3.39
4	3	−6.92	−3.85	−2.32	−0.84	−0.29	−0.04	0.18	0.50	1.06	1.55	2.43
	4	−2.37	−1.50	−0.96	−0.37	−0.08	0.09	0.25	0.55	1.07	1.49	2.15
5	3	−9.35	−5.22	−3.04	−1.22	−0.50	−0.19	0.06	0.40	0.86	1.20	1.76
	4	−3.13	−1.94	−1.24	−0.50	−0.16	0.02	0.18	0.45	0.88	1.22	1.74
	5	−1.63	−1.08	−0.73	−0.31	−0.06	0.08	0.22	0.47	0.89	1.20	1.64
6	3	−10.54	−6.12	−3.72	−1.56	−0.69	−0.32	−0.04	0.33	0.75	1.02	1.39
	4	−3.69	−2.39	−1.59	−0.67	−0.25	−0.05	0.12	0.38	0.76	1.03	1.42
	5	−2.05	−1.36	−0.91	−0.38	−0.11	0.04	0.17	0.40	0.77	1.04	1.41
	6	−1.29	−0.91	−0.64	−0.28	−0.06	0.07	0.19	0.41	0.77	1.04	1.39
7	3	−13.00	−7.39	−4.45	−1.87	−0.89	−0.48	−0.16	0.26	0.68	0.90	1.20
	4	−4.67	−2.95	−1.94	−0.84	−0.36	−0.13	0.05	0.32	0.66	0.89	1.20
	5	−2.48	−1.59	−1.10	−0.48	−0.17	−0.02	0.12	0.34	0.66	0.89	1.21
	6	−1.54	−1.04	−0.73	−0.32	−0.10	0.03	0.15	0.35	0.67	0.90	1.20
	7	−1.09	−0.79	−0.56	−0.26	−0.06	0.05	0.17	0.36	0.68	0.90	1.18
8	3	−14.36	−8.15	−5.01	−2.14	−1.04	−0.58	−0.21	0.24	0.67	0.88	1.12
	4	−5.34	−3.30	−2.18	−0.99	−0.43	−0.19	0.02	0.30	0.64	0.83	1.07
	5	−2.78	−1.86	−1.25	−0.56	−0.22	−0.05	0.10	0.32	0.62	0.82	1.07
	6	−1.80	−1.20	−0.83	−0.36	−0.12	0.01	0.13	0.33	0.63	0.82	1.08
	7	−1.28	−0.88	−0.61	−0.27	−0.07	0.04	0.15	0.33	0.63	0.82	1.08
	8	−0.97	−0.70	−0.50	−0.22	−0.05	0.06	0.16	0.34	0.63	0.82	1.07
9	3	−15.68	−9.12	−5.64	−2.38	−1.17	−0.66	−0.28	0.20	0.66	0.86	1.06
	4	−6.31	−3.78	−2.47	−1.08	−0.50	−0.24	−0.01	0.28	0.61	0.79	1.00
	5	−3.19	−2.10	−1.40	−0.63	−0.26	−0.08	0.08	0.30	0.58	0.76	0.98
	6	−2.01	−1.38	−0.94	−0.41	−0.15	−0.01	0.11	0.30	0.57	0.76	0.99
	7	−1.43	−0.99	−0.70	−0.31	−0.10	0.02	0.13	0.31	0.57	0.76	0.99
	8	−1.08	−0.76	−0.55	−0.25	−0.07	0.04	0.14	0.31	0.58	0.76	0.99
	9	−0.87	−0.64	−0.47	−0.21	−0.05	0.05	0.15	0.32	0.58	0.76	0.98

Example: For $n=6$, $r=4$, we have $V_{0.10}=0.76$.

Percentiles of the distribution of $(\tilde{u}-u)/\tilde{b}$ (*Continued*)

n	r	0.02	0.05	0.10	0.25	0.40	0.50	0.60	0.75	0.90	0.95	0.98
						$1-\alpha$						
10	3	−17.45	−9.98	−6.05	−2.58	−1.29	−0.76	−0.34	0.17	0.66	0.87	1.07
	4	−6.54	−4.17	−2.70	−1.22	−0.58	−0.28	−0.04	0.27	0.60	0.77	0.96
	5	−3.56	−2.37	−1.56	−0.73	−0.31	−0.12	0.05	0.28	0.56	0.72	0.93
	6	−2.21	−1.51	−1.03	−0.48	−0.19	−0.04	0.09	0.28	0.54	0.71	0.92
	7	−1.56	−1.08	−0.77	−0.35	−0.12	−0.00	0.11	0.28	0.54	0.70	0.93
	8	−1.20	−0.86	−0.62	−0.27	−0.08	0.02	0.12	0.28	0.53	0.71	0.93
	9	−0.97	−0.70	−0.50	−0.23	−0.06	0.04	0.13	0.29	0.54	0.71	0.93
	10	−0.80	−0.60	−0.44	−0.20	−0.04	0.04	0.14	0.29	0.54	0.71	0.92
11	3	−18.52	−10.68	−6.42	−2.76	−1.41	−0.85	−0.42	0.13	0.65	0.87	1.07
	4	−7.26	−4.57	−2.95	−1.37	−0.66	−0.36	−0.10	0.24	0.58	0.75	0.92
	5	−4.00	−2.58	−1.75	−0.81	−0.37	−0.16	0.01	0.26	0.54	0.69	0.88
	6	−2.45	−1.67	−1.16	−0.53	−0.22	−0.07	0.06	0.26	0.52	0.66	0.85
	7	−1.70	−1.21	−0.85	−0.40	−0.15	−0.02	0.09	0.26	0.50	0.65	0.86
	8	−1.30	−0.92	−0.66	−0.30	−0.11	0.00	0.10	0.26	0.50	0.65	0.86
	9	−1.06	−0.76	−0.54	−0.25	−0.08	0.02	0.11	0.26	0.50	0.65	0.86
	10	−0.87	−0.63	−0.46	−0.21	−0.06	0.03	0.12	0.27	0.50	0.65	0.86
	11	−0.75	−0.55	−0.42	−0.19	−0.05	0.03	0.12	0.27	0.50	0.65	0.85
12	3	−19.08	−11.23	−6.92	−3.03	−1.58	−0.97	−0.49	0.10	0.64	0.88	1.10
	4	−7.44	−4.81	−3.17	−1.47	−0.74	−0.40	−0.14	0.21	0.58	0.75	0.92
	5	−4.17	−2.72	−1.88	−0.89	−0.42	−0.20	−0.01	0.24	0.53	0.68	0.84
	6	−2.63	−1.83	−1.27	−0.60	−0.26	−0.10	0.05	0.25	0.50	0.64	0.81
	7	−1.91	−1.32	−0.92	−0.42	−0.17	−0.04	0.08	0.25	0.48	0.62	0.80
	8	−1.41	−1.00	−0.71	−0.33	−0.12	−0.01	0.09	0.25	0.48	0.62	0.79
	9	−1.15	−0.80	−0.58	−0.27	−0.09	0.01	0.10	0.25	0.47	0.62	0.80
	10	−0.91	−0.67	−0.48	−0.23	−0.07	0.02	0.11	0.25	0.47	0.62	0.80
	11	−0.78	−0.58	−0.43	−0.20	−0.06	0.03	0.11	0.25	0.47	0.62	0.80
	12	−0.69	−0.53	−0.39	−0.19	−0.05	0.03	0.11	0.25	0.47	0.62	0.79
13	3	−19.77	−11.66	−7.41	−3.21	−1.64	−1.02	−0.54	0.08	0.65	0.88	1.09
	4	−8.22	−5.21	−3.37	−1.60	−0.82	−0.48	−0.19	0.20	0.59	0.76	0.93
	5	−4.44	−2.95	−1.99	−0.96	−0.47	−0.24	−0.04	0.24	0.54	0.68	0.84
	6	−2.86	−1.94	−1.35	−0.66	−0.31	−0.13	0.03	0.25	0.51	0.64	0.79
	7	−2.04	−1.40	−0.98	−0.46	−0.19	−0.06	0.06	0.25	0.47	0.61	0.77
	8	−1.52	−1.06	−0.77	−0.36	−0.14	−0.02	0.08	0.24	0.46	0.59	0.75
	9	−1.18	−0.86	−0.61	−0.29	−0.10	−0.00	0.09	0.24	0.45	0.58	0.74
	10	−1.00	−0.72	−0.52	−0.24	−0.08	0.01	0.10	0.24	0.45	0.58	0.74
	11	−0.85	−0.63	−0.45	−0.21	−0.06	0.02	0.11	0.24	0.45	0.58	0.75
	12	−0.74	−0.56	−0.41	−0.19	−0.05	0.03	0.11	0.25	0.45	0.59	0.75
	13	−0.67	−0.51	−0.38	−0.18	−0.05	0.04	0.11	0.25	0.45	0.59	0.75

Percentiles of the distribution of $(\tilde{u}-u)/\tilde{b}$ (*Continued*)

		1 − α										
n	r	0.02	0.05	0.10	0.25	0.40	0.50	0.60	0.75	0.90	0.95	0.98
14	3	−21.43	−12.49	−7.65	−3.31	−1.71	−1.08	−0.57	0.06	0.65	0.90	1.11
	4	−8.30	−5.38	−3.53	−1.68	−0.87	−0.49	−0.20	0.19	0.59	0.77	0.94
	5	−4.72	−3.13	−2.17	−1.03	−0.51	−0.26	−0.04	0.24	0.54	0.69	0.84
	6	−3.07	−2.10	−1.45	−0.70	−0.32	−0.14	0.02	0.24	0.50	0.63	0.79
	7	−2.16	−1.50	−1.06	−0.50	−0.22	−0.07	0.06	0.24	0.47	0.60	0.75
	8	−1.67	−1.15	−0.81	−0.39	−0.15	−0.04	0.08	0.24	0.45	0.58	0.73
	9	−1.30	−0.93	−0.66	−0.30	−0.11	−0.01	0.09	0.23	0.44	0.56	0.72
	10	−1.07	−0.76	−0.54	−0.26	−0.09	0.00	0.09	0.23	0.43	0.56	0.72
	11	−0.89	−0.65	−0.48	−0.22	−0.07	0.01	0.09	0.23	0.43	0.56	0.72
	12	−0.76	−0.57	−0.42	−0.19	−0.06	0.02	0.10	0.23	0.43	0.56	0.72
	13	−0.68	−0.51	−0.38	−0.18	−0.05	0.02	0.10	0.23	0.43	0.56	0.72
	14	−0.63	−0.47	−0.36	−0.17	−0.05	0.03	0.10	0.23	0.43	0.56	0.72
15	3	−23.14	−13.14	−8.14	−3.63	−1.92	−1.20	−0.65	0.02	0.64	0.89	1.12
	4	−8.79	−5.55	−3.74	−1.78	−0.94	−0.55	−0.23	0.19	0.60	0.78	0.95
	5	−4.88	−3.35	−2.27	−1.10	−0.56	−0.29	−0.07	0.23	0.55	0.70	0.85
	6	−3.21	−2.21	−1.55	−0.75	−0.36	−0.17	−0.00	0.23	0.50	0.64	0.78
	7	−2.29	−1.56	−1.11	−0.55	−0.25	−0.09	0.04	0.23	0.47	0.59	0.74
	8	−1.72	−1.20	−0.86	−0.42	−0.18	−0.05	0.06	0.23	0.45	0.57	0.71
	9	−1.35	−0.96	−0.70	−0.35	−0.13	−0.03	0.07	0.23	0.43	0.56	0.69
	10	−1.10	−0.82	−0.59	−0.28	−0.10	−0.01	0.08	0.23	0.42	0.55	0.68
	11	−0.96	−0.70	−0.51	−0.24	−0.08	0.01	0.09	0.23	0.42	0.54	0.69
	12	−0.83	−0.62	−0.45	−0.21	−0.07	0.01	0.09	0.23	0.41	0.54	0.68
	13	−0.73	−0.55	−0.41	−0.19	−0.06	0.02	0.10	0.23	0.41	0.54	0.68
	14	−0.66	−0.50	−0.37	−0.18	−0.05	0.03	0.10	0.22	0.41	0.54	0.68
	15	−0.59	−0.46	−0.35	−0.17	−0.04	0.03	0.10	0.23	0.42	0.54	0.68
16	3	−22.72	−13.55	−8.42	−3.73	−2.01	−1.27	−0.69	0.00	0.66	0.92	1.13
	4	−9.38	−5.89	−3.92	−1.89	−0.99	−0.58	−0.24	0.18	0.60	0.79	0.97
	5	−5.17	−3.45	−2.35	−1.15	−0.58	−0.32	−0.10	0.21	0.54	0.70	0.85
	6	−3.34	−2.34	−1.64	−0.79	−0.38	−0.19	−0.01	0.22	0.50	0.63	0.77
	7	−2.42	−1.68	−1.19	−0.57	−0.27	−0.11	0.03	0.23	0.46	0.59	0.73
	8	−1.81	−1.30	−0.93	−0.44	−0.19	−0.06	0.05	0.22	0.44	0.56	0.70
	9	−1.44	−1.05	−0.74	−0.36	−0.14	−0.03	0.07	0.22	0.42	0.54	0.68
	10	−1.18	−0.86	−0.61	−0.29	−0.11	−0.01	0.07	0.21	0.41	0.53	0.67
	11	−1.00	−0.72	−0.52	−0.25	−0.08	0.00	0.08	0.21	0.41	0.52	0.66
	12	−0.87	−0.64	−0.46	−0.21	−0.07	0.01	0.09	0.21	0.40	0.52	0.66
	13	−0.76	−0.56	−0.41	−0.19	−0.06	0.02	0.09	0.21	0.40	0.52	0.66
	14	−0.68	−0.51	−0.37	−0.17	−0.05	0.03	0.10	0.21	0.40	0.52	0.66
	15	−0.61	−0.46	−0.35	−0.16	−0.04	0.03	0.10	0.22	0.40	0.52	0.66
	16	−0.56	−0.44	−0.33	−0.15	−0.04	−0.03	0.10	0.22	0.40	0.52	0.66

Percentiles of the distribution of $(\tilde{u} - u)/\tilde{b}$ *(Continued)*

		$1-\alpha$										
n	r	0.02	0.05	0.10	0.25	0.40	0.50	0.60	0.75	0.90	0.95	0.98
17	3	−24.35	−13.91	−8.80	−3.79	−2.01	−1.27	−0.69	0.04	0.69	0.95	1.17
	4	−9.31	−6.05	−4.07	−1.92	−1.00	−0.60	−0.26	0.17	0.62	0.81	0.98
	5	−5.32	−3.60	−2.50	−1.21	−0.62	−0.34	−0.10	0.21	0.55	0.72	0.86
	6	−3.54	−2.43	−1.75	−0.85	−0.42	−0.21	−0.03	0.23	0.50	0.64	0.78
	7	−2.60	−1.82	−1.28	−0.62	−0.29	−0.13	0.02	0.22	0.46	0.59	0.71
	8	−1.94	−1.39	−0.98	−0.48	−0.21	−0.07	0.05	0.22	0.44	0.55	0.68
	9	−1.49	−1.11	−0.78	−0.38	−0.17	−0.05	0.06	0.21	0.42	0.53	0.66
	10	−1.25	−0.92	−0.66	−0.32	−0.13	−0.03	0.07	0.21	0.40	0.51	0.64
	11	−1.07	−0.77	−0.56	−0.27	−0.10	−0.01	0.07	0.21	0.39	0.50	0.63
	12	−0.90	−0.67	−0.50	−0.24	−0.08	−0.00	0.08	0.21	0.39	0.49	0.63
	13	−0.80	−0.59	−0.44	−0.21	−0.07	0.01	0.08	0.20	0.38	0.49	0.63
	14	−0.72	−0.54	−0.40	−0.19	−0.06	0.01	0.08	0.20	0.38	0.49	0.63
	15	−0.65	−0.49	−0.36	−0.18	−0.05	0.02	0.09	0.20	0.38	0.48	0.62
	16	−0.60	−0.46	−0.34	−0.16	−0.05	0.02	0.09	0.21	0.38	0.48	0.63
	17	−0.56	−0.43	−0.32	−0.16	−0.05	0.02	0.09	0.21	0.38	0.49	0.63
18	3	−25.92	−14.29	−8.73	−3.84	−2.01	−1.27	−0.69	0.02	0.69	0.97	1.21
	4	−9.67	−6.23	−4.12	−2.00	−1.04	−0.63	−0.29	0.17	0.62	0.83	1.00
	5	−5.55	−3.74	−2.59	−1.27	−0.66	−0.36	−0.12	0.21	0.56	0.72	0.87
	6	−3.67	−2.56	−1.77	−0.88	−0.43	−0.23	−0.03	0.22	0.51	0.65	0.78
	7	−2.64	−1.87	−1.31	−0.65	−0.31	−0.14	0.01	0.22	0.47	0.60	0.72
	8	−2.02	−1.42	−1.02	−0.51	−0.23	−0.09	0.04	0.22	0.44	0.55	0.68
	9	−1.62	−1.16	−0.83	−0.40	−0.17	−0.06	0.05	0.21	0.42	0.52	0.65
	10	−1.33	−0.95	−0.67	−0.33	−0.13	−0.03	0.06	0.21	0.40	0.50	0.63
	11	−1.13	−0.81	−0.58	−0.28	−0.11	−0.01	0.07	0.21	0.38	0.49	0.62
	12	−0.95	−0.69	−0.50	−0.24	−0.08	−0.00	0.08	0.20	0.37	0.48	0.61
	13	−0.84	−0.61	−0.45	−0.21	−0.07	0.01	0.08	0.20	0.37	0.48	0.61
	14	−0.74	−0.55	−0.40	−0.19	−0.06	0.01	0.08	0.20	0.37	0.48	0.60
	15	−0.67	−0.50	−0.37	−0.18	−0.05	0.02	0.09	0.20	0.37	0.48	0.61
	16	−0.61	−0.46	−0.35	−0.16	−0.05	0.02	0.09	0.20	0.37	0.48	0.60
	17	−0.57	−0.44	−0.32	−0.15	−0.04	0.02	0.09	0.20	0.37	0.48	0.61
	18	−0.54	−0.41	−0.31	−0.15	−0.04	0.02	0.09	0.20	0.37	0.48	0.61

Percentiles of the distribution of $(\tilde{u}-u)/\tilde{b}$ (*Continued*)

		$1-\alpha$										
n	r	0.02	0.05	0.10	0.25	0.40	0.50	0.60	0.75	0.90	0.95	0.98
19	3	−25.46	−14.84	−9.23	−4.11	−2.17	−1.37	−0.77	−0.02	0.68	0.96	1.22
	4	−10.39	−6.56	−4.35	−2.08	−1.13	−0.67	−0.31	0.15	0.62	0.83	1.01
	5	−5.84	−3.94	−2.68	−1.31	−0.69	−0.39	−0.14	0.21	0.57	0.72	0.87
	6	−3.76	−2.62	−1.84	−0.92	−0.46	−0.24	−0.05	0.22	0.51	0.65	0.78
	7	−2.77	−1.94	−1.38	−0.68	−0.33	−0.16	0.00	0.22	0.47	0.59	0.71
	8	−2.11	−1.50	−1.07	−0.53	−0.25	−0.10	0.04	0.22	0.43	0.55	0.67
	9	−1.67	−1.20	−0.87	−0.42	−0.19	−0.06	0.05	0.21	0.41	0.52	0.63
	10	−1.37	−0.98	−0.72	−0.35	−0.14	−0.04	0.06	0.21	0.40	0.50	0.62
	11	−1.13	−0.84	−0.61	−0.29	−0.11	−0.02	0.07	0.20	0.38	0.48	0.61
	12	−0.97	−0.72	−0.53	−0.25	−0.09	−0.01	0.07	0.20	0.37	0.48	0.60
	13	−0.84	−0.63	−0.46	−0.22	−0.08	0.00	0.08	0.20	0.36	0.47	0.59
	14	−0.75	−0.56	−0.41	−0.20	−0.07	0.01	0.08	0.20	0.36	0.47	0.59
	15	−0.68	−0.51	−0.37	−0.18	−0.06	0.01	0.08	0.19	0.36	0.47	0.59
	16	−0.62	−0.47	−0.35	−0.17	−0.05	0.02	0.08	0.19	0.36	0.46	0.59
	17	−0.58	−0.44	−0.33	−0.16	−0.05	0.02	0.08	0.19	0.36	0.46	0.58
	18	−0.54	−0.41	−0.31	−0.15	−0.04	0.02	0.08	0.19	0.36	0.46	0.59
	19	−0.51	−0.39	−0.30	−0.15	−0.04	0.02	0.09	0.20	0.36	0.46	0.59
20	3	−26.67	−15.33	−9.32	−4.12	−2.18	−1.40	−0.79	−0.02	0.71	0.99	1.25
	4	−10.49	−6.64	−4.47	−2.15	−1.16	−0.70	−0.34	0.15	0.62	0.83	1.02
	5	−5.99	−4.00	−2.78	−1.37	−0.71	−0.41	−0.16	0.19	0.56	0.73	0.89
	6	−3.95	−2.73	−1.94	−0.96	−0.49	−0.26	−0.06	0.21	0.51	0.65	0.79
	7	−2.91	−2.04	−1.43	−0.72	−0.35	−0.17	−0.01	0.21	0.46	0.59	0.72
	8	−2.20	−1.55	−1.11	−0.55	−0.26	−0.11	0.02	0.21	0.43	0.55	0.68
	9	−1.72	−1.26	−0.91	−0.44	−0.19	−0.07	0.04	0.21	0.41	0.52	0.63
	10	−1.42	−1.03	−0.75	−0.37	−0.15	−0.05	0.05	0.20	0.39	0.49	0.60
	11	−1.19	−0.88	−0.63	−0.31	−0.13	−0.03	0.06	0.20	0.38	0.47	0.59
	12	−1.02	−0.76	−0.56	−0.26	−0.10	−0.01	0.07	0.20	0.36	0.46	0.59
	13	−0.89	−0.67	−0.49	−0.23	−0.08	−0.00	0.07	0.19	0.35	0.46	0.58
	14	−0.80	−0.59	−0.43	−0.20	−0.07	0.01	0.08	0.19	0.35	0.45	0.57
	15	−0.72	−0.53	−0.39	−0.19	−0.06	0.01	0.08	0.19	0.35	0.45	0.57
	16	−0.65	−0.49	−0.37	−0.17	−0.05	0.02	0.08	0.19	0.35	0.45	0.57
	17	−0.60	−0.46	−0.35	−0.16	−0.05	0.02	0.08	0.19	0.35	0.45	0.57
	18	−0.56	−0.43	−0.33	−0.15	−0.04	0.02	0.08	0.19	0.35	0.45	0.57
	19	−0.53	−0.41	−0.31	−0.14	−0.04	0.02	0.08	0.19	0.35	0.45	0.57
	20	−0.50	−0.40	−0.30	−0.14	−0.04	0.02	0.09	0.19	0.35	0.45	0.57

[a]Taken from: Mann, N. R.; K. W. Fertig and E. M. Scheuer: *Confidence and Tolerance Bounds and a New Goodness-of-Fit Test for Two-Parameter Weibull or Extreme-Value Distribution* (*With Tables for Censored Samples of Size* 3(1)25); Aerospace Research Laboratories, Wright–Patterson Air Force Base, Ohio, ARL 71-0077, May, 1971.

appendix XII percentiles of the distribution of the statistic[a] $V_R = (\tilde{u} - x_R)/\tilde{b}$

Percentiles of the Distribution of $V_{0.90}$

		$1-\alpha$										
n	r	0.02	0.05	0.10	0.25	0.40	0.50	0.60	0.75	0.90	0.95	0.98
3	3	0.75	1.10	1.43	2.18	2.88	3.40	4.06	5.50	8.99	13.16	20.93
4	3	0.78	1.16	1.49	2.18	2.82	3.33	3.96	5.38	9.03	13.07	20.23
	4	0.87	1.16	1.46	2.06	2.60	2.99	3.45	4.40	6.47	8.39	11.66
5	3	0.78	1.18	1.51	2.17	2.79	3.27	3.87	5.24	8.78	12.58	20.38
	4	0.97	1.23	1.51	2.09	2.61	2.99	3.44	4.40	6.49	8.48	11.73
	5	0.97	1.23	1.49	2.02	2.49	2.82	3.20	3.93	5.48	6.73	8.66
6	3	0.73	1.18	1.53	2.15	2.73	3.18	3.74	4.98	8.24	11.74	18.65
	4	1.00	1.28	1.55	2.10	2.60	2.98	3.41	4.30	6.33	8.18	11.39
	5	1.02	1.29	1.54	2.05	2.50	2.82	3.21	3.94	5.42	6.73	8.89
	6	1.02	1.27	1.53	2.01	2.42	2.70	3.04	3.67	4.86	5.83	7.31
7	3	0.64	1.18	1.53	2.13	2.66	3.08	3.60	4.79	7.80	11.12	17.54
	4	1.04	1.31	1.58	2.10	2.57	2.91	3.33	4.21	6.16	7.89	10.90
	5	1.08	1.33	1.57	2.06	2.49	2.80	3.15	3.87	5.36	6.68	8.44
	6	1.08	1.32	1.56	2.03	2.42	2.70	3.01	3.63	4.86	5.82	7.23
	7	1.08	1.32	1.55	2.00	2.37	2.62	2.90	3.44	4.46	5.25	6.37

[a]Taken from: Mann, N. R.; E. W. Fertig and E. M. Scheuer: *Tolerance Bounds and a New Goodness-of-Fit Test for Two-Parameter Weibull or Extreme-Value Distribution (With Tables for Censored Samples of Size 3(1)25)*; Aerospace Research Laboratories, Wright-Patterson Air Force Base, Ohio, ARL 71-0077, May, 1971.

Percentiles of the Distribution of $V_{0.90}$ (*Continued*)

						$1-\alpha$						
n	r	0.02	0.05	0.10	0.25	0.40	0.50	0.60	0.75	0.90	0.95	0.98
8	3	0.49	1.13	1.52	2.11	2.62	3.01	3.48	4.62	7.51	10.67	16.36
	4	1.04	1.33	1.60	2.10	2.56	2.88	3.27	4.10	5.96	7.79	10.76
	5	1.11	1.36	1.60	2.08	2.49	2.78	3.12	3.82	5.28	6.50	8.62
	6	1.13	1.36	1.59	2.05	2.43	2.71	3.02	3.62	4.83	5.83	7.18
	7	1.12	1.36	1.58	2.03	2.38	2.64	2.93	3.46	4.49	5.31	6.40
	8	1.12	1.36	1.58	2.01	2.34	2.57	2.83	3.32	4.21	4.90	5.84
9	3	0.42	1.12	1.51	2.09	2.57	2.95	3.40	4.43	7.14	10.21	15.61
	4	1.06	1.36	1.61	2.10	2.52	2.84	3.21	4.00	5.77	7.39	10.26
	5	1.17	1.41	1.63	2.08	2.47	2.76	3.08	3.76	5.13	6.34	8.13
	6	1.19	1.41	1.62	2.06	2.43	2.70	2.99	3.59	4.74	5.67	7.06
	7	1.19	1.41	1.62	2.04	2.39	2.64	2.91	3.45	4.48	5.28	6.46
	8	1.19	1.40	1.61	2.02	2.36	2.59	2.84	3.34	4.26	4.95	5.94
	9	1.19	1.40	1.60	2.00	2.33	2.55	2.78	3.22	4.04	4.66	5.50
10	3	0.09	0.99	1.46	2.05	2.51	2.84	3.27	4.25	6.75	9.36	14.88
	4	0.99	1.34	1.62	2.08	2.48	2.77	3.13	3.90	5.56	7.17	9.60
	5	1.17	1.42	1.64	2.07	2.45	2.71	3.02	3.67	5.00	6.13	8.02
	6	1.20	1.43	1.64	2.05	2.41	2.66	2.94	3.53	4.67	5.59	6.99
	7	1.21	1.43	1.64	2.04	2.38	2.62	2.88	3.41	4.41	5.18	6.29
	8	1.21	1.43	1.63	2.02	2.35	2.58	2.83	3.31	4.22	4.91	5.83
	9	1.21	1.42	1.63	2.01	2.32	2.54	2.77	3.22	4.03	4.63	5.51
	10	1.21	1.42	1.62	1.99	2.30	2.50	2.72	3.13	3.86	4.41	5.16
11	3	−0.09	0.90	1.42	2.01	2.45	2.77	3.17	4.07	6.41	9.11	14.47
	4	0.97	1.35	1.61	2.06	2.44	2.73	3.06	3.79	5.46	7.04	9.98
	5	1.18	1.43	1.64	2.05	2.41	2.68	2.98	3.60	4.90	6.07	7.83
	6	1.24	1.45	1.64	2.04	2.38	2.63	2.91	3.46	4.58	5.52	6.96
	7	1.25	1.45	1.64	2.03	2.35	2.59	2.86	3.36	4.36	4.16	6.34
	8	1.25	1.45	1.64	2.01	2.33	2.56	2.80	3.28	4.15	4.87	5.82
	9	1.25	1.44	1.64	2.00	2.31	2.53	2.76	3.21	4.01	4.63	5.54
	10	1.25	1.44	1.64	1.99	2.29	2.49	2.71	3.14	3.87	4.44	5.23
	11	1.25	1.45	1.63	1.98	2.28	2.46	2.67	3.06	3.76	4.26	4.94

Percentiles of the Distribution of $V_{0.90}$ (*Continued*)

		1 − α										
n	r	0.02	0.05	0.10	0.25	0.40	0.50	0.60	0.75	0.90	0.95	0.98
12	3	−0.38	0.75	1.37	1.98	2.41	2.71	3.08	3.89	6.00	8.40	12.96
	4	0.95	1.34	1.60	2.05	2.42	2.69	3.00	3.67	5.17	6.60	9.07
	5	1.20	1.44	1.66	2.05	2.40	2.65	2.93	3.52	4.72	5.79	7.35
	6	1.26	1.46	1.67	2.04	2.38	2.62	2.88	3.39	4.41	5.31	6.61
	7	1.28	1.47	1.67	2.03	2.36	2.58	2.82	3.30	4.21	4.98	6.09
	8	1.28	1.47	1.66	2.02	2.34	2.54	2.78	3.22	4.06	4.75	5.71
	9	1.27	1.46	1.66	2.01	2.31	2.52	2.74	3.16	3.94	4.53	5.40
	10	1.27	1.47	1.65	2.00	2.30	2.49	2.70	3.11	3.87	4.37	5.11
	11	1.27	1.46	1.64	2.00	2.28	2.47	2.67	3.05	3.72	4.23	4.88
	12	1.28	1.47	1.64	1.99	2.27	2.44	2.63	3.00	3.62	4.07	4.68
13	3	−0.45	0.72	1.34	1.99	2.40	2.69	3.04	3.85	5.88	8.16	12.45
	4	0.88	1.31	1.60	2.06	2.42	2.68	2.98	3.64	5.10	6.45	8.82
	5	1.20	1.45	1.67	2.07	2.40	2.64	2.92	3.49	4.71	5.75	7.32
	6	1.27	1.48	1.68	2.07	2.38	2.61	2.86	3.38	4.43	5.30	6.49
	7	1.30	1.49	1.68	2.06	2.36	2.58	2.82	3.30	4.23	4.96	6.02
	8	1.30	1.49	1.68	2.04	2.34	2.55	2.78	3.22	4.06	4.73	5.63
	9	1.30	1.49	1.68	2.03	2.33	2.52	2.75	3.16	3.94	4.55	5.32
	10	1.30	1.49	1.68	2.03	2.31	2.50	2.72	3.12	3.83	4.37	5.11
	11	1.30	1.49	1.68	2.02	2.30	2.48	2.69	3.06	3.74	4.23	4.90
	12	1.30	1.49	1.67	2.01	2.28	2.46	2.65	3.02	3.65	4.09	4.73
	13	1.30	1.49	1.67	2.01	2.27	2.44	2.62	2.97	3.57	3.97	4.51
14	3	−0.81	0.57	1.25	1.93	2.34	2.62	2.95	3.70	5.56	7.69	11.56
	4	0.83	1.29	1.59	2.03	2.37	2.62	2.92	3.54	4.93	6.17	8.28
	5	1.18	1.46	1.67	2.05	2.36	2.60	2.86	3.42	4.58	5.54	6.96
	6	1.28	1.51	1.69	2.05	2.35	2.57	2.82	3.34	4.33	5.12	6.27
	7	1.32	1.52	1.69	2.04	2.34	2.55	2.78	3.27	4.15	4.82	5.75
	8	1.33	1.52	1.69	2.03	2.33	2.52	2.74	3.19	4.03	4.61	5.47
	9	1.33	1.52	1.69	2.03	2.31	2.50	2.71	3.14	3.90	4.45	5.18
	10	1.33	1.51	1.68	2.02	2.30	2.48	2.70	3.09	3.78	4.30	4.94
	11	1.33	1.51	1.68	2.01	2.28	2.47	2.67	3.05	3.71	4.20	4.79
	12	1.33	1.51	1.68	2.01	2.27	2.45	2.65	3.01	3.64	4.09	4.67
	13	1.33	1.51	1.68	2.00	2.26	2.43	2.62	2.97	3.55	3.98	4.51
	14	1.33	1.51	1.68	2.00	2.25	2.42	2.60	2.93	3.46	3.85	4.36

Percentiles of the Distribution of $V_{0.90}$ (*Continued*)

		$1-\alpha$										
n	r	0.02	0.05	0.10	0.25	0.40	0.50	0.60	0.75	0.90	0.95	0.98
15	3	−1.05	0.43	1.19	1.91	2.33	2.60	2.91	3.64	5.39	7.23	10.78
	4	0.77	1.26	1.59	2.03	2.37	2.61	2.89	3.49	4.78	5.95	7.94
	5	1.15	1.44	1.67	2.06	2.37	2.59	2.85	3.38	4.43	5.36	6.85
	6	1.29	1.50	1.69	2.06	2.36	2.57	2.81	3.30	4.22	4.97	6.19
	7	1.33	1.52	1.70	2.06	2.35	2.55	2.78	3.23	4.08	4.72	5.77
	8	1.34	1.52	1.70	2.05	2.33	2.53	2.74	3.17	3.95	4.57	5.40
	9	1.35	1.52	1.69	2.04	2.32	2.51	2.72	3.13	3.85	4.40	5.16
	10	1.35	1.52	1.69	2.04	2.31	2.49	2.69	3.09	3.76	4.26	4.95
	11	1.35	1.52	1.69	2.03	2.30	2.48	2.67	3.04	3.69	4.15	4.76
	12	1.34	1.52	1.69	2.02	2.28	2.46	2.64	3.00	3.62	4.08	4.62
	13	1.35	1.52	1.68	2.01	2.27	2.44	2.63	2.96	3.55	3.98	4.51
	14	1.35	1.51	1.69	2.01	2.27	2.43	2.60	2.93	3.49	3.89	4.39
	15	1.35	1.52	1.68	2.01	2.25	2.41	2.58	2.89	3.41	3.77	4.23
16	3	−1.38	0.25	1.10	1.90	2.31	2.57	2.88	3.57	5.22	7.07	10.49
	4	0.74	1.23	1.58	2.03	2.37	2.59	2.87	3.45	4.72	5.90	7.94
	5	1.17	1.45	1.68	2.06	2.37	2.59	2.84	3.36	4.42	5.33	6.73
	6	1.30	1.52	1.71	2.07	2.36	2.57	2.81	3.28	4.20	4.98	6.18
	7	1.35	1.53	1.72	2.06	2.35	2.56	2.77	3.21	4.05	4.74	5.81
	8	1.37	1.54	1.72	2.06	2.34	2.53	2.75	3.16	3.94	4.56	5.38
	9	1.37	1.54	1.72	2.05	2.33	2.51	2.72	3.11	3.84	4.38	5.77
	10	1.37	1.54	1.71	2.04	2.31	2.50	2.70	3.08	3.74	4.24	4.97
	11	1.37	1.54	1.71	2.04	2.30	2.48	2.68	3.04	3.67	4.13	4.79
	12	1.37	1.54	1.71	2.03	2.29	2.46	2.65	3.00	3.60	4.05	4.65
	13	1.37	1.54	1.71	2.03	2.28	2.45	2.63	2.97	3.52	3.94	4.49
	14	1.37	1.54	1.71	2.02	2.27	2.43	2.61	2.94	3.48	3.87	4.38
	15	1.37	1.54	1.71	2.02	2.26	2.42	2.59	2.90	3.41	3.79	4.26
	16	1.38	1.54	1.71	2.02	2.26	2.40	2.56	2.87	3.36	3.71	4.16
17	3	−1.56	0.10	1.03	1.86	2.26	2.50	2.79	3.41	4.96	6.54	9.76
	4	0.61	1.18	1.56	2.01	2.32	2.54	2.80	3.33	4.48	5.56	7.33
	5	1.08	1.43	1.67	2.04	2.33	2.54	2.78	3.27	4.24	5.09	6.35
	6	1.28	1.52	1.71	2.05	2.33	2.52	2.75	3.20	4.07	4.76	5.81
	7	1.34	1.54	1.72	2.05	2.32	2.51	2.72	3.14	3.95	4.56	5.44
	8	1.36	1.55	1.72	2.04	2.31	2.49	2.69	3.10	3.85	4.41	5.19
	9	1.37	1.56	1.72	2.04	2.30	2.48	2.67	3.05	3.76	4.28	4.97
	10	1.38	1.55	1.72	2.03	2.29	2.47	2.65	3.01	3.69	4.16	4.80
	11	1.38	1.55	1.72	2.03	2.28	2.45	2.64	2.98	3.62	4.05	4.66
	12	1.38	1.55	1.71	2.02	2.27	2.44	2.62	2.95	3.56	3.98	4.53
	13	1.38	1.55	1.71	2.02	2.26	2.42	2.60	2.97	3.50	3.90	4.45
	14	1.38	1.55	1.71	2.01	2.25	2.42	2.58	2.90	3.45	3.83	4.31
	15	1.38	1.55	1.71	2.01	2.25	2.40	2.57	2.88	3.40	3.76	4.19
	16	1.38	1.55	1.71	2.00	2.24	2.39	2.55	2.85	3.35	3.70	4.10
	17	1.38	1.55	1.71	2.00	2.24	2.38	2.53	2.82	3.29	3.62	4.04

Percentiles of the Distribution of $V_{0.90}$ (*Continued*)

		1 − α										
n	r	0.02	0.05	0.10	0.25	0.40	0.50	0.60	0.75	0.90	0.95	0.98
18	3	−1.61	0.11	1.01	1.85	2.26	2.50	2.77	3.38	4.81	6.43	9.64
	4	0.66	1.19	1.56	2.01	2.32	2.54	2.79	3.32	4.44	5.52	7.38
	5	1.10	1.45	1.69	2.05	2.34	2.54	2.77	3.26	4.22	5.04	6.40
	6	1.30	1.53	1.73	2.06	2.34	2.53	2.75	3.19	4.06	4.73	5.79
	7	1.37	1.56	1.74	2.06	2.33	2.52	2.73	3.14	3.94	4.55	5.41
	8	1.40	1.57	1.74	2.06	2.32	2.50	2.71	3.09	3.84	4.39	5.15
	9	1.41	1.58	1.74	2.05	2.31	2.49	2.68	3.06	3.75	4.25	4.99
	10	1.41	1.58	1.74	2.05	2.30	2.48	2.67	3.03	3.67	4.14	4.80
	11	1.41	1.58	1.74	2.05	2.29	2.46	2.65	3.00	3.63	4.05	4.63
	12	1.41	1.58	1.74	2.04	2.28	2.45	2.63	2.97	3.56	3.97	4.51
	13	1.41	1.57	1.74	2.04	2.28	2.44	2.61	2.94	3.50	3.91	4.40
	14	1.41	1.57	1.73	2.03	2.27	2.43	2.60	2.92	3.45	3.84	4.30
	15	1.41	1.57	1.73	2.03	2.26	2.42	2.59	2.89	3.40	3.75	4.21
	16	1.41	1.57	1.73	2.02	2.26	2.41	2.57	2.87	3.36	3.71	4.14
	17	1.41	1.58	1.74	2.02	2.25	2.40	2.55	2.84	3.31	3.66	4.04
	18	1.42	1.58	1.74	2.02	2.24	2.39	2.54	2.82	3.27	3.59	3.97
19	3	−2.13	−0.16	0.83	1.81	2.23	2.47	2.72	3.29	4.68	6.09	9.01
	4	0.43	1.11	1.51	1.99	2.30	2.51	2.75	3.25	4.32	5.33	7.04
	5	1.09	1.44	1.67	2.04	2.33	2.52	2.74	3.20	4.13	4.94	6.21
	6	1.30	1.53	1.72	2.05	2.33	2.52	2.72	3.15	3.98	4.67	5.67
	7	1.38	1.56	1.74	2.05	2.32	2.50	2.70	3.11	3.87	4.45	5.33
	8	1.41	1.58	1.74	2.05	2.31	2.49	2.68	3.07	3.79	4.30	5.09
	9	1.42	1.58	1.74	2.05	2.31	2.48	2.67	3.03	3.71	4.23	4.88
	10	1.43	1.58	1.74	2.05	2.30	2.46	2.65	3.00	3.64	4.11	4.73
	11	1.43	1.58	1.74	2.04	2.29	2.45	2.63	2.98	3.57	4.02	4.63
	12	1.43	1.58	1.74	2.04	2.28	2.45	2.62	2.95	3.52	3.94	4.51
	13	1.43	1.58	1.74	2.03	2.27	2.43	2.60	2.93	3.48	3.99	4.41
	14	1.43	1.58	1.73	2.03	2.26	2.42	2.59	2.90	3.43	3.81	4.29
	15	1.42	1.58	1.74	2.03	2.26	2.41	2.57	2.88	3.39	3.75	4.22
	16	1.42	1.58	1.73	2.03	2.25	2.40	2.56	2.86	3.36	3.70	4.10
	17	1.42	1.58	1.74	2.02	2.25	2.40	2.55	2.84	3.32	3.65	4.06
	18	1.42	1.58	1.74	2.02	2.24	2.38	2.54	2.81	3.29	3.59	3.99
	19	1.43	1.59	1.74	2.02	2.24	2.38	2.52	2.78	3.24	3.53	3.90

Percentiles of the Distribution of $V_{0.90}$ (*Continued*)

		$1-\alpha$										
n	r	0.02	0.05	0.10	0.25	0.40	0.50	0.60	0.75	0.90	0.95	0.98
20	3	−2.27	−0.19	0.82	1.79	2.20	2.44	2.69	3.23	4.53	6.04	9.01
	4	0.46	1.10	1.51	1.99	2.29	2.49	2.72	3.21	4.24	5.24	7.00
	5	1.07	1.42	1.67	2.04	2.31	2.50	2.72	3.16	4.04	4.82	6.16
	6	1.30	1.54	1.73	2.06	2.32	2.50	2.71	3.13	3.90	4.55	5.68
	7	1.38	1.58	1.74	2.06	2.31	2.49	2.69	3.08	3.80	4.41	5.33
	8	1.42	1.59	1.75	2.06	2.30	2.48	2.67	3.04	3.71	4.29	5.06
	9	1.43	1.60	1.75	2.06	2.29	2.46	2.65	3.01	3.65	4.15	4.85
	10	1.44	1.60	1.75	2.05	2.29	2.45	2.64	2.98	3.59	4.05	4.71
	11	1.44	1.60	1.75	2.05	2.28	2.44	2.63	2.95	3.54	3.97	4.57
	12	1.44	1.60	1.75	2.04	2.28	2.43	2.61	2.92	3.49	3.92	4.46
	13	1.44	1.60	1.75	2.04	2.27	2.42	2.59	2.90	3.44	3.84	4.39
	14	1.44	1.60	1.75	2.04	2.26	2.41	2.58	2.88	3.40	3.77	4.29
	15	1.43	1.60	1.75	2.03	2.26	2.40	2.56	2.86	3.37	3.72	4.20
	16	1.44	1.60	1.75	2.03	2.25	2.40	2.55	2.84	3.33	3.67	4.12
	17	1.44	1.60	1.75	2.03	2.24	2.39	2.54	2.82	3.29	3.61	4.04
	18	1.44	1.60	1.74	2.02	2.24	2.38	2.53	2.81	3.25	3.57	3.97
	19	1.44	1.60	1.75	2.02	2.24	2.37	2.52	2.78	3.21	3.52	3.92
	20	1.44	1.60	1.75	2.02	2.23	2.36	2.51	2.76	3.17	3.47	3.84

Percentiles of the Distribution of $V_{0.95}$

		$1-\alpha$										
n	r	0.02	0.05	0.10	0.25	0.40	0.50	0.60	0.75	0.90	0.95	0.98
3	3	1.26	1.64	2.04	2.94	3.83	4.49	5.33	7.20	11.85	17.21	27.32
4	3	1.38	1.73	2.11	2.98	3.82	4.47	5.30	7.19	12.17	17.55	27.59
	4	1.36	1.69	2.04	2.78	3.44	3.95	4.51	5.73	8.40	10.88	15.06
5	3	1.44	1.79	2.16	3.00	3.81	4.46	5.27	7.16	12.07	17.36	28.30
	4	1.45	1.76	2.10	2.82	3.49	3.98	4.56	5.80	8.56	11.14	15.51
	5	1.44	1.74	2.06	2.72	3.29	3.70	4.17	5.11	7.06	8.68	11.14
6	3	1.48	1.83	2.20	2.99	3.77	4.37	5.15	6.93	11.53	16.66	26.85
	4	1.52	1.83	2.15	2.84	3.49	3.98	4.55	5.75	8.47	10.95	15.32
	5	1.51	1.81	2.12	2.76	3.33	3.73	4.21	5.17	7.08	8.82	11.58
	6	1.50	1.80	2.10	2.70	3.20	3.56	3.97	4.77	6.27	7.53	9.39
7	3	1.52	1.87	2.22	2.97	3.71	4.30	5.04	6.80	11.20	16.07	25.31
	4	1.59	1.88	2.19	2.86	3.47	3.94	4.49	5.69	8.39	10.80	14.80
	5	1.59	1.86	2.16	2.78	3.32	3.72	4.18	5.13	7.12	8.84	11.18
	6	1.57	1.84	2.14	2.72	3.21	3.57	3.96	4.76	6.33	7.61	9.40
	7	1.58	1.85	2.14	2.68	3.14	3.45	3.79	4.47	5.76	6.73	8.19
8	3	1.53	1.90	2.24	2.96	3.67	4.22	4.95	6.61	11.02	15.76	24.57
	4	1.63	1.91	2.22	2.87	3.46	3.91	4.44	5.60	8.19	10.74	15.22
	5	1.63	1.90	2.20	2.79	3.32	3.71	4.18	5.09	7.07	8.78	11.57
	6	1.62	1.89	2.18	2.75	3.24	3.58	3.99	4.77	6.35	7.67	9.43
	7	1.62	1.89	2.17	2.71	3.16	3.48	3.83	4.52	5.83	6.91	8.38
	8	1.62	1.89	2.17	2.69	3.10	3.38	3.71	4.31	5.44	6.29	7.50
9	3	1.55	1.93	2.26	2.95	3.63	4.18	4.85	6.45	10.71	15.33	23.80
	4	1.67	1.96	2.25	2.88	3.44	3.86	4.40	5.50	8.02	10.40	14.41
	5	1.69	1.95	2.23	2.80	3.33	3.70	4.13	5.07	6.90	8.59	11.05
	6	1.68	1.95	2.22	2.76	3.24	3.58	3.98	4.77	6.27	7.51	9.46
	7	1.67	1.94	2.21	2.73	3.18	3.50	3.85	4.53	5.86	6.91	8.40
	8	1.68	1.94	2.20	2.70	3.12	3.41	3.73	4.36	5.53	6.39	7.73
	9	1.69	1.95	2.19	2.68	3.08	3.35	3.63	4.19	5.72	6.00	7.09

Percentiles of the Distribution of $V_{0.95}$ *(Continued)*

		$1-\alpha$										
n	r	0.02	0.05	0.10	0.25	0.40	0.50	0.60	0.75	0.90	0.95	0.98
10	3	1.51	1.91	2.26	2.91	3.56	4.06	4.70	6.23	10.24	14.50	23.00
	4	1.70	1.98	2.27	2.86	3.39	3.81	4.29	5.42	7.81	10.12	13.69
	5	1.72	1.97	2.24	2.80	3.31	3.65	4.07	4.97	6.87	8.39	11.00
	6	1.71	1.97	2.23	2.76	3.22	3.55	3.93	4.70	6.24	7.50	9.42
	7	1.71	1.96	2.22	2.73	3.18	3.48	3.80	4.50	5.79	6.83	8.29
	8	1.70	1.96	2.21	3.71	3.13	3.41	3.72	4.33	5.52	6.40	7.61
	9	1.71	1.96	2.21	2.69	3.08	3.35	3.64	4.20	5.23	6.01	7.12
	10	1.72	1.96	2.21	2.67	3.04	3.29	3.56	4.08	4.98	5.67	6.65
11	3	1.48	1.93	2.25	2.88	3.49	3.98	4.59	6.07	9.89	14.11	22.60
	4	1.73	2.01	2.27	2.83	3.37	3.76	4.24	5.31	7.71	10.03	14.44
	5	1.75	2.01	2.26	2.78	3.27	3.62	4.03	4.89	6.72	8.34	10.93
	6	1.75	2.00	2.24	2.74	3.19	3.52	3.90	4.63	6.16	7.42	9.39
	7	1.75	1.99	2.23	2.72	3.14	3.45	3.79	4.45	5.79	6.83	8.42
	8	1.75	1.99	2.22	2.70	3.10	3.39	3.70	4.31	5.46	6.38	7.65
	9	1.75	1.98	2.22	2.68	3.07	3.34	3.63	4.21	5.23	6.04	7.23
	10	1.76	2.00	2.22	2.67	3.04	3.28	3.56	4.09	5.03	5.75	6.73
	11	1.77	2.00	2.22	2.66	3.01	3.25	3.50	3.98	4.85	5.49	6.35
12	3	1.42	1.91	2.26	2.87	3.47	3.91	4.50	5.85	9.41	13.40	21.39
	4	1.74	2.02	2.28	2.84	3.34	3.72	4.18	5.16	7.42	9.56	13.27
	5	1.79	2.03	2.28	2.78	3.26	3.61	3.99	4.82	6.54	8.08	10.40
	6	1.79	2.02	2.27	2.75	3.20	3.52	3.86	4.56	5.97	7.22	9.00
	7	1.79	2.02	2.26	2.73	3.14	3.43	3.76	4.40	5.63	6.66	8.08
	8	1.79	2.01	2.25	2.71	3.11	3.38	3.68	4.26	5.36	6.27	7.49
	9	1.78	2.00	2.24	2.69	3.07	3.33	3.62	4.16	5.16	5.95	7.06
	10	1.77	2.01	2.23	2.67	3.04	3.29	3.56	4.07	4.99	5.67	6.63
	11	1.78	2.01	2.24	2.67	3.02	3.25	3.51	3.98	4.84	5.47	6.29
	12	1.80	2.02	2.24	2.66	3.00	3.22	3.46	3.91	4.68	5.26	6.00
13	3	1.44	1.92	2.27	2.88	3.45	3.90	4.47	5.84	9.23	13.11	20.76
	4	1.78	2.05	2.31	2.84	3.35	3.73	4.18	5.18	7.38	9.47	13.09
	5	1.82	2.06	2.30	2.81	3.26	3.60	3.98	4.80	6.57	8.04	10.25
	6	1.82	2.05	2.29	2.78	3.21	3.50	3.86	4.58	6.03	7.24	8.89
	7	1.82	2.05	2.28	2.75	3.15	3.44	3.75	4.41	5.65	6.68	8.10
	8	1.81	2.04	2.27	2.74	3.12	3.39	3.69	4.28	5.40	6.29	7.43
	9	1.81	2.04	2.27	2.72	3.10	3.35	3.63	4.16	5.17	6.00	6.99
	10	1.81	2.04	2.27	2.70	3.07	3.31	3.58	4.09	5.01	5.70	6.66
	11	1.82	2.04	2.27	2.70	3.05	3.27	3.53	4.02	4.87	5.50	6.36
	12	1.82	2.04	2.27	2.69	3.02	3.24	3.48	3.95	4.73	5.30	6.12
	13	1.83	2.05	2.27	2.68	3.00	3.21	3.43	3.87	4.61	5.12	5.80

Percentiles of the Distribution of $V_{0.95}$ (*Continued*)

		$1-\alpha$										
n	r	0.02	0.05	0.10	0.25	0.40	0.50	0.60	0.75	0.90	0.95	0.98
14	3	1.39	1.92	2.26	2.84	3.39	3.83	4.36	5.64	8.84	12.73	19.14
	4	1.77	2.06	2.31	2.82	3.30	3.67	4.09	5.06	7.18	9.10	12.45
	5	1.84	2.08	2.31	2.79	3.22	3.55	3.93	4.72	6.38	7.82	9.93
	6	1.85	2.08	2.30	2.76	3.17	3.46	3.81	4.52	5.91	7.07	8.71
	7	1.85	2.07	2.29	2.73	3.12	3.40	3.72	4.38	5.58	6.53	7.78
	8	1.85	2.07	2.28	2.72	3.09	3.36	3.66	4.25	5.35	6.16	7.30
	9	1.84	2.07	2.28	2.71	3.07	3.32	3.59	4.15	5.14	5.88	6.92
	10	1.85	2.06	2.27	2.70	3.05	3.29	3.56	4.07	4.98	5.65	6.49
	11	1.85	2.07	2.28	2.69	3.03	3.26	3.51	4.00	4.86	5.48	6.26
	12	1.85	2.07	2.28	2.68	3.01	3.23	3.47	3.93	4.73	5.31	6.06
	13	1.85	2.07	2.29	2.68	2.99	3.21	3.44	3.87	4.61	5.14	5.80
	14	1.86	2.08	2.28	2.66	2.98	3.18	3.41	3.81	4.48	4.97	5.60
15	3	1.24	1.88	2.24	2.84	3.38	3.80	4.35	5.56	8.75	12.22	18.38
	4	1.77	2.06	2.31	2.83	3.30	3.65	4.07	4.99	7.00	8.90	11.93
	5	1.84	2.09	2.32	2.80	3.23	3.55	3.92	4.69	6.25	7.64	9.79
	6	1.87	2.09	2.31	2.79	3.18	3.47	3.80	4.48	5.79	6.91	8.55
	7	1.87	2.08	2.30	2.76	3.14	3.41	3.73	4.34	5.50	6.41	7.90
	8	1.86	2.08	2.29	2.74	3.11	3.36	3.66	4.23	5.29	6.10	7.26
	9	1.86	2.07	2.29	2.73	3.09	3.33	3.60	4.14	5.11	5.81	6.87
	10	1.85	2.07	2.28	2.72	3.06	3.30	3.56	4.08	4.96	5.61	6.50
	11	1.85	2.07	2.28	2.70	3.04	3.27	3.52	3.99	4.84	5.43	6.22
	12	1.86	2.07	2.28	2.70	3.02	3.25	3.48	3.94	4.73	5.31	6.00
	13	1.87	2.07	2.28	2.69	3.01	3.22	3.45	3.88	4.63	5.17	5.88
	14	1.87	2.07	2.28	2.68	3.00	3.20	3.41	3.83	4.53	5.02	5.66
	15	1.89	2.08	2.28	2.68	2.98	3.18	3.39	3.77	4.43	4.88	5.46
16	3	1.13	1.85	2.25	2.84	3.38	3.78	4.30	5.47	8.49	11.98	18.76
	4	1.77	2.06	2.33	2.83	3.30	3.64	4.06	4.95	7.00	8.88	12.30
	5	1.87	2.11	2.34	2.81	3.24	3.54	3.89	4.67	6.28	7.67	9.72
	6	1.89	2.11	2.33	2.79	3.19	3.48	3.81	4.47	5.80	6.92	8.63
	7	1.89	2.10	2.33	2.77	3.15	3.42	3.72	4.33	5.51	6.47	7.98
	8	1.89	2.10	2.32	2.75	3.12	3.38	3.66	4.22	5.28	6.12	7.27
	9	1.88	2.09	2.31	2.74	3.10	3.34	3.62	4.13	5.11	5.84	6.92
	10	1.88	2.09	2.31	2.72	3.07	3.32	3.57	4.07	4.95	5.60	6.60
	11	1.89	2.09	2.30	2.71	3.05	3.28	3.54	4.00	4.85	5.44	6.30
	12	1.88	2.09	2.31	2.71	3.03	3.25	3.50	3.94	4.71	5.30	6.06
	13	1.89	2.10	2.31	2.70	3.02	3.23	3.47	3.89	4.60	5.14	5.87
	14	1.90	2.10	2.31	2.69	3.00	3.21	3.43	3.84	4.52	5.02	5.68
	15	1.90	2.11	2.31	2.69	2.99	3.18	3.40	3.79	4.44	4.91	5.53
	16	1.92	2.12	2.33	2.69	2.98	3.16	3.36	3.74	4.35	4.80	5.35

Percentiles of the Distribution of $V_{0.95}$ (*Continued*)

		$1-\alpha$										
n	r	0.02	0.05	0.10	0.25	0.40	0.50	0.60	0.75	0.90	0.95	0.98
17	3	1.07	1.84	2.22	2.80	3.29	3.67	4.15	5.26	8.19	11.24	17.44
	4	1.76	2.08	2.32	2.80	3.24	3.57	3.95	4.81	6.65	8.45	11.48
	5	1.86	2.11	2.33	2.79	3.19	3.48	3.83	4.57	5.98	7.34	9.21
	6	1.90	2.11	2.33	2.77	3.15	3.42	3.75	4.40	5.61	6.67	8.19
	7	1.90	2.11	2.32	2.76	3.11	3.38	3.67	4.26	5.39	6.26	7.56
	8	1.90	2.11	2.32	2.74	3.09	3.34	3.61	4.15	5.17	5.95	7.05
	9	1.89	2.10	2.31	2.73	3.07	3.30	3.56	4.07	5.01	5.72	6.70
	10	1.89	2.10	2.31	2.71	3.05	3.28	3.52	4.00	4.89	5.53	6.37
	11	1.89	2.10	2.31	2.71	3.03	3.25	3.48	3.94	4.77	5.35	6.16
	12	1.89	2.11	2.31	2.70	3.01	3.23	3.45	3.89	4.67	5.23	5.92
	13	1.89	2.11	2.31	2.69	3.00	3.20	3.42	3.84	4.58	5.10	5.79
	14	1.90	2.12	2.31	2.69	2.99	3.19	3.40	3.80	4.51	5.00	5.61
	15	1.90	2.11	2.31	2.68	2.98	3.17	3.38	3.77	4.43	4.88	5.44
	16	1.90	2.12	2.31	2.67	2.97	3.15	3.35	3.72	4.34	4.79	5.30
	17	1.92	2.12	2.32	2.67	2.96	3.14	3.33	3.68	4.27	4.68	5.19
18	3	1.11	1.83	2.23	2.80	3.29	3.66	4.12	5.23	7.96	11.18	17.89
	4	1.81	2.09	2.34	2.82	3.25	3.56	3.95	4.79	6.66	8.46	11.66
	5	1.90	2.14	2.36	2.80	3.20	3.49	3.83	4.56	6.00	7.32	9.43
	6	1.93	2.14	2.36	2.78	3.16	3.43	3.75	4.40	5.63	6.61	8.27
	7	1.94	2.14	2.35	2.77	3.13	3.38	3.68	4.25	5.39	6.23	7.45
	8	1.94	2.14	2.35	2.75	3.10	3.35	3.63	4.15	5.17	5.95	6.97
	9	1.94	2.14	2.34	2.74	3.08	3.32	3.58	4.09	5.03	5.72	6.71
	10	1.93	2.13	2.34	2.73	3.06	3.29	3.54	4.03	4.88	5.51	6.38
	11	1.93	2.13	2.34	2.73	3.04	3.26	3.51	3.98	4.79	5.35	6.15
	12	1.93	2.14	2.34	2.72	3.03	3.24	3.48	3.92	4.68	5.23	5.96
	13	1.93	2.14	2.33	2.71	3.01	3.22	3.45	3.86	4.60	5.12	5.75
	14	1.93	2.13	2.33	2.70	3.01	3.21	3.42	3.83	4.52	5.01	5.60
	15	1.94	2.14	2.34	2.70	3.00	3.19	3.40	3.79	4.44	4.88	5.46
	16	1.94	2.14	2.34	2.70	2.99	3.18	3.38	3.75	4.37	4.83	5.36
	17	1.95	2.15	2.34	2.69	2.97	3.16	3.35	3.71	4.31	4.74	5.22
	18	1.97	2.17	2.35	2.69	2.97	3.15	3.33	3.68	4.24	4.65	5.11

Percentiles of the Distribution of $V_{0.95}$ (*Continued*)

						$1-\alpha$						
n	r	0.02	0.05	0.10	0.25	0.40	0.50	0.60	0.75	0.90	0.95	0.98
19	3	0.96	1.73	2.19	2.79	3.27	3.62	4.08	5.16	7.80	10.74	16.87
	4	1.76	2.08	2.34	2.81	3.24	3.54	3.92	4.72	6.52	8.23	11.18
	5	1.90	2.14	2.36	2.80	3.19	3.47	3.79	4.50	5.95	7.23	9.20
	6	1.94	2.15	2.36	2.78	3.15	3.42	3.72	4.35	5.56	6.60	8.07
	7	1.95	2.16	2.35	2.76	3.13	3.37	3.65	4.23	5.31	6.16	7.45
	8	1.95	2.15	2.35	2.75	3.10	3.34	3.60	4.14	5.14	5.87	6.94
	9	1.95	2.15	2.35	2.74	3.07	3.31	3.56	4.05	4.98	5.70	6.58
	10	1.95	2.41	2.34	2.73	3.05	3.27	3.52	4.00	4.84	5.49	6.35
	11	1.95	2.14	2.34	2.72	3.03	3.26	3.50	3.95	4.73	5.33	6.17
	12	1.95	2.14	2.34	2.72	3.02	3.24	3.47	3.89	4.65	5.22	5.97
	13	1.95	2.14	2.34	2.71	3.01	3.22	3.44	3.86	4.55	5.10	5.80
	14	1.95	2.14	2.33	2.70	3.00	3.20	3.41	3.81	4.49	4.98	5.61
	15	1.95	2.14	2.33	2.70	2.99	3.18	3.39	3.78	4.43	4.88	5.51
	16	1.95	2.15	2.33	2.70	2.98	3.17	3.37	3.74	4.38	4.81	5.35
	17	1.95	2.15	2.34	2.70	2.98	3.16	3.35	3.71	4.31	4.74	5.25
	18	1.96	2.16	2.35	2.70	2.97	3.14	3.33	3.68	4.26	4.65	5.15
	19	1.98	2.17	2.35	2.69	2.96	3.13	3.31	3.64	4.20	4.57	5.02
20	3	0.94	1.75	2.19	2.78	3.24	3.60	4.02	5.00	7.72	10.78	16.96
	4	1.79	2.09	2.34	2.81	3.21	3.52	3.88	4.66	6.49	8.16	11.03
	5	1.92	2.15	2.36	2.80	3.18	3.46	3.79	4.45	5.85	7.09	9.12
	6	1.94	2.17	2.37	2.79	3.15	3.40	3.71	4.31	5.48	6.49	8.11
	7	1.95	2.17	2.36	2.77	3.12	3.37	3.64	4.19	5.25	6.11	7.42
	8	1.96	2.17	2.36	2.76	3.09	3.32	3.59	4.09	5.06	5.88	6.97
	9	1.96	2.16	2.35	2.75	3.07	3.29	3.55	4.02	4.92	5.62	6.57
	10	1.95	2.16	2.35	2.74	3.05	3.27	3.51	3.97	4.79	5.43	6.30
	11	1.95	2.16	2.35	2.73	3.04	3.24	3.49	3.91	4.70	5.29	6.09
	12	1.95	2.16	2.35	2.72	3.02	3.22	3.45	3.87	4.61	5.18	5.92
	13	1.96	2.16	2.35	2.72	3.01	3.21	3.42	3.83	4.53	5.07	5.79
	14	1.96	2.16	2.34	2.71	3.00	3.19	3.40	3.79	4.46	4.96	5.63
	15	1.96	2.16	2.35	2.70	2.99	3.17	3.38	3.76	4.41	4.86	5.49
	16	1.97	2.16	2.35	2.70	2.98	3.16	3.36	3.73	4.35	4.78	5.37
	17	1.98	2.17	2.35	2.70	2.97	3.15	3.34	3.70	4.29	4.70	5.23
	18	1.98	2.18	2.35	2.69	2.96	3.14	3.33	3.67	4.23	4.64	5.14
	19	1.99	2.18	2.36	2.69	2.96	3.13	3.31	3.63	4.18	4.57	5.06
	20	2.00	2.19	2.36	2.70	2.95	3.12	3.29	3.60	4.12	4.49	4.95

Percentiles of the Distribution of $V_{0.99}$

		$1-\alpha$										
n	r	0.02	0.05	0.10	0.25	0.40	0.50	0.60	0.75	0.90	0.95	0.98
3	3	2.24	2.75	3.32	4.63	5.93	6.96	8.19	11.05	18.15	26.71	42.07
4	3	2.38	2.87	3.44	4.74	6.02	7.07	8.38	11.31	19.38	27.97	43.72
	4	2.37	2.82	3.31	4.38	5.34	6.09	6.91	8.74	12.79	16.62	23.12
5	3	2.45	2.95	3.47	4.80	6.09	7.14	8.48	11.55	19.73	28.71	46.68
	4	2.43	2.86	3.37	4.44	5.44	6.20	7.09	9.00	13.31	17.41	24.31
	5	2.45	2.87	3.32	4.28	5.11	5.71	6.39	7.81	10.75	13.23	16.87
6	3	2.55	3.02	3.55	4.80	6.08	7.11	8.39	11.42	19.28	28.02	45.73
	4	2.52	2.96	3.45	4.49	5.50	6.25	7.16	9.01	13.49	17.54	24.27
	5	2.52	2.94	3.38	4.33	5.18	5.79	6.52	7.95	10.96	13.67	17.96
	6	2.56	2.94	3.39	4.25	4.96	5.49	6.08	7.26	9.49	11.41	14.37
7	3	2.64	3.11	3.61	4.81	6.07	7.04	8.37	11.40	19.27	27.76	44.93
	4	2.61	3.04	3.49	4.52	5.48	6.25	7.14	9.08	13.53	17.47	23.96
	5	2.59	3.00	3.44	4.37	5.19	5.80	6.51	8.02	11.20	13.91	17.78
	6	2.61	3.01	3.41	4.26	5.00	5.53	6.14	7.32	9.75	11.74	14.36
	7	2.65	3.02	3.43	4.21	4.87	5.32	5.82	6.83	8.75	10.20	12.38
8	3	2.67	3.12	3.65	4.82	6.01	7.03	8.27	11.28	19.24	27.78	44.29
	4	2.66	3.07	3.53	4.54	5.51	6.23	7.12	9.04	13.42	17.64	25.03
	5	2.64	3.05	3.47	4.39	5.20	5.83	6.55	8.03	11.12	13.92	18.42
	6	2.64	3.04	3.45	4.30	5.04	5.57	6.17	7.41	9.82	11.87	14.78
	7	2.67	3.06	3.47	4.24	4.90	5.38	5.90	6.95	8.92	10.52	12.89
	8	2.71	3.08	3.47	4.20	4.79	5.21	5.69	6.58	8.27	9.52	11.35
9	3	2.75	3.20	3.66	4.81	6.02	6.97	8.17	11.14	19.00	27.91	43.02
	4	2.72	3.12	3.56	4.56	5.49	6.21	7.10	8.99	13.28	17.57	24.34
	5	2.71	3.10	3.52	4.40	5.23	5.83	6.53	8.05	10.98	13.78	17.99
	6	2.72	3.10	3.52	4.32	5.06	5.59	6.20	7.43	9.82	11.78	14.84
	7	2.71	3.10	3.51	4.27	4.96	5.42	5.95	6.99	9.06	10.66	12.96
	8	2.75	3.12	3.50	4.24	4.84	5.27	5.74	6.68	8.44	9.75	11.85
	9	2.79	3.17	3.50	4.20	4.78	5.16	5.59	6.41	7.90	9.04	10.71

Percentiles of the Distribution of $V_{0.99}$ (*Continued*)

		$1-\alpha$										
n	r	0.02	0.05	0.10	0.25	0.40	0.50	0.60	0.75	0.90	0.95	0.98
10	3	2.75	3.20	3.68	4.77	5.91	6.86	8.10	10.93	18.61	27.05	43.21
	4	2.74	3.16	3.60	4.53	5.45	6.15	7.00	8.87	13.12	17.17	23.48
	5	2.75	3.13	3.54	4.41	5.22	5.79	6.47	7.94	11.05	13.70	17.82
	6	2.75	3.11	3.51	4.33	5.04	5.56	6.15	7.38	9.81	11.86	15.01
	7	2.77	3.12	3.50	4.28	4.95	5.47	5.92	6.99	8.99	10.66	13.03
	8	2.77	3.15	3.50	4.23	4.86	5.29	5.75	6.68	8.49	9.88	11.69
	9	2.81	3.16	3.51	4.21	4.78	5.17	5.60	6.43	7.97	9.17	10.82
	10	2.84	3.19	3.53	4.18	4.72	5.07	5.48	6.23	7.57	8.57	10.03
11	3	2.79	3.22	3.67	4.75	5.87	6.79	7.98	10.78	18.19	26.45	43.82
	4	2.79	3.16	3.58	4.53	5.43	6.13	6.96	8.81	13.02	17.25	24.97
	5	2.79	3.15	3.55	4.39	5.19	5.77	6.44	7.88	10.99	13.66	18.13
	6	2.77	3.14	3.52	4.30	5.02	5.53	6.13	7.33	9.79	11.86	15.02
	7	2.78	3.16	3.51	4.26	4.92	5.39	5.92	6.95	9.03	10.72	13.30
	8	2.81	3.16	3.51	4.23	4.83	5.27	5.74	6.69	8.43	9.86	11.88
	9	2.84	3.17	3.51	4.20	4.77	5.16	5.61	6.49	8.02	9.24	11.08
	10	2.86	3.20	3.53	4.18	4.71	5.09	5.49	6.27	7.67	8.73	10.20
	11	2.88	3.24	3.55	4.17	4.66	5.01	5.38	6.09	7.36	8.31	9.57
12	3	2.88	3.28	3.72	4.77	5.85	6.76	7.88	10.59	17.59	25.73	41.44
	4	2.85	3.22	3.62	4.52	5.41	6.08	6.87	8.67	12.72	16.60	23.41
	5	2.82	3.19	3.58	4.39	5.17	5.75	6.41	7.79	10.68	13.37	17.35
	6	2.82	3.19	3.55	4.31	5.03	5.54	6.12	7.26	9.61	11.60	14.59
	7	2.81	3.18	3.54	4.27	4.92	5.39	5.89	6.89	8.85	10.55	12.65
	8	2.84	3.18	3.54	4.24	4.83	5.25	5.73	6.63	8.34	9.75	11.64
	9	2.86	3.18	3.53	4.21	4.77	5.16	5.60	6.44	7.98	9.18	10.90
	10	2.86	3.20	3.54	4.18	4.73	5.10	5.50	6.26	7.66	8.68	10.17
	11	2.90	3.23	3.55	4.17	4.69	5.03	5.41	6.12	7.37	8.31	9.53
	12	2.94	3.27	3.57	4.17	4.64	4.97	5.32	5.97	7.09	7.96	9.03
13	3	2.88	3.30	3.74	4.78	5.87	6.76	7.89	10.68	17.58	25.37	40.28
	4	2.89	3.25	3.66	4.56	5.46	6.11	6.94	8.75	12.81	16.56	23.73
	5	2.87	3.22	3.60	4.43	5.20	5.78	6.43	7.81	10.89	13.47	17.15
	6	2.89	3.21	3.57	4.35	5.05	5.54	6.14	7.32	9.71	11.72	14.47
	7	2.89	3.21	3.57	4.30	4.94	5.38	5.92	6.94	8.97	10.61	13.04
	8	2.89	3.22	3.57	4.28	4.86	5.29	5.75	6.69	8.46	9.88	11.74
	9	2.91	3.23	3.56	4.25	4.81	5.21	5.64	6.47	8.02	9.27	10.88
	10	2.91	3.24	3.57	4.23	4.78	5.14	5.54	6.31	7.72	8.80	10.25
	11	2.94	3.25	3.58	4.22	4.73	5.07	5.44	6.18	7.46	8.42	9.65
	12	2.97	3.27	3.60	4.21	4.68	5.00	5.37	6.05	7.21	8.03	9.27
	13	3.00	3.30	3.61	4.20	4.66	4.95	5.28	5.91	6.99	7.75	8.76

Percentiles of the Distribution of $V_{0.99}$ (*Continued*)

		$1-\alpha$										
n	r	0.02	0.05	0.10	0.25	0.40	0.50	0.60	0.75	0.90	0.95	0.98
14	3	2.90	3.30	3.75	4.75	5.80	6.63	7.76	10.37	17.23	25.09	39.16
	4	2.91	3.28	3.68	4.52	5.39	6.05	6.83	8.61	12.48	16.19	22.41
	5	2.91	3.25	3.62	4.40	5.14	5.71	6.38	7.77	10.64	13.22	16.89
	6	2.88	3.24	3.59	4.33	5.00	5.50	6.07	7.25	9.59	11.55	14.29
	7	2.91	3.23	3.58	4.28	4.90	5.35	5.86	6.92	8.88	10.47	12.51
	8	2.91	3.24	3.58	4.25	4.83	5.25	5.72	6.66	8.40	9.71	11.54
	9	2.93	3.25	3.58	4.23	4.79	5.16	5.59	6.45	8.00	9.17	10.81
	10	2.95	3.26	3.58	4.22	4.74	5.11	5.52	6.31	7.71	8.72	10.06
	11	2.95	3.28	3.60	4.20	4.71	5.06	5.44	6.17	7.46	8.42	9.61
	12	2.98	3.31	3.61	4.20	4.67	5.00	5.36	6.04	7.25	8.11	9.25
	13	3.00	3.31	3.62	4.19	4.65	4.96	5.30	5.93	7.03	7.82	8.81
	14	3.02	3.34	3.63	4.17	4.62	4.92	5.24	5.82	9.81	7.49	8.45
15	3	2.96	3.34	3.77	4.78	5.83	6.72	7.82	10.45	17.15	24.54	38.14
	4	2.94	3.29	3.69	4.55	5.40	6.05	6.82	8.54	12.41	15.96	21.82
	5	2.93	3.28	3.64	4.42	5.18	5.72	6.36	7.72	10.56	12.94	16.99
	6	2.93	3.26	3.61	4.37	5.03	5.52	6.09	7.23	9.45	11.38	14.24
	7	2.93	3.25	3.59	4.31	4.92	5.37	5.90	6.89	8.81	10.31	12.84
	8	2.93	3.25	3.59	4.27	4.86	5.27	5.73	6.64	8.36	9.64	11.58
	9	2.95	3.26	3.58	4.26	4.81	5.19	5.62	6.48	8.00	9.11	10.79
	10	2.94	3.27	3.59	4.24	4.77	5.13	5.54	6.33	7.71	8.72	10.06
	11	2.95	3.26	3.59	4.21	4.73	5.08	5.46	6.18	7.48	8.37	9.61
	12	2.98	3.29	3.61	4.21	4.70	5.04	5.39	6.07	7.29	8.17	9.24
	13	3.01	3.30	3.61	4.20	4.68	4.99	5.33	5.96	7.09	7.89	8.92
	14	3.03	3.32	3.62	4.19	4.65	4.95	5.26	5.86	6.92	7.63	8.59
	15	3.07	3.35	3.62	4.19	4.63	4.91	5.21	5.77	6.70	7.37	8.24
16	3	2.98	3.36	3.78	4.79	5.86	6.70	7.78	10.30	17.06	24.06	38.50
	4	2.96	3.32	3.70	4.57	5.41	6.04	6.82	8.52	12.49	16.24	23.03
	5	2.94	3.30	3.65	4.46	5.18	5.73	6.37	7.71	10.64	13.23	16.90
	6	2.93	3.29	3.63	4.37	5.05	5.53	6.11	7.23	9.53	11.53	14.38
	7	2.94	3.28	3.62	4.32	4.96	5.40	5.89	6.89	8.84	10.54	12.97
	8	2.95	3.28	3.62	4.30	4.88	5.30	5.75	6.65	8.39	9.75	11.59
	9	2.96	3.28	3.61	4.27	4.84	5.22	5.66	6.47	8.03	9.18	10.96
	10	2.97	3.28	3.62	4.25	4.79	5.16	5.55	6.32	7.71	8.75	10.29
	11	2.99	3.29	3.62	4.23	4.74	4.10	5.48	6.19	7.50	8.43	9.76
	12	2.99	3.31	3.63	4.23	4.71	5.04	5.41	6.09	7.27	8.17	9.32
	13	3.01	3.34	3.63	4.21	4.68	5.01	5.36	5.99	7.07	7.86	9.04
	14	3.05	3.35	3.65	4.21	4.66	4.96	5.30	5.90	6.93	7.68	8.67
	15	3.07	3.37	3.66	4.21	4.64	4.92	5.24	5.80	6.78	7.46	8.39
	16	3.12	3.39	3.69	4.20	4.62	4.88	5.17	5.72	6.60	7.26	8.10

Percentiles of the Distribution of $V_{0.99}$ (*Continued*)

		$1-\alpha$										
n	r	0.02	0.05	0.10	0.25	0.40	0.50	0.60	0.75	0.90	0.95	0.98
17	3	2.97	3.37	3.76	4.71	5.69	6.50	7.56	10.06	16.55	23.57	37.77
	4	2.97	3.32	3.70	4.51	5.32	5.94	6.71	8.36	11.93	15.54	21.68
	5	2.96	3.30	3.66	4.42	5.12	5.65	3.27	7.62	10.23	12.58	16.34
	6	2.95	3.28	3.63	4.34	5.01	5.47	6.03	7.14	9.32	11.13	13.79
	7	2.96	3.29	3.63	4.30	4.90	5.34	5.83	6.83	8.70	10.18	12.33
	8	2.97	3.28	3.61	4.27	4.83	5.23	5.69	6.58	8.25	9.56	11.40
	9	2.97	3.29	3.61	4.25	4.79	5.17	5.58	6.39	7.91	9.07	10.65
	10	2.98	3.30	3.61	4.23	4.76	5.12	5.50	6.25	7.66	8.70	10.01
	11	2.99	3.31	3.62	4.22	4.71	5.06	5.42	6.13	7.42	8.33	9.63
	12	3.01	3.33	3.63	4.21	4.69	5.01	5.36	6.02	7.22	8.08	9.21
	13	3.03	3.34	3.63	4.21	4.66	4.97	5.30	5.93	7.05	7.86	8.92
	14	3.04	3.35	3.64	4.19	4.64	4.93	5.25	5.85	6.91	7.67	8.63
	15	3.05	3.37	3.66	4.19	4.62	4.91	5.21	5.79	6.77	7.45	8.29
	16	3.09	3.39	3.67	4.19	4.60	4.87	5.17	5.71	6.62	7.29	8.04
	17	3.12	3.41	3.68	4.18	4.58	5.85	5.12	5.63	6.48	7.11	7.84
18	3	3.03	3.38	3.78	4.72	5.69	6.50	7.53	9.97	16.20	23.42	38.75
	4	3.03	3.35	3.72	4.55	5.33	5.94	6.72	8.32	12.10	15.75	21.93
	5	3.01	3.35	3.68	4.44	5.14	5.67	6.29	7.61	10.29	12.75	16.76
	6	3.00	3.33	3.66	4.37	5.01	5.47	6.04	7.15	9.34	11.10	14.07
	7	3.00	3.33	3.65	4.33	4.93	5.35	5.86	6.84	8.80	10.17	12.25
	8	3.01	3.33	3.65	4.30	5.87	5.28	5.73	6.60	8.28	9.62	11.30
	9	3.03	3.33	3.64	4.27	4.82	5.19	5.61	6.46	7.97	9.12	10.71
	10	3.04	3.34	3.64	4.26	4.77	5.15	5.54	6.31	7.63	8.69	10.04
	11	3.04	3.34	3.64	4.25	4.74	5.08	5.47	6.20	7.45	8.34	9.69
	12	3.05	3.35	3.66	4.25	4.71	5.03	5.39	6.09	7.26	8.11	9.24
	13	3.06	3.37	3.67	4.23	4.68	4.99	5.34	5.99	7.10	7.92	8.84
	14	3.08	3.38	3.67	4.21	4.67	4.97	5.29	5.90	6.95	7.70	8.60
	15	3.11	3.39	3.68	4.21	4.65	4.94	5.25	5.82	6.82	7.50	8.32
	16	3.13	3.41	3.69	4.21	4.64	4.91	5.22	5.76	6.68	7.37	8.15
	17	3.16	3.43	3.70	4.20	4.61	4.88	5.17	5.69	6.57	7.19	7.94
	18	3.19	3.47	3.73	4.20	4.60	4.86	5.13	5.63	6.45	7.04	7.75

Percentiles of the Distribution of $V_{0.99}$ (*Continued*)

		$1-\alpha$										
n	r	0.02	0.05	0.10	0.25	0.40	0.50	0.60	0.75	0.90	0.95	0.98
19	3	3.03	3.40	3.79	4.73	5.72	6.52	7.53	9.96	16.17	23.13	36.55
	4	3.05	3.37	3.73	4.55	5.35	5.95	6.70	8.27	11.98	15.31	21.70
	5	3.03	3.36	3.69	4.43	5.14	5.65	6.27	7.54	10.29	12.70	16.31
	6	3.02	3.34	3.66	4.36	5.01	5.47	6.01	7.10	9.28	11.15	13.95
	7	3.02	3.33	3.66	4.33	4.93	5.36	5.83	6.80	8.65	10.16	12.41
	8	3.02	3.33	3.64	4.29	4.87	5.25	5.69	6.58	8.25	9.52	11.35
	9	3.03	3.33	3.65	4.26	4.80	5.18	5.59	6.40	7.93	9.09	10.61
	10	3.03	3.34	3.64	4.25	4.76	5.11	5.52	6.26	7.63	8.67	10.08
	11	3.06	3.35	3.65	4.24	4.72	5.07	5.44	6.15	7.42	8.34	9.70
	12	3.07	3.37	3.65	4.23	4.70	5.03	5.39	6.05	7.22	8.13	9.34
	13	3.08	3.38	3.65	4.22	4.68	4.99	5.33	5.97	7.06	7.88	8.99
	14	3.09	3.38	3.66	4.21	4.66	4.96	5.28	5.89	6.93	7.69	8.63
	15	3.10	3.40	3.67	4.22	4.64	4.93	5.24	5.81	6.79	7.49	8.46
	16	3.13	3.42	3.68	4.21	4.63	4.90	5.20	5.76	6.70	7.35	8.19
	17	3.14	3.43	3.69	4.21	4.62	4.88	5.16	5.70	6.59	7.23	8.01
	18	3.17	3.44	3.71	4.21	4.60	4.86	5.14	5.64	6.49	7.08	7.84
	19	3.20	3.46	3.72	4.21	4.59	4.83	5.10	5.57	6.38	6.91	7.60
20	3	3.01	3.39	3.79	4.72	5.67	6.42	7.38	9.80	16.16	23.55	37.26
	4	3.02	3.37	3.73	4.54	5.32	5.93	6.66	8.21	11.97	15.38	21.61
	5	3.03	3.36	3.70	4.44	5.13	5.64	6.26	7.53	10.15	12.57	16.40
	6	3.02	3.34	3.67	4.38	5.01	5.46	6.01	7.08	9.17	11.05	14.08
	7	3.02	3.34	3.66	4.33	4.93	5.35	5.82	6.78	8.63	10.10	12.40
	8	3.03	3.35	3.66	4.30	4.85	5.25	5.68	6.53	8.18	9.55	11.47
	9	3.04	3.34	3.65	4.28	4.80	5.17	5.58	6.36	7.86	9.02	10.65
	10	3.04	3.35	3.65	4.26	4.76	5.12	5.50	6.24	7.59	8.61	10.02
	11	3.06	3.36	3.66	4.25	4.73	5.06	5.45	6.12	7.37	8.35	9.57
	12	3.08	3.38	3.66	4.24	4.70	5.03	5.38	6.03	7.20	8.08	9.28
	13	3.09	3.38	3.67	4.23	4.68	4.98	5.33	5.95	7.01	7.86	9.02
	14	3.11	3.40	3.67	4.22	4.65	4.95	5.26	5.87	6.89	7.66	8.68
	15	3.12	3.41	3.69	4.21	4.63	4.92	5.23	5.80	6.79	7.49	8.44
	16	3.14	3.42	3.70	4.21	4.62	4.89	5.20	5.74	6.67	7.34	8.23
	17	3.16	3.44	3.71	4.20	4.61	4.87	5.16	5.69	6.57	7.22	7.98
	18	3.18	3.46	3.72	4.20	4.60	4.86	5.13	5.63	6.47	7.07	7.82
	19	3.21	3.47	3.73	4.21	4.58	4.84	5.11	5.57	6.38	6.96	7.64
	20	3.24	3.49	3.74	4.21	4.57	4.81	5.06	5.52	6.25	6.80	7.48

appendix XIII percentiles of the distribution of the *S*-statistic for goodness-of-fit tests on the two-parameter weibull[a]

Percentiles of the Distribution of S and Expected Values of M_i

n	i	M_i	0.75	0.80	0.85	0.90	0.95	0.99
3	1	1.216395						
	2	0.863046						
	3		0.75	0.79	0.84	0.90	0.95	0.99
4	1	1.150727						
	2	0.706698						
	3	0.679596	0.74	0.79	0.85	0.90	0.95	0.99
	4		0.50	0.55	0.60	0.67	0.76	0.89
5	1	1.115718						
	2	0.645384						
	3	0.532445	0.75	0.80	0.85	0.90	0.95	0.99
	4	0.583273	0.50	0.56	0.61	0.68	0.77	0.89
	5		0.67	0.71	0.75	0.79	0.86	0.94
6	1	1.093929						
	2	0.612330						
	3	0.474330	0.75	0.80	0.85	0.90	0.95	0.99
	4	0.442920	0.50	0.55	0.61	0.68	0.76	0.89
	5	0.522759	0.67	0.71	0.75	0.80	0.86	0.93
	6		0.54	0.57	0.61	0.66	0.73	0.84

[a]Taken from: Mann, N. R.; K. W. Fertig and E. M. Scheuer: Tolerance Bounds and a new Goodness-of-Fit Test for Two-Parameter Weibull or Extreme-Value Distribution (With Tables for Censored Samples of Size 3(1)25); Aerospace Research Laboratories, Wright-Patterson Air Force Base, Ohio, ARL 71-0077, Contract no. AF33(615)-70-C-1216, May,1971.

Percentiles of the distribution of S and expected values of M_i (*Continued*)

n	i	M_i	0.75	0.80	0.85	0.90	0.95	0.99
7	1	1.079055						
	2	0.591587						
	3	0.442789	0.75	0.80	0.85	0.90	0.95	0.99
	4	0.387289	0.50	0.55	0.61	0.68	0.77	0.89
	5	0.387714	0.67	0.71	0.75	0.80	0.86	0.94
	6	0.480648	0.54	0.58	0.62	0.67	0.74	0.85
	7		0.64	0.67	0.70	0.74	0.80	0.88
8	1	1.068252						
	2	0.577339						
	3	0.422889	0.75	0.80	0.85	0.90	0.95	0.99
	4	0.356967	0.50	0.55	0.61	0.68	0.77	0.90
	5	0.334089	0.67	0.71	0.75	0.80	0.86	0.94
	6	0.349907	0.54	0.58	0.62	0.67	0.74	0.85
	7	0.449338	0.64	0.67	0.70	0.74	0.80	0.89
	8		0.55	0.58	0.61	0.65	0.71	0.81
9	1	1.060046						
	2	0.566942						
	3	0.409157	0.75	0.80	0.85	0.90	0.95	0.99
	4	0.337763	0.50	0.55	0.61	0.68	0.77	0.89
	5	0.304777	0.67	0.71	0.75	0.80	0.86	0.94
	6	0.297949	0.54	0.58	0.62	0.67	0.75	0.86
	7	0.322189	0.63	0.67	0.70	0.74	0.80	0.89
	8	0.424958	0.55	0.58	0.61	0.66	0.72	0.82
	9		0.62	0.64	0.67	0.71	0.76	0.85
10	1	1.053606						
	2	0.559013						
	3	0.399100	0.75	0.80	0.85	0.90	0.95	0.99
	4	0.324470	0.50	0.55	0.61	0.68	0.77	0.90
	5	0.286163	0.67	0.71	0.75	0.80	0.86	0.94
	6	0.269493	0.54	0.58	0.62	0.68	0.75	0.85
	7	0.271645	0.63	0.67	0.71	0.75	0.81	0.89
	8	0.300869	0.55	0.58	0.62	0.66	0.72	0.81
	9	0.405316	0.62	0.65	0.68	0.71	0.76	0.85
	10		0.55	0.58	0.61	0.64	0.69	0.79

Percentiles of the distribution of S and expected values of M_i (*Continued*)

n	i	M_i	0.75	0.80	0.85	0.90	0.95	0.99
11	1	1.048411						
	2	0.552769						
	3	0.391410	0.75	0.80	0.85	0.90	0.95	0.99
	4	0.314705	0.49	0.55	0.61	0.68	0.77	0.90
	5	0.273245	0.67	0.71	0.75	0.80	0.86	0.94
	6	0.251386	0.54	0.58	0.63	0.68	0.75	0.86
	7	0.243928	0.64	0.67	0.71	0.75	0.81	0.89
	8	0.251548	0.55	0.58	0.62	0.66	0.72	0.82
	9	0.283879	0.62	0.64	0.68	0.71	0.77	0.85
	10	0.389071	0.55	0.58	0.61	0.64	0.70	0.79
	11		0.60	0.63	0.65	0.69	0.74	0.82
12	1	1.044137						
	2	0.547721						
	3	0.385338	0.75	0.79	0.84	0.90	0.95	0.99
	4	0.307221	0.50	0.55	0.61	0.68	0.78	0.89
	5	0.263737	0.67	0.71	0.75	0.80	0.86	0.94
	6	0.238797	0.54	0.58	0.62	0.67	0.74	0.85
	7	0.226264	0.64	0.67	0.70	0.75	0.81	0.89
	8	0.224477	0.55	0.58	0.62	0.66	0.72	0.82
	9	0.235630	0.62	0.64	0.68	0.71	0.77	0.85
	10	0.269966	0.55	0.58	0.61	0.65	0.70	0.79
	11	0.375356	0.60	0.63	0.66	0.69	0.74	0.82
	12		0.55	0.57	0.60	0.63	0.68	0.76
13	1	1.040555						
	2	0.543556						
	3	0.380417	0.75	0.80	0.85	0.90	0.95	0.99
	4	0.301300	0.50	0.55	0.61	0.68	0.77	0.89
	5	0.256437	0.67	0.71	0.75	0.80	0.86	0.94
	6	0.229515	0.54	0.58	0.63	0.68	0.75	0.86
	7	0.213966	0.64	0.67	0.71	0.75	0.81	0.90
	8	0.207205	0.55	0.58	0.62	0.66	0.72	0.82
	9	0.209131	0.62	0.65	0.68	0.72	0.77	0.85
	10	0.222667	0.55	0.58	0.61	0.65	0.70	0.79
	11	0.258323	0.60	0.63	0.66	0.69	0.74	0.82
	12	0.363582	0.55	0.57	0.60	0.64	0.68	0.76
	13		0.59	0.61	0.64	0.67	0.72	0.79

Percentiles of the distribution of S and expected values of M_i (*Continued*)

n	i	M_i	0.75	0.80	0.85	0.90	0.95	0.99
14	1	1.037513						
	2	0.540059						
	3	0.376352	0.75	0.79	0.85	0.90	0.95	0.99
	4	0.296496	0.49	0.54	0.61	0.68	0.77	0.90
	5	0.250650	0.67	0.71	0.75	0.80	0.86	0.94
	6	0.222377	0.54	0.58	0.62	0.68	0.74	0.86
	7	0.204885	0.64	0.67	0.71	0.75	0.81	0.89
	8	0.195165	0.55	0.58	0.62	0.66	0.73	0.82
	9	0.192209	0.62	0.65	0.68	0.72	0.77	0.85
	10	0.196679	0.55	0.58	0.61	0.65	0.70	0.79
	11	0.211875	0.60	0.63	0.66	0.69	0.74	0.82
	12	0.248409	0.55	0.57	0.60	0.64	0.68	0.77
	13	0.353334	0.59	0.61	0.64	0.67	0.72	0.79
	14		0.55	0.57	0.59	0.62	0.67	0.75
15	1	1.034894						
	2	0.537085						
	3	0.372934	0.75	0.80	0.84	0.90	0.95	0.99
	4	0.292518	0.51	0.56	0.62	0.69	0.78	0.90
	5	0.245947	0.68	0.71	0.76	0.80	0.86	0.94
	6	0.216712	0.54	0.58	0.62	0.67	0.75	0.86
	7	0.197893	0.64	0.67	0.71	0.75	0.81	0.89
	8	0.186266	0.55	0.58	0.62	0.66	0.72	0.82
	9	0.180402	0.62	0.65	0.68	0.72	0.77	0.85
	10	0.180072	0.55	0.58	0.61	0.65	0.70	0.79
	11	0.186347	0.61	0.63	0.66	0.69	0.74	0.82
	12	0.202727	0.55	0.57	0.60	0.64	0.68	0.77
	13	0.239842	0.59	0.62	0.64	0.67	0.72	0.79
	14	0.344309	0.55	0.57	0.60	0.63	0.67	0.75
	15		0.59	0.61	0.63	0.66	0.70	0.77

Percentiles of the distribution of S and expected values of M_i (*Continued*)

n	i	M_i	0.75	0.80	0.85	0.90	0.95	0.99
16	1	1.032617						
	2	0.534521						
	3	0.370021	0.75	0.80	0.85	0.90	0.95	0.99
	4	0.289169	0.51	0.56	0.62	0.69	0.78	0.89
	5	0.242049	0.68	0.72	0.76	0.80	0.86	0.94
	6	0.212103	0.54	0.58	0.63	0.68	0.75	0.86
	7	0.192338	0.64	0.67	0.71	0.75	0.81	0.89
	8	0.179407	0.55	0.58	0.62	0.66	0.72	0.82
	9	0.171667	0.62	0.65	0.68	0.72	0.77	0.85
	10	0.168476	0.55	0.58	0.61	0.65	0.71	0.79
	11	0.170026	0.60	0.63	0.66	0.69	0.74	0.82
	12	0.177619	0.55	0.58	0.60	0.64	0.69	0.77
	13	0.194859	0.60	0.62	0.64	0.68	0.72	0.80
	14	0.232350	0.55	0.57	0.60	0.63	0.67	0.75
	15	0.336283	0.59	0.61	0.63	0.66	0.70	0.77
	16		0.55	0.57	0.59	0.62	0.66	0.73
17	1	1.030618						
	2	0.532290						
	3	0.367507	0.75	0.80	0.85	0.90	0.95	0.99
	4	0.286312	0.50	0.55	0.61	0.69	0.78	0.90
	5	0.238765	0.67	0.71	0.75	0.80	0.87	0.94
	6	0.208278	0.54	0.58	0.62	0.68	0.74	0.85
	7	0.187813	0.64	0.67	0.71	0.75	0.80	0.89
	8	0.173951	0.55	0.58	0.62	0.66	0.72	0.81
	9	0.164928	0.62	0.65	0.68	0.72	0.77	0.85
	10	0.159891	0.55	0.58	0.61	0.65	0.70	0.79
	11	0.158624	0.61	0.63	0.66	0.69	0.74	0.82
	12	0.161559	0.55	0.58	0.61	0.64	0.69	0.77
	13	0.170132	0.59	0.62	0.64	0.67	0.72	0.80
	14	0.188005	0.55	0.57	0.60	0.63	0.68	0.75
	15	0.225729	0.59	0.61	0.63	0.66	0.70	0.77
	16	0.329085	0.55	0.57	0.59	0.62	0.66	0.74
	17		0.58	0.60	0.62	0.65	0.69	0.75

Percentiles of the distribution of S and expected values of M_i (*Continued*)

n	i	M_i	0.75	0.80	0.85	0.90	0.95	0.99
18	1	1.028850						
	2	0.530332						
	3	0.365314	0.75	0.80	0.85	0.90	0.95	0.99
	4	0.283846	0.49	0.55	0.61	0.68	0.77	0.90
	5	0.235958	0.67	0.71	0.75	0.80	0.86	0.94
	6	0.205051	0.54	0.58	0.62	0.67	0.75	0.86
	7	0.184055	0.64	0.67	0.71	0.75	0.81	0.89
	8	0.169504	0.55	0.58	0.62	0.66	0.73	0.82
	9	0.159564	0.62	0.65	0.68	0.72	0.77	0.85
	10	0.153263	0.55	0.58	0.61	0.65	0.71	0.80
	11	0.150176	0.61	0.63	0.66	0.69	0.74	0.82
	12	0.150333	0.55	0.68	0.61	0.64	0.69	0.77
	13	0.154313	0.60	0.62	0.64	0.68	0.72	0.80
	14	0.163630	0.55	0.57	0.60	0.63	0.67	0.76
	15	0.181971	0.59	0.61	0.63	0.66	0.70	0.78
	16	0.219825	0.55	0.57	0.59	0.62	0.66	0.74
	17	0.322580	0.58	0.60	0.62	0.65	0.69	0.76
	18		0.55	0.57	0.59	0.61	0.65	0.72
19	1	1.027277						
	2	0.528594						
	3	0.363389	0.75	0.80	0.85	0.90	0.95	0.99
	4	0.281692	0.50	0.55	0.61	0.69	0.78	0.90
	5	0.233535	0.67	0.71	0.76	0.81	0.86	0.94
	6	0.202291	0.54	0.58	0.62	0.68	0.75	0.86
	7	0.180882	0.64	0.67	0.71	0.75	0.81	0.89
	8	0.165807	0.55	0.58	0.62	0.67	0.72	0.82
	9	0.155189	0.62	0.65	0.68	0.72	0.77	0.85
	10	0.147984	0.55	0.58	0.61	0.65	0.71	0.80
	11	0.143650	0.61	0.63	0.66	0.69	0.74	0.82
	12	0.142012	0.55	0.58	0.60	0.64	0.69	0.77
	13	0.143250	0.60	0.62	0.64	0.68	0.72	0.80
	14	0.148031	0.55	0.58	0.60	0.63	0.68	0.76
	15	0.157921	0.59	0.61	0.63	0.66	0.70	0.78
	16	0.176611	0.55	0.57	0.59	0.62	0.66	0.74
	17	0.214520	0.58	0.60	0.62	0.65	0.69	0.76
	18	0.316666	0.55	0.57	0.59	0.61	0.65	0.72
	19		0.57	0.59	0.61	0.64	0.67	0.74

Percentiles of the distribution of S and expected values of M_i (*Continued*)

n	i	M_i	0.75	0.80	0.85	0.90	0.95	0.99
20	1	1.025866						
	2	0.527046						
	3	0.361682	0.75	0.80	0.85	0.90	0.95	0.99
	4	0.279798	0.50	0.55	0.61	0.68	0.78	0.90
	5	0.231417	0.67	0.71	0.75	0.80	0.86	0.94
	6	0.199905	0.54	0.58	0.62	0.67	0.75	0.86
	7	0.178167	0.64	0.67	0.71	0.75	0.81	0.89
	8	0.162684	0.55	0.58	0.62	0.66	0.73	0.82
	9	0.151549	0.62	0.65	0.68	0.72	0.77	0.85
	10	0.143674	0.55	0.58	0.61	0.65	0.71	0.80
	11	0.138448	0.61	0.63	0.66	0.69	0.74	0.83
	12	0.135580	0.55	0.58	0.61	0.64	0.69	0.77
	13	0.135306	0.60	0.62	0.65	0.68	0.72	0.80
	14	0.137120	0.55	0.57	0.60	0.63	0.68	0.76
	15	0.142527	0.59	0.61	0.63	0.66	0.71	0.78
	16	0.152861	0.55	0.57	0.59	0.62	0.67	0.74
	17	0.171810	0.58	0.60	0.62	0.65	0.69	0.76
	18	0.209721	0.55	0.57	0.59	0.62	0.66	0.72
	19	0.311257	0.58	0.59	0.61	0.64	0.68	0.74
	20		0.55	0.56	0.58	0.61	0.65	0.71
21	1	1.024594						
	2	0.525657						
	3	0.360159	0.75	0.80	0.85	0.90	0.95	0.99
	4	0.278117	0.50	0.56	0.62	0.69	0.78	0.90
	5	0.229551	0.68	0.71	0.76	0.80	0.86	0.94
	6	0.197821	0.54	0.58	0.62	0.67	0.74	0.85
	7	0.175815	0.64	0.67	0.71	0.75	0.80	0.89
	8	0.160009	0.55	0.58	0.62	0.66	0.73	0.82
	9	0.148471	0.62	0.65	0.68	0.72	0.77	0.85
	10	0.140087	0.55	0.58	0.61	0.65	0.70	0.80
	11	0.134200	0.60	0.63	0.66	0.69	0.74	0.82
	12	0.130451	0.55	0.58	0.60	0.64	0.69	0.77
	13	0.128702	0.59	0.62	0.64	0.68	0.72	0.79
	14	0.129025	0.55	0.57	0.60	0.63	0.67	0.75
	15	0.131756	0.59	0.61	0.63	0.66	0.70	0.78
	16	0.137659	0.55	0.57	0.60	0.63	0.67	0.74
	17	0.148341	0.58	0.60	0.62	0.65	0.69	0.76
	18	0.167481	0.55	0.57	0.59	0.62	0.66	0.73
	19	0.205352	0.58	0.60	0.62	0.64	0.68	0.75
	20	0.306285	55	0.56	0.58	0.61	0.65	0.72
	21		0.57	0.59	0.61	0.63	0.67	0.73

Percentiles of the distribution of S and expected values of M_i (*Continued*)

n	i	M_i	0.75	0.80	0.85	0.90	0.95	0.99
22	1	1.023439						
	2	0.524405						
	3	0.358790	0.75	0.80	0.85	0.90	0.95	0.99
	4	0.276618	0.50	0.55	0.61	0.68	0.77	0.90
	5	0.227895	0.67	0.71	0.75	0.80	0.86	0.94
	6	0.195983	0.54	0.58	0.63	0.68	0.75	0.85
	7	0.173760	0.64	0.67	0.71	0.75	0.81	0.89
	8	0.157692	0.55	0.58	0.62	0.66	0.72	0.82
	9	0.145834	0.62	0.65	0.68	0.72	0.77	0.85
	10	0.137052	0.55	0.58	0.61	0.65	0.70	0.80
	11	0.130662	0.61	0.63	0.66	0.69	0.74	0.82
	12	0.126260	0.55	0.58	0.61	0.64	0.69	0.78
	13	0.123640	0.60	0.62	0.65	0.68	0.72	0.80
	14	0.122763	0.55	0.58	0.60	0.63	0.68	0.75
	15	0.123763	0.59	0.61	0.63	0.67	0.71	0.78
	16	0.127019	0.55	0.57	0.60	0.62	0.67	0.74
	17	0.133316	0.58	0.60	0.62	0.65	0.69	0.76
	18	0.144273	0.55	0.57	0.59	0.62	0.66	0.73
	19	0.163552	0.58	0.60	0.62	0.64	0.68	0.75
	20	0.201355	0.55	0.57	0.59	0.61	0.65	0.72
	21	0.301693	0.57	0.59	0.61	0.64	0.67	0.73
	22		0.54	0.56	0.58	0.61	0.64	0.70
23	1	1.022389						
	2	0.523269						
	3	0.357557	0.75	0.80	0.85	0.90	0.95	0.99
	4	0.275268	0.50	0.55	0.61	0.68	0.77	0.89
	5	0.226417	0.67	0.71	0.75	0.80	0.86	0.94
	6	0.194351	0.55	0.59	0.63	0.68	0.76	0.86
	7	0.171948	0.64	0.68	0.71	0.76	0.82	0.89
	8	0.155666	0.56	0.59	0.63	0.67	0.73	0.83
	9	0.143549	0.62	0.65	0.68	0.72	0.78	0.86
	10	0.134451	0.56	0.59	0.62	0.66	0.71	0.80
	11	0.127667	0.61	0.63	0.66	0.70	0.75	0.82
	12	0.122768	0.55	0.58	0.61	0.64	0.69	0.78
	13	0.119503	0.60	0.62	0.65	0.68	0.73	0.80
	14	0.117764	0.55	0.57	0.60	0.63	0.68	0.76
	15	0.117577	0.59	0.61	0.63	0.67	0.71	0.78
	16	0.119120	0.55	0.57	0.60	0.63	0.67	0.75
	17	0.122799	0.58	0.60	0.63	0.65	0.69	0.77
	18	0.129416	0.55	0.57	0.59	0.62	0.66	0.73
	19	0.140590	0.58	0.60	0.62	0.64	0.68	0.75
	20	0.159966	0.55	0.57	0.59	0.61	0.65	0.72
	21	0.197679	0.57	0.59	0.61	0.63	0.67	0.73
	22	0.297435	0.55	0.56	0.58	0.60	0.64	0.70
	23		0.57	0.58	0.60	0.63	0.66	0.72

Percentiles of the distribution of S and expected values of M_i (*Continued*)

n	i	M_i	0.75	0.80	0.85	0.90	0.95	0.99
24	1	1.021431						
	2	0.522233						
	3	0.356436	0.75	0.80	0.85	0.90	0.95	0.99
	4	0.274051	0.50	0.56	0.62	0.69	0.78	0.90
	5	0.225086	0.67	0.71	0.76	0.81	0.86	0.94
	6	0.192892	0.54	0.58	0.62	0.68	0.75	0.85
	7	0.170338	0.64	0.67	0.71	0.75	0.81	0.89
	8	0.153877	0.55	0.58	0.62	0.67	0.73	0.83
	9	0.141549	0.62	0.65	0.68	0.72	0.77	0.86
	10	0.132195	0.56	0.58	0.61	0.66	0.71	0.80
	11	0.125099	0.61	0.63	0.66	0.70	0.75	0.83
	12	0.119811	0.55	0.58	0.61	0.64	0.70	0.78
	13	0.116054	0.60	0.62	0.65	0.68	0.73	0.80
	14	0.113677	0.55	0.58	0.60	0.64	0.68	0.76
	15	0.112638	0.59	0.61	0.64	0.67	0.71	0.78
	16	0.113007	0.55	0.57	0.60	0.63	0.67	0.74
	17	0.114990	0.58	0.60	0.62	0.65	0.69	0.76
	18	0.119014	0.55	0.57	0.59	0.62	0.66	0.73
	19	0.125889	0.58	0.60	0.62	0.64	0.68	0.75
	20	0.137235	0.55	0.57	0.59	0.61	0.65	0.72
	21	0.156679	0.57	0.59	0.61	0.64	0.67	0.73
	22	0.194285	0.55	0.56	0.58	0.61	0.64	0.71
	23	0.293473	0.57	0.59	0.60	0.63	0.66	0.72
	24		0.54	0.56	0.58	0.60	0.64	0.69

Percentiles of the distribution of S and expected values of M_i (*Continued*)

n	i	M_i	0.75	0.80	0.85	0.90	0.95	0.99
25	1	1.020551						
	2	0.521285						
	3	0.355415	0.75	0.80	0.85	0.90	0.95	0.99
	4	0.272945	0.50	0.56	0.61	0.69	0.78	0.91
	5	0.223885	0.67	0.71	0.76	0.81	0.87	0.94
	6	0.191578	0.54	0.58	0.62	0.68	0.75	0.86
	7	0.168899	0.64	0.67	0.71	0.75	0.81	0.89
	8	0.152286	0.55	0.58	0.62	0.66	0.72	0.82
	9	0.139783	0.62	0.65	0.68	0.72	0.77	0.85
	10	0.130219	0.56	0.58	0.61	0.65	0.70	0.80
	11	0.122871	0.61	0.63	0.66	0.70	0.75	0.82
	12	0.117274	0.55	0.58	0.61	0.64	0.69	0.78
	13	0.113132	0.60	0.62	0.65	0.68	0.73	0.81
	14	0.110268	0.55	0.58	0.60	0.63	0.68	0.76
	15	0.108598	0.59	0.61	0.64	0.66	0.71	0.78
	16	0.108124	0.55	0.67	0.60	0.63	0.67	0.74
	17	0.108944	0.58	0.60	0.62	0.65	0.69	0.76
	18	0.111289	0.55	0.57	0.59	0.62	0.66	0.73
	19	0.115596	0.58	0.60	0.62	0.64	0.68	0.75
	20	0.122683	0.55	0.57	0.59	0.61	0.65	0.72
	21	0.134165	0.57	0.59	0.61	0.63	0.67	0.74
	22	0.153650	0.55	0.56	0.58	0.61	0.64	0.71
	23	0.191137	0.57	0.58	0.60	0.63	0.66	0.72
	24	0.289773	0.54	0.56	0.58	0.60	0.63	0.70
	25		0.56	0.58	0.60	0.62	0.65	0.71

Author Index

Subject Index